De Gruyter Expositions in Mathematics

Volume 65

Yakov G. Berkovich and Zvonimir Janko

Groups of Prime Power Order

Volume 6

DE GRUYTER

Mathematics Subject Classification 2010
20-02, 20D15, 20E07

Authors
Prof. Dr. Yakov G. Berkovich
18251 Afula
Israel
berkov@math.haifa.ac.il

Prof. Dr. Zvonimir Janko
Ruprecht-Karls-Universität Heidelberg
Mathematisches Institut
Im Neuenheimer Feld 288
69120 Heidelberg
Germany
janko@mathi.uni-heidelberg.de

ISBN 978-3-11-053097-1
e-ISBN (PDF) 978-3-11-053314-9
e-ISBN (EPUB) 978-3-11-053100-8
ISSN 0938-6572

Library of Congress Control Number: 2018941447

Bibliographic information published by the Deutsche Nationalbibliothek
The Deutsche Nationalbibliothek lists this publication in the Deutsche Nationalbibliografie; detailed bibliographic data are available on the Internet at http://dnb.dnb.de.

Typesetting: Dimler & Albroscheit, Müncheberg
Printing and binding: CPI books GmbH, Leck

MIX
Papier aus verantwortungsvollen Quellen
FSC® C083411

www.degruyter.com

Contents

https://doi.org/10.1515/9783110533149-202

List of definitions and notations

Set theory

- $|M|$ is the cardinality of a set M (if G is a finite group, then $|G|$ is called its order).
- $x \in M$ ($x \notin M$) means that x is (is not) an element of a set M; $N \subseteq M$ ($N \nsubseteq M$) means that N is (is not) a subset of the set M; moreover, if $M \neq N \subseteq M$, we write $N \subset M$.
- $\emptyset$ is the empty set.
- N is called a nontrivial subset of M if $N \neq \emptyset$ and $N \subset M$. If $N \subset M$, we say that N is a proper subset of M.
- $M \cap N$ is the intersection and $M \cup N$ is the union of sets M and N. If M, N are sets, then $N - M = \{x \in N \mid x \notin M\}$ is the difference of N and M.

Number theory and general algebra

- p is always a prime number.
- π is a set of primes; π' is the set of all primes not contained in π.
- m, n, k, r, s are, as a rule, natural numbers.
- $\pi(m)$ is the set of prime divisors of m; then m is a π-number.
- n_p is the p-part of n, n_π is the π-part of n.
- $\mathrm{GCD}(m, n)$ is the greatest common divisor of m and n.
- $\mathrm{LCM}(m, n)$ is the least common multiple of m and n.
- $m \mid n$ should be read as: m divides n.
- $\mathrm{GF}(p^m)$ is the finite field containing p^m elements.
- F^* is the multiplicative group of a field F.
- $\mathcal{L}(G)$ is the lattice of subgroups of a group G.
- $\mathcal{L}_N(G)$ is the lattice of normal subgroups of a group G.
- If $n = p_1^{\alpha_1} \dots p_k^{\alpha_k}$ is the standard prime decomposition of n, then $\lambda(n) = \sum_{i=1}^{k} \alpha_i$.

Groups

- We consider only finite groups which are denoted, with a pair exceptions, by upper case Latin letters.
- If G is a group, then $\pi(G) = \pi(|G|)$.
- G is a p-group if $|G|$ is a power of p; G is a π-group if $\pi(G) \subseteq \pi$.
- G is, as a rule, a finite p-group.
- $H \leq G$ means that H is a subgroup of G.

https://doi.org/10.1515/9783110533149-203

- $H < G$ means that $H \leq G$ and $H \neq G$ (in that case H is called a *proper* subgroup of G); $\{1\}$ denotes the group containing only one element.
- H is a nontrivial subgroup of G if $\{1\} < H < G$.
- H is a maximal subgroup of G if $G > \{1\}$, $H < G$ and it follows from $H \leq M < G$ that $H = M$.
- If H is a proper normal subgroup of G, then we write $H \triangleleft G$. The expressions "normal subgroup of G' and 'G-invariant subgroup" are synonyms.
- A normal subgroup H of G is nontrivial provided $G > H > \{1\}$.
- H is a minimal normal subgroup of G if (a) H normal in G; (b) $H > \{1\}$; (c) $N \triangleleft G$ and $N < H$ implies $N = \{1\}$. Thus, the group $\{1\}$ has no minimal normal subgroup.
- A group G is *metabelian* if G/A is abelian for some abelian $A \triangleleft G$. Groups of class 2 are metabelian but the converse is not true.
- A p-group G is said to be *Dedekindian* if all its subgroups are normal.
- A group G is said to be *minimal nonabelian* if it is nonabelian but all its proper subgroups are abelian.
- Let $\mathcal{A}_1(G)$ be the set of all minimal nonabelian subgroups of a p-group G.
- A group H is said to be *capable* if there exists a group G such that $G/Z(G) \cong H$.
- A p-group G is said to be *metahamiltonian* if all its minimal nonabelian (so all nonabelian) subgroups are normal.
- $H \leq G$ is quasinormal if it is permutable with all subgroups of G.
- A p-group is said to be *modular* if all its subgroups are quasinormal.
- G is simple if it is a minimal normal subgroup of G (so $|G| > 1$).
- H is a maximal normal subgroup of G if $H < G$ and G/H is simple.
- The subgroup generated by all minimal normal subgroups of G is called the *socle* of G and denoted by $\mathrm{Sc}(G)$. We put, by definition, $\mathrm{Sc}(\{1\}) = \{1\}$.
- $\mathrm{N}_G(M) = \{x \in G \mid x^{-1}Mx = M\}$ is the normalizer of a subset M in G.
- $\mathrm{C}_G(x)$ is the centralizer of an element x in G: $\mathrm{C}_G(x) = \{z \in G \mid zx = xz\}$.
- (G) is the number of nonidentity cyclic subgroups of G.
- $\mathrm{C}_G(M) = \bigcap_{x \in M} \mathrm{C}_G(x)$ is the centralizer of a subset M in G.
- If $A \leq B$ and A, B are normal in G, then $\mathrm{C}_G(B/A) = H$, where $H/A = \mathrm{C}_{G/A}(B/A)$.
- $H < G$ is a TI-subgroup if $H \cap H^x = \{1\}$ for all $x \in G - \mathrm{N}_G(H)$.
- $A \operatorname{wr} B$ is the wreath product of the "passive" group A and the transitive permutation group B (in what follows we assume that B is a regular permutation group, as a rule, a p-group); B is called the active factor of the wreath product). Then the order of that group is $|A|^{|B|} \cdot |B|$.
- $\mathrm{Aut}(G)$ is the group of automorphisms of G (the automorphism group of G).
- $\mathrm{Inn}(G)$ is the group of all inner automorphisms of G.
- $\mathrm{Out}(G) = \mathrm{Aut}(G)/\mathrm{Inn}(G)$ is the outer automorphism group of G.
- $\mathcal{N}(G)$ is the norm of G, the intersection of normalizers of all subgroups of G.
- If $a, b \in G$, then $a^b = b^{-1}ab$.
- $a \in G$ is real if it is conjugate with a^{-1}.
- An element $x \in G$ inverts a subgroup $H \leq G$ if $h^x = h^{-1}$ for all $h \in H$.

- If $M \subseteq G$, then $\langle M \rangle = \langle x \mid x \in M \rangle$ is the subgroup of G generated by M.
- $M^x = x^{-1}Mx = \{y^x \mid y \in M\}$ for $x \in G$ and $M \subseteq G$.
- $[x, y] = x^{-1}y^{-1}xy = x^{-1}x^y$ is the *commutator* of elements x, y of G. If $M, N \subseteq G$, then $[M, N] = \langle [x, y] \mid x \in M, y \in N \rangle$ is a subgroup of G.
- $\mathrm{o}(x)$ is the order of an element x of G.
- An element $x \in G$ is a π-element if $\pi(\mathrm{o}(x)) \subseteq \pi$.
- G is a π-group if $\pi(G) \subseteq \pi$. Obviously, G is a π-group if and only if all of its elements are π-elements.
- G' is the subgroup generated by all commutators $[x, y]$, $x, y \in G$ (i.e., $G' = [G, G]$), $G^{(2)} = [G', G'] = G'' = (G')'$, $G^{(3)} = [G'', G''] = (G'')'$ and so on; G' is called the *commutator* (or *derived*) subgroup of G.
- $\mathrm{Z}(G) = \bigcap_{x \in G} \mathrm{C}_G(x)$ is the center of G.
- $\mathrm{Z}_i(G)$ is the i-th member of the upper central series of G; in particular, $\mathrm{Z}_0(G) = \{1\}$, $\mathrm{Z}_1(G) = \mathrm{Z}(G)$; $\mathcal{N}(G)$ is the *norm* of G, i.e., the intersection of normalizers of all subgroups of G.
- $\mathrm{K}_i(G)$ is the i-th member of the lower central series of G; in particular, $\mathrm{K}_2(G) = G'$. We have $\mathrm{K}_i(G) = [G, \dots, G]$ ($i \geq 1$ times). We set $\mathrm{K}_1(G) = G$.
- If G is nonabelian, then $\eta(G)/\mathrm{K}_3(G) = \mathrm{Z}(G/\mathrm{K}_3(G))$.
- $\mathcal{M}(G) = \langle x \in G \mid \mathrm{C}_G(x) = \mathrm{C}_G(x^p) \rangle$ is the Mann subgroup of a p-group G.
- $\mathrm{Syl}_p(G)$ is the set of p-Sylow subgroups of an arbitrary finite group G.
- S_n is the *symmetric* group of degree n.
- A_n is the *alternating* group of degree n.
- Σ_{p^n} is a Sylow p-subgroup of S_{p^n}.
- $\mathcal{H}_{2,p}$ is a nonabelian metacyclic p-group of order p^4 and exponent p^2.
- $\mathrm{GL}(n, F)$ is the set of all nonsingular $n \times n$ matrices with entries in a field F, the n-dimensional *general linear* group over F.
- $\mathrm{SL}(n, F) = \{A \in \mathrm{GL}(n, F) \mid \det(A) = 1 \in F\}$, the n-dimensional *special linear* group over F.
- If $H \leq G$, then $H_G = \bigcap_{x \in G} x^{-1}Hx$ is the *core* of the subgroup H in G and H^G, the intersection of all normal subgroups of G containing H, is the *normal closure* or *normal hull* of H in G. Obviously, H_G is normal in G.
- If G is a p-group, then $p^{b(x)} = |G : \mathrm{C}_G(x)|$; $b(x)$ is said to be the *breadth* of $x \in G$, where G is a p-group; $b(G) = \max\{b(x) \mid x \in G\}$ is the *breadth* of G.
- If $H \leq G$ and $|G : \mathrm{N}_G(H)| = p^{\mathrm{sb}(H)}$, then $\mathrm{sb}(H)$ is said to be the *subgroup breadth* of H. Next, $\mathrm{sb}(G) = \max\{\mathrm{sb}(H) \mid H \leq G\}$.
- $\Phi(G)$ is the Frattini subgroup of G (i.e., the intersection of all maximal subgroups of G), $\Phi(\{1\}) = \{1\}$, $p^{\mathrm{d}(G)} = |G : \Phi(G)|$.
- $\Gamma_i = \{H < G \mid \Phi(G) \leq H, |G : H| = p^i\}$, $i = 1, \dots, \mathrm{d}(G)$, where $G > \{1\}$.
- If $H < G$, then $\Gamma_1(H)$ is the set of all maximal subgroups of H.
- $\exp(G)$ is the exponent of G (the least common multiple of the orders of elements of G). If G is a p-group, then $\exp(G) = \max\{\mathrm{o}(x) \mid x \in G\}$.
- $\mathrm{k}(G)$ is the number of conjugacy classes of G (= G-classes), the class number of G.

- K_x is the G-class containing an element x (sometimes we also write $\mathrm{ccl}_G(x)$).
- C_m is the cyclic group of order m.
- $A \times B$ is the direct product of groups A and B.
- $G^m = G \times \cdots \times G$ (m times) is the direct product of m copies of a group G.
- $A * B$ is a central product of groups A and B, i.e., $A * B = AB$ with $[A, B] = \{1\}$. In particular, the direct product $A \times B$ is a central product of groups A and B.
- $E_{p^m} = C_p^m$ is the elementary abelian group of order p^m. G is an elementary abelian p-group if and only if it is a p-group $> \{1\}$ and G coincides with its socle. Next, $\{1\}$ is elementary abelian for each prime p, by definition.
- A group G is said to be *homocyclic* if it is a direct product of isomorphic cyclic subgroups (obviously, elementary abelian p-groups are homocyclic).
- $ES(m, p)$ is an *extraspecial* group of order p^{1+2m} (a p-group G is said to be extraspecial if $G' = \Phi(G) = Z(G)$ is of order p). Note that for each positive integer m, there are exactly two non-isomorphic extraspecial groups of order p^{2m+1}.
- $S(p^3)$ is a nonabelian group of order p^3 and exponent $p > 2$.
- A *special* p-group is a nonabelian p-group G such that $G' = \Phi(G) = Z(G)$ is elementary abelian. Direct products of extraspecial p-groups are special.
- D_{2m} is the *dihedral* group of order $2m$, $m > 2$. Some authors consider E_{2^2} as the dihedral group D_4.
- Q_{2^m} is the *generalized quaternion* group of order $2^m \geq 2^3$.
- SD_{2^m} is the *semidihedral group* of order $2^m \geq 2^4$.
- M_{p^m} is a nonabelian p-group containing exactly p cyclic subgroups of index p (see Theorem 1.2).
- $\mathrm{cl}(G)$ is the *nilpotence class* of a p-group G.
- $\mathrm{dl}(G)$ is the *derived length* of a p-group G.
- $\mathrm{CL}(G)$ is the set of all G-classes.
- A p-group of *maximal class* is a nonabelian group G of order p^m with $\mathrm{cl}(G) = m - 1$.
- A p-group is s-self-dual if every its subgroup is isomorphic to a quotient group.
- A p-group is q-self-dual if every its quotient group is isomorphic to a subgroup.
- $GL(n, F)$ ($SL(n, p)$) is the general (special) linear group of degree n over the field F.
- $\Omega_m(G) = \langle x \in G \mid o(x) \leq p^m \rangle$, $\Omega_m^{\#}(G) = \langle x \in G \mid o(x) = p^m \rangle$, $\mho_m(G) = \langle x^{p^m} \mid x \in G \rangle$.
- A p-group G is said to be *regular* if for any $x, y \in G$ there exists $z \in \langle x, y \rangle'$ such that $(xy)^p = x^p y^p z^p$.
- A p-group is *absolutely regular* if $|G/\mho_1(G)| < p^p$. Absolutely regular p-groups are regular.
- A p-group is *thin* if it is either absolutely regular or of maximal class.
- $G = A \cdot B$ is a *semidirect product* with *kernel* B and *complement* A.
- A group G is an extension of a normal subgroup N by a group H if $G/N \cong H$. A group G splits over N if $G = H \cdot N$ with $H \leq G$ and $H \cap N = \{1\}$ (in that case, G is a semidirect product of H and N with kernel N).
- $H^{\#} = H - \{e_H\}$, where e_H is the identity element of the group H. If $M \subseteq G$, then $M^{\#} = M - \{e_G\}$.

- An automorphism α of G is *regular* (i.e., *fixed-point-free*) if it induces a regular permutation on $G^{\#}$ (a permutation is said to be *regular* if it has no fixed points).
- An *involution* is an element of order 2 in a group.
- A group G is said to *metacyclic* if it contains a normal cyclic subgroup C such that G/C is cyclic.
- A group G is said to be *minimal nonmetacyclic* if it is nonmetacyclic but all its proper subgroups are metacyclic.
- A subgroup A of a group G is said to be *soft* if $\mathrm{C}_G(A) = A$ and $|\mathrm{N}_G(A) : A| = p$.
- A *section* of a group G is an epimorphic image of some subgroup of G.
- If $F = \mathrm{GF}(p^n)$, we usually write $\mathrm{GL}(m, p^n)$, $\mathrm{SL}(m, p^n)$, etc., instead of $\mathrm{GL}(m, F)$, $\mathrm{SL}(m, F)$, etc.
- $\mathrm{c}_n(G)$ is the number of cyclic subgroups of order p^n in a p-group G.
- $\mathrm{c}(G)$ is the number of nonidentity cyclic subgroups of G.
- $\mathrm{s}_n(G)$ is the number of subgroups of order p^n in a p-group G.
- $\nu_n(G)$ is the number of normal subgroups of order p^n in a p-group G.
- $e_n(G)$ is the number of subgroups of order p^n and exponent p in G.
- A group G is said to be s-self-dual (q-self-dual) if any its subgroup (quotient group) is isomorphic to some quotient group (subgroup).
- $M_p(m, n) = \langle a, b \mid \mathrm{o}(a) = p^m,\ \mathrm{o}(b) = p^n,\ a^b = a^{1+p^{m-1}} \rangle$ is the metacyclic minimal nonabelian group of order p^{m+n}.
- $M_p(m, n.1) = \langle a, b \mid \mathrm{o}(a) = p^m,\ \mathrm{o}(b) = p^n,\ [a, b] = c,\ [a, c] = [b, c] = c^p = 1 \rangle$ is the nonmetacyclic minimal nonabelian group of order p^{m+n+1}.
- An $\mathcal{A}_n$-group is a p-group G all of whose subgroups of index p^n are abelian but G contains a nonabelian subgroup of index p^{n-1}. In particular, $\mathcal{A}_1$-group is a minimal nonabelian p-group for some p.
- $\alpha_n(G)$ is the number of $\mathcal{A}_n$-subgroups in a p-group G.
- $\mathcal{D}_n$-group is a 2-group all of whose subgroups of index p^n are Dedekindian, containing a non-Dedekindian subgroup of index p^{n-1} and which is not an $\mathcal{A}_n$-group.
- $\mathcal{MA}(G)$ is the set of minimal nonabelian subgroups of a p-group G.
- $\mathcal{MA}_k(G) = \{H \in \mathcal{MA}(G) \mid \Omega_k(H) = H\}$.
- $\mathrm{D}_k(G) = \langle \mathcal{MA}_k(G) \rangle = \langle H \mid H \in \mathcal{MA}_k(G) \rangle$.
- $L_n = |\{x \in G \mid x^n = 1\}|$.
- If G is a metacyclic p-group and $w(G) = \max\{i \mid |\Omega_i(G)| = 2^{2i}\}$, then $R(G) = \Omega_{w(G)}(G)$. In that case, $G/R(G)$ is either cyclic or a 2-group of maximal class.

Characters and representations

- $\mathrm{Irr}(G)$ is the set of all *irreducible* characters of G over complex numbers.
- A character of degree 1 is said to be *linear*.
- $\mathrm{Lin}(G)$ is the set of all *linear* characters of G (obviously, $\mathrm{Lin}(G) \subseteq \mathrm{Irr}(G)$).

- $\mathrm{Irr}_1(G) = \mathrm{Irr}(G) - \mathrm{Lin}(G)$ is the set of all *nonlinear* irreducible characters of G; $\mathrm{n}(G) = |\mathrm{Irr}_1(G)|$.
- $\chi(1)$ is the *degree* of a character χ of G.
- χ_H is the *restriction* of a character χ of G to $H \leq G$.
- χ^G is the character of G induced from the character χ of some subgroup of G.
- $\overline{\chi}$ is a character of G defined as follows: $\overline{\chi}(x) = \overline{\chi(x)}$ (here $\overline{w}$ is the complex conjugate of a complex number w).
- $\mathrm{X}(G)$ is a character table of G.
- $\mathrm{Irr}(\chi)$ is the set of irreducible constituents of a character χ of G.
- 1_G is the principal character of G.
- $\mathrm{Irr}^{\#}(G) = \mathrm{Irr}(G) - \{1_G\}$.
- If χ is a character of G, then $\ker(\chi) = \{x \in G \mid \chi(x) = \chi(1)\}$ is the *kernel* of a character χ.
- $\mathrm{Z}(\chi) = \{x \in G \mid |\chi(x)| = \chi(1)\}$ is the *quasikernel* of χ.
- If N is normal in G, then $\mathrm{Irr}(G \mid N) = \{\chi \in \mathrm{Irr}(G) \mid N \nleq \ker(\chi)\}$.
- $\langle \chi, \tau \rangle = |G|^{-1} \sum_{x \in G} \chi(x)\tau(x^{-1})$ is the *inner product* of characters χ and τ of G.
- $\mathrm{I}_G(\phi) = \langle x \in G \mid \phi^x = \phi \rangle$ is the *inertia subgroup* of $\phi \in \mathrm{Irr}(H)$ in G, where $H \triangleleft G$.
- 1_G is the *principal character* of G ($1_G(x) = 1$ for all $x \in G$).
- $\mathrm{M}(G)$ is the *Schur multiplier* of G.
- $\mathrm{cd}(G) = \{\chi(1) \mid \chi \in \mathrm{Irr}(G)\}$.
- $\mathrm{mc}(G) = \mathrm{k}(G)/|G|$ is the *measure of commutativity* of G.
- $\mathrm{T}(G) = \sum_{\chi \in \mathrm{Irr}(G)} \chi(1)$.
- $\mathrm{f}(G) = \mathrm{T}(G)/|G|$.

Preface

This is the sixth volume of the series devoted to elementary parts of finite p-group theory. As in the previous volumes, numerous characterizations of different classes of finite p-groups are presented. Minimal nonabelian subgroups and minimal nonnilpotent groups play an important role in numerous sections. Also counting theorems which were proved in Volume 1 are used here. Most of the presented material appears in book form for the first time.

Below we list some new results devoted to characterizations of certain classes of p-groups and results presented in this volume:

(1) classification of p-groups in which the normal closures of nonnormal subgroups have index p (this result was obtained in § 62, but the proof presented in this volume is different),

(2) classification of p-groups, $p > 2$, in which the intersection of any two nonincident subgroups is normal,

(3) classification of nonabelian p-groups with exactly one minimal nonabelian subgroup of exponent $> p$,

(4) description of nonabelian p-groups, $p > 2$, all of whose minimal nonabelian subgroups are isomorphic to M_{p^3},

(5) description of nonabelian p-groups such that $\mathrm{Z}(S) \leq \mathrm{Z}(G)$ for any minimal nonabelian subgroup $S \leq G$,

(6) description of nonabelian p-groups such that $\mathrm{Z}(S) \leq \mathrm{Z}(T)$ for any two minimal nonabelian subgroups $S, T \leq G$,

(7) investigation of nonabelian p-groups all of whose subgroups are powerful,

(8) classification of nonabelian p-groups all of whose maximal abelian subgroups are isolated,

(9) classification of irregular p-groups all of whose maximal absolutely regular subgroups (metacyclic subgroups) are isolated,

(10) alternate classification of nonabelian p-groups all of whose minimal nonabelian subgroups are isolated,

(11) alternate proof of Theorem 109.1, Theorem 1.23 and Lemma 57.1.

(12) alternate proof of the theorem of Passman on p-groups all of whose subgroups of order $\leq p^s$ are normal,

(13) proof of Thompson's $A \times B$ lemma,

(14) classification of minimal non-p-central p-groups,

(15) classification of p-groups all of whose regular subgroups are either absolutely regular or of exponent p,

(16) proof that any p-group has a critical subgroup,

(17) a number of results on action of p'-subgroups on p-groups,

(18) groups G such that $H < G$ and $|H|^2 < |G|$ imply that $H \lhd G$,

https://doi.org/10.1515/9783110533149-204

(19) nonabelian p-groups in which any nonabelian subgroup contains its centralizer,
(20) p-groups whose center contains the centers of their minimal nonabelian subgroups,
(21) p-groups with exactly two conjugate classes of non-normal maximal cyclic subgroups,
(22) p-groups all of whose maximal subgroups, except one, are minimal nonmetacyclic,
(23) p-groups, $p > 2$, with the metacyclic Frattini subgroup,
(24) description of metacyclic p-groups with abelian subgroup of index p,
(25) classification of p-groups all of whose maximal metacyclic (absolutely regular) subgroups have index p,
(26) description of p-groups all of whose subgroups of order p^4 are isomorphic,
(27) description of nonabelian p-groups all of whose proper nonabelian subgroups have exponent p,
(28) description of the non-Dedekindian p-groups in which each nonnormal subgroup has a cyclic complement in its normalizer,
(29) proof that any irregular p-group contains a non-isolated maximal regular subgroup,
(30) new proof of some known counting theorems,
(31) description of the nonabelian p-groups in which any two nonincident subgroups have an abelian intersection,
(32) study of the prime divisors of the automorphism groups of the 2-groups without elementary abelian subgroup of order 8,
(33) classification of nonabelian metacyclic p-groups with abelian Frattini subgroup,
(34) exposition of the number of elementary results from the Odd Order paper and Thompson's paper on groups all of whose local subgroups are solvable,
(35) description of the p-groups all of whose maximal subgroups of exponent p are normal and have order p^p,
(36) description of nonabelian p-groups with exactly p normal closures of minimal nonabelian subgroups,
(37) proof of an analog of Thompson's dihedral lemma.

For further information, see Contents.

The problems from the section “Research problems and themes VI” are posed, if it is not stated otherwise, by the first author. That section contains more than 700 items some of which were solved by the second author. Many problems from the lists in the previous five volumes were solved mainly by the second author and mathematicians from Shanxi Normal University. Most of the presented problems were discussed with Avinoam Mann and, of course, with the second author. This list also contains information on solved problems.

In the book there are many hundreds exercises most of which are solved (those exercises, if it is not stated otherwise, were written by the first author). Many exercises contain new results which supplement the main text. Most of the exercises are concentrated in § 293. Only few exercises are cited in the main text (among of them are Exercises 1.6, 9.1, 9.13. 10.10, A.101.7 from the previous volumes).

Sections and appendices written by the authors are listed in the author index.

We added in the bibliography of the previous volumes those items that are cited in this volume and omitted most of those ones that are not cited here. In that list we have also included those papers which we suggest to read for obtaining more full information.

We are grateful to the publishing house of Walter de Gruyter and all its workers supporting and promoting the publication of this and the five previous volumes.

§ 257 Nonabelian p-groups with exactly one minimal nonabelian subgroup of exponent $> p$

It is well known that a nonabelian p-group G is generated by its minimal nonabelian subgroups (Proposition 10.28). Therefore any information about the structure of some minimal nonabelian subgroups influences very strongly the structure of the whole group.

Mann proved that if any minimal nonabelian subgroup of a nonabelian p-group G, $p > 2$, has exponent p (i.e., it is isomorphic to $\mathrm{S}(p^3)$), then G possesses an abelian subgroup of index p.

Here we shall determine in Theorem 257.1 the structure of the title p-groups G (solving Problem 3887), where we assume that G is not minimal nonabelian. In that case we must have, as in the previous paragraph, $p > 2$ and so the problem is quite difficult. The main tool will be a result of Hogan–Kappe [HogK] stating that the index of a nontrivial Hughes subgroup $\mathrm{H}_p(G)$ of a metabelian p-group G is equal at most p. Also we use very often some lemmas which we reproduce here in a simpler form:

Lemma 57.1. *Let A be a maximal normal abelian subgroup of a nonabelian p-group G. Then for each $x \in G - A$, there is $a \in A$ such that $\langle x, a\rangle$ is minimal nonabelian.*

Lemma 65.1. *Let G be a minimal nonabelian p-group with $p > 2$ which has an elementary abelian maximal subgroup. Then G is isomorphic to $\mathrm{S}(p^3)$ (a nonabelian group of order p^3 and exponent p), M_{p^3} (a nonabelian group of order p^3 and exponent p^2) or $\mathrm{M}_p(2, 1, 1)$ (a nonmetacyclic minimal nonabelian group of order p^4), where*

$$\mathrm{M}_p(2,1,1) = \langle g, a \mid g^{p^2} = a^p = 1,\ [g, a] = z,\ z^p = [z, g] = [z, a] = 1\rangle.$$

Lemma 65.2 (a). *A two-generator p-group G with $|G'| = p$ is minimal nonabelian.*

We shall prove here the following general result.

Theorem 257.1. *Let G be a nonabelian p-group which is not minimal nonabelian but possesses exactly one minimal nonabelian subgroup M of exponent $> p$. Then $p > 2$ and $|G : M| = p$ and one of the following holds:*

(a) *$M = \mathrm{H}_p(G)$,*

(b) *G is an irregular 3-group of class 3 and order $\leq 3^5$ with exactly one abelian maximal subgroup having exponent 9.*

Proof. Let G be a nonabelian p-group which is not minimal nonabelian but possesses exactly one minimal nonabelian subgroup M of exponent $> p$. As all other minimal nonabelian subgroups of G have exponent p and so are isomorphic to $\mathrm{S}(p^3)$ (the nonabelian group of order p^3 and exponent p), we have $p > 2$ and $M \trianglelefteq G$.

(i) Suppose that G has no abelian subgroup of index p. Let A be a maximal normal abelian subgroup in G of maximal possible exponent. Let C/A be any subgroup of

https://doi.org/10.1515/9783110533149-001

order p^2 in G/A. Suppose for a moment that $C/A \cong C_{p^2}$ and let D/A be the subgroup of order p in C/A. If all elements of the set $C - D$ lie in M, then $C \leq M$ (indeed, $\langle C - D\rangle = C$). Since M is minimal nonabelian, D would be abelian, contrary to $C_G(A) = A$. Hence there is $g \in C - D$ such that $g \notin M$. By Lemma 57.1, there is $a \in A$ so that $\langle g, a\rangle$ is minimal nonabelian. Since $o(g) \geq p^2$, we get $\langle g, a\rangle = M$, a contradiction. Hence we must have $C/A \cong E_{p^2}$ implying that $\exp(G/A) = p$.

(i1) Assume that A is elementary abelian. Then $\exp(G) = p^2$ *and* each normal abelian subgroup in G is elementary abelian. Let $g \in G$ with $o(g) = p^2$ so that $g \nleq A$ and $1 \neq g^p \in A$. By Lemma 57.1, there is $a \in A$ such that $\langle g, a\rangle$ is minimal nonabelian. By hypothesis, $\langle g, a\rangle = M \trianglelefteq G$. Because of the uniqueness of M (and using again Lemma 57.1), we see that all elements in $G - M$ are of order p so that $M = H_p(G)$. If $\langle [g, a]\rangle = \langle g^p\rangle$, then $M = M_{p^3}$ and, if $\langle [g, a]\rangle = z \notin \langle g\rangle$, then

$$M \cong M_p(2, 1, 1) = \langle g, a \mid g^{p^2} = a^p = 1,\ [g, a] = z,\ z^p = [z, g] = [z, a] = 1,$$

which is nonmetacyclic minimal nonabelian of order p^4. Set $AM = D$ so that D is nonabelian with $|D : A| = p$, by the product formula. Also, $D < G$ and, as we have noted, each normal abelian subgroup in G is elementary abelian.

Suppose that $M \cong M_{p^3}$. Since in this case M has exactly p cyclic subgroups of order p^2 and no one of them is normal in G (indeed. all normal abelian subgroups of G have exponent p), it follows that they form a conjugate class in G so that $|G : N_G(\langle g\rangle)| = p$ (about g see the previous paragraph). Suppose that there is $x \in G - M$ such that $o(x) = p$ and x normalizes $\langle g\rangle$. Then there are elements of order p^2 in the nonabelian subgroup $\langle g, x\rangle - \langle g\rangle$, a contradiction. We have proved that $N_G(\langle g\rangle) = M$ implying that $|G : M| = p$. Then $M = D$ and so $A \cong E_{p^2}$. Since $C_G(A) = A$, we must have $|G : A| = p$, a contradiction.

We have proved that $M \cong M_p(2, 1, 1)$. In this case,

$$Z(M) = \Phi(M) = \langle z, g^p\rangle \cong E_{p^2},$$

where $\langle z\rangle = M'$ and $\langle g^p\rangle = \mho_1(G)$ so $Z(M) \leq Z(G)$. Since $M = H_p(G)$, we have

$$Z(G) \leq Z(M)$$

implying that

$$Z(M) = Z(G) = M' \times \mho_1(M) \cong E_{p^2}.$$

Set $D = AM$ and let $D < C \leq G$ with $|C : D| = p$ so that $C/A \cong E_{p^2}$ and therefore C is metabelian with $M = H_p(G)$. It follows (by a result of Hogan–Kappe [HogK]) that $M = D$ so that $|C : M| = p$ and $A \cong E_{p^3}$.

Suppose that $C \neq G$ and let $C < F \leq G$ with $|F : C| = p$. We have $F/M \cong E_{p^2}$ and we consider $\overline{F} = F/\langle z\rangle$ (with bar convention) so that $M/\langle z\rangle$ is abelian. Hence $\overline{F}$ is metabelian with $\overline{M} = H_p(\overline{F})$ and this contradicts the result of Hogan–Kappe applied to the group $\overline{F}$. We have proved that $C = G$ and so $|G| = p^5$ and then $|G : M| = p$ and $M = H_p(G)$ so that we have obtained a group from part (a) of Theorem 257.1.

(i2) Suppose that A is of exponent $> p$. Let C/A be a normal subgroup of order p^2 in G/A so that $C/A \cong E_{p^2}$ (recall that $\exp(G/A) = p$). Let C_i/A, $i = 1, 2, \dots, p+1$, be $p+1$ subgroups of order p in C/A. Since M is contained in at most one of the subgroups $C_1, C_2, \dots, C_{p+1}$, it follows that we may assume that M is not contained in any of $C_1, C_2, \dots, C_p$. Then all minimal nonabelian subgroups in any of the groups $C_1, C_2, \dots, C_p$ are isomorphic to $S(p^3)$ so that (by Lemma 57.1) all elements in $C_i - A$ are of order p ($i = 1, 2, \dots, p$). If $M \nleq C_{p+1}$, then $A = H_p(C)$ and so we have a contradiction by the result of Hogan–Kappe [HogK]. Hence $M \leq C_{p+1}$ and so $D = C_{p+1} = AM \trianglelefteq G$ since $A, M \triangleleft G$.

Suppose that $G > C$. If $N_G(C_1) > C$, then take an element $k \in N_G(C_1) - C$ so that $o(k) = p$, $(C_1\langle k\rangle)/A \cong E_{p^2}$ and all elements of the set $(C_1\langle k\rangle) - A$ are of order p. Since $H_p(C_1\langle k\rangle) = A$, we get again a contradiction by the result of Hogan–Kappe applied on the metabelian group $C_1\langle k\rangle$. We have proved that $N_G(C_1) = C$. But C/A contains exactly $p+1$ subgroups of order p with $C_{p+1} \trianglelefteq G$ and so $|G : C| = p$. Since G/A is nonabelian, we get $G/A \cong S(p^3)$. We have proved that either $G = C$ and then $G/A \cong E_{p^2}$ or $G/A \cong S(p^3)$ with $(G/A)' = D/A$, where $D = C_{p+1} \trianglelefteq G$.

Assume for a moment that $M < D$, where $D = C_{p+1} = MA \trianglelefteq G$. Set $S = D - (A \cup M)$ so that S is a nonempty set of elements of order p (by Lemma 57.1). Since we have $\Phi(M) = Z(M) \leq Z(D)$, it follows that each $s \in S$ centralizes $Z(M) < A \cap M$ implying that $Z(M)$ is elementary abelian and so $\exp(M) = p^2$. If $s \in S$ would centralize $A \cap M$, then $A \cap M \leq Z(D)$ and so $A \cap M \leq Z(M)$, a contradiction. Hence $(A \cap M)\langle s\rangle$ is nonabelian with $((A \cap M)\langle s\rangle) \cap M = A \cap M$ so that each minimal nonabelian subgroup of $(A \cap M)\langle s\rangle$ is isomorphic to $S(p^3)$. For each $a \in (A \cap M) - Z(M)$, $[s, a] \in Z(M) \leq Z(D)$ and so (by Lemma 65.2), $\langle s, a\rangle$ is minimal nonabelian. Therefore all elements in $(A \cap M) - Z(M)$ are of order p so that $A \cap M = \langle (A \cap M) - Z(M)\rangle$ is elementary abelian. By Lemma 65.1, we have $M \cong M_{p^3}$ or $M \cong M_p(2, 1, 1)$ and in any case all elements in $M - A$ are of order p^2. It follows that M has exactly p maximal subgroups of exponent p^2 and let M_0 be one of them. If $N_D(M_0) > M$, then there is an element $s_0 \in S$ normalizing M_0. If all elements in $(M\langle s_0\rangle) - M$ centralize M_0, then $M\langle s_0\rangle$ centralizes M_0 implying that $M_0 \leq Z(M)$, a contradiction. Hence there is an element $y \in (M\langle s_0\rangle) - M$ which does not centralize M_0 and so if $x \in M_0 - Z(M)$, then $[x, y] \neq 1$ and $o([x, y]) = p$ (since $[x, y] \in Z(M)$). By Lemma 65.2, $\langle x, y\rangle$ is minimal nonabelian. Since $\langle x, y\rangle \neq M$, we get $\langle x, y\rangle \cong S(p^3)$ and therefore $o(x) = p$, a contradiction (because all elements in $M_0 - Z(M)$ are of order p^2). We have proved that $N_D(M_0) = M$ implying that $|D : M| = p$ and $N_G(M_0)$ covers G/D. If $M \cong M_{p^3}$, then $M_0 \cong C_{p^2}$ and so there is $c \in C - D$ with $o(c) = p$ normalizing M_0. But then there is an element of order p^2 in $(M_0\langle c\rangle) - M_0$, a contradiction (because all elements in $C - D$ are of order p). Suppose that $M \cong M_p(2, 1, 1)$ so that in this case $M_0 = \langle g\rangle \times \langle h\rangle$, where $o(g) = p^2$ and $o(h) = p$. By the structure of $M_p(2, 1, 1)$, $\langle g\rangle$ is not normal in M and so M fuses all p subgroups isomorphic to C_{p^2} in M_0 and since $N_G(M_0)$ covers C/D, there is $c \in C - D$ of order p which normalizes $\langle g\rangle \cong C_{p^2}$. But this gives a contradiction as above. We have proved that $M = D$.

Suppose that $G/A \cong S(p^3)$ so that $G/M \cong E_{p^2}$ and then all elements in $G - M$ are of order p. If $\exp(M/M') = p$, then $\Phi(M) = M' \cong C_p$ implies $M \cong M_{p^3}$ so that $A \cong C_{p^2}$. But $C_G(A) = A$ and so $G/A \cong S(p^3)$ must act faithfully on A, a contradiction. Hence we must have $\exp(M/M') > p$ and then $H_p(G/M') = M/M'$, where $|G : M| = p^2$. This contradicts the result of Hogan–Kappe [HogK] applied on the metabelian group G/M'. We have proved that $G/A \cong E_{p^2}$ so that $|G : M| = p$ and $H_p(G) = M$ and hence we have obtained the groups in part (a) of Theorem 257.1.

(ii) Finally, assume that G has an abelian subgroup A of index p. Then $G = AM$, where $M \trianglelefteq G$, $M < G$, $G/M \cong A/M_0$, is abelian; here $M_0 = A \cap M$ is a maximal subgroup in M. Also, $Z(M) = \Phi(M) \le Z(G)$ and since $G' \le A \cap M = M_0$, we get $[G, G'] \le Z(M)$ and so G is of class at most 3. Set $S = G - (A \cup M)$, where the set S is nonempty. By Lemma 57.1, for each $s \in S$, there is $a \in A$ so that $\langle s, a\rangle$ is minimal nonabelian. This gives $o(s) = p$ and so S consists of elements of order p.

Since for each $z \in Z(M) \le Z(G)$ and $s \in S$, we have $sz \in S$, we get $o(sz) = p$ and so $o(z) \le p$ and therefore $Z(M)$ is elementary abelian.

Let $m \in M_0 - Z(M)$ and $s \in S$, where $|M_0 : Z(M)| = p$. If $[s, m] = 1$, then $m \in Z(G)$ and so $m \in Z(M)$, a contradiction. Hence $1 \ne [s, m] \in Z(M) \le Z(G)$. By Lemma 65.2, $\langle s, m\rangle$ is minimal nonabelian distinct from M and so $\langle s, m\rangle \cong S(p^3)$. In particular, $o(m) = p$ and so we have proved that M_0 is elementary abelian. By Lemma 65.1, $M \cong M_{p^3}$ or $M \cong M_p(2, 1, 1)$, where all elements in $M - M_0$ are of order p^2.

(ii1) First assume $M \cong M_{p^3}$. Let $g \in M - M_0$, where

$$M \cap A = M_0 \cong E_{p^2}, \quad o(g) = p^2, \quad \langle g^p\rangle = Z(M) = M'.$$

Note that M has exactly p cyclic subgroups of order p^2. Suppose, by way of contradiction, that $N_G(\langle g\rangle) > M$ and let $M < X \le N_G(\langle g\rangle)$ with $|X : M| = p$. If all elements in $X - M$ centralize $\langle g\rangle$, then M centralizes $\langle g\rangle$, a contradiction. Hence there is an element $x \in X - M$ such that $\langle x\rangle$ acts nontrivially on $\langle g\rangle$. But then $\langle g, x\rangle$ is minimal nonabelian of exponent $> p$ and is distinct from M, a contradiction. We have proved that $N_G(\langle g\rangle) = M$ and so $|G : M| = p$, $|G| = p^4$ and G fuses all p cyclic subgroups of order p^2 in M. We know that $G' \le M_0$. But if $G' = \langle g^p\rangle = M'$, then $\langle g\rangle \trianglelefteq G$, a contradiction. Hence $G' = M_0 \cong E_{p^2}$ and so G is of class 3.

We show that $\Omega_1(G) = G$. For an $s \in S = G - (A \cup M)$, we have $|M_0\langle s\rangle| = p^3$ and $\exp(M_0\langle s\rangle) = p$. If $\Omega_1(G) < G$, then $\Omega_1(G) = M_0\langle s\rangle$ and so $|S| + |M_0| = p^3$ giving

$$(p^4 - p^3 - (p^3 - p^2)) + p^2 = p^3 \quad \text{or} \quad p^4 - 3p^3 + 2p^2 = 00 \quad \text{or} \quad p^2 - 3p + 2 = 0,$$

a contradiction since $p > 2$. Hence we have proved that $\Omega_1(G) = G$. If $p > 3$, then $cl(G) = 3 < p$ and so G is a regular p-group (Theorem 7.1 (b)) and then $G = \Omega_1(G)$ is of exponent p, a contradiction. Hence $p = 3$ and G is irregular.

If A is elementary abelian, then $M = H_3(G)$ and so we get a group from part (a) of Theorem 257.1. If $\exp(A) > 3$, then $\exp(A) = 9$ and so we have obtained a group from part (b) of Theorem 257.1.

(ii2) Now assume $M \cong M_p(2, 1, 1)$. Here we have $M \cap A = M_0 \cong E_{p^3}$ and the other p maximal subgroups of M are abelian of type (p^2, p); let M_1 be one of them. We have $Z(M) = \mho_1(M) \times M' \cong E_{p^2}$ and $Z(M) \le Z(G)$ so that $|M_1 : Z(M)| = p$.

Suppose, by way of contradiction, that $N_G(M_1) > M$ and let $M < X \le N_G(M_1)$ with $|X : M| = p$. If all elements in $X - M$ centralize M_1, then X centralizes M_1, a contradiction. Hence there is $x \in X - M$ such that x does not centralize M_1. In particular, for an $m \in M_1 - Z(M)$, we have $1 \ne [x, m] \in Z(M) \le Z(G)$ and so by Lemma 65.2, $\langle x, m\rangle$ is minimal nonabelian. But $o(m) = p^2$ and $\langle x, m\rangle \ne M$ and this is a contradiction. We have proved that $N_G(M_1) = M$ and so $|G : M| = p$ and $|G| = p^5$. If $G' \le Z(M) \le Z(G)$, then M_1 is normal in G, a contradiction. Hence $G' \nleq Z(G)$ and therefore $cl(G) = 3$.

We show that $\Omega_1(G) = G$. For an $s \in S = G - (A \cup M)$, we have $|M_0\langle s\rangle| = p^4$ and $\exp(M_0\langle s\rangle) = p$. If $\Omega_1(G) < G$, then $\Omega_1(G) = M_0\langle s\rangle$ and so $|S| + |M_0| = p^4$ giving

$$(p^5 - p^4 - (p^4 - p^3)) + p^3 = p^4 \quad \text{or} \quad p^5 - 3p^4 + 2p^3 = 0 \quad \text{or} \quad p^2 - 3p + 2 = 0,$$

a contradiction. Hence we have proved that $\Omega_1(G) = G$. If $p > 3$, then $cl(G) = 3 < p$ and so G is a regular p-group (Theorem 7.1 (b)) and then $G = \Omega_1(G)$ is of exponent p, a contradiction. Hence $p = 3$ and G is irregular.

If A is elementary abelian, then $M = H_3(G)$ and so we get a group from part (a) of Theorem 257.1. If $\exp(A) > 3$, then $\exp(A) = 9$ and so we have obtained a group from part (b) of Theorem 257.1. □

Problem 1. Classify the nonabelian p-groups containing $\le p$ minimal nonabelian subgroups of exponent $> p$.

Problem 2. Classify the $\mathcal{A}_n$-group, $n > 2$, that are p-groups and containing exactly one (i) $\mathcal{A}_2$-subgroup, (ii) $\mathcal{A}_2$-subgroup of exponent $> p$.

Problem 3. Classify the irregular p-groups containing $\le p$ maximal regular subgroups of exponent $> p$.

§ 258 2-groups with some prescribed minimal nonabelian subgroups

In § 57 nonabelian 2-groups are classified if all minimal nonabelian subgroups are isomorphic and have exponent 4. In Theorem 10.33 it was proved that if G is a nonabelian 2-group all of whose minimal nonabelian subgroups are isomorphic to D_8, then G is quasidihedral. In Corollary A.17.3 it was shown that if G is a nonabelian 2-group all of whose minimal nonabelian subgroups are isomorphic to Q_8, then $G = Q \times V$, where $Q \cong Q_{2^n}$ ($n \geq 3$) and $\exp(V) \leq 2$.

Here we modify the above two results concerning a nonabelian finite 2-group G. In Theorem 258.1, we assume that all minimal nonabelian subgroups of G, except one, are isomorphic to D_8 and in Theorem 258.2, we suppose that all minimal nonabelian subgroups of G, except one, are isomorphic to Q_8. The idea to consider such a type of problems is due to the first author. In both cases we determine such 2-groups up to isomorphism and we obtain only 2-groups of small order $\leq 2^5$. This was somewhat surprising since we assume nothing about the structure of that exceptional minimal nonabelian subgroup.

Note that a quasidihedral group is a 2-group G with an abelian subgroup H of exponent > 2 and index 2 and there is an involution $i \in G - H$ which inverts each element in H (in that case, all elements of the set $G - H$ are involutions). Finally, the Hughes subgroup $H_p(G)$ of a p-group G is the subgroup of G generated by all elements of order $\neq p$.

Theorem 258.1 (Janko). *Let G be a nonabelian 2-group which is not minimal nonabelian but all its minimal nonabelian subgroups, except one, denoted with M, are isomorphic to D_8. Then one of the following holds:*

(a) *$M \cong Q_8$ and then either $G \cong SD_{16}$ or $G = Q * Z$ (a central product), where $Q \cong Q_8$, $Z \cong C_4$ (a cyclic group of order 4) and $Q \cap Z = Z(Q)$.*

(b) *M is the metacyclic minimal nonabelian group of order 2^4 and exponent 4 (i.e., $M \cong M_2(2,2)$) and*

$$G = \langle a, c \mid a^8 = c^4 = [a, c^2] = 1,\ [a, c] = a^6 c^2,\ [a^2, c] = a^4 \rangle,$$

where

$$|G| = 2^5, \quad cl(G) = 3, \quad G' = \langle a^2 c^2 \rangle \cong C_4, \quad Z(G) = \langle a^4, c^2 \rangle \cong E_4$$

and $\langle a, c^2 \rangle \cong C_8 \times C_2$ is a unique abelian maximal subgroup in G.

Conversely, all the above 2-groups satisfy the assumptions of the theorem.

Proof. Let G be a nonabelian 2-group satisfying the assumptions of Theorem 258.1. Then a unique minimal nonabelian subgroup M which is not isomorphic to D_8 is normal in G.

Suppose, by way of contradiction, that G has no abelian subgroup of index 2. Let A be a maximal normal abelian subgroup in G and assume in addition that $\exp(A) > 2$ if

https://doi.org/10.1515/9783110533149-002

such subgroups exist. Let C/A be a subgroup of order 4 in G/A. Suppose that $C/A \cong C_4$ and let D/A be the subgroup of order 2 in C/A. If all elements in $C - D$ lie in M, then $C \leq M$. But M is minimal nonabelian and so in that case D would be abelian, a contradiction. Hence there is $g \in C - D$ so that $g \notin M$. Then Lemma 57.1 implies that there is $a \in A$ such that $\langle g, a\rangle$ is minimal nonabelian. Hence $\langle g, a\rangle \cong D_8$ and so $o(g) = 4$, $\langle g\rangle \cap A = \{1\}$ and $\langle g, a\rangle \cap A = \langle a\rangle \cong C_2$. It follows that $\langle g, a\rangle = \langle g\rangle \times \langle a\rangle$ is abelian, a contradiction. We have proved that $C/A \cong E_4$ and so G/A is elementary abelian.

Let C_1/A, C_2/A and C_3/A be the three subgroups of order 2 in C/A. Since M is contained in at most one of $\{C_1, C_2, C_3\}$, we may assume that M is not contained in C_1 and C_2.

Assume at the moment that A is elementary abelian. Since C_1 is nonabelian and each minimal nonabelian subgroup of C_1 is isomorphic to D_8, Theorem 10.33 implies that C_1 is quasidihedral and so setting $H = H_2(C_1)$, we have that H is abelian of exponent > 2 and $|C_1 : H| = 2$. But H is characteristic in C_1 and $C_1 \trianglelefteq G$ and so $H \trianglelefteq G$. Hence abelian normal subgroups of exponent > 2 exist and so, by our choice of A, we must have $\exp(A) > 2$, a contradiction. We have proved that $\exp(A) > 2$. It follows then (by Theorem 10.33) that $A = H_2(C_1) = H_2(C_2)$. Let $c_i \in C_i - A$, $i = 1, 2$, so that c_1 and c_2 invert each element in A. But then $c_1c_2 \notin A$ and c_1c_2 centralizes A, a final contradiction. We have proved that G possesses an abelian subgroup A of index 2.

We have $M < G$, $G = AM$, $M \trianglelefteq G$, the group G/M is abelian and so $G' \leq M \cap A = M_0$, $Z(M) = \Phi(M) \leq Z(G)$ and $|M_0 : Z(M)| = 2$. Since $[G, M_0] \leq Z(M) \leq Z(G)$, we get that G is of class at most 3.

We claim that for each $x \in G - A$ we have $x^2 \in Z(M)$. Indeed, if $x \in M$, then $x^2 \in \Phi(M) = Z(M)$. If $x \in G - (A \cup M)$, then Lemma 57.1 implies that there is $y \in A$ such that $\langle x, y\rangle$ is minimal nonabelian. But $x \notin M$ and so $F = \langle x, y\rangle \cong D_8$. Then

$$x^2 \in F' \cap Z(F) \leq Z(G) \cap G' \leq Z(M).$$

Applying a result of Burnside on $G/Z(M)$, we get that each $x \in G - A$ inverts each element in $A/Z(M)$. Since $G' \leq M_0$, it follows that G/M_0 is abelian and so x centralizes A/M_0 and therefore A/M_0 is elementary abelian. Hence G/M_0 is elementary abelian.

Let $x \in G - (A \cup M)$, where $x^2 \in Z(M)$. Noting that $M_0 \trianglelefteq G$, we see that $K = M_0\langle x\rangle$ is nonabelian with $Z(K) = Z(M)$. Since $M \not\leq K$, Theorem 10.33 implies that K is quasidihedral and then $Z(M)$ is elementary abelian, where $Z(M) < H_2(K)$. It follows that $\exp(M) = 4$.

(i) Suppose that G is metacyclic. Since G has a subgroup isomorphic to D_8, Proposition 10.19 (a) implies that G is of maximal class. But we know that $cl(G) \leq 3$ and so $|G| = 2^4$, $cl(G) = 3$ and $G \cong SD_{16}$. We have obtained a group stated in part (a) of our theorem. From now on we assume that G is nonmetacyclic.

(ii) Assume that $Z(G) > Z(M)$. Then we have $Z(G) < A$. From the fact that G/M_0 is elementary abelian we get that $Z(G)/Z(M) \neq \{1\}$ is elementary abelian. We set $M = \langle m, n\rangle$. If there is an involution $j \in Z(G) - Z(M)$, then $\langle jm, n\rangle \cong M$ and $\langle jm, n\rangle \neq M$, a contradiction. Hence all elements in $Z(G) - Z(M)$ are of order 4 (noting that $Z(M)$ is elementary abelian). Let $v \in Z(G) - Z(M)$ so that $1 \neq v^2 \in Z(M)$. Then $\langle mv, n\rangle$ is minimal nonabelian and is distinct from M and so must be isomorphic to D_8.

Suppose for a moment that M is not isomorphic to Q_8. By Lemma 65.1 (and noting that $\exp(M) = 4$), we have exactly three remaining possibilities for the structure of M:

- If
$$M = \langle m, n \mid m^4 = n^4 = 1,\ [m, n] = m^2\rangle,$$
then we consider $S = \langle mv, n\rangle$, where $[mv, n] = [m, n] = m^2$. But then we have $\Phi(S) \geq \langle m^2, n^2\rangle \cong E_4$, a contradiction.
- If
$$M = \langle m, n \mid m^2 = n^4 = 1,\ [m, n] = k,\ k^2 = [m, k] = [n, k] = 1\rangle,$$
then we consider $S = \langle mv, n\rangle$, where $[mv, n] = [m, n] = k$. But $\Phi(S) \geq \langle k, n^2\rangle \cong E_4$ (noting that M is nonmetacyclic and so k is not a square in M), and this is a contradiction.
- Finally, if
$$M = \langle m, n \mid m^4 = n^4 = 1,\ [m, n] = k,\ k^2 = [m, k] = [n, k] = 1\rangle,$$
then we consider $S = \langle mv, n\rangle$, where $[mv, n] = [m, n] = k$. But $\Phi(S) \geq \langle k, n^2\rangle \cong E_4$ (noting that again M is nonmetacyclic and so k is not a square in M), and this is a contradiction.

We have proved that we must have $M \cong Q_8$. Set $Z(M) = \langle z\rangle \cong C_2$ and we have $M_0 \cong C_4$. Since for each $v \in Z(G) - Z(M)$, $v^2 = z$ and there are no involutions in $Z(G) - Z(M)$, we get $Z(G) = \langle v\rangle \cong C_4$.

Suppose that $M\langle v\rangle < G$. Take an element $x \in G - (A \cup M\langle v\rangle)$. We know that $x^2 \in \langle z\rangle = Z(M)$. If $o(x) = 2$, then we replace x with xv so that $o(xv) = 4$. Hence we may assume from the start that $x \in G - (A \cup M\langle v\rangle)$ with $o(x) = 4$ and so $x^2 = z$. Then considering $M_0\langle x\rangle$, we see that $M_0 \trianglelefteq G$ and $M_0 \nleq Z(G)$ so that $M_0\langle x\rangle \cong Q_8$ and $M_0\langle x\rangle \neq M$, a contradiction. Hence we must have $M * \langle v\rangle = G$ and so we have obtained a group stated in part (a) of our theorem. From now on we assume that $Z(M) = Z(G)$.

(iii) Assume that G is of class 2. Then $G' \leq Z(G) = Z(M)$ and $Z(M)$ is elementary abelian and so any two non-commuting elements in G generate a minimal nonabelian subgroup (see Lemma 65.2). Since M is not generated by its involutions (otherwise, $M \cong D_8$, a contradiction), there is an element $m \in M - M_0$ with $o(m) = 4$. We have $m^2 \in Z(M)$ and $C_G(m) = \langle m\rangle C_A(m) = \langle m\rangle Z(M)$. If all elements in $G - M$ are involutions, then (by a result of Burnside) M would be abelian, a contradiction. Hence there is $g \in G - M$ with $o(g) \geq 4$. By the above, $[g, m] \neq 1$ and so $\langle g, m\rangle \neq M$ is minimal nonabelian not isomorphic to D_8 (since $\langle g, m\rangle$ is generated by elements of order ≥ 4), a contradiction. We have proved that $cl(G) = 3$.

(iv) Suppose, by way of contradiction, that there is $a \in A - M_0$ such that $a^2 \in Z(M)$. Set $M^* = M\langle a\rangle$ so that $|M^* : M| = 2$. Note that $Z(M^*) = Z(M) = Z(G)$ and set $M_0^* = M^* \cap A$.

Since $\mho_1(M_0^*) \leq Z(M)$, we have $\mho_1(M^*) \leq Z(M)$ and so M^* is of class 2. There is an element $m \in M - M_0$ with $o(m) = 4$ and $C_{M^*}(m) = \langle m \rangle Z(M)$. Also, there is $m^* \in M^* - M$ with $o(m^*) = 4$ and $[m, m^*] \neq 1$. Then $\langle m, m^* \rangle \neq M$ is minimal nonabelian and is generated by elements of order 4 and so is not isomorphic to D_8, a contradiction. We have proved that for each $x \in A - M_0$, we have $x^2 \in M_0 - Z(M)$ and then the fact that A/M_0 is elementary abelian implies that $|A : M_0| = |G : M| = 2$.

(v) Suppose that M_0 is elementary abelian. Then A is abelian of type $(4, 2, \ldots, 2)$ and $\mho_1(A) \cong C_2$ with $M_0 = Z(M) \times \mho_1(A)$. But then $\mho_1(A) \leq Z(G)$ and so $M_0 \leq Z(G)$, a contradiction. We have proved that M_0 is not elementary abelian and so all elements in $M_0 - Z(M)$ are of order 4. Since there is $a \in A - M_0$ with $a^2 \in M_0 - Z(M)$, we have $o(a) = 8$ and $\exp(A) = 8$. But all elements in $G - A$ are of order ≤ 4 and so $\exp(G) = 8$.

(vi) Now we can determine the structure of G. We have $A = \langle a \rangle Z(M)$, where $o(a) = 8$. Set $a^4 = z$ so that $\mho_1(M_0) = \langle z \rangle$, $\mho_1(A) = \langle a^2 \rangle \trianglelefteq G$, $\langle a^2 \rangle \nleq Z(G)$. For each $x \in G - (A \cup M)$, $x^2 \in Z(G)$ and $\langle a^2, x \rangle \neq M$ is minimal nonabelian and so $\langle a^2, x \rangle \cong D_8$ so that $o(x) = 2$. Since $\langle a^2 \rangle \nleq Z(G)$, we have $M' = \langle z \rangle$. Because z is a square in M, it follows (by Lemma 65.1) that M is metacyclic. Let $c \in M - M_0$ with $o(c) = 4$ (noting that all elements in $M - M_0$ cannot be involutions). If $c^2 = z$, then $M \cong Q_8$ and so $|G| = 2^4$. But $\exp(G) = 8$ and so G would be metacyclic, contrary to our assumption in (i). Hence $c^2 \neq z$ and then $M = \langle a^2, c \rangle$ and this is the metacyclic minimal nonabelian group of order 2^4 and exponent 4 and then $|G| = 2^5$.

By Lemma 1.1,

$$2^5 = |G| = 2|Z(G)||G'| = 2^3|G'| \quad \text{and so} \quad |G'| = 4.$$

But $\mathrm{cl}(G) = 3$ and so $G' \nleq Z(G)$ and therefore G' covers $M_0/Z(G)$ implying that $G' \cong C_4$. Since G is nonmetacyclic, $\langle a \rangle$ is not normal in G and so $G' \nleq \langle a \rangle$. Since $G' \leq M_0$, we get

$$[a, c] = a^2c^2z^\epsilon, \quad \epsilon = 0, 1, \quad \text{and so} \quad a^c = a^3c^2z^\epsilon.$$

We know that all elements in $G - (A \cup M)$ are involutions and so $o(ca) = 2$. We compute

$$1 = (ca)^2 = caca = c^2a^ca = c^2(a^3c^2z^\epsilon)a = c^4a^4z^\epsilon = a^4z^\epsilon,$$

and so we get $\epsilon = 1$. It follows that $[a, c] = a^2c^2z = a^6c^2$. The structure of G is completely determined as given in part (b) of our theorem. Since the converse is clear, our theorem is proved. □

Theorem 258.2. *Let G be a nonabelian 2-group which is not minimal nonabelian but all its minimal nonabelian subgroups, except one denoted with M, are isomorphic to Q_8. Then G has an abelian subgroup of index 2, G is of class 3, and one of the following holds:*

(a) *If G is metacyclic, then $G \cong SD_{16}$ and $M \cong D_8$.*

(b) *If G is nonmetacyclic, then*

$$G = \langle a, c \mid a^8 = c^4 = [a, c^2] = 1,\ [a, c] = a^2c^2,\ [a^2, c] = a^4 \rangle,$$

where

$$|G| = 2^5, \quad \exp(G) = 8, \quad G' = \langle a^2c^2\rangle \cong C_4, \quad \Omega_1(G) = Z(G) = \langle a^4, c^2\rangle \cong E_4,$$

$\langle a, c^2\rangle \cong C_8 \times C_2$ *is a unique abelian maximal subgroup in* G, $M = \langle a^2, c\rangle$ *and this is the metacyclic minimal nonabelian group of order* 2^4 *and exponent* 4.

Conversely, the above 2-groups satisfy the assumptions of our theorem.

Proof. Let G be a nonabelian 2-group which is not minimal nonabelian but all its minimal nonabelian subgroups, except one denoted with M, are isomorphic to Q_8. Then M is characteristic in G and so $M \trianglelefteq G$.

Suppose, by way of contradiction, that G does not have an abelian subgroup of index 2. Let A be a maximal normal abelian subgroup in G, where $|G : A| > 2$. Let C/A be a subgroup of order 4 in G/A. First suppose that $C/A \cong C_4$ and let D/A be the subgroup of order 2 in C/A. If all elements in $C - D$ lie in M, then $C \le M$. But M is minimal nonabelian and so in this case $D > A$ would be abelian, a contradiction. Hence there is $g \in C - D$ so that $g \notin M$ and $o(g) \ge 4$. By Lemma 57.1, there is $a \in A$ such that $\langle g, a\rangle$ is minimal nonabelian and so $\langle g, a\rangle \cong Q_8$. In this case $o(g) = 4$ and so g^2 is an involution in $D - A$. But $\langle a\rangle \le A$ contains another involution so that $\langle g, a\rangle$ possesses at least two involutions, a contradiction. We have proved that $C/A \cong E_4$.

Let C_i/A, $i = 1, 2, 3$, be three pairwise distinct subgroups of order 2 in C/A. Since M is contained in at most one of $\{C_1, C_2, C_3\}$, we may assume that M is not contained in C_1 and C_2. Consider a subgroup C_j ($j \in \{1, 2\}$) and take an element $c_j \in C_j - A$. Since C_j does not contain M, it follows that all minimal nonabelian subgroups in the nonabelian group C_j are isomorphic to Q_8. By Corollary A.17.3, $C_j \cong Q_{2^n} \times E_{2^m}$, where $n \ge 3$ and $m \ge 0$. Noting that A is an abelian maximal subgroup in C_j, we see that $o(c_j) = 4$ and c_j inverts each element in A. But then $c_1c_2 \notin A$ and c_1c_2 centralizes A, a contradiction. We have proved that G has an abelian subgroup A of index 2.

We have $M < G$, $G = AM$, $G/M \cong A/(M \cap A)$ is abelian and so $G' \le M \cap A = M_0$, where M_0 is a maximal subgroup in M containing $Z(M)$, $Z(M) = \Phi(M) \le Z(G)$ and $|M_0 : Z(M)| = 2$. Since $[G, M_0] \le Z(M) \le Z(G)$, we get that G is of class at most 3.

We claim that for each $x \in G - A$, we have $x^2 \in Z(M)$. Indeed, if $x \in M$, then $x^2 \in \Phi(M) = Z(M)$. If $x \in G - (A \cup M)$, then Lemma 57.1 implies that there is $y \in A$ such that $\langle x, y\rangle$ is minimal nonabelian. But $x \notin M$ and so $F = \langle x, y\rangle \cong Q_8$. Then

$$x^2 \in Z(F) \cap F' \le Z(G) \cap G' \le Z(G) \cap M_0 = Z(M).$$

Applying a result of Burnside on $G/Z(M)$, we see that each $x \in G - A$ inverts each element in $A/Z(M)$. Since $G' \le M_0$, it follows that G/M_0 is abelian and so x centralizes A/M_0 and therefore A/M_0 is elementary abelian. We have proved that G/M_0 is elementary abelian.

Let $x \in G - (A \cup M)$, where $x^2 \in Z(M)$. Noting that $M_0 \trianglelefteq G$, we see that $K = M_0\langle x\rangle$ is nonabelian with $Z(K) = Z(M)$. Indeed, assume that K is abelian. Then $M_0 = K \cap A \le Z(G)$, a contradiction. Since $M \nleq K$, Corollary A.17.3 implies that $K \cong Q_{2^n} \times E_{2^m}$, where $n \ge 3$

and $m \geq 0$, and so $Z(K)$ is elementary abelian. Since $Z(K) = Z(M)$, we see that $Z(M)$ is elementary abelian but (noting that $|K : Z(M)| = 4$) M_0 is not elementary abelian. Finally, $M/Z(M) \cong E_4$ and so $\exp(M) = 4$.

All above preliminary facts will be used in the rest of the proof without explicit quotations.

(i) Assume that G has non-commuting involutions. Since any two non-commuting involutions generate a dihedral subgroup, it follows that G has a minimal nonabelian subgroup isomorphic to D_8 and so $M \cong D_8$. Then $\langle v \rangle = M_0 \cong C_4$ and so $K = M_0\langle x \rangle \cong Q_8$, where $x \in G - (A \cup M)$. In particular, all elements in $M - M_0$ are involutions. Suppose that there is an involution $j \in A$ which does not commute with an involution $i \in M - M_0$. Then $\langle i, j \rangle \cong D_{2^k}$, $k \geq 3$. Since D_{2^k} is generated with its subgroups isomorphic to D_8, there is $D \cong D_8$ with $D \leq \langle i, j \rangle$ and $D \nleq M$, contrary to the uniqueness of $M \cong D_8$. Hence $\Omega_1(A)$ centralizes M. Assume that $\Omega_1(A) > \langle v^2 \rangle = Z(M)$ and let $j \in \Omega_1(A) - \langle v^2 \rangle$. Then for an $i \in M - M_0$, ij inverts $\langle v \rangle$ and so $\langle ij, v \rangle \cong D_8$ but $\langle ij, v \rangle \neq M$, a contradiction. Hence $\Omega_1(A) = \langle v^2 \rangle$ and so A is cyclic. Since $A > M_0$ and A/M_0 is elementary abelian, we get $A \cong C_8$. Since $i \in M - M_0$ inverts $\langle v \rangle$ and Q_8 is a subgroup in G, it follows that $G \cong SD_{16}$ and this is a group from part (a) of our theorem. Therefore from now on we assume that $\Omega_1(G)$ is elementary abelian. Then M is not isomorphic to D_8 or Q_8 and this implies that $Z(M) \cong E_4$ or E_8.

(ii) Suppose that $\Omega_1(A) > Z(M)$. Set $M = \langle m, n \rangle$, where $\Omega_1(A) \cap M_0 = Z(M)$, and let $c \in \Omega_1(A) - Z(M)$. If c centralizes M, then $M = \langle m, n \rangle \cong \langle mc, n \rangle$ but $M \neq \langle mc, n \rangle$, a contradiction. Hence c does not centralize M and so there is $a \in M - M_0$ such that $[a, c] \neq 1$ and so $o(a) = 4$. We consider the subgroup $X = \langle a, c \rangle$, where we set $a^2 = b$ and note that $[b, c] = 1$. Since D_8 is not a subgroup in X, Theorem A.105.1 gives that X is a nonmetacyclic minimal nonabelian group of order 2^4 with $X \neq M$ (since $c \notin M$), a contradiction. We have proved that $\Omega_1(A) = Z(M)$ and all elements in $G - (A \cup M)$ are of order 4.

(iii) Assume that there is an involution $i \in M - M_0$. Let $x \in G - M$ such that $x^2 \in Z(M)$. Then $o(x) = 4$ and x centralizes i. Indeed, suppose that $[x, i] \neq 1$. Theorem A.105.1 together with the fact that D_8 is not a subgroup in G implies that $\langle x, i \rangle$ is a nonmetacyclic minimal nonabelian group of order 2^4 distinct from M (since $x \notin M$), a contradiction. If $x \in A$, then x centralizes $M = \langle m, i \rangle$, where $m \in M_0 - Z(M)$. But then $\langle mx, i \rangle$ is minimal nonabelian (since $[mx, i] = [m, i]$) distinct from M and is not isomorphic to Q_8, a contradiction. It follows that $A = \langle a \rangle Z(M)$, where $a^2 \in M_0 - Z(M)$ and $o(a) = 8$. Then $C_A(i) = Z(M)$ and so $C_G(i) = \langle i \rangle \times Z(M)$ is elementary abelian. But (by the above) if $x \in G - (A \cup M)$, then $1 \neq x^2 \in Z(M)$, $o(x) = 4$ and x centralizes i, a contradiction. We have proved that there are no involutions in $M - A$ and so $\Omega_1(G) = Z(M) \cong E_4$ or E_8. By Lemma 65.1, we have one of the following:

$$M = \langle m, n \mid m^4 = n^4 = 1,\ [m, n] = m^2 \rangle,$$
$$M = \langle m, n \mid m^4 = n^4 = 1,\ [m, n] = s,\ s^2 = [s, m] = [s, n] = 1 \rangle.$$

(iv) We show that $Z(G) = Z(M) = \Omega_1(G)$. Indeed, suppose that there is $v \in Z(G)$ with $o(v) = 4$. Then $v \in A - M_0$ and $v^2 \in Z(M)$. If v^2 is a square in M, then there is $y \in M$ with $v^2 = y^2$. But then vy is an involution and $vy \notin M$, a contradiction. Hence $v^2 \neq m^2$ and $v^2 \neq n^2$. The subgroup $\langle vm, vn\rangle$ is minimal nonabelian distinct from M since $vm \notin M$. But then $(vm)^2 = v^2m^2 \neq 1$, $(vn)^2 = v^2n^2 \neq 1$ and $(vm)^2 \neq (vn)^2$ since $m^2 \neq n^2$ and so $\langle vm, vn\rangle$ is not isomorphic to Q_8, a contradiction.

(v) Suppose that for an element $x \in A - M$, we have $x^2 \in Z(M) = Z(G) = \Omega_1(G)$. Then we set $G_0 = M\langle x\rangle$ and $G_1 = G_0 \cap A$ and note that $Z(G_0) = Z(G)$. Since $G_0/Z(G_0)$ is elementary abelian, G_0 is of class 2. In this case we find $b \in G_1 - M$ and $c \in M - A$ such that $b^2 \neq c^2$. Then $1 \neq [b, c] \in Z(G)$ so that (by Lemma 65.2) $\langle b, c\rangle$ is minimal nonabelian. But $\langle b, c\rangle \neq M$ (since $b \notin M$) and $\langle b, c\rangle$ is not isomorphic to Q_8, and so we get a contradiction.

We have proved that for each $a \in A - M$, $a^2 \in M_0 - Z(M)$ and so $o(a) = 8$, $G_0 = G$, $A = \langle a\rangle Z(M)$ and therefore $|A : M_0| = |G : M| = 2$.

(vi) Now it is easy to determine the structure of G. By Lemma 1.1, $|G| = 2|Z(G)||G'|$ and so from $|G/Z(G)| = 2^3$ we get $|G'| = 4$. All elements in $G - A$ are of order 4 and they generate G. Therefore, if G is of class 2, then $\exp(G) = 4$, a contradiction. Hence G is of class 3 and noting that $G' \leq M_0$ and $|M_0 : Z(M)| = 2$, we see that G' covers $M_0/Z(M)$ so that $G' \cong C_4$.

Suppose that $|M| = 2^5$ and in this case M is nonmetacyclic. But $G' \trianglelefteq M$ and $G' \nleq Z(M)$ so that $M' < G'$ implying that an involution in M' is a square in M, a contradiction. We have proved that $|M| = 2^4$ and so M is the metacyclic minimal nonabelian group of order 2^4 and exponent 4 and $|G| = 2^5$. If G would be metacyclic, then the fact that Q_8 is a subgroup in G implies (by Proposition 10.19 (a)) that G is of maximal class, contrary to $|G| = 2^5$ and $cl(G) = 3$. Hence G is nonmetacylic. In particular, $G' \nleq \langle a\rangle$. Let $c \in M - M_0$ so that $o(c) = 4$ and $a^4 \neq c^2$. Then $[a, c^2] = 1$ and $[a^2, c] = a^4$. Set $a^4 = z$ and then $[a, c] = a^2c^2z^\epsilon$, where $\epsilon = 0, 1$, because $G' \nleq \langle a\rangle$. This gives $a^c = a^3c^2z^\epsilon$. We know that all elements in $G - (A \cup M)$ are of order 4 and so $o(ca) = 4$. We compute

$$(ca)^2 = caca = c^2a^ca = c^2(a^3c^2z^\epsilon)a = c^4a^4z^\epsilon = a^4z^\epsilon,$$

and so we get $\epsilon = 0$. It follows that $[a, c] = a^2c^2$. The structure of G is completely determined as given in part (b) of our theorem. Since the converse is clear, our theorem is proved. □

Problem 1. Classify the 2-groups all of whose minimal nonabelian subgroups, except one, are isomorphic to Q_8.

Problem 2. Classify the 2-groups all of whose minimal nonabelian subgroups, except one, have order 8.

Problem 3. Classify the p-groups containing at most one minimal nonabelian subgroup not isomorphic to M_{p^n}.

§ 259 Nonabelian p-groups, $p > 2$, all of whose minimal nonabelian subgroups are isomorphic to M_{p^3}

It is known (see Mann's comment to #115) that if all minimal nonabelian subgroups of a nonabelian p-group G of exponent $p > 2$ are isomorphic to $S(p^3)$, then G has an abelian subgroup of index p. Below we consider the nonabelian p-groups of exponent $> p^2$, $p > 2$, all of whose minimal nonabelian subgroups are isomorphic to M_{p^3}. In what follows the following result from [HogK] is used: If G is a metabelian p-group, then its H_p-subgroup has index $\le p$.

Theorem 259.1. *Let G be a nonabelian p-group, $p > 2$, of exponent $> p^2$ all of whose minimal nonabelian subgroups are isomorphic to M_{p^3}. Then G has an abelian subgroup A of index p and all elements in $G - A$ are of order p^2.*

Proof. Let G be a nonabelian p-group, $p > 2$, of exponent $> p^2$ all of whose minimal nonabelian subgroups are isomorphic to M_{p^3}.

(i) Assume, by way of contradiction, that G has no abelian subgroup of index p. Let A be a maximal abelian subgroup of exponent $> p^2$. Let X be any subgroup such that $A < X \le G$ and $|X : A| = p$. By Lemma 57.1, for any $x \in X - A$, there is $a \in A$ such that the subgroup $\langle x, a\rangle$ is minimal nonabelian and so $\langle x, a\rangle \cong M_{p^3}$. In particular, $o(x) \le p^2$ and so $x^p \in \Omega_1(A)$. Thus, all elements in the set $G - A$ have orders $\le p^2$. Since all elements of order $\ge p^3$ in X lie in A and they generate A, it follows that A is characteristic in X. Hence if C is such that $X < C \le G$ and $|C : X| = p$, then $A \trianglelefteq C$.

First assume that $C/A \cong C_{p^2}$ and let $c \in C - X$ so that $\langle c\rangle$ covers C/A and therefore $o(c) \ge p^2$. By Lemma 57.1, there is $b \in A$ such that the subgroup $\langle b, c\rangle$ is minimal nonabelian and so $\langle b, c\rangle \cong M_{p^3}$. In particular,

$$o(c) = p^2, \quad \langle c\rangle \cap A = \{1\}, \quad \langle b, c\rangle \cap A = \langle b\rangle \cong C_p.$$

But in this case, $\langle b\rangle \trianglelefteq \langle b, c\rangle$ and so $\langle b, c\rangle = \langle b\rangle \times \langle c\rangle$ is abelian, a contradiction. Thus, we have $\exp(G/A) = p$.

We have proved that, if $|C/A| = p^2$, then $C/A \cong E_{p^2}$ and then all elements in the set $C - A$ are of order $\le p^2$ so that for each $g \in C - A$, $g^p \in \Omega_1(A)$. Since $\exp(A/\Omega_1(A)) \ge p^2$, we have

$$H_p(C/\Omega_1(A)) = A/\Omega_1(A) > \{1\}. \tag{1}$$

Note that $C/\Omega_1(A)$ is metabelian and so, applying a result of Hogan–Kappe [HogK] to the group $C/\Omega_1(A)$, we see that $A/\Omega_1(A)$ has index p in $C/\Omega_1(A)$. But this contradicts equality (1). Thus, the group G/A has no subgroup of order p^2. We have proved that G has an abelian subgroup A of index p. By Lemma 57.1, all elements in the set $G - A$ are of order $\le p^2$ and $\exp(A) \ge p^3$.

https://doi.org/10.1515/9783110533149-003

(ii) Let $M \cong M_{p^3}$ be a minimal nonabelian subgroup in G so that $A \cap M = M_0 \trianglelefteq G$ is of order p^2 and

$$Z(M) = M' = \mho_1(M) = \langle z \rangle \cong C_p, \quad \text{where } \langle z \rangle \leq Z(G),\ C_G(M_0) = A. \tag{2}$$

Indeed, $C_G(z) \geq AM = G$.

If $G' = \langle z \rangle$ (and so G is of class 2 so regular since $p > 2$), then the fact that G is generated by its elements of order $\leq p^2$ (since G is generated by its subgroups isomorphic to M_{p^3}, by Proposition 10.28) would imply that $\exp(G) = p^2$, a contradiction. Hence

$$|G'| > p. \tag{3}$$

Let $g \in G - A$ so that $o(g) \leq p^2$ and $\langle g \rangle$ acts nontrivially on M_0 (here we use the last equality in (2)). Hence, the subgroup $M_0\langle g \rangle$ is minimal nonabelian and so $M_0\langle g \rangle \cong M_{p^3}$ implying that $g^p \in \langle z \rangle$ (indeed, $g^p \in Z(G)$). Hence, for each $g \in G - A$, $g^p \in \langle z \rangle$. But, by (3), $G/\langle z \rangle$ is nonabelian with an abelian maximal subgroup $A/\langle z \rangle$ of exponent $> p$ and all elements in $G/\langle z \rangle - A/\langle z \rangle$ are of order p. Then each minimal nonabelian subgroup in $G/\langle z \rangle$ is generated by elements of order p and so is isomorphic to $S(p^3)$, by Lemma 65.1.

Let $K/\langle z \rangle$ be a subgroup in $G/\langle z \rangle$ isomorphic to $S(p^3)$. Then K covers G/A and $K_0 = K \cap A$ is of order p^3 since $|K| = p^4$. Since $S(p^3)$ is not a subgroup in G, it follows that K does not split over $\langle z \rangle$ and so $d(K) = 2$. If $|K'| = p$, then by Lemma 65.2 (a), K would be minimal nonabelian of order p^4, a contradiction. Hence $|K'| = p^2$ and so by Lemma 1.1, $Z(K) = \langle z \rangle$ is of order p. Hence K is of class 3 and K_0 is an abelian maximal subgroup in K. By Proposition 145.4, we get $K' \cong E_{p^2}$. But $C_G(K') = A$ and so for each $g \in G - A$, $\langle g \rangle K'$ is minimal nonabelian and so $\langle g \rangle K' \cong M_{p^3}$. It follows that $o(g) = p^2$ and $1 \neq g^p \in \langle z \rangle$. The proof is complete. □

Problem 1. Classify the nonabelian p-groups, $p > 2$, all of whose minimal nonabelian subgroups are isomorphic M_{p^3}.

Problem 2. Classify the nonabelian p-groups, $p > 2$, all of whose minimal nonabelian subgroups have order p^3. (See § 90, where a similar problem is solved for 2-groups.)

§ 260 *p*-groups with many modular subgroups M_{p^n}

A modular p-group M_{p^n} of order p^n, $n > 3$, is defined with

$$M_{p^n} = \langle g, t \mid g^{p^{n-1}} = t^p = 1,\ [g, t] = g^{p^{n-2}} \rangle.$$

Since this is a minimal nonabelian group, it is clear that if a p-group G has many such modular subgroups, then the structure of G must be very restricted.

The first author has proposed to consider the case, where a p-group G is of exponent $> p$ and for each cyclic subgroup M of order p^2 we have $C_G(M) \cong M_{p^n}$ for some $n > 3$. It is surprising that in that case we must have $p = 2$ and G is a unique group of order 2^5 with $\Omega_2(G) \cong D_8 \times C_2$. In fact, we shall prove the following result.

Theorem 260.1. *Let G be p-group of exponent $> p$ such that for each cyclic subgroup M of order p^2 we have $C_G(M) \cong M_{p^n}$ for some $n > 3$. Then $p = 2$ and G is a faithful and non-splitting extension of an elementary abelian group of order 8 by C_4 given by*

$$G = \langle a, e_1 \mid a^8 = e_1^2 = 1,\ a^4 = z,\ [a, e_1] = e_2,\ e_2^2 = [e_1, e_2] = 1,\ [a, e_2] = z \rangle.$$

Here $E = \langle e_1, e_2, z \rangle$ is a self-centralizing normal elementary abelian subgroup of order 8 with

$$G/E \cong C_4, \quad Z(G) = \langle z \rangle \cong C_2, \quad G' = \langle z, e_2 \rangle \cong E_4$$

and $\Omega_2(G) = \langle E, a^2 \rangle \cong D_8 \times C_2$.

Conversely, the above group of order 2^5 satisfies the assumptions of the theorem.

Proof. Let G be p-group of exponent $> p$ such that for each cyclic subgroup M of order p^2 we have $C_G(M) \cong M_{p^n}$ for some $n > 3$ depending of M.

Let $\langle a \rangle \cong C_{p^2}$ be any cyclic subgroup of order p^2 in G. Then $A = C_G(a) \cong M_{p^{m+1}}$ with $m \geq 3$. Let $\langle z \rangle$ be a cyclic subgroup of index p in A so that $\langle z \rangle \cong C_{p^m}$ and $\langle z \rangle \geq \langle a \rangle$. Let t be an element of order p in $A - \langle z \rangle$ so that

$$\Omega_1(A) = \langle a^p, t \rangle \cong E_{p^2} \quad \text{and} \quad \Omega_2(A) = \langle t, a \rangle = \langle t \rangle \times \langle a \rangle \cong C_p \times C_{p^2}.$$

Then

$$A_0 = \langle t \rangle \times \langle z^p \rangle \cong C_p \times C_{p^{m-1}}$$

is an abelian maximal subgroup in A and $a \in A_0$. Hence A_0 is also a maximal abelian subgroup in G since $C_G(a) = A$.

Set $b = at (\in A_0)$ so that $o(b) = p^2$, $C_A(b) = A_0$ and $B = C_G(b) \cong M_{p^{m'+1}}$ with $m' \geq 3$. We also have $b \nleq Z(A)$ and $B < A$ is maximal cyclic. Note that A_0 is a maximal abelian subgroup in G and so A_0 is a maximal abelian subgroup in B. In particular, $Z(B) \leq A_0$ and $Z(B) \geq \langle b \rangle$. Hence $Z(B) = \langle b \rangle \cong C_{p^2}$ and so $m' = 3$ and $B \cong M_{p^4}$. As A_0 is maximal in A and B, we get $m = 3$. We have proved that for each cyclic subgroup $\langle a \rangle$ of order p^2 in G we have $C_G(a) \cong M_{p^4}$ and so $\exp(G) = p^3$.

If G has no normal subgroup isomorphic to E_{p^2}, then G is cyclic or a 2-group of maximal class (Lemma 1.4). But then M_{p^4} is not a subgroup in G, a contradiction. Hence G has a normal subgroup U isomorphic to E_{p^2}.

https://doi.org/10.1515/9783110533149-004

If a is any element of order p^2 in G, then a is a p-th power of an element $g \in G$ so that a centralizes U (see the previous sentence and consider the nonabelian subgroup $\langle g \rangle U$) and so $C_G(a) = U\langle g \rangle \cong M_{p^4}$, where $\Omega_1(C_G(a)) = U$, $\langle g \rangle \cap U \cong C_p$ and $\langle g \rangle$ acts nontrivially on U. In particular, $G_0 = C_G(U)$ is of index p in G, $\exp(G_0) = p^2$ and all elements in $G - G_0$ are of order p^3. If $h \in G - G_0$, then $\langle h \rangle \cap U = \langle z \rangle \cong C_p$, where $\langle z \rangle$ is a fixed subgroup of order p in U. Also, U is a unique normal subgroup isomorphic to E_{p^2}. If G has a cyclic subgroup $\langle x \rangle \cong C_{p^2}$ contained in $Z(G)$, then, by hypothesis, $G \cong M_{p^4}$. But then G has another cyclic subgroup $\langle y \rangle$ of order p^2 with $C_G(y) \cong C_p \times C_{p^2}$, a contradiction. It follows that $Z(G) = \langle z \rangle \cong C_p$.

Maximal abelian subgroups in G are isomorphic to C_{p^3} or to $C_p \times C_{p^2}$ or they are elementary abelian. Hence if A is a maximal normal abelian subgroup in G containing U, then either $A \cong C_p \times C_{p^2}$ or A is elementary abelian of order $\geq p^3$. Indeed, if $A \cong E_{p^2}$, then $|G| = p^3$, a contradiction.

(i) Suppose that a maximal normal abelian subgroup A containing U is isomorphic to $C_p \times C_{p^2}$. Let $a \in A$ with $\langle a \rangle \cong C_{p^2}$ so that setting $a^p = z$ we have $\langle z \rangle = Z(G) \cong C_p$. Set $B = C_G(a)(\cong M_{p^4})$, where $A = \Omega_2(B)$ so that all elements in $B - A$ are of order p^3 and for each $b \in B - A$, we have $\langle b^p \rangle = \langle a \rangle$ since $b^i nZ(B) = \langle a \rangle$. Let $t \in U - \langle z \rangle$ so that $a' = at$ is another element of order p^2 with $\langle a' \rangle \neq \langle a \rangle$ and, setting $C = C_G(a')(\cong M_{p^4})$, we have for each $c \in C - A$, $\langle c^p \rangle = \langle a' \rangle$. This implies $B \cap C = A$ and so $G > B$. Set $G_0 = C_G(U)$ so that $|G : G_0| = p$, $\Omega_2(G) = G_0$ and $G_0 > A$. Let $A < D \leq G_0$ with $|D : A| = p$ so that $Z(D) = U$. By Lemma 1.1, $|D'| = p$ and so D is of class 2. In fact, we have $C_{G_0}(a) = A$ and $C_{G_0}(a') = A$ since $\exp(G_0) = p^2$.

Suppose that $p > 2$. Then G/A is isomorphic to a subgroup of $S(p^3)$ and so $\exp(G/A) = p$. If there is $d \in D - A$ with $o(d) = p^2$, then there is $g \in G$ with $g^p = d$, contrary to $\exp(G/A) = p$. Hence all elements in $D - A$ are of order p. Let $d \in D - A$. Then

$$(da)^p = d^p a^p [a, d]^{\binom{p}{2}} = a^p = z$$

and so $da \in D - A$ and $o(da) = p^2$, a contradiction.

We have proved that we must have $p = 2$. In this case G/A is isomorphic to a subgroup of D_8. Suppose that $G/A \cong D_8$ and in this case $|G_0/A| = 4$ and note that $\exp(G_0) = 4$. If $G_0/A \cong C_4$, then there is $v \in G_0 - A$ such that $\langle v \rangle$ covers G_0/A and so $o(v) = 4$ and $\langle v \rangle \cap A = \{1\}$. But then

$$C_G(v) \geq \langle v, U \rangle \cong C_4 \times E_4,$$

a contradiction. Hence we have $G_0/A \cong E_4$. If $x \in G_0 - A$, then x is either an involution or $o(x) = 4$ with $x^2 \in U$. In the second case there is $g \in G - G_0$ with $g^2 = x$. Since $C_G(U) = G_0$, we have in that case $x^2 = z$, where $\langle z \rangle = Z(G)$. We have proved that $\mho_1(G_0) = \langle z \rangle$ and so $G_0/\langle z \rangle$ is elementary abelian implying $G_0' = \langle z \rangle \cong C_2$. But we know that $C_{G_0}(a) = A$ and so a has four conjugates in G_0, a contradiction.

We have proved that $|G_0/A| = 2$ and so $G_0 = D$ and $|G/A| = 4$ implying $|G| = 2^5$. But B/A, C/A and D/A are three pairwise distinct subgroups of order 2 in G/A and so $G/A \cong E_4$. If there is an element $d \in D - A$ of order 4, then there is $g \in G$ with $g^2 = d$, contrary to $G/A \cong E_4$. Hence all elements in $D - A$ are involutions and so (by a result of Burnside) if $d \in D - A$, then d inverts each element in A implying that D is quasi-dihedral and so $D \cong D_8 \times C_2$. But $D = G_0 = \Omega_2(G)$ and so by the results of § 52 and by Theorem 52.2 (a) for $n = 2$, G is isomorphic to the group of order 2^5 stated in our theorem.

(ii) Suppose that a maximal normal abelian subgroup A containing U is elementary abelian of order $\geq p^3$. Let $U < E \leq A$ with $E \trianglelefteq G$ and $|E : U| = p$ with $E \cong E_{p^3}$. Set $B = C_G(E) \trianglelefteq G$ so that $\exp(B) = p$. If $p > 2$, then G/B is isomorphic to s subgroup of $S(p^3)$ and then $\exp(G) = p^2$, a contradiction. It follows $p = 2$ and in this case B is elementary abelian implying $B = A$ and G/A is isomorphic to a subgroup of D_8. For an element $x \in G - A$ of order 4 we have $x^2 = z$, $1 \neq z \in U$, $\langle z \rangle = Z(G)$ and $C_A(x) = U \cong E_4$. This implies that either $A = E$ or $A \cong E_{16}$. Set $C = A\langle g \rangle$ so that $Z(C) = U$.

First assume that $A \cong E_{16}$. Then the fact that $C_A(g) = U \cong E_4$ implies (with an elementary argument) that setting $U = \langle z, t \rangle$, the set $C - A$ consists of four involutions, four elements whose squares are equal to z, four elements whose squares are equal to t and four elements whose squares are equal to tz. Let $c \in C - A$ with $c^2 = t$. Then there is $g \in G - G_0$ so that $g^2 = c$, where $G_0 = C_G(U)$ and $|G : G_0| = 2$. But then $U \leq Z(G)$, a contradiction. We have proved that $A = E \cong E_8$. Since all elements in $G - G_0$ are of order 8, it follows that $G/A \cong C_4$ and so $|G| = 2^5$.

Let $a \in G - G_0$ so that $o(a) = 8$ and $\langle a \rangle \cap A = \langle z \rangle = Z(G)$. Then

$$M = C_G(a^2) = \langle a \rangle U \cong M_{2^4}, \quad \text{where } \langle a^2 \rangle = Z(M).$$

But $|G : M| = 2$ and so $\langle a^2 \rangle \trianglelefteq G$. Let $e \in A - U$ and $t \in U - \langle z \rangle$. Then

$$\langle a^2, e \rangle \cong D_8 \quad \text{and} \quad A\langle a^2, e \rangle = \langle a^2, e \rangle \times \langle t \rangle \cong D_8 \times C_2.$$

Since

$$\Omega_2(G) = A\langle a^2 \rangle \cong D_8 \times C_2,$$

the results of § 52 and Theorem 52.2 (a) for $n = 2$ imply that G is isomorphic again to the group of order 2^5 stated in our theorem and we are done. □

Problem 1. Study the nonabelian p-groups, $p > 2$, all of whose minimal nonabelian subgroups have order p^3.

Problem 2. Study the p-groups all of whose $\mathcal{A}_2$-subgroups are nonmetacyclic. (See § 270 where the p-groups with all metacyclic $\mathcal{A}_2$-subgroups are classifies.)

Problem 3. Classify the p-groups all of whose minimal nonabelian subgroups have exactly p cyclic subgroups of index p.

Problem 4. Classify the p-groups all of whose minimal nonabelian subgroups, except ONE, have exactly p cyclic subgroups of index p.

Problem 5. What will be if $n \geq 3$ in Theorem 260.1?

§ 261 Nonabelian p-groups of exponent $> p$ with a small number of maximal abelian subgroups of exponent $> p$

Let G be a nonabelian p-group of exponent $> p$. If S is a minimal nonabelian subgroup in G, then S has exactly $p + 1$ maximal subgroups $S_1, S_2, \dots, S_{p+1}$ and they are abelian and lie in $p + 1$ pairwise distinct maximal abelian subgroups in G. If at least two of the subgroups S_i are elementary abelian, then S is generated by its elements of order p and then, by Lemma 65.1, $S \cong \mathrm{D}_8$ or $S \cong \mathrm{S}(p^3)$ (the nonabelian group of order p^3 and exponent $p > 2$) or $S \cong \mathrm{Q}_8$. If all minimal nonabelian subgroups of G are generated by elements of order p, then G has only one maximal abelian subgroup A of exponent $> p$, where A is of index p in G and $A = \mathrm{H}_p(G)$ (Hughes subgroup). However, if a minimal nonabelian subgroup of G has at most one elementary abelian maximal subgroup, then G has at least p maximal abelian subgroups of exponent $> p$.

From the above it follows that a nonabelian p-group G of exponent $> p$ has either exactly one maximal abelian subgroup of exponent $> p$ or G has at least p of them. Therefore the first author has proposed to classify nonabelian p-groups of exponent $> p$ which have exactly p maximal abelian subgroups of exponent $> p$ and this is done here in Theorem 261.1 for $p = 2$ and in Theorem 261.2 for $p > 2$. By the above, such a group G possesses a minimal nonabelian subgroup S which is not isomorphic to D_8 or $\mathrm{S}(p^3)$. Also, such an S has exactly one maximal subgroup X which is elementary abelian so that $\Phi(S) = \mathrm{Z}(S)$ is elementary abelian and $|S : \Phi(S)| = p^2$. Let $a \in S - X$ and $b \in X - \Phi(S)$ so that $o(a) \le p^2$, $o(b) = p$ and $S = \langle a, b\rangle$, where $\Phi(S) = \langle a^p, [a, b]\rangle$. If $|\Phi(S)| = p$, then $|S| = p^3$ and $S \cong \mathrm{M}_{p^3}$ (the nonabelian group of order p^3 and exponent p^2, where $p > 2$). If $|\Phi(S)| = p^2$, then $S \cong \mathrm{M}_p(2, 1, 1)$, where

$$\mathrm{M}_p(2, 1, 1) = \langle a, b \mid a^{p^2} = b^p = 1,\ [a, b] = c,\ c^p = [c, a] = [c, b] = 1\rangle.$$

Suppose that G possesses a nonnormal maximal abelian subgroup H of exponent $> p$. Set $K = \mathrm{N}_G(H)$ so that $|G : K| = p$, $H < K$ and $H^G \le K$. All elements in $G - K$ are of order p. If $p = 2$, then K is abelian (by a result of Burnside), a contradiction. Hence in this case we must have $p > 2$. For any $g \in G - K$, $H^g \le K$ and so H and H^g normalize each other.

The first author has also proposed to consider the next critical case, where G has exactly $p + 1$ maximal abelian subgroups of exponent $> p$. However, we have been able to classify such p-groups only in case $p = 2$ in Theorem 261.3.

Theorem 261.1. *Let G be a nonabelian 2-group with exactly two maximal abelian subgroups of exponent > 2. Then $G = M \times V$, where*

$$M \cong \mathrm{M}_2(2, 1, 1) = \langle a, b \mid a^4 = b^2 = 1,\ [a, b] = c,\ c^2 = [c, a] = [c, b] = 1\rangle$$

and $\exp(V) \le 2$.

https://doi.org/10.1515/9783110533149-005

Proof. Let G be a nonabelian 2-group with exactly two maximal abelian subgroups of exponent > 2. Let H_1 and H_2 be the two maximal abelian subgroups of exponent > 2, where we know that H_1 and H_2 are normal in G. If $H_1H_2 < G$, then all elements in $G - (H_1H_2)$ are involutions and then (by a result of Burnside) H_1H_2 would be abelian, a contradiction. Hence $H_1H_2 = G$ and $H_1 \cap H_2 = Z(G)$ so that G is of class 2 and all elements in $G - (H_1 \cup H_2)$ are involutions. Indeed, all elements of order > 2 lie in H_1 or H_2 (by our hypothesis). If $g \in G - (H_1 \cup H_2)$, then a maximal abelian subgroup H containing $\langle g\rangle$ is elementary abelian implying that $Z(G)$ is elementary abelian. Since $H \trianglelefteq G$ (noting that G is of class 2), Lemma 57.1 implies that for any $x \in G - H$ there is $h \in H$ such that $\langle x, h\rangle$ is minimal nonabelian. Since $\langle x, h\rangle \cong D_8$ or $M_2(2, 1, 1)$, it follows that $\exp(\langle x, h\rangle) = 4$ and so $o(x) \leq 4$. We have proved that $\exp(G) = 4$. For any $x, y \in G$, $[x^2, y] = [x, y]^2 = 1$ and so we get $\mho_1(G) \leq Z(G)$.

Suppose that both H_1 and H_2 are not maximal subgroups in G. Then $|H_i : Z(G)| \geq 4$ for $i = 1, 2$ and let $h_i \in H_i - Z(G)$ be an element of order 4 ($i = 1, 2$) so that $1 \neq h_i^2 \in Z(G)$. Let H_i^* be a maximal subgroup of H_i which contains $Z(G)\langle h_i\rangle$, $i = 1, 2$. Then $M_1 = H_1H_2^*$ and $M_2 = H_2H_1^*$ are distinct maximal subgroups of G containing H_1 and H_2, respectively. Since all elements in $G - (H_1 \cup H_2)$ are involutions, it follows that all elements in $G - (M_1 \cup M_2)$ are involutions. Let $g \in G - (M_1 \cup M_2)$ and $m \in M_1 \cap M_2$. Then g and $gm \in G - (M_1 \cup M_2)$ are involutions and so we get

$$1 = (gm)^2 = gmgm = g^2m^gm = m^gm \quad \text{and so} \quad m^g = m^{-1}.$$

Hence g inverts each element in $M_1 \cap M_2$ so that a result of Burnside implies that $M_1 \cap M_2$ is abelian. In particular, $\langle h_1, h_2\rangle$ is abelian. Let Y be a maximal abelian subgroup in G containing $\langle h_1, h_2\rangle$. By our hypothesis, $Y = H_1$ or $Y = H_2$, a contradiction. We have proved that we may assume $|G : H_1| = 2$ and so H_1 is a maximal subgroup in G.

Let H_1^* be a maximal subgroup of H_1 containing $\Omega_1(H_1)$. Then $M_2 = H_2H_1^*$ is a maximal subgroup of G and all elements in $G - (H_1 \cup M_2)$ are involutions. If $g \in G - (H_1 \cup M_2)$, then for any $x \in H_1^* = H_1 \cap M_2$, $gx \in G - (H_1 \cup M_2)$ is an involution. This implies $x^g = x^{-1}$ and so g inverts each element in H_1^*. In particular, g centralizes $\Omega_1(H_1)$. It follows that $\Omega_1(H_1) \leq Z(G)$ and so $\Omega_1(H_1) = Z(G) = H_1 \cap H_2$ and therefore all elements in $H_1 - Z(G)$ are of order 4.

Suppose that $Z(G)$ is not a maximal subgroup in H_1. Note that all elements in $G - (H_1 \cup H_2)$ are involutions and all elements in $H_2 - H_1$ and in $H_1 - Z(G)$ are of order 4. Let $v \in H_1 - Z(G)$ so that $v^2 \in Z(G)$ and let H_1^{**} be a maximal subgroup of H_1 containing $Z(G)\langle v\rangle$ so that $M_2^* = H_1^{**}H_2$ is a maximal subgroup in G. If $g \in G - (H_1 \cup M_2^*)$, then g and $gv \in G - (H_1 \cup M_2^*)$ are involutions implying that $v^g = v^{-1}$. Then each element in $G - H_1$ also inverts $\langle v\rangle$. Indeed, for any $h \in H_1$, $v^{hg} = v^g = v^{-1}$. Hence each element in $G - H_1$ inverts each element of order 4 in H_1 and since it also centralizes $Z(G)$, it follows that each element in $G - H_1$ inverts each element in H_1. But then G is quasidihedral (since there are involutions in $G - (H_1 \cup H_2)$) and so all elements in $G - H_1$ must be involutions, a contradiction. We have proved that $Z(G) = H_1 \cap H_2$ is a maximal subgroup in H_1 and so H_2 is also a maximal subgroup in G.

If each minimal nonabelian subgroup in G is isomorphic to D_8, then by Theorem 10.33 our group G is quasidihedral and so G has only one maximal abelian subgroup of exponent > 2, a contradiction. Hence G possesses a minimal nonabelian subgroup

$$M \cong M_2(2,1,1) = \langle a, b \mid a^4 = b^2 = 1,\ [a,b] = c,\ c^2 = [c,a] = [c,b] = 1\rangle.$$

Then M covers G/H_1 and $H_1/Z(G)$ and $M \cap H_1$ is abelian of type $(4,2)$, where we have $M \cap Z(G) \cong E_4$. Indeed, if M does not cover G/H_1 or $H_1/Z(G)$, then M would be abelian, a contradiction. Let V be a complement of $M \cap Z(G)$ in $Z(G)$. Then $G = M \times V$ and our theorem is proved. □

Theorem 261.2. *Let G be a nonabelian p-group of exponent $> p$, where $p > 2$. Suppose that G has exactly p maximal abelian subgroups $H_1, H_2, \ldots, H_p$ of exponent $> p$. Then* $\exp(G) = p^2$, $Z(G)$ *is elementary abelian, each H_i normalizes each H_j $(i, j = 1, 2, \ldots, p)$, $H = H_1H_2 \ldots H_p = H_p(G)$ (Hughes subgroup) and $\mho_1(G) \le Z(H) = H_1 \cap H_2 \cap \cdots \cap H_p$.*

Proof. Let G be a p-group, $p > 2$, satisfying the assumptions of Theorem 261.2. It is easy to see that G possesses at least one minimal nonabelian subgroup M which is isomorphic to M_{p^3} or $M_p(2,1,1)$. Suppose that this is false. Then all minimal nonabelian subgroups of G are isomorphic do $S(p^3)$ and so G has an abelian subgroup A of exponent $> p$ and index p such that $A = H_p(G)$. But then G has only one maximal abelian subgroup of exponent $> p$, a contradiction. Hence there is such M as above. Any two maximal subgroups of M lie in two distinct maximal abelian subgroups in G. In this way we get p pairwise distinct maximal abelian subgroups in G of exponent $> p$ and one maximal abelian subgroup which is elementary abelian. In particular, $Z(G)$ is elementary abelian.

We want to show that $\exp(G) = p^2$. Let $H_1, H_2, \ldots, H_p$ be the set of all p maximal abelian subgroups in G which are of exponent $> p$. Set $\exp(G) = p^e$, where $e \ge 2$ and let g be an element of order p^e so that $g \in H = H_1H_2 \ldots H_p$, where we know that each H_i normalizes each H_j. If g is not contained in all H_i $(i = 1, 2, \ldots, p)$, say $g \notin H_1$, then by Lemma 57.1, there is $h_1 \in H_1$ such that $\langle g, h_1\rangle$ is minimal nonabelian. Since all minimal nonabelian subgroups of G are of exponent $\le p^2$, we get $e = 2$. So suppose that $g \in H_i$ for all $i = 1, 2, \ldots, p$. In particular, $g \in H_1 \cap H_2$. Since $\langle H_2 - H_1\rangle = H_2$, there is $h \in H_2 - H_1$ such that $o(h) = p^e$. By Lemma 57.1, there is $k \in H_1$ such that $\langle h, k\rangle$ is minimal nonabelian. This implies again $e = 2$. We have proved that $\exp(G) = p^2$. If $H < G$, then all elements in $G - H$ are of order p and so $H = H_p(G)$. Now, $Z(H)$ centralizes all H_i and so $Z(H) \le H_1 \cap H_2 \cap \cdots \cap H_p$. But $H_1 \cap H_2 \cap \cdots \cap H_p \le Z(H)$ and so we get $Z(H) = H_1 \cap H_2 \cap \cdots \cap H_p$.

Let g be any element of order p^2 in G. Then $g \in H = H_1H_2 \ldots H_p$, where $H_i \trianglelefteq H$ for all $i = 1, 2, \ldots, p$. We have either $g \in H_i$ (and then also $g^p \in H_i$) or (by Lemma 57.1) there is $h_i \in H_i$ such that $M = \langle g, h_i\rangle$ is minimal nonabelian, where $M \cong M_{p^3}$ or $M \cong M_p(2,1,1)$. Then we know that M contains exactly one maximal subgroup X of exponent p^2 such that $X \le H_i$. This implies that $g^p \in X \le H_i$. Hence in any case we get $g^p \in H_i$ for all $i = 1, 2, \dots, p$. It follows that $g^p \in H_1 \cap H_2 \cap \cdots \cap H_p = Z(H)$ and so $\mho_1(G) \leq Z(H)$. Our theorem is proved. □

Theorem 261.3. *Let G be a nonabelian 2-group with exactly three maximal abelian subgroups H_1, H_2, H_3 of exponent > 2. Then $G = H_1H_2H_3$ and $Z(G) = H_1 \cap H_2 \cap H_3$.*

(a) *If H_1 is conjugate in G to (say) H_2, then $\exp(H_1) = 4$, H_3 is of index 2 in G with $\exp(H_3) \leq 8$, $Z(G)$ is elementary abelian and G has a maximal subgroup which is quasidihedral of exponent 4. (A smallest example of such groups is SD_{16}.)*

(b) *If all H_i are normal in G, $i = 1, 2, 3$, then G is of class 2, $\mho_1(G) \leq Z(G)$ and so G' is elementary abelian and we have the following two possibilities:*
   (1) *If $G \neq H_1 \cup H_2 \cup H_3$, then $Z(G)$ is elementary abelian.*
   (2) *If $G = H_1 \cup H_2 \cup H_3$, then H_i are maximal subgroups in G, $i = 1, 2, 3$, $G/Z(G) \cong E_4$ and $|G'| = 2$.*

Proof. Let G be a nonabelian 2-group with exactly three maximal abelian subgroups H_1, H_2, H_3 of exponent > 2. Set $H = \langle H_1, H_2, H_3 \rangle$ so that $H \trianglelefteq G$. If $H < G$, then all elements in $G - H$ are involutions. But then (by a result of Burnside) H is abelian, a contradiction. Hence we have $G = \langle H_1, H_2, H_3 \rangle$ and then obviously $Z(G) = H_1 \cap H_2 \cap H_3$.

(i) First we consider the case where some H_i are not normal in G. Then we may assume that H_1 and H_2 are conjugate in G and then $H_3 \trianglelefteq G$. We set $K = N_G(H_1)$ so that $|G : K| = 2$, $H_1 < K$ and $K = N_G(H_2)$. For any $g \in G - K$, $H_2 = H_1^g$ and $H_1H_2 = H_1^G$. Then H_3 covers $G/(H_1H_2)$ so that $G = (H_1H_2)H_3$. All elements in $G - (K \cup H_3)$ are involutions and so for each involution $i \in G - (K \cup H_3)$, a maximal abelian subgroup in G containing i is elementary abelian. In particular, $Z(G)$ is elementary abelian.

Set $G_1 = H_1H_3$ and let $g \in H_3 - K$ so that $H_2 = H_1^g \leq G_1$. It follows $G_1 = G$ and set $H_3^* = H_3 \cap K$ so that H_3^* normalizes H_1. We have that $H_1 \cap H_3 = Z(G)$ is elementary abelian and also $H_2 \cap H_3 = Z(G)$. Then $K = H_1H_3^*$ and

$$K' \leq H_1 \cap H_3^* = Z(G) \leq Z(K)$$

so that K is of class 2 and K' is elementary abelian. For any $k_1, k_2 \in K$ it follows that $[k_1^2, k_2] = [k_1, k_2]^2 = 1$ and so $\mho_1(K) \leq Z(K)$. We have $Z(K) < H_1$ and if $Z(K) > Z(G)$, then $Z(K)H_3^*$ is contained in a maximal abelian subgroup in G distinct from H_1, H_2 and H_3 and so $Z(K)H_3^*$ must be elementary abelian. We have proved that in any case $Z(K)$ is elementary abelian and so $\exp(K) = 4$ and $4 \leq \exp(H_3) \leq 8$.

Assume, by way of contradiction, that $Z(K) > Z(G)$. Since $Z(K) < H_1$, it follows that $L = Z(K)H_3$ is a proper subgroup of G. We know that all elements in $G - (K \cup L)$ are involutions. Let $i \in G - (K \cup L)$ and $x \in K \cap L$. Then $ix \in G - (K \cup L)$ and so

$$1 = (ix)^2 = ixix \quad \text{implying} \quad x^i = x^{-1}.$$

Since i inverts each element in $K \cap L$, it follows that i centralizes $Z(K)$ (noting that $Z(K)$ is elementary abelian). But then $Z(K) \leq Z(G)$, a contradiction. We have proved that $Z(K) = Z(G)$ and so, in particular, $\mho_1(H_1) \leq Z(G)$.

Suppose, by way of contradiction, that H_3 is not a maximal subgroup in G. Let v be an element of order 4 in H_1 so that $v^2 \in Z(K) = Z(G)$ and we set $R = H_3\langle v\rangle$. Since $|R : H_3| = 2$, it follows that R is a proper subgroup of G and all elements in $G - (K \cup R)$ are involutions. If $i \in G - (K \cup R)$ and $y \in K \cap R$, then $iy \in G - (K \cup R)$ so that iy is an involution implying $y^i = y^{-1}$. Thus i inverts each element in $K \cap R = \langle v\rangle H_3^*$ implying that $K \cap R$ is abelian. Let X be a maximal abelian subgroup of G containing $K \cap R$. Since X is obviously distinct from each H_i, $i = 1, 2, 3$, and $\exp(X) > 2$, we have a contradiction. We have proved that H_3 is a maximal subgroup in G.

All elements in $G - (K \cup H_3)$ are involutions, where K and H_3 are two distinct maximal subgroups in G. Then each involution $i \in G - (K \cup H_3)$ inverts each element in $K \cap H_3 = H_3^*$. In particular, i centralizes $\Omega_1(H_3^*)$ and so $\Omega_1(H_3^*) = H_1 \cap H_3 = Z(G)$. Since $H_1 \cap H_3 < H_3^*$, it follows that $\exp(H_3^*) = 4$. Then $H_3^*\langle i\rangle$ is quasidihedral of exponent 4 and $H_3^*\langle i\rangle$ is a maximal subgroup in G. Finally, $H_1 \cap H_3^* = Z(G)$ is a maximal subgroup of H_1 and so $\exp(H_1) = 4$ and $G = H_1H_3 = H_1H_2H_3$. We have proved all properties of G stated in part (a) of our theorem.

(ii) Assume that all H_i are normal in G, $i = 1, 2, 3$. Then we have again $G = H_1H_2H_3$.

(ii1) First suppose that H_1, H_2 and H_3 do not cover G. Then $G - (H_1 \cup H_2 \cup H_3)$ is not empty so that all elements in $G-(H_1 \cup H_2 \cup H_3)$ are involutions. Let $i \in G-(H_1 \cup H_2 \cup H_3)$ and let A be a maximal abelian subgroup in G containing i so that A is distinct from H_1, H_2 and H_3 implying that A must be elementary abelian. Since $Z(G) < A$, it follows that $Z(G)$ is elementary abelian.

It is easy to see that $\exp(G) = 4$. Suppose that $g \in G$ with $o(g) \geq 8$. For any $i \in \{1, 2, 3\}$, we have either $g \in H_i$ (and then also $g^2 \in H_i$) or $g \in G - H_i$. In the second case Lemma 57.1 implies that there is $h \in H_i$ such that $M = \langle g, h\rangle$ is minimal nonabelian. Since $\exp(M) \geq 8$, each of the three maximal subgroups M_i $(i = 1, 2, 3)$ of M are of exponent > 2 and they lie in three pairwise distinct maximal abelian subgroups H_1, H_2, H_3 of exponent > 2 in G. Hence for an $j \in \{1, 2, 3\}$, we have $M_j \leq H_i$ and then $g^2 \in M_j \leq H_i$. We have proved that in any case $g^2 \in H_i$ for each $i \in \{1, 2, 3\}$ and so $g^2 \in H_1 \cap H_2 \cap H_3 = Z(G)$. But $Z(G)$ is elementary abelian and so $o(g^2) \leq 2$, a contradiction. We have proved that $\exp(G) = 4$.

Suppose that there is $h \in G$ of order 4 such that $h^2 \notin Z(G)$. Since all elements of order 4 in G are contained in $H_1 \cup H_2 \cup H_3$, we may assume that $h \in H_1$. Then interchanging H_2 and H_3 (if necessary), we may assume that $h^2 \notin H_2$. Set $K_0 = H_1H_2$ so that $Z(K_0) = H_1 \cap H_2$ and $h^2 \notin Z(K_0)$. We have $K_0' \leq H_1 \cap H_2 = Z(K_0)$ and so K_0 is of class 2. Suppose, by way of contradiction, that $\exp(Z(K_0)) = 4$. Let $k \in K_0 - (H_1 \cup H_2)$ and let B be a maximal abelian subgroup of G containing $Z(K_0)\langle k\rangle$ so that we must have $B = H_3$. But then $H_3 \geq Z(K_0)$ and so $Z(K_0) = H_1 \cap H_2 \cap H_3 = Z(G)$, a contradiction. Hence $Z(K_0)$ is elementary abelian. But then for all $x \in K_0$, $[h^2, x] = [h, x]^2 = 1$ and so $h^2 \in Z(K_0)$, a final contradiction. We have proved that $\mho_1(G) \leq Z(G)$ implying that G' is elementary abelian and so we have obtained some 2-groups from part (b1) of our theorem.

(ii2) Now assume that $G = H_1 \cup H_2 \cup H_3$, i.e., H_1, H_2, H_3 cover G. By Lemma 116.3 (b), H_i is a maximal subgroup of G for $i = 1, 2, 3$. Then $H_1 \cap H_2 = Z(G)$ and so $G/Z(G) \cong E_4$. By Lemma 1.1, $|G'| = 2$ and our theorem is proved. □

§ 262 Nonabelian p-groups all of whose subgroups are powerful

We recall that a p-group G is powerful if either $p > 2$ and $G' \leq \mho_1(G)$ or $p = 2$ and $G' \leq \mho_2(G)$.

A metacyclic 2-group G is called ordinary metacyclic (with respect to a subgroup H) if G has a cyclic normal subgroup H such that G/H is cyclic and $[G, H] \leq \mho_2(H)$. Then each subgroup U of G is ordinary metacyclic with respect to $H \cap U$.

For convenience we state some known results which we need here.

Lemma 1 (Theorem 26.29). *All subgroups of a 2-group G are powerful if and only if G is D_8-free and Q_8-free.*

Lemma 2 (Theorem 26.27). *A two-generator 2-group is powerful if and only if it is ordinary metacyclic.*

Lemma 3 (Exercise 26.1). *All subgroups in a p-group G with $p > 2$ are powerful if and only if each two-generator subgroup in G is metacyclic.*

Lemma 4 (Theorem 80.1). *If G is a minimal non-Q_8-free 2-group, then G has a normal subgroup N such that $G/N \cong Q_8$ and $d(G) = 2$.*

Lemma 5 (see paragraph preceding Remark 1 in § 78). *If G is a minimal non-D_8-free 2-group, then G has a normal subgroup N such that $G/N \cong D_8$ and $d(G) = 2$.*

Now we prove the following result presenting the solution of Problem 2824 (i).

Theorem 262.1. *Let G be a nonabelian p-group such that each two-generator subgroup is metacyclic when $p > 2$ and ordinary metacyclic when $p = 2$. Then all subgroups of G are powerful.*

Proof. If $p > 2$, then our result follows from Lemma 3.

From now on assume that $p = 2$. First suppose that G is not Q_8-free. Let H be a minimal non-Q_8-free subgroup in G. Then $d(H) = 2$ and H possesses a normal subgroup H_0 such that $H/H_0 \cong Q_8$ (by Lemma 4). By our hypothesis, H is ordinary metacyclic and then each subgroup of H is also ordinary metacyclic. By Lemma 2, each subgroup of H is powerful. But then by Lemma 1, H is Q_8-free, a contradiction. We have proved that G is Q_8-free.

Similarly, suppose that G is not D_8-free. Let K be a minimal non-D_8-free subgroup in G. Then $d(K) = 2$ (by Lemma 5) and K possesses a normal subgroup K_0 such that $K/K_0 \cong D_8$. By our hypothesis, K is ordinary metacyclic and then each subgroup of K is also ordinary metacyclic. By Lemma 2, each subgroup of K is powerful. But then by Lemma 1, K is D_8-free, a contradiction. We have proved that G is D_8-free.

Since G is Q_8-free and D_8-free, Lemma 1 implies that each subgroup of G is powerful and we are done. □

https://doi.org/10.1515/9783110533149-006

§ 263 Nonabelian 2-groups G with $C_G(x) \le H$ for all $H \in \Gamma_1$ and $x \in H - Z(G)$

In this section we solve Problem 3702 for $p = 2$ and prove the following result.

Theorem 263.1. *Let G be a nonabelian 2-group with $C_G(x) \le H$ for all $H \in \Gamma_1$ and $x \in H - Z(G)$. Then we have $\Phi(G) = Z(G)$ and each maximal abelian subgroup of G has order $2|Z(G)|$.*

Conversely, all such nonabelian 2-groups satisfy the assumptions of our theorem.

Proof. Let G be a nonabelian 2-group with $C_G(x) \le H$ for all $H \in \Gamma_1$ and $x \in H - Z(G)$.

Let H be a fixed maximal subgroup in G. Then $Z(G) < H$ so that $Z(G) \le \Phi(G)$. For each $g \in G - H$, $g^2 \in Z(G)$ (indeed, $C_G(g^2) \not\le H$) and so by a result of Burnside applied to $G/Z(G)$, we have that $H/Z(G)$ is abelian and g inverts each element of $H/Z(G)$ (indeed, all elements in the set $G/Z(G) - H/Z(G)$ have order 2). Hence each maximal subgroup of $G/Z(G)$ is abelian and so we have either $G' \le Z(G)$ (and then $\mathrm{cl}(G) = 2$) or $G/Z(G)$ is minimal nonabelian. Suppose that we have the second case. Since minimal nonabelian (two generator) group $G/Z(G)$ is generated with involutions, we get $G/Z(G) \cong D_8$ and since g inverts each element of $H/Z(G)$, it follows that $H/Z(G) \cong C_4$ so that H is abelian and G is not of class 2. But H was an arbitrary maximal subgroup in G so that each maximal subgroup of G is abelian. This implies that $|G'| = 2$ and so $G' \le Z(G)$ and G is of class 2, a contradiction.

We have proved that G is of class 2. Since $G/Z(G)$ is abelian and is generated by involutions, $G/Z(G)$ is elementary abelian and so $\Phi(G) = \mho_1(G) \le Z(G)$. But we have also $Z(G) \le \Phi(G)$ and so $\Phi(G) = Z(G)$.

Suppose that G possesses an abelian subgroup $A > Z(G)$ such that $A/Z(G) \cong E_4$. Then $A < G$. Let K be a maximal subgroup of G which does not contain A. Then $(A \cap K) > Z(G)$ and if $x \in (A \cap K) - Z(G)$, then $C_G(x) \not\le K$ since $A \le C_G(a)$, a contradiction. We have proved that each maximal abelian subgroup in G is of order $2|Z(G)|$. The theorem is proved since the converse is clear. □

https://doi.org/10.1515/9783110533149-007

§ 264 Nonabelian 2-groups of exponent ≥ 16 all of whose minimal nonabelian subgroups, except one, have order 8

In § 90 we have determined the nonabelian 2-groups all of whose minimal nonabelian subgroups are of order 8, i.e., they are isomorphic to D_8 or Q_8. According to an idea of the first author, we study here such 2-groups G, where we allow one exception, i.e., G has exactly one minimal nonabelian subgroup of order > 8. In studying such 2-groups we shall use often the well-known Lemma 57.1 and the following:

Lemma 57.2. *Let G be a nonabelian 2-group of exponent ≥ 8 all of whose minimal nonabelian subgroups are of exponent 4. Then for any maximal normal abelian subgroup A in G we have either G/A is cyclic of order ≤ 4 or $G/A \cong Q_8$.*

Here we prove the following result:

Theorem 264.1. *Let G be a nonabelian 2-group of exponent ≥ 16 which is not minimal nonabelian but all its minimal nonabelian subgroups, except one denoted with M, have order 8. Then one of the following holds:*

(a) *G has an abelian subgroup A of index 2 and all elements in $G - (A \cup M)$ are of order ≤ 4.*

(b) *We have $|G : M| = 2$ and all elements in $G - M$ are of order ≤ 4.*

Proof. Let G be a nonabelian 2-group of exponent ≥ 16 which is not minimal nonabelian but all its minimal nonabelian subgroups, except one denoted with M, have order 8.

If A is any maximal normal abelian subgroup in G, then G/A is elementary abelian. Indeed, let $A < C \leq G$ with $C/A \cong C_4$ and let D/A be the subgroup of order 2 in C/A. If all $x \in C - D$ lie in M, then $C \leq M$ and since M is minimal nonabelian, D would be abelian, contrary to $C_G(A) = A$. Hence there is $g \in C - D$ such that $g \notin M$. By Lemma 57.1, there is $a \in A$ such that $\langle a, g\rangle$ is minimal nonabelian. But $\langle a, g\rangle \neq M$ and so $\langle a, g\rangle \cong D_8$ or Q_8. Noting that $\langle g\rangle$ covers $C/A \cong C_4$, we get $o(g) = 4$, $\langle g\rangle \cap A = \{1\}$ and $\langle a, g\rangle = \langle a\rangle \times \langle g\rangle$ with $o(a) = 2$, a contradiction.

If A is an abelian subgroup of index 2 in G, then $G = AM$ and all elements in $G - (A \cup M)$ are of order ≤ 4. Indeed, if $x \in G - (A \cup M)$, then by Lemma 57.1, there is $a \in A$ such that $\langle x, a\rangle$ is minimal nonabelian. But $\langle x, a\rangle \neq M$ and so $\langle x, a\rangle \cong D_8$ or Q_8 implying $o(x) \leq 4$.

From now on we assume that G has no abelian subgroup of index 2 and A will always denote a maximal normal abelian subgroup of the maximal possible exponent. Then $\exp(A) > 2$. Indeed, let $A < C \leq G$ with $|C/A| = 4$ so that (by the above) $C/A \cong E_4$. Let C_i/A, $i = 1, 2, 3$, be the three subgroups of order 2 in C/A. Since M is contained in at most one of $\{C_1, C_2, C_3\}$, we may assume that M is not contained in C_1 and C_2.

https://doi.org/10.1515/9783110533149-008

Suppose for a moment that A is elementary abelian. Then $C_1 \trianglelefteq G$ and M and Q_8 are not contained in C_1. Hence all minimal nonabelian subgroups of the nonabelian group C_1 are isomorphic to D_8. Then by Theorem 10.33, C_1 is quasidihedral, i.e., C_1 has an abelian subgroup H of exponent > 2 and index 2 and $H = H_2(C_1)$. But H is characteristic in C_1 and so $H \trianglelefteq G$. Let K be a maximal normal abelian subgroup in G containing H so that $\exp(K) > 2$. This is a contradiction to the assumption that A was a maximal normal abelian subgroup of the maximal possible exponent.

Now assume that setting $M_0 = M \cap A$, we have $|M/M_0| > 2$ so that $M/M_0 \cong E_4$ and then $M_0 = \Phi(M) = Z(M)$ and $|M_0| \geq 4$. Set $C = AM$ and let C_i/A, $i = 1, 2, 3$, be the three subgroups of order 2 in C/A. We have $M < C$ and M covers $C/A \cong E_4$. Indeed, if $M = C$, then C_i would be abelian, a contradiction. Note that $M_0 = Z(M) \leq Z(C)$ and so for each $m \in M - M_0$ (using Lemma 57.1) there is $a \in A - M_0$ such that $\langle m, a\rangle$ is minimal nonabelian. But $\langle m, a\rangle \neq M$ and therefore $\langle m, a\rangle \cong D_8$ or Q_8 and so $o(m) \leq 4$. Hence if $m_1, m_2 \in M - M_0$ are such that $M = \langle m_1, m_2\rangle$, then

$$M_0 = \Phi(M) = \langle m_1^2, m_2^2, M'\rangle$$

implying that M_0 is elementary abelian and so $\exp(M) = 4$. But then each minimal nonabelian subgroup in G is of exponent 4, $\exp(G) \geq 16$ (by our basic hypothesis) and G/A is elementary abelian of order ≥ 4 and so by Lemma 57.2 we get a contradiction.

We have proved that setting $M_0 = M \cap A$, we have $|M : M_0| = 2$. Set $H = MA$ so that $Z(M) \leq Z(H)$. If $\exp(M) = 4$, then Lemma 57.2 again gives a contradiction (since G/A is elementary abelian of order ≥ 4). Hence $\exp(M) \geq 8$ and so there are elements of order ≥ 8 in $M - M_0$.

Assume that $M < H = AM$ so that the set $S = H - (A \cup M)$ is nonempty and (by Lemma 57.1) it consists of elements of order ≤ 4. For any $h \in S$ and $m_0 \in M_0 - Z(M)$, we have $[h, m_0] \neq 1$. Indeed, if $[h, m_0] = 1$, then $m_0 \in Z(H)$, a contradiction. Set $X = M_0\langle h\rangle$ so that $X' \leq Z(M) \leq Z(H)$ and therefore $\mathrm{cl}(X) = 2$. We get

$$[h, m_0] \in Z(M) \quad \text{and} \quad [h, m_0]^2 = [h^2, m_0] = 1$$

and so

$$\langle h, m_0\rangle' = \langle [h, m_0]\rangle \cong C_2.$$

Therefore by Lemma 65.2 (a), $\langle h, m_0\rangle$ is minimal nonabelian. Since $\langle h, m_0\rangle \neq M$, we get $o(m_0) \leq 4$. It follows that all elements in $M_0 - Z(M)$ are of order ≤ 4 and so $\exp(M_0) \leq 4$ and $\exp(M) = 8$.

Let $m \in M - M_0$ with $o(m) = 8$ and set $v = m^2$ and $v^2 = z$. Since

$$C_4 \cong \langle v\rangle \leq Z(M) \leq Z(H),$$

it follows that there are elements of order 4 in $S = H - (A \cup M)$ and let h be one of them. Let k be any element of order 4 in $M_0 - Z(M)$. By the above, $\langle h, k\rangle$ is minimal nonabelian distinct from M. Since $\langle h\rangle$ and $\langle k\rangle$ are distinct cyclic subgroups of order 4 in $\langle h, k\rangle$, we get $\langle h, k\rangle \cong Q_8$. In particular, we get $h^2 = k^2 \in Z(M)$ for a fixed element h of order 4 in S and an arbitrary element k of order 4 in $M_0 - Z(M)$. It follows that $\mho_1(M_0) = \langle h^2\rangle$

and $h^2 = z$. Then kv is an involution in $M_0 - Z(M)$. Hence in any case there is an in involution t in $M_0 - Z(M)$. It follows that $M = \langle m, t\rangle$ and so we have either

$$M \cong M_{2^4} = \langle m, t \mid m^8 = t^2 = 1,\ [m, t] = m^4 = z\rangle$$

or

$$M \cong M_2(3, 1, 1) = \langle m, t \mid m^8 = t^2 = 1,\ [m, t] = z',\ (z')^2 = [z', m] = [z', t] = 1\rangle.$$

By the structure of M, in both cases M has exactly two conjugate classes of cyclic subgroups of order 8 and let $Y = \langle m\rangle$ be one of them. If $C_G(Y) \nleq M$, then there is $c \in G - M$ such that c centralizes Y. Then $ct \notin M$ and $\langle ct, Y\rangle \neq M$ is minimal nonabelian of order $\geq 2^4$, a contradiction. It follows that $C_G(Y) \leq M$.

Assume that $|A : M_0| = 2$ and since $\exp(M_0) = 4$, it follows that $\exp(A) \leq 8$. Note that (using Lemma 57.1) for each $g \in G - H$, there is a minimal nonabelian subgroup L containing g and so $L \cong D_8$ or Q_8 implying that $o(g) \leq 4$. It follows that $\exp(G) = 8$, contrary to our basic hypothesis. Hence $|A : M_0| = |H : M| > 2$ and note that $M \trianglelefteq G$ so that we can act on the two conjugate classes of cyclic subgroups of order 8 in M. It follows that

$$|N_G(Y)/N_M(Y)| = 4$$

so that (noting that $C_G(Y) \leq M$) there is $b \in N_G(Y) - N_M(Y)$ such that $\langle Y, b\rangle$ is minimal nonabelian of order $\geq 2^4$ and $\langle Y, b\rangle \neq M$, a contradiction.

We have proved that $A < M$ and so $|M : A| = 2$. All elements in $G - M$ are of order ≤ 4 (where we are using again Lemma 57.1) and so by our basic hypothesis we have $\exp(M) \geq 16$ implying that $\exp(A) \geq 8$. Let V be a maximal subgroup of G containing A with $M \nleq V$ (where we are using the fact that G/A is elementary abelian). Then all minimal nonabelian subgroups of V are of exponent 4. By Lemma 57.2 applied to V, we get $|V : A| = 2$ and so $|G : M| = 2$, where all elements in $G - M$ are of order ≤ 4. Our theorem is proved. □

§ 265 p-groups all of whose regular subgroups are either absolutely regular or of exponent p

In some cases, knowledge of abelian subgroups of a p-group G allows us to recover the structure of G. For example, if any abelian subgroup of G is two-generator, there is a good description of G (see Theorem 13.7 and § 50). In subsection 1° we describe completely the p-groups all of whose abelian subgroups are either elementary or cyclic. In subsection 2° we describe the p-groups in which all regular subgroups are either absolutely regular or have exponent p. We classify the p-groups of maximal class in terms of their regular subgroups (Theorem 265.2 and Corollary 265.3). Since regular 2-groups are abelian, the partial case of Theorem 265.2 is considered in Lemma 265.1 and used in what follows. The offered proofs are based on some results of §§ 12–13.

1°. We begin with the following:

Definition 1. A p-group G all of whose noncyclic abelian subgroups have exponent p is said to be a CE-*group* and we write $G \in$ CE.

The property CE is inherited by subgroups. Each 2-group of maximal class is a CE-group: all its abelian subgroups are either cyclic or of type $(2, 2)$. Also, all groups of exponent p are CE-groups. Next we assume that our CE-group G is neither cyclic nor of exponent p. If $p > 2$, then any p-group G of maximal class and order $> p^{p+1}$ is not a CE-group (indeed, their fundamental subgroup G_1 contains a noncyclic abelian subgroup of exponent $> p$). Note that Σ_{p^2} (of order p^{p+1}) is a CE-group. A p-group G is a CE-group if and only if it has no abelian subgroup of type (p^2, p). It follows that the group M_{p^n} is not a CE-group for $n > 3$.

Let G be a nonabelian CE-group, and let L be a maximal abelian subgroup of G of exponent $> p$; then L is cyclic. By Lemma 1.4, if G is not a 2-group of maximal class, there is in G a noncyclic maximal abelian subgroup A of exponent p (indeed, if $R < G$ is abelian of type (p, p), then any maximal abelian subgroup A of G containing R, is elementary abelian, by hypothesis). As $Z(G) \le L \cap A$, we get $|Z(G)| = p$. It follows that if $H \le G$ is nonabelian and $\exp(H) > p$, then $|Z(H)| = p$. Therefore, if $S \le G$ is minimal nonabelian, then $|Z(S)| = p$ implies $|S| = p^3$ (Lemma 65.1). In that case,

$$S \in \{D_8, Q_8, M_{p^3}, S(p^3)\}. \tag{1}$$

Any extraspecial p-group G of exponent $> p$ and order $\ge p^5$ is not a CE-group since each such group contains an abelian subgroup of type (p^2, p).

Now we are ready to prove the following result.

Lemma 265.1. *A nonabelian p-group G of order $> p^3$ is a* CE*-group if and only if one of the following holds:*

(i) *G is of exponent p.*

(ii) *G is a 2-group of maximal class.*

https://doi.org/10.1515/9783110533149-009

(iii) $p > 2$, *G is of maximal class with the fundamental subgroup of exponent p (in that case,* $|G| \le p^{p+1}$*).*

Proof. One may assume that G is nonabelian and $\exp(G) = p^e > p$. Let $L < G$ be cyclic of order p^e and $L < M \le G$, where $|M : L| = p$. As M is a CE-group, it follows that either $M \cong \mathrm{M}_{p^3}$, $p > 2$, or $p = 2$ and M is of maximal class. In both cases, M is of maximal class. By Exercise 10.10, G is a p-group of maximal class. Any 2-group of maximal class satisfies the hypothesis.

Next we assume that G is a p-group of maximal class, $p > 2$. If the above chosen $L < G_1$ and $\mathrm{M}_{p^3} \cong M < G_1$, then $\mathrm{C}_{G_1}(L) = L$ so that G_1 is of maximal class (Proposition 1.8), which is a contradiction. Thus, $\exp(G_1) = p$ so that $|G| \le p^{p+1}$. All members of the set Γ_1 that have exponent p^2 are of maximal class. □

If G is a p-group of maximal class with an abelian subgroup of index p, then each its nonabelian subgroup is of maximal class. The converse is also true (see Remark 1).

Remark 1. Let G be a nonabelian p-group in which any nonabelian subgroup has center of order p. Let $R \lhd G$ be of order p^2. Then $A = \mathrm{C}_G(R) \in \Gamma_1$ is abelian, by hypothesis. By Lemma 1.1, $|G : G'| = p|\mathrm{Z}(G)| = p^2$. We prove, by induction on $|G|$, that G is of maximal class. This is a case if $|G| = p^3$. Let $|G| > p^3$ and set $\overline{G} = G/\mathrm{Z}(G)$. Note that $\mathrm{Z}(G) < G'$. One has

$$|\mathrm{Z}(\overline{G})| = \frac{1}{p}|\overline{G} : \overline{G}'| = p.$$

Therefore, by induction, $\overline{G}$ is of maximal class. As $|\mathrm{Z}(G)| = p$, it follows that G is also of maximal class. This argument allows us to give another proof of Lemma 265.1. First of all, any 2-group of maximal class satisfies the condition of that theorem. Let $p > 2$ and $|G| > p^3$. By Theorem 1.2, A is noncyclic so it is elementary abelian. Let $A = \mathrm{Z}(G) \times L$. Then $L_G = \{1\}$ and, since

$$|G : L| = |G : A||A : L| = |G : A||\mathrm{Z}(G)| = p^2,$$

we conclude that G is a subgroup of $\Sigma_{p^2} \in \mathrm{Syl}_p(\mathrm{S}_{p^2})$. The proof is complete.

Exercise 1. Classify the p-groups which are minimal non-CE-groups.

Solution. Our group G has no proper abelian subgroup of type (p^2, p) and $\exp(G) > p$. One may assume that $|G| > p^3$. Therefore, if G is abelian, it is of type (p^2, p). Now let G be nonabelian. As it is not a 2-group of maximal class, it has a normal subgroup $R \cong \mathrm{E}_{p^2}$ (Lemma 1.4). Set $C = \mathrm{C}_G(R)$. If $C = G$, it has no cyclic subgroups of order p^2 so $\exp(G) = p$, a contradiction. Thus, $C < G$. Again C has no cyclic subgroup of order p^2 so that $\exp(C) = p$ and $\exp(G) = p^2$. Let $L < G$ be cyclic of order p^2. As G has no abelian subgroup of type (p^2, p), we get $\mathrm{C}_G(L) = L$ so that G is of maximal class (Proposition 1.8). As above, the fundamental subgroup G_1 of G has exponent p so G is a CE-group, a contradiction. Thus, G is abelian of type (p^2, p). □

Exercise 2. Study the nonabelian p-groups all of nonabelian subgroups of exponent $> p$ have cyclic centers.

2°. Recall that a p-group X is said to be absolutely regular if $|X/\mho_1(X)| < p^p$. By Theorem 9.8 (a), absolutely regular p-groups are regular.

Definition 2. A p-group G is said to be an AE-group if each its regular subgroup is either absolutely regular or has exponent p.

It follows that a 2-group which is an AE-group is also a CE-group (indeed, regular 2-groups are abelian and absolutely regular 2-groups are cyclic). The p-groups that are either absolutely regular or of exponent p are AE-groups. A p-group of order $\leq p^p$ is an AE-group so any irregular group of order p^{p+1} is an AE-group. Moreover, any p-group of maximal class is an AE-group (see Theorems 9.5, 9.6, 13.19). Indeed, let a p-group G be of maximal class and order $> p^{p+1}$. To prove our assertion, we use induction on $|G|$. All irregular maximal subgroups of G are of maximal class so AE-groups, by induction. Next, G contains exactly one regular maximal subgroup and it is absolutely regular (Theorem 9.6) so an AE-group. It follows that G is an AE-group.

Theorem 265.2. *If a p-group G is an* AE*-group, then one and only one of the following holds:*
(a) *G is absolutely regular.*
(b) $\exp(G) = p$.
(c) *G is irregular of maximal class.*

Proof. Up to the end of the proof we assume that we are given a p-group $G \in$ AE. If G is regular, it is either absolutely regular or of exponent p, by definition. In what follows we assume that G is irregular of order $> p^{p+1}$. By the above remark, in view of Lemma 265.1, one may assume that $p > 2$.

Let $C < G$ be cyclic of order $> p$ and let $C < S < G$, where S is a maximal regular subgroup of G. Then, by hypothesis, S is absolutely regular. If $S < R \leq G$, where $|R : S| = p$, then R is irregular. In that case, R is either of maximal class or $R = S\Omega_1(R)$, where $\Omega_1(R)$ is of order p^p and exponent p (Theorem 12.1 (b)).

(i) Suppose that for any choice, the subgroup R is of maximal class. It follows from Exercise 10.10 that the group G is also of maximal class. Such a G, by the above, is an AE-group.

(ii) Suppose that G is not of maximal class. Then, for some choice, R is not of maximal class, by (i). In that case, $|R| > p^{p+1}$ (indeed, any irregular group of order p^{p+1} is of maximal class, by Theorem 7.1 (b)) and $R = S\Omega_1(R)$, where $\Omega_1(R)$ is of order p^p and exponent p (Theorem 12.1 (b)). Let $\Omega_1(R) < T < R$, where $|T : \Omega_1(R)| = p$ (T exists since $|R| > p^{p+1} = p|\Omega_1(G)|$). In that case, $\exp(T) > p$ and T is not absolutely regular. Besides, $\exp(T) = p^2 > p$. Since $|T| = p^{p+1}$, it follows that T is of maximal class. Thus, all subgroups of R containing $\Omega_1(R)$ as a subgroup of index p are of maximal class. It follows from Exercise 10.10 that the subgroup R is of maximal class, which is a contradiction. □

The following corollary yields a subgroup characterization of irregular p-groups of maximal class.

Corollary 265.3. *An irregular p-group is of maximal class if and only if all its regular subgroups are either absolutely regular or of exponent p.*

Exercise 3. Classify the irregular p-groups all of whose maximal subgroups, except one, are absolutely regular. (Example: M_{2^n}.)

Exercise 4. Any nonabelian metacyclic p-group G of order $> p^3$ is not a CE-group, unless it is a 2-group of maximal class.

Hint. Consider $C_G(\Omega_1(G))$. □

Exercise 5. Given $n > 1$, study the nonabelian p-groups all of whose abelian subgroups of exponent $> p^n$ are cyclic.

Exercise 6. Given $n > 1$, study the irregular p-groups all of whose regular subgroups of exponent $> p^n$ are absolutely regular.

Problem 1. Study the p-groups all of whose maximal abelian subgroups are homocyclic.

Problem 2. Study the nonabelian p-groups all of whose maximal abelian subgroups, except one, are either cyclic or elementary abelian.

Problem 3. Study the irregular p-groups all of whose maximal regular subgroups, except one, are either absolutely regular or of exponent p.

Problem 4. Study the irregular p-groups all of whose minimal irregular subgroups have order p^{p+1}. (For $p = 2$, see § 90.)

Problem 5. Study the irregular p-groups containing at most p maximal regular subgroups of exponent $> p$. Describe in detail irregular p-groups containing only one maximal regular subgroup of exponent $> p$.

Problem 6. Study the irregular p-groups containing exactly $p + 1$ maximal regular subgroups. (Example: $\Sigma_{p^2} \times C_p$, where $\Sigma_{p^2} \in \mathrm{Syl}_p(S_{p^2})$.)

§ 266 Nonabelian p-groups in which any two distinct minimal nonabelian subgroups with a nontrivial intersection are non-isomorphic

Here we solve Problem 3702 by proving the following result:

Theorem 266.1. *Let G be a nonabelian p-group in which any two distinct minimal nonabelian subgroups with a nontrivial intersection are non-isomorphic. Then $\Omega_1(Z(G))$ is contained in each minimal nonabelian subgroup and therefore all minimal nonabelian subgroups are pairwise non-isomorphic so G-invariant. In particular, G is metahamiltonian (see § 228 and* [FA, AZ]*).*[1]

Proof. Let G be a nonabelian p-group in which any two distinct minimal nonabelian subgroups with nontrivial intersection are non-isomorphic. Let $M = \langle x, y\rangle$ be a minimal nonabelian subgroup in G and assume that $t \in \Omega_1(Z(G))$ with $t \notin M$; then t is an element of order p. Write $H = M \times \langle t\rangle$. Then all p^2 minimal nonabelian subgroups of H are isomorphic and any two distinct of them have nontrivial intersection, contrary to the hypothesis. Thus, $\Omega_1(Z(G))$ is contained in all minimal nonabelian subgroups of G and therefore all minimal nonabelian subgroups of G must be pairwise non-isomorphic, by hypothesis. In particular, each minimal nonabelian subgroup is normal (even characteristic) in G and this implies that each nonabelian subgroup is normal in G (indeed, any nonabelian p-group is generated by minimal nonabelian subgroups, by Theorem 10.28). This means that the group G is metahamiltonian. □

Problem 1. Does there exist a nonabelian p-group all of whose maximal abelian subgroups are pairwise non-isomorphic? Study the nonabelian p-groups all of whose maximal abelian subgroups are normal. (If $p > 2$, such p-groups are regular.)

Problem 2. Does there exist the nonmetacyclic p-groups all of whose maximal metacyclic subgroups are pairwise non-isomorphic? Study the nonmetacyclic p-groups all of whose maximal metacyclic subgroups are normal.

Problem 3. Study the non-absolutely regular p-groups all of whose maximal absolutely regular subgroups are normal.

Problem 4. Study the nonabelian p-groups all of whose minimal nonabelian subgroups are normal.

1 Recall that a p-group is said to be *metahamiltonian* provided it is nonabelian and all its nonabelian subgroups are normal. In particular, minimal nonabelian p-groups are metahamiltonian.

https://doi.org/10.1515/9783110533149-010

Problem 5. Study the p-groups G such that, whenever $S, T < G$ are minimal nonabelian, then the subgroup $\langle S, T\rangle$ is regular.

Problem 6. Study the nonabelian p-groups G all of whose minimal nonabelian subgroups not contained in $\Phi(G)$ are pairwise non-isomorphic.

§ 267 Thompson's $A \times B$ lemma

A number of results obtained with using the Three Subgroups Lemma are fairly unexpected. One of such results is the following one.

Theorem 267.1 (Thompson's $A \times B$ lemma; see [Gor1, Theorem 5.3.4]). *Let $A \times B$ be a group of automorphisms of a p-group P with A a p'-group and B a p-group. If A centralizes the centralizer $\mathrm{C}_P(B)$, then $A = \{1\}$.*

Proof. It suffices to prove that A centralizes P. We proceed by induction on $|P|$. The subgroup $\mathrm{C}_P(B)$ is $A \times B$-invariant. Let Q be an $A \times B$-invariant subgroup of P of maximal order containing $\mathrm{C}_P(B)$ and such that A centralizes Q. One may assume that $Q < P$ (otherwise, $A = \{1\}$). By induction, any proper $A \times B$-invariant subgroup of P, containing Q is centralized by A. As the $A \times B$-invariant subgroup $N = \mathrm{N}_P(Q) > Q$, we get $N = P$ so that $Q \lhd P$. Set $\overline{P} = P/Q$. The subgroup $\overline{L} = \mathrm{C}_{\overline{P}}(B) > \{1\}$ is $A \times B$-invariant so is its inverse image $L > Q$, and therefore, by induction, $L = P$ hence $[P, B] \leq Q$ and, since $[Q, A] = \{1\}$, by a choice of Q, we get $[P, B, A] \leq [Q, A] = \{1\}$. Since $[B, A] = \{1\}$, we obtain $[B, A, P] = \{1\}$. Then the Three Subgroups Lemma yields $[A, P, B] = \{1\}$. Thus, $[A, P] \leq \mathrm{C}_P(B) \leq Q$ hence the p'-group A stabilizes the normal series $P > Q > \{1\}$ and therefore, by Exercise 268.7, $A = \{1\}$. □

Corollary 267.2. *Let P be a p-group. Let $B = \mathrm{Inn}(P)$ be the group of inner automorphisms of P. If $A \times B$ is a group of automorphisms of P, where A is a p'-group centralizing $\mathrm{Z}(P)$, then $A = \{1\}$.*

Proof. Since A centralizes $\mathrm{C}_P(B) = \mathrm{Z}(P)$, the result follows from Theorem 267.1. □

Problem. Is the following assertion true? Let $A \times B$ be a group of automorphisms of a π-group P with A a π'-group and B a π-group. Suppose, in addition, that the last member of the upper central series of the group BP contains P. If A centralizes $\mathrm{C}_P(B)$, then $A = \{1\}$.

https://doi.org/10.1515/9783110533149-011

§ 268 On automorphisms of some p-groups

To describe the actions of a p'-automorphism on a p-group, we analyze the minimal nonnilpotent subgroups of its holomorph. It is known that a number of the results proved below are used in papers devoted the classification of finite simple groups (CFSG).

Proposition 268.1. *Suppose that a 2-group P has no subgroup isomorphic to* E_8. *If $q > 2$ is a prime divisor of* $|\mathrm{Aut}(P)|$, *then* $q \in \{3, 5\}$.

Proof. Suppose that $\phi \in \mathrm{Aut}(P)$ has a prime order $q > 2$. Let G be a natural semidirect product of $A = \langle\phi\rangle$ and P, and let $S \leq G$ be minimal nonnilpotent such that $A < S$. Then $P_1 \in \mathrm{Syl}_2(S)$ is either elementary abelian or special, A acts irreducibly on $P_1/\Phi(P_1)$ (Theorem A.22.1). Taking in mind our aim, one may assume that $G = S$; then $P_1 = P$. If P is elementary abelian, then $P \cong \mathrm{E}_4$, by hypothesis, and $q = 3$, by Sylow. Now let P be special. Then $|\mathrm{Z}(P)| \leq 4$, by hypothesis. One may assume that $|P| > 2^3$ (if $|P| = 2^3$, then $P \cong \mathrm{Q}_8$ and $q = 3$).

Assume that $|\mathrm{Z}(P)| = 2$; then P is extraspecial of order, say 2^{2m+1}, $m > 1$. In that case P has a subgroup isomorphic to E_{2^m} and therefore $m = 2$. Then, by Lemma 4.3, $P \cong \mathrm{Q}_8 * \mathrm{D}_8$ (note that the group $\mathrm{Q}_8 * \mathrm{Q}_8 \cong \mathrm{D}_8 * \mathrm{D}_8$ has a subgroup isomorphic to E_8). Then $q = |A|$ divides $2^4 - 1$ which implies $q \in \{3, 5\}$. By Example 76.1, there is $\alpha_1(P) = 20$ of minimal nonabelian subgroups in P, it is easy to check that there is in P exactly ten subgroups isomorphic to D_8. In fact, $q \neq 3$ (otherwise, A fixes a subgroup isomorphic to D_8, contrary to irreducibility of action of ϕ on $P/\Phi(P)$).

Now let P be special with noncyclic center; then $\mathrm{Z}(P) \cong \mathrm{E}_4 \cong \Omega_1(P)$, by hypothesis, and this implies that $|P| > 2^4$. Let $|P/P'| = 2^{2m}$ (Theorem A.22.1 since, in the case under consideration, P/L is extraspecial for each $L < \mathrm{Z}(P)$ of index 2). Note that $|\mathrm{Z}(G) : \ker(\chi)| = 2$ for each $\chi \in \mathrm{Irr}_1(G))$. By the Frobenius–Schur formula,

$$4 = 1 + t(P) = 2^{2m} + \sum_{\chi\in\mathrm{Irr}_1(P)} \nu_2(\chi)\chi(1), \tag{1}$$

where $t(P)$ is the number of involutions in the group P, the Frobenius–Schur indicator $\nu_2(\chi) \in \{-1, 1\}$ if $\bar{\chi} = \chi$ and $\nu_2(\chi) = 0$ if $\bar{\chi} \neq \chi$. In the case under consideration, $\nu_2(\chi) = 1$ if χ is linear. If $\chi \in \mathrm{Irr}_1(P)$, then $P/\ker(\chi)$ is extraspecial of order 2^{1+2m} so that $\chi(1) = 2^m$. Therefore, by (1), since

$$\mathrm{Irr}_1(P/\ker(\chi)) = \{\chi\}, \qquad |\mathrm{Irr}_1(G)| = 3,$$
$$\mathrm{c}_1(\mathrm{Z}(G)) = t(G) = 3, \quad |\mathrm{Lin}(G)| = |G/G'| = 2^{2m},$$

one has

$$2^{2m} + 2^m \sum_{\chi\in\mathrm{Irr}_1(P)} \nu_2(\chi) = 2^m(2^m + \tau) = t(G) = 4, \tag{2}$$

where $\tau = \sum_{\chi\in\mathrm{Irr}_1(P)} \nu_2(\chi)$. It follows from (2) that $m = 2$ and so $\tau = -3$, and we conclude that $\nu_2(\chi) = -1$ for each $\chi \in \mathrm{Irr}_1(P)$. Then

$$|P| = |\mathrm{Lin}(G)| + 3\chi(1)^2 = 2^4 + 3 \cdot 4^2 = 2^6. \tag{3}$$

https://doi.org/10.1515/9783110533149-012

Applying to $G/\ker(\chi)$ ($\chi \in \mathrm{Irr}_1(G)$) again the Frobenius–Schur formula, we get

$$t(G/\ker(\chi)) = |G/G'| - \chi(1) - 1 = 11 \implies G/\ker(\chi) \cong \mathrm{Q}_8 * \mathrm{D}_8.$$

Since ϕ acts irreducible in $G/\ker(\phi)$, we get $q = 5$. □

The 2-groups of Proposition 268.1 are described in § 50. We suggest the reader to use that result to produce a new proof of Proposition 268.1.

There is in the Atlas [HS] a special group No. 187 (denote it by X) of order 2^6, which has exactly three involutions generating $Z(X) \cong \mathrm{E}_4$. The group X has automorphisms of orders 3 and 5 and possesses all properties described in the second part of the proof of Proposition 268.1, in particular, all three epimorphic images of X of order 2^5 are extraspecial isomorphic to $\mathrm{Q}_8 * \mathrm{D}_8$.

Exercise 1. The following conditions hold:

(a) Let A, an abelian group of type (p^2, p), be normal in a p-group $G = C \cdot A$, a semidirect product with kernel A and a cyclic complement $C = \langle c \rangle$ of order p^2. If $p > 2$, then $\mathrm{C}_G(A) > A$. In particular, p^2 does not divide $|\mathrm{Aut}(A)|$. Moreover, a Sylow p-subgroup of $\mathrm{Aut}(A)$ is isomorphic to $\mathrm{S}(p^3)$.

(b) Let $E \cong \mathrm{E}_{p^3}$, $p > 3$. Prove that p^2 does not divide $|\mathrm{Aut}(E)|$.

Solution. (a) Since p^2 does not divide $|\mathrm{Aut}(\mathrm{E}_{p^2})|$, c^p centralizes $\Omega_1(A) \cong \mathrm{E}_{p^2}$. One has $\mathrm{C}_G(c^p) \geq H = C\Omega_1(A)$. In that case, $\mho_1(H) = \langle c^p \rangle \triangleleft G$ since $H \triangleleft G$. Since $c^p \in Z(G)$, we get $\mathrm{C}_G(A) > A$ and hence p^2 does not divide $\exp(\mathrm{Aut}(A))$. Thus, a Sylow p-subgroup of $\mathrm{Aut}(A)$, being nonabelian, is isomorphic to $\mathrm{S}(p^3)$.

(b) Assume that $\phi \in \mathrm{Aut}(E)$ is of order p^2 and consider $G = \langle \phi \rangle \cdot E$, the natural semidirect product with kernel E. Take $R < E$, a G-invariant subgroup of order p^2; then $H = \langle \phi \rangle \cdot R \triangleleft G$ since $|G : H| = p$. In that case, $\langle \phi^p \rangle = \mho_1(H) \triangleleft G$ implies $\phi^p \in Z(G)$ (see the corresponding place in the solution of (a); here we use Theorem 7.2 (b)) hence ϕ does not act faithfully on E. Thus, p^2 does not divide $|\mathrm{Aut}(E)|$. As a Sylow p-subgroup of $\mathrm{Aut}(E)$ is isomorphic to a Sylow p-subgroup of $\mathrm{GL}(3, p)$, it is nonabelian hence isomorphic to $\mathrm{S}(p^3)$.[1] □

We suggest the readers to consider in Exercise 1 (b) the case $p = 3$ in detail.

Theorem 268.2 (Theorem 109.1). *Suppose that $p > 2$ and L is a metacyclic p-group without cyclic subgroup of index p. An element $a \in \mathrm{Aut}(L)$ of order p does not centralize $\Omega_1(L)$ if and only if $p = 3$, the partial holomorph $G = \langle a \rangle \cdot L$ is a 3-group of maximal class and one of the following holds:*

(a) *L is abelian of type $(3^m, 3^n)$, $m > 1$, $n > 1$ and $|m - n| \leq 1$.*

(b) $L = \langle a, b \mid \mathrm{o}(a) = 3^m,\ \mathrm{o}(b) = 3^n,\ a^b = a^{1+3^{m-1}} \rangle \cong \mathrm{M}_3(m, n)$, $|m - n| \leq 1$.

1 It is easy to prove that if A is noncyclic abelian group of order 2^3, then a Sylow 2-subgroup of $\mathrm{Aut}(A)$ is isomorphic to D_8.

Proof. By hypothesis, $|L| \geq p^4$. One has

$$R = \Omega_1(L) \cong \mathrm{E}_{p^2}, \quad M = \langle a, R \rangle \cong \mathrm{S}_{p^3} \quad \text{and} \quad L \cap M = R.$$

By hypothesis, $|G : \mathrm{C}_G(R)| = p$.

If G is a 3-group of maximal class, then either (a) or (b) hold (see Theorem 9.6). Next we assume that G is not a 3-group of maximal class. As $p > 2$, (the metacyclic p-group) L is absolutely regular.

Assume that G is regular. In this case, $\Omega_1(G)$ is of order p^3 and exponent p, $\Omega_1(G) \in \{\mathrm{S}(p^3), \mathrm{E}_{p^3}\}$. If $\Omega_1(G) \cong \mathrm{S}(p^3)$, then, by Theorem 13.7 (c), L has a cyclic subgroup of index p, contrary to the hypothesis. Thus, $\Omega_1(G) \cong \mathrm{E}_{p^3}$. As $a \in \Omega_1(G)$ and $R < \Omega_1(G)$, a centralizes R, contrary to the hypothesis.

Thus, the group G is irregular. In that case, $p = 3$ (here we use Theorem 9.8 (a), Theorem 12.1 (b) and Theorem 7.2 (d)). Then, by Theorem 12.1 (b), $G = L\Omega_1(G)$, where $\Omega_1(G) \in \{\mathrm{S}(3^3), \mathrm{E}_{p^3}\}$. In the first case, all members of the set Γ_1 not containing $\Omega_1(G)$, have a cyclic subgroup of index p (Theorem 13.7). As $L \in \Gamma_1$ and $\Omega_1(G) \nleq L$, we get a contradiction (by hypothesis, L has no cyclic subgroup of index p). Thus, $\Omega_1(G) \cong \mathrm{E}_{p^3}$. Again, as in the last sentence of the previous paragraph, $a \in \Omega_1(G) - \Omega_1(L)$ centralizes $R = \Omega_1(L)$, contrary to the assumption. Thus, G is a 3-group of maximal class, as in the statement. □

Compare the above proof with the proof of Theorem 109.1. The obtained result improves essentially the corresponding result in [MS] (see also [AS, Lemma A.1.30]) stated in Corollary 109.2. Note that in Theorem 268.2, the metacyclic subgroup L need not necessarily be abelian, however, its structure is restricted essentially.

Let, for a prime p, $\epsilon(p) = 1$ for $p > 2$ and $\epsilon(2) = 2$. Recall that if a p'-automorphism ϕ of a p-group P induces the identity automorphism on $\Omega_{\epsilon(p)}(P)$, then $\phi = \mathrm{id}_P$ (this is true, by Theorem A.52.2). We use this fact in what follows.

Exercise 2 ([GLS, Lemma 10.34] for the case $p = 2$). Suppose that a group $A \cong \mathrm{E}_{p^2}$ acts on a p-group P which is neither cyclic nor a p-group of maximal class, and let $R = \langle \mathrm{C}_P(a) \mid a \in A^\# \rangle$. Then:

(a) R has an AP-invariant subgroup isomorphic to E_{p^2}.

(b) If some element from A of order p induces a nonidentity inner automorphism on P, then $|R| \geq p^3$.

Solution. Let $G = A \cdot P$ be the natural semidirect product of A and P. By Lemma 1.4, there is in P a G-invariant subgroup $U \cong \mathrm{E}_{p^2}$. Since $|G : \mathrm{C}_G(U)| \leq p$, the subgroup $A \cap \mathrm{C}_G(U)$ contains an element $a \in A$ of order p hence $U \leq R$, and (a) holds. Now suppose that $b \in A^*$ induces a nonidentity inner automorphism of P. Then $\mathrm{C}_P(b) = \mathrm{C}_P(x)$ for some $x \in P$. Assume that $|R| < p^3$; then $|\mathrm{C}_P(x)| = p^2$ (see the definition of R). In that case the nonabelian p-group P is of maximal class (Proposition 1.8), contrary to the hypothesis. □

If P is a p-group, then $r(P)$ is a maximal rank of abelian subgroups of P.

Definition 1. If P is a p-group, then:
(i) $\mathcal{E}^{(p)}(P) = \{A \le P \mid A \cong \mathrm{E}_{p^n},\ n > 1\}$, the set of elementary abelian subgroups of order $> p$ in P.
(ii) $\mathcal{E}_*^{(p)}(P) = \{A \in \mathcal{E}^{(p)}(P) \mid \mathrm{d}(A) = r(P)\}$.
(iii) $J(P) = \langle A \in \mathcal{E}_*^{(p)}(P)\rangle$.

The subgroup $J(P)$ is characteristic in P. If $J(P)$ is abelian, then $J(P) \in \mathcal{E}_*^{(p)}(P)$. If G is regular, then $\exp(J(P)) = p$ (Theorem 7.2 (b)). If $|J(P)| > p^{r(p)}$, then $J(P)$ is nonabelian. If $P = \Sigma_{p^2}$ and $p > 2$, then $|P : J(P)| = p$.

Exercise 3 ([GLS, Lemmas 10.29 and 10.30]). The following conditions hold:
(i) $J(P)$ is the unique subgroup of P of its isomorphism type.
(ii) Suppose that $Q \le P$ and $A \in \mathcal{E}_*^{(p)}(P)$. If $A \le Q \le P$, then $J(Q) \le J(P)$.
(iii) If $P = P_1 \times P_2$, then $J(P) = J(P_1) \times J(P_2)$.
(iv) $\Omega_1(\mathrm{Z}(J(P)) = \bigcap_{A \in \mathcal{E}_*(P)} A$.

Exercise 4 ([GLS, Lemma 11.9]). If ϕ is a nontrivial p'-automorphism of a p-group P, $p > 2$, then $\phi_{J(P)} \neq \mathrm{id}_{J(P)}$.

Solution. Let $G = \langle\phi\rangle \cdot P$ be the natural semidirect product of groups $A = \langle\phi\rangle$ and P with kernel P; then $J(P) \lhd G$. Assume, by way of contradiction, that ϕ centralizes $J(P)$. We claim that $\Omega_1(\mathrm{C}_P(J(P)) \le J(P)$. Assume, however, that $x \in \mathrm{C}_P(J(P)) - J(P)$ is of order p. If $A \in \mathcal{E}_*^{(p)}(P)$, then the elementary abelian subgroup $\langle x\rangle \times A > A$, which is a contradiction. Thus, ϕ centralizes $\Omega_1(\mathrm{C}_P(J(P))$ so $\mathrm{C}_P(J(P))$ (indeed, $\langle\phi\rangle\mathrm{C}_P(J(P))$ has no minimal nonnilpotent subgroup, by Theorem A.22.1). In that case, ϕ centralizes $H = J(P)\mathrm{C}_P(J(P))$. One has $H \lhd G$ implies $\mathrm{C}_P(H) \lhd G$ and $\mathrm{C}_P(H) \le \mathrm{C}_P(J(P)) \le H$. Then we have

$$\mathrm{C}_G(H) = \langle\phi\rangle \times \mathrm{C}_P(H) \lhd G \implies \langle\phi\rangle \lhd G \implies \phi = \mathrm{id}_P,$$

contrary to the assumption. □

Exercise 5. Classify the p-groups P such that $J(P)$ is metacyclic.

Exercise 6 ([GLS, Lemma 11.18]). Let an involution t act on a p-group P, $p > 2$. If P has a subgroup isomorphic to E_{p^3}, then t normalizes a subgroup $Q \le P$ with $Q \cong \mathrm{E}_{p^3}$.

Solution. Let P be a counterexample of minimal order; then $|P| > p^3$ and t acts nontrivially on P. By Theorem 10.4, there is $R \lhd P$ with $R \cong \mathrm{E}_{p^3}$. As $(RR^t)^t = RR^t$, one has $P = RR^t$, by induction. As the natural semidirect product of $\langle t\rangle$ and P is supersolvable (Exercise 293.335), one may assume that $P = RR^t$ is nonabelian (otherwise, the subgroup P, being elementary abelian, possesses a t-invariant subgroup of order p^3). It follows that $R \cap R^t > \{1\}$ (note that $R, R^t \lhd P$). By Fitting's lemma, $\mathrm{cl}(P) = 2$ so that $\exp(P) = p$ (Theorem 7.2 (b)). One has $\mathrm{Z}(P)^t = \mathrm{Z}(P)$. Write $K = \langle t, \mathrm{Z}(P)\rangle$. By the product formula, $|P| \le p^5$ and $|\mathrm{Z}(P)| \le p^2$. Since K is supersolvable, it follows that t normalizes a subgroup L_1 of order p in $\mathrm{Z}(P)$; then $L_1 \lhd \langle t.P\rangle$. Let L_2/L_1 be a t-invariant subgroup of order p in $\mathrm{Z}((P/L_1)$; then $\mathrm{E}_{p^2} \cong L_2 \lhd P$ (recall that $\exp(P) = p$).

Write $C = C_P(L_2)$; then $|P : C| \le p$, $C^t = C > L_2$. If L_3/L_2 is a t-invariant subgroup of order p in $C/L_2 \cap Z(P/L_2)$, then $L_3 \cong E_{p^3}$ is a t-invariant normal subgroup of P.[2] □

If a p-group X acts on a p-group P containing a subgroup isomorphic to E_{p^k}, $k \le 4$, $p > 2$, then P contains an X-invariant subgroup isomorphic to E_{p^k}. This follows from Lemma 1.4 and Theorems 10.4 and 10.5. It follows from Theorem 1.17 (b) and Lemma 1.4 that if the 2-group X acts on the 2-group P which is neither cyclic nor a 2-group of maximal class, then P contains an X-invariant subgroup isomorphic to E_4.

Exercise 7 **([GLS, Lemma 11.5]).** Assume that X is a p'-group acting on the p-group P. If

$$\{1\} = P_1 \lhd P_2 \lhd \cdots \lhd P_{n-1} \lhd P_n = P$$

is the series of X-invariant subgroups of P such that X acts trivially on P_{i+1}/P_i (i.e., X stabilizes that series), then X acts trivially on P.

Solution. Let $W = X \cdot P$ be the natural semidirect product of X and P with kernel P. Assume, by way of contradiction, that W is non-p-nilpotent. Then there is in W a minimal nonnilpotent subgroup $S = Q \cdot P_0$, where $Q \le X$ is cyclic of prime power order, $P_0 = S' = S \cap P \in \mathrm{Syl}_p(S)$. Let k be minimal such that $P_0 \le P_k$; then $P_0 \nleq P_{k-1}$. In that case, P_0P_{k-1}/P_{k-1} is the nonidentity subgroup of P_k/P_{k-1} hence Q centralizes $P_0P_{k-1}/P_{k-1} \cong P_0/(P_0 \cap P_{k-1})$, the nonidentity epimorphic image of P_0, which is a contradiction (note that S has no normal subgroup of index p). □

Exercise 8 **([GLS, Lemma 11.17]).** Let Q be a normal subgroup of exponent p of a p-group P. If $\phi \in \mathrm{Aut}(P)$ stabilizes the series $\{1\} < Q < P$, then $\phi^p = \mathrm{id}_P$.

Solution. Let $a \in P - Q$. Then $a^\phi = ab$ for some $b \in Q$. In that case, $a^{\phi^p} = ab^p = a$, and we are done since $\langle P - Q\rangle = P$.[3] □

Remark 1 **([Tho3, Lemma 5.26]; see also [Gor1, Exercise 5.2]).** Let P be a group of order 2^6 in which $Z(P) = P'$. If P possesses an automorphism ϕ of order 5, then P is special. Indeed, write $A = \langle\phi\rangle$ and let $G = A \cdot P$ be the natural semidirect product of A and P. By hypothesis, P is nonabelian. Let $S \le G$ be minimal nonnilpotent. One may assume that $A < S$ (Sylow). Then $S = A \cdot P_1$, where $P_1 = S \cap P \in \mathrm{Syl}_2(S)$ is of order $\ge 2^4$ hence $|G : S| \le 2^2$. By Sylow, $N = N_G(A) = A \times L$, where $|L| = 4$ and N is maximal in G. As $N_P(L) > L$ is A-invariant, it follows that $L \lhd G$. Considering $C_G(L)$ and taking into account that G/L has no subgroups of index 2, we get $C_G(L) = G$ hence $L = Z(G) = Z(P)$, and we conclude that $L = P'$. It follows that

$$P_1P' = P \implies P_1 = P \implies G = S.$$

In that case, P is special (Theorem A.22.1).

2 The last assertion is stated relative to the case when $P = RR^t$. But in the general case, we do not know if it is possible to choose L_3 to be normal in P.

3 If $\exp(Q) = p^e$, then the same argument shows that $\phi^{p^e} = \mathrm{id}_P$.

Proposition 268.3 ([Gor1, Theorem 5.6.6]). *Let ϕ be an automorphism of the 2-group P and assume that $o(\phi)$ is a Mersenne prime > 3. Then ϕ acts nontrivially on $\Omega_1(P)$.*

Solution. Assume, by way of contradiction, that ϕ acts trivially on $\Omega_1(P)$. Write $A = \langle\phi\rangle$ and let $G = A \cdot P$ be the natural semidirect product of A and P. Let $S \leq G$ be minimal nonnilpotent. One may assume that $A < S$. Then $S = A \cdot P_1$, where $P_1 = S \cap P$. By hypothesis, $\exp(P_1) > 2$. In that case, $|P_1/P_1'| = 2^{2b}$ (this follows easily from Theorem A.22.1). Let $o(\phi) = 2^q - 1$, where $q > 2$ is a prime number ($o(\phi)$ is a Mersenne prime). Then

$$2^q - 1 \mid (2^{2b} - 1) = (2^b - 1)(2^b + 1) \implies 2^q - 1 \mid 2^b + 1$$

since $o(\phi)$ does not divide $2^b - 1$, by the above theorem. Let $b = kq + s$, where $s < q$. Then

$$\begin{aligned}\frac{2^b + 1}{2^q - 1} &= \frac{2^{kq+s} + 1}{2^q - 1}\\ &= \frac{2^s(2^{kq} - 1) + (2^s + 1)}{2^q - 1}\\ &= 2^s(2^{(k-1)q} + 2^{(k-2)q} + \cdots + 2^q + 1) + \frac{2^s + 1}{2^q - 1}\end{aligned}$$

is not a natural number since $2^q - 1 > 2^s + 1$, which is a contradiction. Thus, ϕ acts nontrivially on $\Omega_1(P)$, as required. □

The group $SL(2, 3)$ shows that assumption $q > 3$ in Proposition 268.3 cannot be omitted.

Some of the above results were proved in [Gor1, Chapter 5] using Lie rings methods. In contrast, our arguments are pure group theoretical.

Exercise 9 ([Gor1, Lemma 5.2.1 (ii)]). Given $n > 1$, prove that the order of 5 modulo 2^n is equal to 2^{n-2}. It follows from this that $\mathrm{Aut}(C_{2^n})$ is abelian of type $(2^{n-2}, 2)$.

Hint. This fact is used in the book many times. To prove it, use induction on n and Newton's binomial formula. □

Exercise 10 ([Gor1, Corollary 5.6.4]). Let $P \cong E_{p^n}$. If P admits a p'-automorphism ϕ of order m, which acts irreducibly on P, then $m \mid p^n - 1$.

Solution. Write $A = \langle\phi\rangle$ and let G be the natural semidirect product of A and P. Then $N_G(A) = A$, by hypothesis. Let $x \in P$ and set $A \cap A^x = D$ and assume that $D > \{1\}$. Then $N_G(D) = A \cdot P_1$, where $\{1\} < P_1 < P$ and P_1 is A-invariant, contrary to irreducibility of A-group P. Thus, A is a self-normalizing TI-subgroup of G, i.e., $G = (A, P)$ is a Frobenius group with kernel P and complement A. It follows that $m = |A| \mid |P| - 1 = p^n - 1$, as required. □

It follows that if ϕ is a p'-automorphism of $P \cong E_{p^n}$, then $o(\phi) \mid p^k - 1$ for some $k \leq n$ (to prove this, use Maschke' theorem). Note that $|\mathrm{Aut}(P)| = \prod_{i=0}^{n-1}(p^n - p^i)$, but this does not imply the stated result.

The group $P = E_{p^n}$ admits an automorphism of order $p^n - 1$, acting irreducibly on P (see [Gor1, Corollary 6.4]). Indeed, let us consider P as the additive group of the Galois field $GF(p^n)$. Let the mapping $P \to P$ be defined by $\phi : a \mapsto a\xi$, where $a \in P$ and ξ is a generator of the multiplicative group of the field $GF(p^n)$. Then ϕ is an automorphism of P of order $o(\phi) = p^n - 1$.

Exercise 11 (Hall–Higman; see [GLS, Proposition 11.10]). If α is a nontrivial p'-automorphism of the p-group P that acts trivially on every proper α-invariant subgroup of P, then P is either elementary abelian or special. In the second case, $C_P(\alpha)$ is elementary abelian and, if $p > 2$, then $\exp(P) = p$.

Solution. Let $G = \langle\alpha\rangle \cdot P$ be the natural semidirect product of $\langle\alpha\rangle$ and P. Take in G a minimal nonnilpotent subgroup S. One may assume that a nonnormal Sylow q-subgroup of S is generated by β, a power of α. As β acts nontrivially on S', it follows, by hypothesis, that $S' = P$. Assume that S' is nonabelian. Then it is special (Theorem A.22.1). By hypothesis, $C_P(\beta) = Z(S) > \{1\}$ is elementary abelian. If $p > 2$, then we have $\exp(P) = p$ (Theorem A.22.1). □

It follows from the solution that P is isomorphic to a normal Sylow subgroup of some minimal nonnilpotent group. If P from Exercise 11 is special, one can say about P and G more (see [Gol]). For example, $|Z(P)|^2 \leq |P : P'|$.

Exercise 12 ([GLS, Lemma 11.6] for the case $p > 2$). Let P be a p-group. Suppose that a p'-group A acts trivially on $\Omega_\mu(P)$, where $\mu = 1$ if $p > 2$ and $\mu = 2$ if $p = 2$. Then A acts trivially on P.

Solution. Let G be a natural semidirect product of A and P. Without loss of generality one may assume that A is abelian. Assume that $S \leq G$ is minimal nonnilpotent. Set $P_1 = P \cap S$. One may assume that the p'-Hall subgroup Q of S is contained in A (Schur–Zassenhaus). By Theorem A.22.1, $P_1 \leq \Omega_\mu(P)$ so Q acts trivially on P and we conclude that S is nilpotent, a contradiction. Thus, the group G has no minimal nonnilpotent subgroup so it is nilpotent hence $G = A \times P$, as desired. □

Exercise 13. Suppose that a p'-group A acts on the nonabelian p-group P and centralizes a maximal normal abelian subgroup Q of P. Then A centralizes P.

Solution. Let G be a natural semidirect product of A and P. Clearly,

$$Q \lhd G,\ C_G(Q) = A \times Q \lhd G \implies A \lhd G,$$

and the result follows. □

Exercise 14. Suppose that a p'-group A acts on the p-group P and centralizes a normal subgroup Q of P such that $C_P(Q) \leq Q$. Then A centralizes P.[4]

4 This is a partial case of [GLS, Lemma 11.8].

Solution. Let $G = A \cdot P$ be a semidirect product. Then

$$C_G(Q) = A \times Z(Q) \lhd G \implies A \lhd G$$

so that A centralizes P. □

Exercise 15. Find $t = |\mathrm{Aut}(G)|$, where

$$G = \langle a, b \mid o(a) = o(b) = 2^2,\ a^b = a^3 \rangle = M_2(2, 2)$$

is the unique nonabelian metacyclic group of order 2^4 and exponent 2^2.

Solution. As we know, $t \mid (2^2 - 1)(2^2 - 2)|\Phi(G)|^2 = 3 \cdot 2^5$. Assume that $\phi \in \mathrm{Aut}(G)$ is of order 3. Let $W = \langle \phi \rangle \cdot G$. As all subgroups of W of order 2 are characteristic in W, there is no in W a minimal nonnilpotent subgroup (Theorem A.22.1) so W is nilpotent, a contradiction. Thus, $t \mid 2^5$. Let $\{u, v\}$ be a minimal basis of G such that $\langle v \rangle \lhd G$. Then $v^u = v^3$, so the mapping $a \to v$ and $b \to u$ is an automorphism of G. It is possible to choose v in four ways and u in eight ways so that the numbers of minimal bases of G is $4 \cdot 8 = 2^5$ hence $t = 2^5$ since in the case under consideration $\mathrm{Aut}(G)$ acts on the set of minimal bases transitively and faithfully. □

Exercise 16 (Burnside). Let ϕ be a p'-automorphism of a p-group G which induces the identity on $G/\Phi(G)$. Then $\phi = \mathrm{id}_G$.

Solution. Without loss of generality, one may assume that $o(\phi) = q$, a prime. Let

$$G/\Phi(G) = \langle x_1\Phi(G) \rangle \times \cdots \times \langle x_d\Phi(G) \rangle.$$

As $(x_i\Phi(G))^\phi = x_i\Phi(G)$ and $\mathrm{GCD}(q, |x_i(\Phi(G)|) = 1$, it follows that there is $y_i \in x_i\Phi(G)$ such that $y_i^\phi = y_i$, $i = 1, \ldots, d$. Since $G = \langle y_1, \ldots, y_d \rangle$, we get $\phi = \mathrm{id}_G$. □

Exercise 17 ([Tho3, Lemma 5.27]). Suppose that G is a metacyclic 2-group and $\mathrm{Aut}(G)$ is not a 2-group. Then G is either isomorphic to Q_8 or abelian of type $(2^n, 2^n)$.

Solution. Clearly, G is noncyclic. One may assume that $\phi \in \mathrm{Aut}(G)$ is of prime order $q > 2$. Let $W = \langle \phi \rangle \cdot G$ be the natural semidirect product and let $S \le W$ be minimal nonnilpotent so that $\phi \in S$. Suppose that G is abelian of type $(2^m, 2^n)$. If $m > n$, then ϕ centralizes $\mho_{m-1}(S')$; then ϕ centralizes $\Omega_1(S')$ so S', a contradiction. Thus, if G is abelian, it is of type $(2^n, 2^n)$. Suppose that G is nonabelian. Then $S' \in \{E_4, Q_8\}$ (Theorem A.22.1). In the second case, G is of maximal class (Proposition 10.17) so $G \cong Q_8$ (Theorem 34.8). If $S' \cong E_4$, then $S' \cap G'$ of order 2 is centralized by ϕ so S is nilpotent, a contradiction. Thus, if G is nonabelian it is isomorphic to Q_8. □

Exercise 18. Let G be a regular p-group of exponent $p^e > p$, and let $\phi \in \mathrm{Aut}(G)$ be a p'-automorphism such that $\phi(x) = x$ for each $x \in G$ of order p^2. Prove that $\phi = \mathrm{id}_G$.

Hint. One has $\Omega_2^\#(G) = \Omega_2(G)$ (see Theorem 7.2 (b)). The holomorph $\langle \phi \rangle \cdot G$ has no minimal nonnilpotent subgroup, by Theorem A.22.1. □

Exercise 19. Let ϕ be an automorphism of order 2 of a noncyclic p-group G, $p > 2$. Show that there is in G a normal subgroup R of order p^2 such that $\phi(R) = R$.

Solution. One has $\phi(Z(G)) = Z(G)$. Put $H = \langle\phi\rangle \cdot Z(G)$. If H is nilpotent, then $\phi(L) = L$ for some $L \leq Z(G)$ of order p. Now let H be nonnilpotent and let $\phi \in S \leq H$, where S is minimal nonnilpotent. If $L = S \cap Z(G)$, then L of order p (Theorem A.22.1) is ϕ-invariant. One has $L \lhd G$. By induction, there is in G/L a normal ϕ-invariant subgroup R/L of order p; then R is a ϕ-invariant normal subgroup of G of order p^2. (In fact, the natural semidirect product of $\langle\phi\rangle$ and G, by Exercise 293.335, is supersolvable and this can be used.) □

Exercise 20. Let ϕ be an automorphism of order 2 of a nonmetacyclic p-group G, $p > 3$. Show that there is in G a ϕ-invariant subgroup R of order p^3 and exponent p.

Exercise 21 ([FT, Lemma 8.4]). Let G be a p-group, $p > 2$.

(a) If G has no normal subgroup isomorphic to E_{p^3}, then each its abelian subgroup is generated by two elements.

(b) If G is as in (a) and $\alpha \in \mathrm{Aut}(G)$ is of prime order $q \neq p$, then $q \mid p^2 - 1$.

Solution. (a) By Theorem 10.4, G has no subgroup isomorphic to E_{p^3}. It follows that any abelian subgroup of G is of rank ≤ 2, by Theorem 10.4 (see also Theorem 13.7).

(b) Take in the natural semidirect product $\langle\alpha\rangle \cdot G$ a minimal nonnilpotent subgroup S and use Theorem 13.7 and Theorem A.22.1 (one has $S' \in \{C_p, E_{p^2}, S(p^3)\}$, where $S' \in \mathrm{Syl}_p(S)$). □

Exercise 22 ([FT, Lemma 8.7]). If a p'-group A of automorphisms of a p-group G is such that $[A, G] = G$ and $[A, \Phi(G)] = \{1\}$, then $\Phi(G) \leq Z(G)$. Is it possible to replace $\Phi(G)$ by a characteristic in G subgroup N?

Solution. One has

$$[A, \Phi(G), G] = [\Phi(G), G, A] = \{1\}.$$

Therefore, by the Three Subgroups Lemma, $[G, A, \Phi(G)] = \{1\}$. But $[G, A] = G$ so that $[G, \Phi(G)] = \{1\}$ implies $\Phi(G) \leq Z(G)$. □

Exercise 23. Suppose that all critical subgroups in the p-groups A and B are known. Is it possible to describe the critical subgroups in $A \times B$?

Exercise 24. Let $P \cong C_{p^m} \times C_{p^n}$, $p > 3$, $m \geq 2$, $n \geq 2$. Suppose that $a \in \mathrm{Aut}(P)$ is of order p. Then a centralizes $\Omega_1(P)$.

Solution. Let $G = \langle a\rangle \cdot P$ be the natural semidirect product with kernel P. As

$$|G/\mho_1(G)| = p^3 < p^{p-1},$$

it follows that G is absolutely regular (Theorem 9.8 (a)) so regular. By Theorem 7.2 (d), we have $|\Omega_1(G)| = p^3$; then $G/\Omega_1(G)$ is noncyclic since $m > 1$, $n > 1$. Therefore, by Theorem 13.7, $E = \Omega_1(G) \cong E_{p^3}$. As $a \in E$ and $E \cap P = \Omega_1(P)$, the automorphism a centralizes $\Omega_1(P)$. The condition $m, n \geq 2$ is essential (see Theorem 13.7 (c)). In [MS] and [AS, Lemma A.30.1], a similar result has proved for $m = n = 2$. (For a stronger result, see Theorem 268.2.) □

The following result is a partial case of Theorem 1.17 (a).

Exercise 25. Suppose that a nonabelian p-group G has no normal subgroup isomorphic to E_{p^2}. Then G has a cyclic subgroup of index p (see Theorem 1.2).

Solution (compare with [Gor1, Theorem 5.4.10]). Let $A < G$ be maximal normal abelian subgroup; then A is cyclic. Let B/A of order p is normal in G/A. Then B has no characteristic subgroup isomorphic to E_{p^2}. By Theorem 1.2, $p = 2$ and B is a 2-group of maximal class. Now G is a 2-group of maximal class, by Exercise 10.10. □

Exercise 26 ([Gor1, Theorem 5.3.7 (ii)]). Let A be a p'-group of automorphisms of a p-group P and let $\phi \in A^{\#}$ act trivially on every proper A-invariant subgroup of P. Then P/P' is elementary abelian (so that $P' = \Phi(P)$), A acts irreducibly on P/P' and ϕ acts nontrivially on P/P'.

Solution. As $P' < P$ is A-invariant, ϕ acts trivially on P'. By Burnside, ϕ acts nontrivially on P/P'. Therefore, ϕ acts nontrivially on $L/P' = \Omega_1(P/P')$ so on L, and we conclude that $L = P$, i.e., P/P' is elementary abelian and we get $P' = \Phi(P)$. As P/P' has no nontrivial A-invariant subgroup (otherwise, using Maschke's theorem, we obtain $\phi_P = \mathrm{id}_P$)), A acts irreducibly on P/P'. By Burnside, ϕ acts nontrivially on P/P'. □

Exercise 27 ([FT, Lemma 8.12]). Let $E > \{1\}$ be an elementary abelian subgroup of a p-group P, $p > 2$. If ϕ is a p'-automorphism of P which centralizes $\Omega_1(C_P(E))$, then $\phi = \mathrm{id}_P$.

Solution. Write $A = \langle\phi\rangle$. As the natural semidirect product $A \cdot C_P(E)$ has no minimal nonnilpotent subgroup, it is nilpotent so A centralizes $C_P(E)$. Therefore, one may assume that $C_P(E) < P$; then $C_P(E)\Phi(P) < P$ so, by induction, ϕ centralizes $\Phi(P)$. Now

$$[P, E] \le [P, P] = P' \le \Phi(P) \implies [P, E, A] \le [\Phi(P), A] = \{1\}.$$

By hypothesis, $[E, A] = \{1\}$ implies $[E, A, P] = \{1\}$. Then $[A, P, E] = \{1\}$, by the Three Subgroups lemma, so that $[P, A] \le C_P(E)$, and (the p'-group) A stabilizes the chain $P > C_P(E) > \{1\}$, which implies $A = \{1\}$ (Exercise 7). □

Exercise 28 ([Tho3, Lemma 5.17]). Assume that $G = P \cdot Q$, where $Q \in \mathrm{Syl}_2(G)$ is G-invariant, $P \in \mathrm{Syl}_p(G)$ and $Q = [P, Q] > \{1\}$. If P centralizes every proper characteristic subgroup of Q, then Q is either elementary abelian or special.

Solution. Assume that $m = \mathrm{cl}(Q)$ and $m > 2$. Now $K_{m-1}(Q)$ is a characteristic abelian subgroup of Q so $[K_{m-1}(Q), P] = \{1\}$, by hypothesis. It follows that $[K_{m-1}(Q), P, Q] = \{1\}$. Next, $[Q, K_{m-1}(Q), P] = [K_m(Q), P] = \{1\}$, by assumption. From the Three Subgroups Lemma one has $[P, Q, K_{m-1}(Q)] = \{1\}$. As $[P, Q] = Q$, this implies $[Q, K_{m-1}(Q)] = \{1\}$, i.e., $\mathrm{cl}(Q) < m$, a contradiction. Thus, we get $m \le 2$. Now let $m = 2$ (if $m = 1$, then, by Frobenius' Theorem A.52.2 on normal 2-complement, Q is elementary abelian). Assume that $Q' < \Phi(Q)$. If $H/Q' = \Omega_1(Q/Q')$, then H is a proper characteristic subgroup in Q hence P centralizes H and so $H/Q'(= \Omega_1(Q/Q'))$, and we conclude that P

centralizes Q (Burnside), a contradiction. Thus, $\exp(Q/Q') = p$ hence $Q' = \Phi(Q)$. One has $Q' \leq Z(Q)$ so $Q' = Z(Q)$ since P acts irreducibly on Q/Q', and hence Q is special. □

Exercise 29. Suppose that each two-generator subgroup H of a p-group G, $p > 2$, satisfies $|H' : \mho_1(H')| < p^{p-1}$. Then G is regular.[5]

Solution. Assume that G is a counterexample of minimal order. If a two-generator $H < G$, then $|H' : \mho_1(H')| < p^{p-1}$, so H is regular (Theorem 9.8 (c)). By induction, all maximal subgroups of G are regular so that G is minimal irregular. It follows that $d(G) = 2$. Then $|G' : \mho_1(G')| < p^{p-1}$, by hypothesis, so G is regular (Theorem 9.8 (c) again), contrary to the assumption. □

Exercise 30. Let A be a G-invariant subgroup of a p-group G and let $B < A$ be abelian G-invariant of exponent $p^n > 2$. If F is a maximal G-invariant abelian subgroup of exponent p^n of A containing B, then $\Omega_n(C_A(F)) = F$.

Solution. Let $C = C_A(F)$; then C is G-invariant since it is equal to $C_G(F) \cap A$. Assume that $F < C$ and $C - F$ contains an element x of order $\leq p^n$ such that $x^p \in F$. Put $H = \langle x, F\rangle$. As $p^n > 2$, the number of containing F abelian subgroups of order $p|F|$ and exponent p^n in C is $\equiv 1 \pmod p$ (Theorem 10.1). It follows that one of the above subgroups is G-invariant, so that F is not a maximal G-invariant abelian subgroup of exponent p^n of A containing B. Thus, $\Omega_n(C_A(F)) = F$. □

Exercise 31 (Exercise 11). Let a p'-group A act nontrivially on a nonabelian p-group P. Assume that A centralizes all proper A-invariant normal subgroups of P. Then P is special, A acts irreducibly on $P/\Phi(P)$.

Solution. Obviously, P is critical[6] so that $\mathrm{cl}(P) = 2$, $\Phi(P) \leq Z(P)$. By hypothesis and Maschke's theorem, A acts irreducibly on $P/\Phi(P)$. It follows that $Z(P) \geq \Phi(P)$ hence $Z(P) = \Phi(P)$. Assume that $P' < \Phi(P)$. In that case $\overline{P} = P/P'$ is abelian of exponent $> p$. Then $\overline{P} = \overline{H}_1 \times \cdots \times \overline{H}_k$ is a direct product of homocyclic A-invariant subgroups $\overline{H}_i$ (Corollary 6.2). Let $k > 1$. Then A centralizes all H_i so G, and we get $A = \{1\}$, a contradiction. Now let $k = 1$; then $\overline{G}$ is homocyclic of exponent $> p$. Let $\overline{F} = \Omega_1(\overline{G})$. Then A centralizes F and $\overline{G}$ so $G/\Phi(G)$, contrary to Burnside's theorem. Thus, $P' = \Phi(P) = Z(P)$ so that P is special, as desired. □

---

5 In particular, if all two-generator subgroups of a p-group G, $p > 2$, have cyclic derived subgroups, then G is regular. By [Alp1, Theorem 2], G in that case is metabelian.

6 If not and $L < P$ is critical, then A acts trivially on L, a contradiction.

§ 269 On critical subgroups of p-groups

We begin with the following:

Definition 1. A subgroup H of a p-group G is said to be *critical* if it is characteristic and
(a) $\Phi(H) \leq Z(H) \geq [G, H]$,
(b) $C_G(H) = Z(H)$.

If H is a critical subgroup of a p-group G, then

$$\mathrm{cl}(H) \leq 2 \quad \text{and} \quad \exp(H/Z(H)) = p.$$

A self-centralizing characteristic abelian subgroup of a p-group is critical. A nonabelian p-group G is critical if and only if $\Phi(G) \leq Z(G)$; in that case, $\exp(G') = p$ (indeed, if $x, y \in G$, then $[x, y]^p = [x^p, y] = 1$, by (a)). A two-generator nonabelian critical p-group is minimal nonabelian (check!) so that a critical subgroup of a metacyclic p-group is either abelian or minimal nonabelian. Also, special p-groups are critical. Direct product of critical p-groups is critical. It is possible that a critical p-group contains a proper critical subgroup (for example, the cyclic subgroup of order 4 in a critical group D_8 is critical).

Recall that $\mathrm{Sc}(G)$, the socle of a group G, is the product of all minimal normal subgroups of G. Obviously, $\mathrm{Sc}(G)$ is characteristic in G. If G is a p-group, then $\mathrm{Sc}(G) = \Omega_1(Z(G))$ is elementary abelian. If X is characteristic in a group G so is $C_G(X)$.

The following result is basic.

Theorem 269.1 ([FT, Lemma 8.2]). *Each p-group possesses a critical subgroup.*

Proof. One may assume that a p-group $G > \{1\}$. Let A be an abelian characteristic subgroup of G maximal by inclusion and let $C = C_G(A)$; then C is characteristic in G. If $C = A$, then A is critical in G. Next we assume that $C > A$; then C is nonabelian. As $Z(C) \geq A$ is characteristic abelian in G, we get $Z(C) = A$. Set

$$H/A = \mathrm{Sc}(Z(C/A))(> \{1\});$$

then the subgroup H is characteristic in G and $[G, H] \leq A = Z(H)$, $\Phi(H) \leq A = Z(H)$, by the definition of H. We claim that the subgroup H is critical in G. It remains to show that $C_G(H) = Z(H)(= A)$. Assume that $L = C_G(H) > A$. As L is characteristic in G, it is nonabelian, by maximality of A, hence $H \cap L = Z(H) = A < L$. Next, $L = C_G(H) \leq C_G(A) = C$. The subgroup $M/A = \mathrm{Sc}(L/A) > \{1\}$ is characteristic in C/A hence $M > A$ is characteristic in C so in G. We have

$$A \leq M \cap H \leq L \cap H = A \implies M \cap H = A$$

(recall that H/A is elementary abelian). However, $HM/A = (H/A(\times(M/A)$ is characteristic elementary abelian subgroup of C/A, and $HM/A > H/A$, contrary to the definition of H/A. Thus, $C_G(H) = A = Z(H)$ so that H is critical in G. □

https://doi.org/10.1515/9783110533149-013

Exercise 1. Let G be a p-group of maximal class and H its critical subgroup.
(i) If H is nonabelian, then $|G| = p^3$.
(ii) If $H < G$, then H is abelian.

Solution. (i) As H is nonabelian, the quotient group $H/Z(H)$ is a noncyclic central subgroup of $G/Z(H)$. It follows that $G/Z(H)$ of order $> p$, not being of maximal class, is isomorphic to E_{p^2}, and we conclude that $H = G$ and $Z(H) = Z(G)$. As G is of maximal class, we get $|Z(G)| = p$ so that $|G| = |Z(G)||G/Z(G)| = p^3$.
(ii) As $H < H$, the subgroup H is abelian, by (i). □

It follows that if a p-group of maximal class has order $> p^3$, then the unique critical subgroup of G coincides with its maximal abelian normal subgroup.

Exercise 2 ([FT, Lemma 8.2 (ii)]). The following conditions hold:
(a) Let H be a critical subgroup of a p-group G and $K = C_{\mathrm{Aut}(G)}(H)$. Then K is a p-group.
(b) For odd p, a p-group G possesses a characteristic subgroup D of class ≤ 2 and exponent p such that any nontrivial p'-automorphism of G induces a nontrivial automorphism of D.

Solution. (a) In commutator notation, we get $[K, H] = \{1\}$ so that $[K, H, G] = \{1\}$. Since $[H, G] \leq H$, we also have

$$[H, G, K] \leq [H, K] = \{1\}.$$

By the Three Subgroups Lemma, we get $[G, K, H] = \{1\}$ so that

$$[G, K] \leq C_G(H) = Z(H).$$

Thus, K stabilizes the chain $G \geq H \geq \{1\}$ so it is a p-group.
(b) See the solution of Exercise 3, below. □

Let, for a prime p, $\epsilon(p) = 1$ for $p > 2$ and $\epsilon(2) = 2$. Recall that if a p'-automorphism ϕ of a p-group P induces the identity automorphism on $\Omega_{\epsilon(p)}(P)$, then $\phi = \mathrm{id}_P$ (indeed, the natural semidirect product of $\langle\phi\rangle$ and G has no minimal nonnilpotent subgroup). We use this fact in the proof of the following theorem.

Exercise 3 (see [Gor1, Theorem 5.3]). Any p-group G possesses a characteristic subgroup $D > \{1\}$ of class ≤ 2 and exponent $\leq p^e$, where $e = \epsilon(p)$, such that any nontrivial p'-automorphism of G induces a nontrivial automorphism of D.

Solution. Let H be a critical subgroup of G. Set $D = \Omega_e(H)$. Then $\mathrm{cl}(D) \leq \mathrm{cl}(H) \leq 2$. Let ϕ be a nontrivial p'-automorphism of G; then $\phi_H \neq \mathrm{id}_H$ (Exercise 2 (a)). As the natural semidirect product $\langle\phi\rangle \cdot H$ is nonnilpotent, it contains a minimal nonnilpotent subgroup S. By Theorem A.22.1, $S' \leq \Omega_e(H) = D$. Let $\pi(S) = \{q, p\}$. Then one may assume that a q-Sylow subgroup of S is generated by some power of ϕ. It follows that $\phi_{S'} \neq \mathrm{id}_{S'}$, and this implies that $\phi_D \neq \mathrm{id}_D$. □

Note that the subgroup D from Exercise 3 need not necessarily be critical in G (generally speaking, $C_G(D) \nleq D$).

Theorem 269.2. *Suppose that a p-group P admits a p′-automorphism ϕ of a prime order q and $p \equiv 1 \pmod q$. Then there is in P a normal ϕ-invariant abelian subgroup A of exponent $\leq p^{\epsilon(p)}$ on which ϕ acts nontrivially.*

Proof. Assume, by way of contradiction, that ϕ centralizes all ϕ-invariant normal abelian subgroups of P of exponent $\leq p^{\epsilon(p)}$; then ϕ centralizes all ϕ-invariant abelian subgroups (Theorem A.12.1 and Frobenius' normal p-complement theorem) and hence P is nonabelian. Let H be a critical subgroup of P. By Exercise 2 (a), ϕ induces on H a nontrivial automorphism so H is nonabelian. Then ϕ acts nontrivially on $H/\Phi(H)$ (Burnside). The subgroup $Z(H)$ is a ϕ-invariant normal (even characteristic) abelian subgroup of P. By assumption, ϕ acts trivially on $Z(H)$. Consider the action of ϕ on the elementary abelian p-group $H/Z(H)$. As $[P, H] \leq Z(H)$, all subgroups of $H/Z(H)$ are P-invariant (indeed, in the case under consideration, $H/Z(G)$ is a central subgroup of $P/Z(H)$). As the natural semidirect product of groups $\langle\phi\rangle$ and $H/Z(H)$ is supersolvable [BZ, Exercise 3.19], there is in $H/Z(H)$ a ϕ-invariant subgroup $U/Z(H)$ of order p. By the above, $U \lhd P$. Thus, the abelian subgroup U is normal in P and ϕ-invariant. Therefore, by assumption, ϕ centralizes the subgroup U. By the theorem of Maschke, $H/Z(H) = (U/Z(H)) \times (V/Z(H))$, where V is ϕ-invariant (and, by the above, normal in P). Applying to $V/Z(H)$ the above argument again, we at last decompose $H/Z(H))$ in a direct product of ϕ-invariant subgroups $U/Z(G)) = U_1/Z(G)), \dots, U_d/Z(G))$ of order p. The abelian subgroups $U_1, \dots, U_d$ are ϕ-invariant and normal in P so ϕ centralizes them, by assumption. As $U_1 \dots U_d = H$, the automorphism ϕ centralizes H, a contradiction. Thus, there is in P a normal ϕ-invariant abelian subgroup A on which ϕ induces a nontrivial automorphism. Then ϕ induces a nontrivial automorphism in $\Omega_{\epsilon(p)}(A)$. □

In particular, if a p-group P of odd order admits an automorphism ϕ of order 2, there is in P a normal ϕ-invariant abelian subgroup E of exponent p such that ϕ induces on E a nonidentity automorphism. Note that condition $p \equiv 1 \pmod q$ is essential in Theorem 269.2.

Note that the subgroup D from Exercise 3 need not necessarily be critical in G (generally speaking, $C_G(D) \nleq D$).

Exercise 4. Let F, H be distinct critical subgroups of a p-group G such that $Z(A) = Z(H)$. Is it true that FH is critical in G?

§ 270 p-groups all of whose $\mathcal{A}_k$-subgroups for a fixed $k > 1$ are metacyclic

Given a positive integer n, a p-group G is said to be an *$\mathcal{A}_n$-group* if all its subgroups of index p^n are abelian but it has a nonabelian subgroup of index p^{n-1}. In particular, $\mathcal{A}_1$-groups are minimal nonabelian. If G is an $\mathcal{A}_n$-group, we write $\alpha(G) = n$. In particular, G is an $\mathcal{A}_1$-group if and only if $\alpha(G) = 1$.

If $n > 1$, then an $\mathcal{A}_n$-group G contains an $\mathcal{A}_{n-1}$-subgroup of index p. Indeed, there is in G a nonabelian subgroup B of index p^{n-1}. Let $B \le H < G$, where H is maximal in G; then $|H : B| = p^{n-2}$. As all subgroups of index p^{n-1} in H are abelian (those subgroups have index p^n is G), it follows that H is an $\mathcal{A}_{n-1}$-subgroup. Moreover, given positive integers $n > k$, any primary $\mathcal{A}_n$-group contains an $\mathcal{A}_k$-subgroup. Next, given $n > 1$, there is a maximal chain $G > H_{n-1} > H_{n-2} > \cdots > H_1$, where H_i is an $\mathcal{A}_i$-subgroup for $i = 1, \dots, n-1$.

If a p-group $G > \{1\}$ is metacyclic, it has a cyclic $C \lhd G$ such that G/C is cyclic. If, in addition, G is noncyclic, then C is a maximal cyclic subgroup of G. Indeed, assume that $C < C_1 < G$, where C_1 is cyclic. Then $C \le \Phi(C_1) \le \Phi(G)$ so that

$$\mathrm{d}(G/C) \ge \mathrm{d}(G/\Phi(G)) = 2,$$

a contradiction.

In this section we consider, for a fixed natural integer k, the p-groups G, all of whose $\mathcal{A}_k$-subgroups are metacyclic. We call such a group G an $\mathcal{M}_k$-group if it is an $\mathcal{A}_n$-group for some $n \ge k$. For example, any 3-group G of maximal class and order $> 3^3$ with $\Omega_1(G) \cong \mathrm{E}_{3^2}$ is a nonmetacyclic $\mathcal{M}_1$-group (such a group G is minimal nonmetacyclic if and only if $|G| = 3^4$). Classification of $\mathcal{M}_1$-groups, i.e., p-groups all of whose minimal nonabelian subgroups are metacyclic, is a very difficult open problem. Therefore, in what follows we assume that $k > 1$.

Our main result is the following:

Theorem 270.1. *If $k > 1$, then any prime power $\mathcal{M}_k$-group is metacyclic.*

In Lemma 270.2 some known results, used in what follows, are gathered.

Lemma 270.2. *Suppose that G is a p-group.*

(a) (§§ 66, 69) *Let G be a minimal nonmetacyclic p-group. Then one and only one of the following holds:*
   (i) $G \in \{\mathrm{E}_{p^3}, \mathrm{S}(p^3)\}$,
   (ii) *G is a 3-group of maximal class and order 3^4 with $\Omega_1(G) \cong \mathrm{E}_{3^2}$,*
   (iii) $G = \mathrm{Q}_8 \times \mathrm{C}_2$,
   (iv) *$G = \mathrm{Q}_8 * \mathrm{C}_4 \cong \mathrm{D}_8 * \mathrm{C}_4$ is the central product of order 2^4,*
   (v) *G is the special group of order 2^5, $\Omega_1(G) = \mathrm{Z}(G) \cong \mathrm{E}_{2^2}$.*

   The groups from (ii)–(v) are $\mathcal{A}_2$-groups of exponent p^2.

https://doi.org/10.1515/9783110533149-014

(b) (Theorem 72.1) *G is a metacyclic $\mathcal{A}_n$-group if and only if $|G'| = p^n$.*
(c) (Exercise 1.8a = Lemma 65.1) *The following hold:*
  (i) *If G is an $\mathcal{A}_1$-group, then $|\Omega_1(G)| \leq p^3$.*
  (ii) *If $|G| = |\Omega_1(G)| = p^3$, then $G \in \{\mathrm{S}(p^3), \mathrm{D}_8\}$.*
  (iii) *If $|G| > p^3$ and $|\Omega_1(G) = p^3$, then $G = \mathrm{M}_p(m, n, 1)$ and $m + n > 2$.*
(d) (Lemma 57.1) *If A is a maximal normal abelian subgroup of a nonabelian p-group G, then for any $x \in G - A$ there is $a \in A$ such that $\langle a, x\rangle$ is an $\mathcal{A}_1$-subgroup.*
(e) *Let A be a subgroup of a p-group G. If A is characteristic in $\mathrm{N}_G(A)$, then $A \lhd G$.*

Let us prove Lemma 270.2 (e). Assume that $\mathrm{N}_G(A) < G$. Let $\mathrm{N}_G(A) < N_1 \leq G$, where $|N_1 : \mathrm{N}_G(A)| = p$. Being characteristic in $\mathrm{N}_G(A) \lhd N_1$, the subgroup A is normal in N_1, a contradiction. Thus, $A \lhd G$.

From Lemma 270.2 (a) the following characterizations of metacyclic p-groups G follow:

(1) A p-group G is metacyclic if and only if $\Omega_2(G)$ is. Indeed, the exponent of a minimal nonmetacyclic p-group does not exceed p^2.
(2) If $p > 2$, then G is metacyclic if and only if all subgroups of G of order $\leq p^4$ are metacyclic. Indeed, the order of minimal nonmetacyclic p-group, $p > 2$, is at most p^4.
(3) If $p = 2$, then G is metacyclic if and only if all subgroups of G of order $\leq 2^5$ are metacyclic. Indeed, such a group G has no minimal nonmetacyclic subgroup.
(4) A 2-group G is metacyclic if and only if all its subgroups of order $\leq 2^5$ are two-generator.

The following characterization follows from Theorem 13.7: A p-group G of order $> p^6$, $p > 3$, is metacyclic if and only if any its subgroup of order $> p^3$ is two-generator. Indeed, this G has no subgroup isomorphic to E_{p^3}. For $p = 3$ the above result is not true as the 3-groups of maximal class and order $> 3^4$ show.

Remark 1. If a nonabelian p-group G, $p > 2$, is an $\mathcal{M}_1$-group (i.e., all $\mathcal{A}_1$-subgroups of G are metacyclic), then the subgroup $\Omega_1(G)$ is elementary abelian. Indeed, let E be a G-invariant elementary abelian subgroup of maximal order in G. Assume that there is $x \in G - E$ of order p. By the choice of E and Exercise 10.10, the subgroup $\langle x, E\rangle$ is nonabelian. By Lemma 270.2 (d), there is $a \in A$ such that $S = \langle x, a\rangle$ is an $\mathcal{A}_1$-subgroup. As $\Omega_1(S) = p$, it follows from Lemma 270.2 (c), that $S \cong \mathrm{S}(p^3)$ so that S is nonmetacyclic, a contradiction. Thus, x does not exist hence $\Omega_1(G) = E$ is elementary abelian, as claimed.

If G is an $\mathcal{A}_n$-group, then by order considerations, $|G| \geq p^{2+n}$, and this estimate is best possible (indeed, a p-group of maximal class and order p^{2+n} is an $\mathcal{A}_n$-group, by Theorem 9.6 (f)).

If H is of minimal possible order among $\mathcal{A}_k$-subgroups of G, then H is said to be an $\mathcal{CA}_k$-subgroup.

Remarks 2–4 are obvious and some of then we use freely in what follows.

Remark 2. If $H < G$ is nonabelian, then $\alpha(H) < \alpha(G)$.

Remark 3. Let $H < G$ be a $\mathcal{CA}_k$-subgroup, $H < F \le G$ with $|F:H| = p^s$. Then F is a $\mathcal{CA}_{k+s}$-subgroup.

Remark 4. Let $H < G$. If $R \lhd G$ is minimal such that $R \nleq H$, then $|RH:H| = p$.

Remark 5. Let $H < G$ be nonabelian metacyclic. If $R \lhd G$ be elementary abelian of minimal order such that $R \nleq H$, then the subgroup RH is nonmetacyclic. This is clear if $|R| \neq p^2$. Let $|R| = p^2$. Then $|R \cap H| = p$ so that $H/(R \cap H)$ is noncyclic since H is nonabelian. It follows that $RH/(R \cap H) = (R/(R \cap H) \times (H/(R \cap H)$ is not two-generator hence RH is nonmetacyclic.

Remark 6. Let G be a nonabelian metacyclic p-group of order $> p^3$. If $|G:G'| = p^2$, then $p = 2$ and G is of maximal class. Indeed, let a cyclic $C \lhd G$ be such that G/C is cyclic. Then $|G:C| = p$ since $G' < C$ and $G/G' \cong \mathrm{E}_{p^2}$. Now the result follows from classification of p-groups with cyclic subgroup of index p and Taussky's theorem (see Theorem 1.2, Proposition 1.6).

To prove that a primary $\mathcal{M}_k$-group, $k > 1$, is metacyclic, it suffices to show that it has no minimal nonmetacyclic subgroup. The minimal nonmetacyclic subgroups are classified in Lemma 270.2 (a).

To clear up our path, we offer an easy proof of a partial case of Theorem 270.1 for $k = 2$ (see § 202 and [BZhang]); that result is used in the proof of Theorem 270.1.

Proposition 270.3. *Any p-group G which is a $\mathcal{M}_2$-group is metacyclic.*

Proof. By definition, all $\mathcal{A}_2$-subgroups of G are metacyclic. Let a p-group G, which is an $\mathcal{M}_2$-group, be a counterexample of minimal order and let $\alpha(G) = n$ (i.e., G is an $\mathcal{A}_n$-group); then $n > 2$. There is in G an $\mathcal{A}_{n-1}$-subgroup H of index p. If $n-1 = 2$, then H is metacyclic, by hypothesis. If $n-1 > 2$, then, since $H < G$ is an $\mathcal{M}_2$-group, it is metacyclic, by induction. Thus, all $\mathcal{A}_{n-1}$-subgroups of G are metacyclic. Moreover, all $\mathcal{A}_s$-subgroups, $s \ge 2$, are metacyclic. Therefore, if a maximal subgroup of G is nonmetacyclic, it is either abelian or an $\mathcal{A}_1$-subgroup.

Being nonmetacyclic, the group G possesses a minimal nonmetacyclic subgroup E. Then, by Lemma 270.2 (a), $E \in \{\mathrm{E}_{p^3}, \mathrm{S}(p^3)\}$ since groups of parts (ii)–(v) in (a) of that lemma are nonmetacyclic $\mathcal{A}_2$-groups so they cannot be subgroups of G.

Let $E \cong \mathrm{S}(p^3)$ and $E < E_1 < G$, where $|E_1 : E| = p$; then E_1, by order considerations, is an $\mathcal{CA}_2$-subgroup. As E_1 is nonmetacyclic, this contradicts to the hypothesis. Thus, $E \cong \mathrm{E}_{p^3}$. As the set Γ_1 has a metacyclic member (see the first paragraph of the proof), E is a maximal elementary abelian subgroup of G. Indeed, G has no subgroup isomorphic to E_{p^4}: if $S \cong \mathrm{E}_{p^4}$ is a subgroup of G and $H \in \Gamma_1$ is a (metacyclic) $\mathcal{A}_{n-1}$-subgroup, then $H \cap S$ is a nonmetacyclic subgroup of the metacyclic subgroup H, a contradiction.

Write $N = \mathrm{N}_G(E)$. We claim that E is characteristic in N. Assume that this is false. As N is nonmetacyclic, then, as we have noted, $\alpha(N) < 2$ so that N is either abelian or minimal nonabelian. Since E is a maximal elementary abelian subgroup in G, it follows that $E = \Omega_1(N)$ is characteristic in N. Thus, $N = G$ (Lemma 270.2 (e)) hence $E \lhd G$.

Let $M < G$ be an $\mathcal{A}_1$-subgroup of minimal order (i.e., M is an $\mathcal{CA}_1$-subgroup of G) and let $M < M_1 < G$, where $|M_1 : M| = p$ (we have $M_1 < G$ since M_1 is an $\mathcal{A}_2$-subgroup and $\alpha(G) > 2$). In that case M_1, being $\mathcal{A}_2$-subgroup, is metacyclic; then M is also metacyclic so that $E \nleq M$ since $E \cong \mathrm{E}_{p^3}$ is nonmetacyclic. Let $R \leq E$ be G-invariant of minimal order such that $R \nleq M$. Then $|RM : M| = p$ (Remark 4) and RM is an $\mathcal{A}_2$-subgroup (Remark 3). As, by Remark 5, RM is nonmetacyclic, this contradicts to the hypothesis. Thus, E does not exist so G has no minimal nonmetacyclic subgroup. It follows that G is metacyclic. □

Now we are ready to prove Theorem 270.1.

Proof of Theorem 270.1. Let G be a counterexample of minimal order; then G is a nonmetacyclic p-group and all its proper $\mathcal{A}_s$-subgroups for $s \geq k$ are metacyclic either by hypothesis or by induction. Write $\alpha(G) = n$; then $n > k$. In view of Proposition 270.3, we get $k > 2$; then $n > 3$. Let $H < G$ be an $\mathcal{A}_{n-1}$-subgroup of index p; then H, being an $\mathcal{M}_k$-subgroup, is metacyclic since $n - 1 \geq k$. It follows that G has no subgroup of order p^4 and exponent p (let us consider its intersection with H). If $X < G$ is nonabelian nonmetacyclic, then $\alpha(X) < k$ since all subgroups $Y < G$ with $\alpha(Y) \geq k$ are metacyclic.

Being nonmetacyclic, the group G possesses a minimal nonmetacyclic subgroup E; then E is one of the groups of Lemma 270.2 (a) hence either $E \in \{\mathrm{E}_{p^3}, \mathrm{S}(p^3)\}$ or $\alpha(E) = 2$. Since $\alpha(G) = n > k > 2$, it follows that $E < G$. If $|E| \neq 2^5$, then E is a $\mathcal{CA}_i$-group, $i \in \{1, 2\}$. In that case, E is contained in some $\mathcal{A}_{n-1}$-subgroup $L < G$. As $n - 1 \geq k$, the subgroup L must be metacyclic, by condition. But L contains a nonmetacyclic subgroup E, which is a contradiction.

Thus, E is either isomorphic to E_{p^3} or minimal nonmetacyclic group of order 2^5.

Let $p = 2$ and $S \in \{\mathrm{D}_8, \mathrm{Q}_8\}$. By Proposition 10.17, $\mathrm{C}_G(S) \nleq S$. If $S < T \leq S\mathrm{C}_G(S)$ such that $|T : S| = 2$, then T is a nonmetacyclic CA_2-subgroup, which is impossible as in the second paragraph of the group. Thus, if $p = 2$, then all minimal nonabelian subgroups of G have order $> 2^3$. Thus, if $E \leq G$ is a nonabelian minimal nonmetacyclic, then $|E| = 2^5$. By the above, E is a $\mathcal{CA}_2$-subgroup of G so $E < G$ since $k > 2$. In that case, $|G| > 2^{5+(k-2)}$. Therefore, if $S < T < G$, where $|T| = 2^{5+(k-2)}$, then T is a nonmetacyclic $\mathcal{A}_k$-subgroup, a contradiction. Thus, G has no minimal nonmetacyclic subgroup of order 2^5.

Suppose that $E \cong \mathrm{E}_{p^3}$ and set $N = \mathrm{N}_G(E)$. Assume that $E < \Omega_1(N)$. Then there is $x \in N - E$ of order p. The subgroup $K = \langle x, E\rangle$ is nonabelian since G has no subgroup isomorphic to E_{p^4}. One has $\Omega_1(K) = K$ so K is not minimal nonabelian (Lemma 65.1). It follows that K is a CA_2-group so it is contained in $\mathcal{A}_{k-1}$-subgroup of G which must be metacyclic, a contradiction. Thus, x does not exist so that $E = \Omega_1(N)$ is characteristic in N which implies that $E \trianglelefteq G$. Let $U < G$ be a minimal nonabelian subgroup of minimal order. Then $|G : U| \geq p^{k+1}$. If $U < V < G$, where $|V : U| = p^k$, then V, being an $\mathcal{A}_k$-subgroup, is metacyclic. Thus, the subgroup U is metacyclic. Let $E_0 \leq E$ be a G-invariant subgroup of minimal order not contained in U. Then $|UE_0 : U| = p$

(Remark 4). It follows that a nonmetacyclic subgroup U (Remark 5) is a $\mathcal{CA}_2$-subgroup. This is a contradiction.

It follows that G has no minimal nonmetacyclic subgroup so it is metacyclic. □

Problem 1. Study the p-groups G such that, whenever $H < G$ and $|H|^2 \le G$, then either H is cyclic or $H \triangleleft G$.

Problem 2. Suppose that a 2-group G has no normal subgroup isomorphic to E_8. Describe the $\mathcal{A}_1$-subgroups of G. (See § 50.)

Problem 3. Study the p-groups all of whose $\mathcal{A}_1$-subgroups are metacyclic. In particular, describe the p-groups all of whose two-generator subgroups are metacyclic.

Problem 4. Study the p-groups in which the intersection of any two distinct maximal metacyclic subgroups is cyclic.

Problem 5. Study the p-groups with metacyclic normal closure of any cyclic subgroup.

Problem 6. Given n, study the p-groups G in which $\Omega_n(G)$ is minimal nonabelian.

Problem 7. Study the non-Dedekindian p-groups G such that $HG' = H^G$ for any nonnormal $H < G$.

Problem 8. Study the two-generator p-groups G with cyclic $G' \le Z(G)$. Describe the representation groups of such G.

Problem 9. Study the p-groups with at most three classes of nonnormal conjugate abelian subgroups.

Problem 10. Classify the p-groups all of whose $\mathcal{A}_2$-subgroups, except one, are metacyclic.

Problem 11. Study the p-groups all of whose minimal nonabelian subgroups are normal.

§ 271 Two theorems of Blackburn

In this section we prove two Blackburn's results.

Theorem 271.1 ([Bla6]). *Let G be a p-group and let $E < G$ be abelian of exponent $p^n > 2$ maximal subjecting these restrictions. If $\alpha \in \mathrm{Aut}(G)$ induces the identity automorphism on E, then $\mathrm{o}(\alpha)$ is a power of p.*

Proof. Let $W = \langle\alpha\rangle \cdot G$ be the natural semidirect product of $\langle\alpha\rangle$ and G. Suppose that $\mathrm{o}(\alpha) = p^k m$, where m is a p'-number. Then α^{p^k} of order m induces the identity automorphism on E, and it suffices to show that $m = 1$. Therefore, in what follows we assume, that $\mathrm{o}(\alpha) = m$, a p'-number. We have to show that $m = 1$. Assume, by way of contradiction, that $m > 1$. By Corollary 10.2,

$$\Omega_n(\mathrm{C}_G(E)) = E. \tag{1}$$

Write $A = \langle\alpha\rangle$ and let $W = A \cdot G$ be the natural semidirect product of A and G.

(i) First assume that $E \lhd G$; then $\mathrm{N}_W(E) \geq AG = W$ so that $E \lhd W$. Write $C = \mathrm{C}_W(E)$; then $A < C \lhd W$. Assume that the conclusion is false. One has $P = C \cap G = \mathrm{C}_G(E) \in \mathrm{Syl}_p(C)$; then P is A-invariant and $C = A \cdot P$, the semidirect product with kernel P. By what we have said in the previous paragraph, $E = \Omega_n(P)$. Therefore, by Theorem A.22.1, $C = A \cdot P$ has no minimal nonnilpotent subgroup (indeed, if $S \leq C$ is minimal nonnilpotent such that p'-Sylow subgroup of S is contained in A, then $S' \in \mathrm{Syl}_p(S)$, being of exponent $\leq p^n$, is contained in $\Omega_n(P) = E \leq \mathrm{C}_G(A)$, which is impossible since A centralizes E); therefore C is nilpotent hence $C = A \times P$. Since A is characteristic in $C \lhd W$, it follows that $A \lhd W$, i.e., $W = A \times G$, and we conclude that α is the identity automorphism of G, a contradiction, since, by assumption, $\mathrm{o}(\alpha) = m > 1$. Thus, $m = 1$ and $\mathrm{o}(\alpha) = p^k$, as asserted.

(ii) Now assume that the abelian subgroup E from the statement is not G-invariant and α, a nonidentity p'-automorphism of G of order $m > 1$, induces the identity automorphism on E. We have to prove that this is impossible. We proceed by induction on $|G|$. Write, as above, $A = \langle\alpha\rangle$. Let $R < G$ be a maximal proper A-invariant subgroup of G containing E and write $N = \mathrm{N}_G(R)$; then $N > R$. Then A normalizes R so N is also A-invariant, and, by induction and the choice of R, we must have $N = G$ so that $R \lhd W$ since $\mathrm{N}_G(R) = AG = W$, i.e., $R \lhd W$. Write $C = \mathrm{C}_G(R)$; then $C \lhd W$. Set $C_1 = A \cdot R$. We have

$$\mathrm{C}_G(R) \leq \mathrm{C}_G(E) \implies \Omega_n(\mathrm{C}_G(R)) \leq \Omega_n(\mathrm{C}_G(E)) = E$$

(the last equality coincides with (1)). Therefore, by Theorem A.22.1, C_1 has no minimal nonnilpotent subgroup so it is nilpotent hence $C_1 = A \times R$. In that case, A is a characteristic subgroup of $C_W(R) \leq \mathrm{C}_W(E) \lhd W$ so that $A \lhd W$, and therefore, α, a generator of A, induces on G the identity automorphism, contrary to the assumption. The proof is complete. □

https://doi.org/10.1515/9783110533149-015

If E from Theorem 271.1 is a maximal abelian subgroup of G, then $o(\alpha)$ is a power of p.

Theorem 271.2 ([Bla12]). *Let G be a p-group of class k. Suppose that $s \in G$ and $|C_G(s)| = p^r$. Then $|G| \le p^{kr}$.*

Proof. One may assume that $s \in G - Z(G)$. Let

$$G = Z_1 > Z_2 > \cdots > Z_k > Z_{k+1} = \{1\}$$

be the lower central series of G. For $i \le k$ set $H_i = \langle s, Z_i\rangle$. For $h \in H_i$, one has $s^h = s[s,h]$ so the number of conjugates of s in H_i is at most $|H_i'|$. Therefore,

$$|H_i'| \ge |H_i : C_{H_i}(s)| \ge \frac{|H_i|}{|C_G(s)|} = \frac{|H_i|}{p^r} \ge \frac{|Z_i|}{p^r}. \tag{2}$$

As $H_i' \le Z_{i+1}$, we get, by (2), that

$$|Z_i : Z_{i+1}| \le |Z_i : H_i'| \le p^r \quad \text{for all } i \le k. \tag{3}$$

Therefore, by (3), one obtains

$$|G| = |Z_1 : Z_2||Z_2 : Z_3| \cdots |Z_k : Z_{k+1}| \le p^{kr},$$

and we are done. □

If a p-group G of maximal class has order p^n, Theorem 271.1 yields $|G| \le p^{2(n-1)}$.

Problem. Given k and r as in Theorem 271.2, does there exist a nonabelian p-group G for which $|G| = p^{kr}$?

§ 272 Nonabelian p-groups all of whose maximal abelian subgroups, except one, are either cyclic or elementary abelian

The knowledge of all maximal abelian subgroups of a nonabelian p-group allows us to obtain an additional information on the structure of G. For example, if $p > 2$ and all maximal abelian subgroups of G are normal, then G is regular. Here we study the p-groups all of whose maximal abelian subgroups are either cyclic or elementary abelian.

We initiate here a study of Problem 3843, by proving the following result.

Theorem 272.1 (Janko). *Let G be a nonabelian p-group all of whose maximal abelian subgroups, except one, are either cyclic or elementary abelian. Then G is non-Dedekindian and G possesses a maximal normal abelian subgroup A of index $\le p^2$ which is neither cyclic nor elementary abelian and G/A is elementary abelian so that G is metabelian.*

Proof. Let G be a nonabelian p-group all of whose maximal abelian subgroups, except one, are either cyclic or elementary abelian. Since Hamiltonian 2-groups do not satisfy our hypothesis, it follows that G is non-Dedekindian.

Let A be a maximal abelian subgroup which is neither cyclic nor elementary abelian. Because of the uniqueness of such a subgroup, $A \lhd G$. Then there is a subgroup $U \cong \mathrm{E}_{p^2}$ which is contained in $\Omega_1(A)$ and which is normal in G. Assume that $|G : A| \ge p^3$. Then there is a subgroup H such that $A < H < G$, $|H : A| = p^2$ and $U \le \mathrm{Z}(H)$ (take $H \le \mathrm{C}_G(U)$). For each $h \in H - A$, $\langle U, h\rangle$ is abelian and noncyclic. Therefore a maximal abelian subgroup X of G containing $\langle U, h\rangle$ must be elementary abelian. In particular, $\mathrm{o}(h) = p$ and so all elements in the set $H - A$ are of order p. It follows that $\{1\} \ne \mathrm{H}_p(H) = A$ and note that H is metabelian. This contradicts a result of Hogan–Kappe [HogK] stating that in a metabelian p-group a nontrivial Hughes subgroup is of index at most p. We have proved that $|G : A| \le p^2$ and so G is metabelian.

Suppose that G/A is not elementary abelian. Then $|G : A| = p^2$ and $G/A \cong \mathrm{C}_{p^2}$. Let $g \in G - A$ be such that the cyclic subgroup $\langle g\rangle$ covers G/A. Since $\langle g\rangle$ acts on $U \cong \mathrm{E}_{p^2}$ (where $U \le \Omega_1(A)$ and $U \trianglelefteq G$), it follows that $1 \ne g^p \in G - A$ centralizes U and so $\mathrm{o}(g^p) = p$ and $\mathrm{o}(g) = p^2$ with $A \cap \langle g\rangle = \{1\}$. Let $1 \ne z \in U \cap \mathrm{Z}(G)$. Then $\langle g, z\rangle$ is abelian of type (p^2, p). Let Y be a maximal abelian subgroup in G containing $\langle g, z\rangle$. Then Y is neither cyclic nor elementary abelian and $Y \ne A$, contrary to our hypothesis. We have proved that G/A is elementary abelian of order $\le p^2$, completing the proof. □

Note that in Lemma 265.1 the p-groups all of whose noncyclic abelian subgroups have exponent p are classified. Also § 265 contains a number of related results.

https://doi.org/10.1515/9783110533149-016

§ 273 Nonabelian p-groups all of whose noncyclic maximal abelian subgroups are elementary abelian

In § 216, the p-groups all of whose nonnormal subgroups are elementary abelian were classified.

In this section we solve Problem 3732 offering to classify the p-groups in which all noncyclic maximal abelian subgroups are elementary abelian. We prove the following:

Theorem 273.1. *Let G be a nonabelian p-group all of whose noncyclic maximal abelian subgroups are elementary abelian. Then one of the following holds:*

(a) *G is a 2-group of maximal class.*

(b) *$p > 2$ and $\exp(G) = p$.*

(c) *$p > 2$, $\exp(G) = p^2$ and G is of maximal class with $\mho_1(G) = Z(G) \cong C_p$. In particular, $|G| \le p^{p+1}$, the fundamental subgroup of G is of exponent p.*

Proof. Obviously, any nonabelian group of exponent p satisfies the hypothesis. As any noncyclic abelian subgroup, say A, is contained in maximal abelian subgroup, say B, of G, then $\exp(B) = p$ so B is elementary abelian, and this implies that A is also elementary abelian. Thus, the hypothesis is inherited by nonabelian subgroups of G.

Let G be a nonabelian p-group of exponent $> p$ all of whose noncyclic maximal abelian subgroups are elementary abelian. If G has no normal abelian subgroup of type (p, p), then by Lemma 1.4, we have $p = 2$ and G is of maximal class. Each 2-group of maximal class satisfies the hypothesis. Next we also assume that $|G| > p^3$.

Now assume that G has a normal abelian subgroup U of type (p, p) and $\exp(G) > p$. Set $H = C_G(U)$. If a cyclic subgroup $L < H$ is of order $> p$, then UL is noncyclic abelian of exponent $p^2 > p$, a contradiction. Thus, $\exp(H) = p$. By assumption, $H < G$ so that $|G : H| = p$ and therefore $U \nleq Z(G)$ and $\exp(G) = p^2$. We have $U \cap Z(G) = \langle z \rangle \cong C_p$. If $g \in G - H$ is of order p^2, then $\langle g, z \rangle$, being abelian of exponent $p^2 > p$, must be cyclic, and hence $\langle g^p \rangle = \langle z \rangle$. It follows that from $\exp(G) = p^2 = o(g)$ that $\langle g \rangle$ is a maximal abelian subgroup of G. Therefore, G is of maximal class, by Proposition 1.8. As each cyclic subgroup of G of order $p^2 = \exp(G)$ contains z, we get $\mho_1(G) = \langle z \rangle$. By Theorems 9.5 and 9.6, $|G/\mho_1(G)| \le p^p$. If $p = 2$, then $|G| = 2^3$ and $G \cong D_8$, the dihedral group of order 8. Now let $p > 2$. Let G_1 be the fundamental subgroup of G. Since any cyclic subgroup of G of order p^2 coincides with its centralizer, it follows that $\exp(G_1) = p$ since $Z(G_1)$ is noncyclic in view of $p > 2$. This group G satisfies the hypothesis. □

Problem 1. Classify the non-Dedekindian p-groups all of whose nonnormal noncyclic abelian subgroups are elementary abelian.

Problem 2. Classify the irregular p-groups all of whose maximal nonnormal regular subgroups are absolutely regular.

https://doi.org/10.1515/9783110533149-017

§ 274 Non-Dedekindian p-groups in which any two nonnormal subgroups normalize each other

There are in our book a number of characterizations of p-groups close to Dedekindian groups. One of such results is contained in Theorem 274.1.

We solve here Problem 4037 by proving the following result.

Theorem 274.1 (Janko). *Let G be a non-Dedekindian p-group in which any two nonnormal subgroups normalize each other. If G_0 is the subgroup of G generated by all nonnormal subgroups, then $G_0 < G$ is abelian, G is of class 2 and the derived subgroup G' is cyclic.*

Proof. Let G be a non-Dedekindian p-group in which any two nonnormal subgroups normalize each other. Let H be a nonnormal subgroup in G so that $N_G(H) < G$. By our hypothesis, all nonnormal subgroups of G normalize H and so they all lie in $N_G(H)$. This implies that the subgroup G_0 generated by all nonnormal subgroups of G is a proper subgroup of G. In that case, by Theorem 231.1, $\mathrm{cl}(G) = 2$, G/G_0 is cyclic and for each $g \in G - G_0$,

$$\{1\} \neq \langle g\rangle \cap G_0 \trianglelefteq G \quad \text{and} \quad G/(\langle g\rangle \cap G_0) \text{ is abelian}$$

so that $G' > \{1\}$ is cyclic.

It remains to determine the structure of G_0. Let $X < G$ be nonnormal. Then $X \leq G_0$ and since G_0 is generated by all nonnormal subgroups of G and they all normalize X, it follows that $X \trianglelefteq G_0$. But then each subgroup of G_0 is normal in G_0 and so G_0 is Dedekindian.

Assume, by way of contradiction, that G_0 is nonabelian. Then, by Theorem 1.20, $p = 2$ and $G_0 = QZ(G_0)$, where $Q \cong Q_8$ and $Z(G_0)$ is elementary abelian. Also we have

$$\mho_1(G_0) = Z(Q) = \langle z\rangle \cong C_2.$$

Let $g \in G - G_0$ be such that $\langle g\rangle$ covers G/G_0 (see Theorem 231.1) and then $\{1 \neq \langle g\rangle \cap G_0$ is cyclic of order $\leq \exp(G_0) = 4$ and $G' \leq \langle g\rangle \cap G_0$. If $G' \cong C_4$, then the fact that $\mathrm{cl}(G) = 2$ implies $G' \leq Z(G)$, contrary to the fact that $Z(G_0)$ is elementary abelian. Hence we have $G' \cong C_2$ and so $G' = Z(Q) = \langle z\rangle$. It follows that each subgroup of G_0 which is not elementary abelian is normal in G (indeed, every cyclic subgroup of order > 2 in G_0 contains $Z(Q) = G'$). Hence, all nonnormal subgroups of G lie in $Z(G_0) = \Omega_1(G_0)$, contrary to the fact that they generate G_0. We have proved that G_0 must be abelian and this completes the proof. □

Proposition 274.2. *If any two nonnormal cyclic subgroups of a non-Dedekindian minimal nonabelian p-group G normalize one other, then*

$$G = G = \langle a, b \mid o(a) = p^m,\ o(b) = p^n,\ a^b = a^{p^{m-1}},\ m > 1\rangle \cong M_p(m, n), \quad m > n.$$

Proof. Clearly, G must be metacyclic and $G \ncong D_8$ (see Lemma 65.1). Any noncyclic subgroup of G contains G' so G-invariant. If $|G| = p^3$, then $G \in \{M_{p^3}, Q_8\}$. Next we

https://doi.org/10.1515/9783110533149-018

assume that $|G| > p^3$. It follows from Lemma 65.1 that G is metacyclic. Let

$$G = \langle a, b \mid o(a) = p^m,\ o(b) = p^n,\ a^b = a^{p^{m-1}},\ m > 1\rangle \cong M_p(m, n).$$

Let $A = \langle A\rangle$. Any non-G-invariant cyclic subgroup has trivial intersection with A. Let $G_0 < G$ be generated by all nonnormal (cyclic) subgroups of G. If $n = 1$, then $G \cong M_p(m, 1) = M_{p^{m+1}}$ satisfies the hypothesis, $G_0 = \Omega_1(G) \cong E_{p^2}$. Now let $n > 1$. It follows that $G \ncong M_2(2, 2)$ (for that group $G_0 = G$ since $c_2(G) = 6$ and there is in G exactly four nonnormal cyclic subgroups).

If $n < m$, all nonnormal subgroups of G have order p^n (they have trivial intersection with $\langle a\rangle$ and all cyclic subgroups of order $< p^n$ lie in $Z(G)$). In that case, $G_0 \le \Omega_n(G) < G$, and therefore G satisfies the hypothesis since G_0 is abelian of type (p^n, p^n).

Now assume that $n = m$. Then $\Omega_{m-1}(G) = Z(G) = \Phi(G)$ so all nonnormal cyclic subgroups of G have order p^n. Let $A < H \in \Gamma_1$; then $H = \langle a\rangle \times \langle b^p\rangle$. All cyclic subgroups of G not contained in H are non-G-invariant of order p^n so that $G_0 = G$. Therefore G is generated by two nonnormal cyclic subgroups, say L_1 and L_2, of order p^n. Clearly, L_1 does not normalize L_2 so $M_p(n, n)$ does not satisfy the hypothesis.

Let $m < n$. If $L < G$ is cyclic of order p^n, then $\mho_1(L) \le Z(G)$. The subgroup $\Omega_{n-1}(G)$ is abelian. Let $M < G$ be cyclic of order p^{n-1} such that $G' \nleq M$. Then $AM = A \times M = H \in \Gamma_1$. Note that $H = \Omega_{m-1}(G)$. Thus, all nonnormal subgroups of order p^{p-1} contained in H so any two such subgroups normalize one other. Any cyclic subgroup, say B, of order p^n is not G-invariant since it does not contain G' and is not contained in $Z(G)$ (of exponent p^{n-1}). Let $C \neq B$ be cyclic of order p^n such that $\langle B, C\rangle = G$. Then $C \triangleleft G$ and G/C is cyclic so $G' < C$, a contradiction. In that case, G does not satisfy the condition. □

Remark 1. The group of Theorem 274.1 is modular (indeed, it is D_8-free and $S(p^3)$-free. Therefore, it is possible to use this fact to produce its new proof.

Problem 1. Classify the non-Dedekindian p-groups such that any two nonnormal cyclic (abelian) subgroups normalize each other.

Problem 2. Classify the non-Dedekindian p-groups such that any two distinct conjugate subgroups normalize each other.

Problem 3. Study the non-Dedekindian p-groups G such that, whenever $A, B < G$ are nonnormal, then one of these subgroups normalizes another.

Problem 4. Study the non-Dedekindian p-groups G such that any its two nonnormal subgroups generate a subgroup of class ≤ 2.

Problem 5. Classify the non-Dedekindian p-groups all of whose minimal nonabelian subgroups are isomorphic either Q_8 or M_{p^n}. (A knowledge of such groups allows us to produce a new proof of Theorem 274.1.)

Problem 6. Study the non-Dedekindian p-groups in which any two nonnormal subgroups are permutable.

§ 275 Nonabelian p-groups with exactly p normal closures of minimal nonabelian subgroups

Suppose that a p-group G is neither abelian nor minimal nonabelian. Then G contains p pairwise distinct minimal nonabelian subgroups (see § 76). It appears that such a group G contains p minimal nonabelian subgroups whose normal closures are pairwise distinct. In this section we characterize the p-groups containing exactly p pairwise distinct normal closures minimal nonabelian subgroups (Theorem 275.3).

Lemma 275.1. *Suppose that H is a unique normal subgroup of a nonabelian p-group G of index p^2. If $H = Z(G)$, then $|G| = p^3$.*

Proof. Since $|G : \Phi(G)| \geq p^2$ and the elementary abelian group $G/\Phi(G)$ contains only one subgroup of index p^2, we get $H = \Phi(G)$. Assume that $H = Z(G)$. Then all members of the set Γ_1 are abelian, and we conclude that G is minimal nonabelian hence $|G'| = p$ (Lemma 65.1). Since a noncyclic abelian p-group G/G' has only one subgroup of index p^2, we get $|G/G'| = p^2$. Thus, $|G| = |G/G'||G'| = p^3$. □

Lemma 275.2. *Suppose that $|G : Z(G)| > p^2$ for a nonabelian p-group G. Then there are in G minimal nonabelian subgroups $S_1, \dots, S_p$ such that their normal closures are pairwise distinct.*

Proof. Let $H \triangleleft G$ be such that $G/H \cong E_{p^2}$. Let $M_1/H, \dots, M_{p+1}/H$ be all subgroups of order p in G/H; then all $M_i \in \Gamma_1$. As $H \neq Z(G)$, at most one of the subgroups M_i, say M_{p+1}, is abelian; then $M_1, \dots, M_p$ are nonabelian. By Theorem 10.28, for any $i \leq p$, the subgroup M_i contains a minimal nonabelian subgroup S_i such that $S_iH = M_i$. Then $S_i \nleq M_j$ for $j \neq i$ and $S_i^G \leq M_i$. Therefore, we get $S_i^G \neq S_j^G$, $j \neq i$. Thus, the normal closures $S_1^G, \dots, S_p^G$ are pairwise distinct. □

Theorem 275.3. *Let a p-group G be such that $|G : Z(G)| > p^2$. Then G contains exactly p minimal nonabelian subgroups $S_1, \dots, S_p$ whose normal closures $S_1^G, \dots, S_p^G$ are pairwise distinct if and only if $\Gamma_1 = \{M_1, \dots, M_p, A\}$, where all M_i are nonabelian and $S_i^G = M_i$ for all $i \leq p$ and A is abelian.*

Proof. Let $\Phi(G) \leq H < G$, where $|G : H| = p^2$. Let M_i/H, $i = 1, \dots, p+1$, be pairwise distinct subgroups of order p in G/H. As above, one may assume that M_i, $i \leq p$, are nonabelian. If $S_i \leq M_i$ is minimal nonabelian for $i \leq p$ such that $S_i \nleq H$ (Theorem 10.28), then $S_1^G, \dots, S_p^G$ are pairwise distinct (see the proof of the previous lemma). Clearly, $M_{p+1} = A$ is abelian (otherwise, if $S_{i+1} \leq M_{i+1}$ is such that $S_{i+1} \nleq H$, then $S_{i+1}^G \notin \{S_1, \dots, S_p\}$). In particular, the subgroup H is abelian. If $S_i^G < M_i$ for some $i \leq p$, there is a minimal nonabelian $S < M_i$ with $S \nleq S_i^G$; then $S^G \neq S_i^G$. In that case the $p+1$ normal closures S_i^G, $i \leq p$, and S^G are pairwise distinct, contrary to the hypothesis. Thus, $S_i^G = M_i$ for $i \leq p$. Assume that $d(G) > 2$. Let $M \in \Gamma_1 - \{M_1, \dots, M_p, A\}$ be nonabelian (such a subgroup M exists since, otherwise, $|G : Z(G)| = p^2$, contrary

https://doi.org/10.1515/9783110533149-019

to the assumption). Let $S \leq M$ be minimal nonabelian; then $S^G \leq M$. All S_i^G, $i \leq P$, contain H but S^G does not contain H (indeed, H is contained in exactly $p+1$ members $M_1, \dots, M_p, A$ of the set Γ_1 and $M \notin \{M_1, \dots, M_p, A\}$) so that

$$S^G \notin \{S_1^G, \dots, S_p^G\} = \{M_1, \dots, M_p\},$$

contrary to the hypothesis. Thus, $d(G) = 2$. The proof is complete. □

Problem 1. Classify the p-groups G containing exactly $p+1$ minimal nonabelian subgroups $S_1, \dots, S_{p+1}$ whose normal closures $S_1^G, \dots, S_{p+1}^G$ are pairwise distinct.

Exercise 1. Is it true that, if G is from Problem 1, it is metabelian?

Exercise 2. Let G be a non-Dedekindian p-group. Prove that there is in G maximal cyclic subgroups L_i, $i \leq p+1$, such that their normal closures L_i^G, $i \leq p$, are pairwise distinct.

Hint. The number of subgroups covering a noncyclic p-group is at least $p+1$ (see §76). □

Problem 2. Classify the p-groups G containing exactly $p+1$ maximal cyclic subgroups $S_1, \dots, S_{p+1}$ whose normal closures $S_1^G, \dots, S_{p+1}^G$ are pairwise distinct.

§ 276 2-groups all of whose maximal subgroups, except one, are Dedekindian

In this section we consider a partial case of the following:

Problem 1. A p-group X is said to be a T-group if $X = S \times E$ whenever S is nonabelian of order p^3 and E is abelian of exponent $\leq p$ (i.e., either $E = \{1\}$ or E is elementary abelian). Classify the p-groups all of whose maximal subgroups, except one, are T-groups.

Namely, we assume that $p = 2$ and $S \cong Q_8$. Thus, we intend to classify the non-Dedekindian 2-groups all of whose maximal subgroups, except one, are Dedekindian. This solves Problem 3773.

Recall that a 2-group G is termed *Hamiltonian* is it is nonabelian Dedekindian. By § 245, if all maximal subgroups of a non-Dedekindian 2-group G are Dedekindian, then G is either minimal nonabelian or isomorphic to Q_{2^4}.

Theorem 276.1 (Janko). *Let G be a 2-group all of whose maximal subgroups, except one, are Dedekindian. Then one of the following holds:*

(a) $G \cong SD_{16}$.

(b) *The group G is of order 2^5,*

$$G = \langle g, w \mid g^8 = w^4 = 1,\ g^2 = v,\ w^2 = t,\ [w, g] = vt,\ [t, g] = 1,\ v^w = v^{-1} \rangle,$$

where

$$\Phi(G) = \langle v, t \rangle \cong C_4 \times C_2, \quad G' = \langle vt \rangle \cong C_4, \quad \Omega_1(G) = Z(G) = \langle v^2, t \rangle \cong E_4$$

and the three maximal subgroups of G are

$$\begin{aligned} A &= \langle g, t \rangle \cong C_8 \times C_2, \\ H &= \langle w, vt \rangle \times \langle t \rangle \cong Q_8 \times C_2, \\ M &= \langle v, w \rangle = \langle v^4 = w^4 = 1, \ , w^v = w^{-1} \rangle \cong M_2(2, 2). \end{aligned}$$

Proof. Let G be a 2-group with exactly one maximal subgroup M which is not Dedekindian.

(i) First assume that $|M'| > 2$ so that we may use Theorem 145.8 about p-groups with exactly one maximal subgroup whose commutator group is of order $> p$. By that theorem, $d(G) = 2$ and G' is abelian of type $(4, 2)$. Further,

$$[G', G] = \Omega_1(G') \leq Z(G), \ \Phi(G) = C_G(G') \text{ is abelian and } \mho_2(G) \leq Z(G).$$

Also, if $\Gamma_1 = \{H_1, H_2, M\}$, then $H_1' = \langle z_1 \rangle$ and $H_2' = \langle z_2 \rangle$ are both of order 2,

$$\langle z_1, z_2 \rangle = \Omega_1(G') = M' \cong E_4, \quad d(M) = 3 \quad \text{and} \quad \mho_1(G') = \langle z_1 z_2 \rangle$$

and we have the following two possibilities.

https://doi.org/10.1515/9783110533149-020

(i1) $d(H_1) = 2$ and $d(H_2) = 2$ in which case H_1 and H_2 are minimal nonabelian, by Lemma 65.2 (a). Since H_1 and H_2 are Hamiltonian (nonabelian Dedekindian), we get $H_1 \cong H_2 \cong Q_8$. But then $|G| = 2^4$ and so $|M| = 2^3$ and this contradicts to the fact that $M' \cong E_4$.

(i2) $d(H_1) = 3$ and $d(H_2) = 3$ so that $H_1 \cong H_2 \cong C_2 \times Q_8$ and (by Theorem 145.8) $|G| \geq 2^7$. This is a contradiction since a maximal subgroup H_1 is of order 2^4.

(ii) We have proved that $|M'| = 2$ noting that M is non-Dedekindian and so nonabelian.

(ii1) First assume that G' is noncyclic. Then we may use Theorem 139.A implying that either $d(G) = 2$, $cl(G) = 3$ and either $G' \cong E_4$ or $d(G) = 3$, $cl(G) = 2$ and either $G' \cong E_4$ or E_8 and also $\Phi(G) = Z(G)$.

(ii1a) Assume that $d(G) = 2$, $cl(G) = 3$ and $G' \cong E_4$. Let $\Gamma_1 = \{H_1, H_2, M\}$, where H_1 and H_2 are Dedekindian and M is non-Dedekindian with $|M'| = 2$. Since the number of abelian maximal subgroups of G is 0 or 1 or 3 (Exercise 1.6 (a)), it follows that one may assume that $|H_2'| = 2$ and so H_2 is Hamiltonian, i.e., $H_2 = Q \times V$, where $Q \cong Q_8$ and $\exp(V) \leq 2$. But H_2' and M' are central in G and since $cl(G) = 3$ (and so $E_4 \cong G' \nleq Z(G)$), we must have $H_2' = M' \geq H_1'$ (where H_1 is either abelian or Hamiltonian). Also, since $\exp(H_2) = 4$, we have $\exp(G) \leq 8$. Note that $G' \cong E_4$ lies in each maximal subgroup of G (since $G' \leq \Phi(G)$). Since $\Omega_1(H_2) \leq Z(H_2)$, we have $G' \leq Z(H_2)$ implying that $C_G(G') = H_2$. But $G' \leq H_1$ and H_1 is either abelian or Hamiltonian and so in any case $C_G(G') \geq H_1$ and therefore $C_G(G') \geq \langle H_1, H_2\rangle = G$, a contradiction. We have proved that the case (ii1a) cannot occur and so we must be here in the following case:

(ii1b) Let $d(G) = 3$, $cl(G) = 2$, $G' \cong E_4$ or E_8 and $\mho_1(G) = \Phi(G) = Z(G)$. Then G has at least five Hamiltonian maximal subgroups $H_1, H_2, \ldots, H_5$ and H_6 is also Dedekindian (but possibly abelian) and H_7 is non-Dedekindian with $|H_7'| = 2$. We have $|G : H_1| = 2$ and so we get $H_1 = Z(G)Q_1$ (noting that $Z(G) = \Phi(G)$ gives $Z(G) < H_1$ and therefore $|H_1 : Z(G)| = 4$ so that $Z(G) = Z(H)$), where $Q_1 \cong Q_8$ and $Q_1 \cap Z(G) = \langle z_1\rangle \cong C_2$. It follows that $Z(G)$ is elementary abelian and

$$a_1^2 = b_1^2 = 1,$$

where $\langle a_1, b_1\rangle = Q_1$ and $\mho_1(H_1) = \langle z_1\rangle$ with $\Omega_1(H_1) = Z(G)$. Consider $H_i \neq H_1$ with $i \in \{2, \ldots, 5\}$ so that H_i is Hamiltonian with $H_i \cap H_1 > Z(G)$ hence $H_i' = \mho_1(H_i) = \langle z_1\rangle$. But then in the nonabelian group $G/\langle z_1\rangle$ (noting that $|G'| \geq 4$) we have at least five abelian maximal subgroups $H_i/\langle z_1\rangle$, $i = 1, 2, \ldots, 5$, a contradiction (Exercise 1.6 (a)). We have proved that a noncyclic G' is not possible. Hence we must have:

(ii2) The group G' is cyclic and so by Theorem 137.7, we have either $G' \cong C_2$ or $d(G) = 2$ and $G' \cong C_4$.

(ii2a) First suppose that $G' \cong C_2$. If in addition $d(G) = 2$, then Lemma 65.2 (a) implies that G is minimal nonabelian, contrary to our hypothesis. Hence $d(G) \geq 3$ so that G has at least three maximal subgroups which are Hamiltonian and let H be one of them. Then $H = QZ(H)$, where $Q \cong Q_8$ and $\exp(Z(H)) = 2$ with $Z(Q) = \langle z \rangle = G' \cong C_2$. As $\langle z \rangle = G'$, we have $Q \trianglelefteq G$ and $C = C_G(Q)$ covers G/Q so that $Q \cap C = \langle z \rangle$, $C \cap H = Z(H)$ with $|C : Z(H)| = 2$ and $C \trianglelefteq G$. Assume for a moment that C is nonabelian. Then $G/C \cong E_4$ and so there are exactly three nonabelian maximal subgroups of G containing C and so at least two of them are Hamiltonian. In particular, the nonabelian subgroup C contains a subgroup $Q_1 \cong Q_8$. This is not possible since $Z(H)$ is an elementary abelian subgroup of index 2 in C. We have proved that C is abelian and so $C = Z(G)$.

If C is elementary abelian, then G is Hamiltonian and then all maximal subgroups of G are Dedekindian, a contradiction. Hence C is not elementary abelian and then $\Omega_1(C) = Z(H)$ and all elements in the set $C - Z(H)$ are of order 4 and let $c \in C - Z(H)$ be one of them. We set $Q = \langle a, b \rangle$ and let V be a complement of $\langle z, c^2 \rangle$ in $Z(H)$.

First assume that $c^2 = z$ and consider the maximal subgroups $D_1 = \langle bc, a \rangle \cong D_8$ and $D_2 = \langle b, ca \rangle \cong D_1$ of $Q * \langle c \rangle$ so that $D_1 \times V$ and $D_2 \times V$ are distinct maximal subgroups of G. Since D_1 and D_2 are not Dedekindian, it follows that also $D_1 \times V$ and $D_2 \times V$ are non-Dedekindian. But this contradicts our basic hypothesis.

Now we suppose that $c^2 \notin \langle z \rangle$ and then we consider the maximal subgroups $H_1 = \langle bc, a \rangle$ and $H_2 = \langle b, ca \rangle \cong H_1$ of $Q \times \langle c \rangle$ so that $H_1 \times V$ and $H_2 \times V$ are distinct maximal subgroups of G. Since $H_1 \cong H_2$ is a metacyclic minimal nonabelian group of order 16 and exponent 4, it follows that H_1 and H_2 are non-Dedekindian and then also $H_1 \times V$ and $H_2 \times V$ are non-Dedekindian, which again contradicts our basic hypothesis. Finally, we have proved that we must be in the remaining case:

(ii2b) Let $d(G) = 2$ and $G' \cong C_4$. Since G has at most one abelian maximal subgroup, it follows that G has at least one maximal subgroup H which is Hamiltonian. By Schreier's inequality (Theorem A.25.1), we have $d(H) \leq 3$. This implies that either $H \cong Q_8$ or $H \cong Q_8 \times C_2$. If $H \cong Q_8$, then $|G| = 2^4$ and since $G' \cong C_4$, a result of O. Taussky (Lemma 1.6) implies that G is of maximal class. It follows that in this case $G \cong SD_{16}$.

From now on assume that $H = Q \times \langle t \rangle$, where $Q \cong Q_8$, $Z(Q) = \langle z \rangle \cong C_2$ and $\langle t \rangle \cong C_2$. It follows that $|G| = 2^5$ and since $G' \cong C_4$ and $G' < H$, we may assume that $G' = \langle a \rangle < Q$, where $Q = \langle a, b \rangle \cong Q_8$. We have

$$\Phi(G) = \mho_1(G) < H, \quad \Phi(G) > G' \quad \text{and} \quad |\Phi(G)| = 2^3.$$

By a result of Burnside, $\Phi(G) \ncong Q_8$ and so $\Phi(G)$ is abelian containing $\langle a \rangle = G'$. Hence $\Phi(G) = \langle a \rangle \times \langle t \rangle$ is abelian of type $(4, 2)$. Since $\Phi(G) = \mho_1(G) < H$, there is $g \in G - H$ such that $g^2 \in \Phi(G) - \langle z, t \rangle$ and so $o(g) = 8$. Also, $G' \nleq Z(G)$ and so G is of class 3

Assume for a moment that G has another Hamiltonian maximal subgroup H_1 so that $H \cap H_1 = \Phi(G)$. In that case $\Omega_1(H_1) = \langle z, t \rangle = Z(H_1)$ implying that $t \in Z(G)$. But then $g \in G - H$ centralizes $g^2 \in \Phi(G) - \langle z, t \rangle$ and t and so $\Phi(G)\langle g \rangle$ is an abelian maximal subgroup in G. In that case all three maximal subgroups of G would be Dedekindian, a contradiction.

We have proved that G has only one Hamiltonian maximal subgroup H and so G must possess an abelian maximal subgroup A so that $A \cap H = \Phi(G)$ and in particular, $\langle z, t \rangle \leq Z(G)$. On the other hand, $Z(G) \leq \Phi(G)$ and $G' \nleq Z(G)$ implying that $Z(G) = \langle z, t \rangle$.

In this case $[t, g] = 1$ and so $A = \langle g \rangle \times \langle t \rangle$ is abelian of type $(8, 2)$, where $g^4 = z$ and we set $g^2 = v$.

Since one has, for each $x \in G - A$, $x^2 \in Z(G)$ and $Z(G) = \langle z, t \rangle \cong E_4$, it follows that all elements in $G - A$ are of order ≤ 4. We note that $\mho_1(A) = \langle v \rangle \cong C_4$ and so if for each $x \in G - A$, $x^2 \in \langle z \rangle$, then $\mho_1(G) = \Phi(G) = \langle v \rangle$, a contradiction. It follows that there is $w \in G - A$ such that $w^2 = t$, where we have replaced t with tz, if necessary.

We have $C_4 \cong G' \leq A$. However, if $G' = \langle v \rangle$, then $\langle g \rangle \trianglelefteq G$ and since $G = \langle g \rangle \langle w \rangle$, G would be metacyclic. This is not possible since G has a nonmetacyclic maximal subgroup $H \cong Q_8 \times C_2$. It follows that $G' = \langle vt \rangle$ and so we may set $[w, g] = vt$. Since $\langle vt \rangle \trianglelefteq G$ and $\Phi(G) = \langle v, t \rangle \trianglelefteq G$, it follows that $\langle v \rangle \trianglelefteq G$, where we note that $\langle v, t \rangle$ has exactly two cyclic subgroups of order 4. Since $v \notin Z(G)$, we get $v^w = v^{-1}$ and so

$$M = \langle v, w \mid v^4 = w^4 = 1,\ v^w = v^{-1} \rangle$$

is a metacyclic minimal nonabelian group of order 16 and exponent 4 and this is a unique non-Dedekindian maximal subgroup in G.

The Hamiltonian maximal subgroup H in G is equal to

$$H = \Phi(G)\langle wg \rangle = \Phi(G) \cup (G - (A \cup M))$$

and so all elements in $G - A$ are of order 4. This implies that

$$\Omega_1(G) = Z(G) = \langle z, t \rangle \cong E_4,$$

and so the structure of G is completely determined as stated in our theorem. □

Problem 2. Study the 2-groups all of whose non-Dedekindian subgroups are conjugate.

Problem 3. Study the 2-groups all of whose non-Dedekindian maximal subgroups are isomorphic.

§ 277 p-groups with exactly two conjugate classes of nonnormal maximal cyclic subgroups

We solve here Problem 3577 by proving the following result.

Theorem 277.1 (Janko). *Let G be a p-group with exactly two conjugate classes of nonnormal maximal cyclic subgroups. Then G' is cyclic and either G is metacyclic or* $\mathrm{cl}(G) = 2$.

Remark 1. Examples of such groups are all 2-groups of maximal class and the metacyclic group $\mathrm{M}_2(2, 2)$.

In proving this theorem we use essentially Theorem 231.1 which we quote here in somewhat simpler form.

Theorem 231.1. *Let G be a non-Dedekindian p-group which is not generated by its nonnormal subgroups. Then G' is cyclic and G is of class 2.*

Proof of Theorem 277.1. Let G be a p-group with exactly two conjugate classes of nonnormal maximal cyclic subgroups with representatives $\langle x_1\rangle$ and $\langle x_2\rangle$. Then for each maximal cyclic subgroup $\langle x\rangle$ in G, we have either $\langle x\rangle$ is conjugate to $\langle x_1\rangle$ or $\langle x\rangle$ is conjugate to $\langle x_2\rangle$ or $\langle x\rangle \trianglelefteq G$.

(i) Assume that $\langle x_1, x_2\rangle = G$ and set $H = \Phi(G)$ so that $G/H \cong \mathrm{E}_{p^2}$. Set $X_1 = (\langle x_1\rangle H)/H$ and $X_2 = (\langle x_2\rangle H)/H$. Then X_1 and X_2 are two distinct maximal subgroups (of order p) in G/H and we note that subgroups (they have index p in G) $\langle x_1\rangle H \trianglelefteq G$ and $\langle x_2\rangle H \trianglelefteq G$. Let $H < K < G$ with $|K : H| = p$ be such that $K/H \neq X_1$ and $K/H \neq X_2$ (where we note that G/H has exactly $p + 1 \geq 3$ pairwise distinct subgroups of order p).

Consider an element $x_3 \in K - H$ so that the subgroup $\langle x_3\rangle$ is maximal cyclic in G. Since $K/H \trianglelefteq G/H$, it follows that $\langle x_3\rangle$ cannot be conjugate to $\langle x_1\rangle$ or $\langle x_2\rangle$ (indeed, $K \trianglelefteq G$ and $x_1, x_2 \notin K$) and so $\langle x_3\rangle \trianglelefteq G$. But $x_1 \in G - K$ and so $G = \langle x_1, x_3\rangle$ implying that G is metacyclic. In particular, G' is cyclic.

(ii) Now assume that $\langle x_1, x_2\rangle < G$. Let M be a maximal subgroup of G containing $\langle x_1, x_2\rangle$ so that $\langle x_1\rangle^G \leq M$ and $\langle x_2\rangle^G \leq M$. Let g be any element in $G - M$. Then the subgroup $\langle g\rangle$ is maximal cyclic in G since $|G : M| = p$ (here we use the product formula). Since g cannot be conjugate to $\langle x_1\rangle$ or $\langle x_2\rangle$, it follows that $\langle g\rangle \trianglelefteq G$.

Let B be any subgroup in G which is not contained in M. Then $B = \langle B - M\rangle$ and for all elements $x \in B - M$, $\langle x\rangle \trianglelefteq G$ implying $B \trianglelefteq G$. It follows that all nonnormal subgroups of G lie in M.

We have proved that G is not generated by its nonnormal subgroups. Since G possesses nonnormal subgroups, it follows by Theorem 231.1 that G' is cyclic and G is of class 2 and we are done. □

Problem. Classify the p-groups G containing a maximal subgroup M such that the set $G - M$ is the union of p conjugacy classes.

https://doi.org/10.1515/9783110533149-021

§ 278 Nonmetacyclic p-groups all of whose maximal metacyclic subgroups have index p

In minimal nonmetacyclic p-groups all maximal metacyclic subgroups have index p. In this section we describe the nonmetacyclic p-groups G all of whose maximal metacyclic subgroups have index p. General speaking, such p-groups G are not necessarily minimal nonmetacyclic (however, this is a case provided $p = 2$).

In this section we prove the following theorems.

Theorem 278.1. *If all maximal metacyclic subgroups of a nonmetacyclic 2-group G have index 2, then G is minimal nonmetacyclic.*

Theorem 278.2. *Suppose that a p-group G is nonmetacyclic, $p > 2$. If any maximal metacyclic subgroup of G has index p, then one of the following holds:*

(a) $|G| = p^3$ *and* $\exp(G) = p$.

(b) *G is minimal nonmetacyclic group of order 3^4 (this group is of maximal class and $\Omega_1(G) \cong E_9$).*

(c) $\Omega_1(G) = S(p^3)$, $G = \Omega_1(G) * C$, *where $C = Z(G)$ is cyclic of order $> p$.*

Recall that if a 2-group G is minimal nonmetacyclic, then $d(G) = 3$. If $p > 2$ and a p-group G is minimal nonmetacyclic, then $G \in \{S(p^3), E_{p^3}, M\}$, where M is a 3-group of maximal class and order 3^4 with $\Omega_1(M) \cong E_{3^2}$.

Proof of Theorem 278.1. Let $H \leq G$ be minimal nonmetacyclic. Then $d(H) = 3$ (Theorem 66.1). Let U, V be distinct maximal subgroups of H and let $U \leq A$, $V \leq B$, where A, B are metacyclic subgroups of index 2 in G. As $\langle U, V\rangle = H$ is nonmetacyclic, we get $A \neq B$. Therefore, the set Γ_1 contains $1+2+2^2$ pairwise distinct metacyclic subgroups of index 2 (here $1+2+2^2$ is the number of maximal subgroups of H), and we conclude that $d(G) \geq 3$. Since $d(G) \leq d(A) + 1 = 3$, we get $d(G) = 3$ so that all maximal subgroups of the (nonmetacyclic) group G are metacyclic. Hence, G is minimal nonmetacyclic. □

Proof of Theorem 278.2. Suppose that there is in G a normal subgroup $E \cong E_{p^3}$. Let U, V be distinct maximal subgroups of H and let $U \leq A$, $V \leq B$, where A, B are metacyclic subgroups of index p in G. As $\langle U, V\rangle = H$ is nonmetacyclic, we get $A \neq B$. Therefore, the set Γ_1 contains $1 + p + p^2$ metacyclic subgroups of index p (here $1 + p + p^2$ is the number of maximal subgroups of H), and we conclude that $d(G) \geq 3$. Since $d(G) \leq d(A) + 1 = 3$, it follows that $d(G) = 3$ so that all maximal subgroups of the (nonmetacyclic) group G are metacyclic. Therefore, G is minimal nonmetacyclic. In the case under consideration, $G = E \cong E_{p^3}$ (see Theorem 69.4).

In what follows we assume that G has no subgroup isomorphic to E_{p^3}. One may assume that $|G| > p^3$ (if $|G| = p^3$, then $G \cong S(p^3)$ is minimal nonmetacyclic). By Theorem 13.7, G is either a 3-group of maximal class or $G = \Omega_1(G)C$, where $\Omega_1(G) \cong S(p^3)$ and C is cyclic of order $> p$.

https://doi.org/10.1515/9783110533149-022

Let G be a 3-group of maximal class. By Theorem 9.6, $|G| < 3^5$ so that $|G| = 3^4$. As G has no subgroup isomorphic to E_{3^3}, their abelian member of the set Γ_1 is of type $(9, 3)$. If there is in G a subgroup isomorphic to $S(3^3)$, G contains $3 + 1(= |\Gamma_1|)$ pairwise distinct metacyclic members so it is minimal nonmetacyclic, a contradiction). Thus, $\Omega_1(G) \cong E_{3^2}$ so that G is minimal nonmetacyclic. In what follows we assume that G is not a 3-group of maximal class.

Then, by Theorem 13.7, we have $G = \Omega_1(G)C$, where $\Omega_1(C) \cong S(p^3)$ and C is cyclic of order $> p$. All $p + 1$ maximal subgroups of $\Omega_1(G)$ are contained in $p + 1$ distinct metacyclic members of the set Γ_1. Therefore, if $d(G) = 2$, then G must be minimal nonmetacyclic, which is a contradiction. Thus, $d(G) = 3$. Then $|G'| = p$ (here we use Lemma 1.4) so that $G = \Omega_1(G) * C$, where C_1 is cyclic of order $|C|$ (Lemma 4.2). If $\Omega_1(G) \le H \in \Gamma_1$, then H is the unique nonmetacyclic member of the set Γ_1. It follows that G satisfies the hypothesis. □

Exercise 1. Classify the nonabelian p-groups all of whose maximal abelian subgroups have index p.

Answer. We have $|G : Z(G)| = p^2$ so that $G = SZ(G)$, where the subgroup $S \le G$ is minimal nonabelian. □

Exercise 2. Classify the irregular p-groups all of whose maximal absolutely regular subgroups have index p.

Problem 1. Classify the nonmetacyclic p-groups G such that, whenever $C < G$ is cyclic, then C is contained in a minimal nonmetacyclic subgroup of G. (In other words, G is covered by minimal nonmetacyclic subgroups.)

The following problem is old (it coincides with Problem 860).

Problem 2. Classify the nonabelian p-groups G such that, whenever $C < G$ is cyclic, then C is contained in a minimal nonabelian subgroup of G.

Problem 3. Classify the p-groups G such that, whenever $C < G$ is cyclic, then $C < M \le G$, where M is a subgroup of maximal class.

Problem 4. Classify the nonmetacyclic p-groups G all of whose maximal metacyclic subgroups, except one, have index p in G.

§ 279 Subgroup characterization of some p-groups of maximal class and close to them

In this section we prove the following four characterization theorems.

Theorem 279.1. *Suppose that for any two distinct maximal abelian subgroups A, B of a nonabelian p-group G one has $|A \cap B| = p$. Then G is a p-group of maximal class with an abelian subgroup of index p, and the reverse assertion is also true.*

Theorem 279.2. *Suppose that for any nonabelian subgroup H of a nonabelian p-group G one has $|Z(H)| = p$. Then G is a p-group of maximal class with an abelian subgroup of index p, and the reverse assertion is also true.*

Theorem 279.3. *Suppose that for any nonabelian subgroup H of a nonabelian p-group G one has $|H : H'| = p^2$. Then G is a p-group of maximal class with an abelian subgroup of index p, and the reverse assertion is also true.*

Theorem 279.4. *Suppose that for any two distinct maximal cyclic subgroups A, B of a nonabelian 2-group G one has $|A \cap B| = 2$. Then $G = Q \times E$, where Q is a generalized quaternion group and* $\exp(E) = 2$.

Proof of Theorem 279.1. Since any maximal abelian subgroup of G contains $Z(G)$, we get $|Z(G)| = p$. Let $H < G$ be nonabelian and let U, V be distinct maximal abelian subgroups of H. Let $U \leq B < G$, $V \leq B < G$, where A, B are maximal abelian subgroups of G; then $A \neq B$ since $\langle U, V\rangle$ is nonabelian. One has

$$|U \cap V| \leq |A \cap B| = p \implies |U \cap V| = p$$

since $U \cap V \geq Z(H) > \{1\}$, so that any proper nonabelian subgroup of our group G satisfies the same hypothesis. Working by induction on G, we see that any proper nonabelian subgroup of G is of maximal class. If G has no proper nonabelian subgroups, i.e., it is minimal nonabelian, then $|Z(G)| = |A \cap B| = p$ (indeed, $\{1\} < Z(G) \leq A \cap B$) which implies that

$$|G| = |Z(G)||G : Z(G)| = p \cdot p^2 = p^3,$$

and we conclude that G is of maximal class. Next we assume that G is not minimal nonabelian hence $|G| \geq p^4$. Then it contains nonabelian maximal subgroups which are of maximal class, by induction. All remaining maximal subgroups of G are abelian, by hypothesis. Therefore, if G itself is of maximal class, we are done since not all members of the set Γ_1 are of maximal class hence there is in the set Γ_1 an abelian member.

Suppose that G is not of maximal class. Then, by Theorem 12.12 (b),

$$\Gamma_1 = \{M_1, \dots, M_{p^2}, A_1, \dots, A_{p+1}\}, \tag{1}$$

where all M_i, $i \leq p^2$, are of maximal class and all A_j, $j \leq p + 1$, not being of maximal

https://doi.org/10.1515/9783110533149-023

class, are abelian. In that case, since $|G| > p^3$, we get

$$||A_1 \cap A_2| = |Z(G)| = \frac{1}{p^2}|G| > p,$$

which is a contradiction.

The reverse assertion is obvious. □

Proof of Theorem 279.2. We proceed by induction on $|G|$. By hypothesis, we have $|Z(G)| = p$. One may assume that G is not minimal nonabelian (otherwise, we would have $|G| = |G : Z(G)||Z(G)| = p^2 \cdot p = p^3$ so G is of maximal class). Then $|G| > p^3$. Obviously, any proper nonabelian subgroup of G satisfies the hypothesis so it is of maximal class, by induction. In particular, all nonabelian members of the set Γ_1 are of maximal class. If G is of maximal class, then its fundamental subgroup G_1, which is not of maximal class, must be abelian, and in this case we are done since, in the case under consideration, such a group G indeed satisfies the hypothesis (note that all maximal abelian subgroups of G not contained in G_1 have order p^2).

Next we assume that the group G is not of maximal class. Then, by Theorem 12.12 (b), one has (1), where all A_i are abelian. In that case, since $|G| \geq p^4$, we get

$$|Z(G)| = |A_1 \cap A_2| = \frac{1}{p^2}|G| \geq p^2 > p,$$

and this is a contradiction. □

Proof of Theorem 279.3. We proceed by induction on $|G|$. The property is inherited by proper nonabelian subgroups so they are of maximal class, by induction. If G is minimal nonabelian, then

$$|G| = |G : G'||G'| = p^2 \cdot p = p^3$$

and G is of maximal class. Next we assume that G is not minimal nonabelian; then $|G| > p^3$ and there is in the set Γ_1 a nonabelian member which is of maximal class.

Assume that G is of maximal class. Then its fundamental subgroup G_1 is abelian since it is not of maximal class (see § 9). The group satisfies the hypothesis since all its nonabelian subgroups B, being of maximal class, satisfy $|B : B'| = p^2$.

Now assume that G is not of maximal class. Then $d(G) = 3$, by Theorem 12.12 (a), so that $|G : G'| \geq |G : \Phi(G)| = p^3 > p^2$, a final contradiction. □

Proof of Theorem 279.4. Let $S \leq G$ be minimal nonabelian. If

$$S \in \{M_2(m, n, 1), M_2(m, n)\},$$

then S contains two distinct maximal cyclic subgroups of orders 2^m and 2^n, respectively, whose intersection is equal to $\{1\}$ (Lemma 65.1), a contradiction. It follows from the classification of minimal nonabelian 2-groups (Lemma 65.1) that $S \cong Q_8$. Therefore, Corollary A.17.3 implies that $p = 2$, $G = Q \times E$, where Q is a generalized quaternion group and $\exp(E) \leq 2$. □

Note that if all proper nonabelian subgroups of a p-group G are of maximal class, then G need not be of maximal class (simplest example yields the group $G = D \times C$, where D is nonabelian of order p^3 and $|C| = p$).

It follows from the classification of minimal nonmetacyclic 2-groups that if all subgroups of a 2-group G of order $\leq 2^5$ are two-generator, then G is metacyclic (see Theorem 66.1). A similar result is not true for 3-groups. Indeed, all subgroups of a (nonmetacyclic) 3-group of maximal class and order $\geq 3^5$ are two-generator.

Problem 1. Study the p-groups all of whose proper normal nonabelian subgroups are two-generator.

Problem 2. Classify the p-groups all of whose proper nonabelian subgroups H satisfy (i) $|Z(H)| = p$, (ii) $|H : H'| = p^2$.

§ 280 Nonabelian p-groups all of whose maximal subgroups, except one, are minimal nonmetacyclic

We solve here Problem 3708 (iii) by proving the following result.

Theorem 280.1. *Let G be a nonabelian p-group all of whose maximal subgroups, except one denoted with M, are minimal nonmetacyclic. Then $p = 3$, G is an irregular group of order 3^4 or 3^5 and $|M'| \leq 3$.*

In proving this theorem we use very heavily the following known results.

Theorem 69.1. *Let G be a minimal nonmetacyclic p-group. Then one of the following holds:*

(a) *G is of order p^3 and exponent p.*

(b) *G is the following group of maximal class and order 3^4:*

$$G = \langle a, c \mid a^9 = c^9 = 1,\ a^3 = c^3,\ [a, c] = b,\ b^3 = [a, b] = 1,\ [b, c] = a^{-3} \rangle.$$

Here

$$G' = \Omega_1(G) = \langle a^3, b \rangle \cong \mathrm{E}_9, \quad Z(G) = \mho_1(G) = \langle a^3 \rangle \cong \mathrm{C}_3,$$

and $A = \langle a \rangle \times \langle b \rangle \cong \mathrm{C}_9 \times \mathrm{C}_3$ is a unique abelian maximal subgroup of G.

(c) *$p = 2$ and*

$$G \cong \mathrm{C}_2 \times \mathrm{Q}_8 \quad \text{or} \quad G \cong \mathrm{Q}_8 * \mathrm{C}_4 \text{ with } Z(\mathrm{Q}_8 * \mathrm{C}_4) = Z(\mathrm{Q}_8)$$

or

$$G \cong G^* = \langle a, b, c \mid a^4 = b^4 = [a, b] = 1,\ c^2 = a^2,\ a^c = ab^2,\ b^c = ba^2 \rangle,$$

where G is special of order 2^5 with $\exp(G) = 4$,

$$\Omega_1(G) = G' = Z(G) = \Phi(G) = \langle a^2, b^2 \rangle \cong \mathrm{E}_4$$

and $\langle a \rangle \times \langle b \rangle \cong \mathrm{C}_4 \times \mathrm{C}_4$ is a unique abelian maximal subgroup of G and all other six maximal subgroups of G are isomorphic to

$$\mathrm{M}_2(2, 2) = \langle x, y \mid x^4 = y^4 = 1,\ x^y = x^{-1} \rangle.$$

Theorem 145.8. *Let G be a nonabelian p-group with exactly one maximal subgroup M such that $|M'| > p$. Then $\mathrm{d}(G) = 2$, $p = 2$ and G' is abelian of type $(4, 2)$.*

Theorem 276.1. *Let G be a 2-group all of whose maximal subgroups, except one, are Dedekindian. Then one of the following holds:*

(a) *$G \cong \mathrm{SD}_{16}$.*

(b) *The group G is of order 2^5,*

$$G = \langle g, w \mid g^8 = w^4 = 1,\ g^2 = v,\ w^2 = t,\ [w, g] = vt,\ [t, g] = 1,\ v^w = v^{-1} \rangle,$$

https://doi.org/10.1515/9783110533149-024

where

$$\Phi(G) = \langle v, t\rangle \cong C_4 \times C_2, \quad G' = \langle vt\rangle \cong C_4, \quad \Omega_1(G) = Z(G) = \langle v^2, t\rangle \cong E_4$$

and the three maximal subgroups of G are

$$A = \langle g, t\rangle \cong C_8 \times C_2,$$
$$H = \langle w, vt\rangle \times \langle t\rangle \cong Q_8 \times C_2,$$
$$M = \langle v, w\rangle = \langle v^4 = w^4 = 1, \, , w^v = w^{-1}\rangle \cong M_2(2, 2).$$

Lemma A. *Let G be a minimal non-Dedekindian 2-group which is not minimal nonabelian. Then* $G \cong Q_{16}$.

Proof. Let X be any minimal nonabelian subgroup in G. Since $X < G$, X is Dedekindian and so $X \cong Q_8$. By Corollary A.17.3, $G = Q \times V$, where $Q \cong Q_{2^n}$, $n \geq 3$, and we have $\exp(V) \leq 2$. If $V \neq \{1\}$, then $Q \cong Q_8$ and so G is Dedekindian, a contradiction. Hence $V = \{1\}$ and each maximal subgroup of Q is Dedekindian and so $G = Q \cong Q_{16}$. □

Theorem 137.7. *A 2-group G has the property that each proper subgroup of G has a derived subgroup of order at most 2 if and only if one of the following holds:*
(a) $|G'| \leq 2$,
(b) $d(G) = 2, |G'| = 4$,
(c) $d(G) = 3$, $cl(G) = 2$, $G' \cong E_4$ *or* E_8 *and* $\Phi(G) = Z(G)$.

Theorem 90.1. *Let G be a nonabelian 2-group all of whose minimal nonabelian subgroups are isomorphic to* D_8 *or* Q_8. *Then G is one of the following groups:*
(a) *G is generalized dihedral (i.e.,* $|G : H_2(G)| = 2$*),*
(b) $G = HZ(G)$, *where H is of maximal class and* $\mho_1(Z(G)) \leq Z(H)$,
(c) $G = HZ(G)$, *where H is extraspecial and* $\mho_1(Z(G)) \leq Z(H)$.

Theorem 258.2. *Let G be a nonabelian 2-group which is not minimal nonabelian but all its minimal nonabelian subgroups, except one denoted with M, are isomorphic to* Q_8. *Then G has an abelian subgroup of index 2, G is of class 3, and one of the following holds:*
(a) *If G is metacyclic, then* $G \cong SD_{16}$ *and* $M \cong D_8$.
(b) *If G is nonmetacyclic, then*

$$G = \langle a, c \mid a^8 = c^4 = [a, c^2] = 1, \, [a, c] = a^2c^2, \, [a^2, c] = a^4\rangle,$$

where

$$|G| = 2^5, \quad \exp(G) = 8, \quad G' = \langle a^2c^2\rangle \cong C_4, \quad \Omega_1(G) = Z(G) = \langle a^4, c^2\rangle \cong E_4,$$

$\langle a, c^2\rangle \cong C_8 \times C_2$ *is a unique abelian maximal subgroup in G,* $M = \langle a^2, c\rangle$ *and this is the metacyclic minimal nonabelian group of order* 2^4 *and exponent 4.*

Conversely, the above 2-groups satisfy the assumptions of our theorem.

Theorem 57.3. *Let G be a nonabelian 2-group all of whose minimal nonabelian subgroups are isomorphic to*

$$\mathrm{M}_2(2,2) = \langle a, b \mid a^4 = b^4 = 1,\ a^b = a^{-1}\rangle.$$

Then the following holds:

(a) *If G is of exponent ≥ 8, then G has a unique abelian maximal subgroup A. We have $\exp(A) \geq 8$ and $E = \Omega_1(A) = \Omega_1(G) = \mathrm{Z}(G)$ is of order ≥ 4. All elements in the set $G - A$ are of order 4 and if v is one of them, then $\mathrm{C}_A(v) = E$ and v inverts $\Phi(A)$ and on A/E.*

(b) *If G is of exponent 4, then $G = K \times V$, where $\exp(V) \leq 2$ and for the group K we have one of the following possibilities:*

(1) *$K \cong \mathrm{M}_2(2,2)$,*

(2) *K is the minimal nonmetacyclic group of order 2^5,*

(3) *K is a unique special group of order 2^6 with $\mathrm{Z}(K) \cong \mathrm{E}_4$ in which every maximal subgroup is minimal nonmetacyclic of order 2^5 (from (2)):*

$$K = \langle a, b, c, d \mid a^4 = b^4 = 1,\ c^2 = a^2b^2,\ [a,b] = 1,\ a^c = a^{-1},\ b^c = a^2b^{-1},$$
$$d^2 = a^2,\ a^d = a^{-1}b^2,\ b^d = b^{-1},\ [c,d] = 1\rangle,$$

(4) *K is a splitting extension of $B = B_1 \times \cdots \times B_m$, $m \geq 2$, with a cyclic group $\langle b\rangle$ of order 4, where $B_i \cong \mathrm{C}_4$, $i = 1, 2, \ldots, m$, and b inverts each element of B (and b^2 centralizes B).*

Theorem 92.6. *Let G be a nonabelian 2-group all of whose minimal nonabelian subgroups are isomorphic to $\mathrm{M}_2(2,2)$ or Q_8. Then one of the following holds:*

(a) *If G is of exponent ≥ 8, then G has a unique abelian maximal subgroup A, $|G'| > 2$, and all elements in $G - A$ are of order 4. We have $\Omega_1(A) = \Omega_1(G) \leq \mathrm{Z}(G)$ and if $x \in G - A$, then x inverts each element in $A/\Omega_1(A)$;*

(b) *If G is of exponent 4, then $G = K \times V$, where $\exp(V) \leq 2$ and for the group K we have one of the following possibilities:*

(1) *$K \cong \mathrm{M}_2(2,2)$ or $K \cong \mathrm{Q}_8$,*

(2) *K is the minimal nonmetacyclic group of order 2^5,*

(3) *K is a unique special group of order 2^6 with $\mathrm{Z}(K) \cong \mathrm{E}_4$ in which every maximal subgroup is minimal nonmetacyclic of order 2^5 (from (2)),*

(4) *$K \cong \mathrm{Q}_8 \times \mathrm{C}_4$,*

(5) *$K \cong \mathrm{Q}_8 \times \mathrm{Q}_8$,*

(6) *$G = K \times V$ has an abelian maximal subgroup B of exponent 4 and an element $v \in G - B$ of order 4 which inverts each element in B,*

(7) *$K = Q * C$ is a central product of $Q = \langle a, b\rangle \cong \mathrm{Q}_8$ and $C = \langle c, d \mid c^4 = d^4 = 1, c^d = c^{-1}\rangle$ with $Q \cap C = \langle c^2d^2\rangle = \mathrm{Z}(Q)$, where K is special of order 2^6 and $\mathrm{Z}(K) = \Omega_1(K) \cong \mathrm{E}_4$.*

Proof of Theorem 280.1. Let G be a nonabelian p-group all of whose maximal subgroups, except one denoted with M, are minimal nonmetacyclic (see Theorem 69.1).

(i) Suppose that $p = 2$.

(i1) Assume that G has a maximal subgroup $H_1 \cong E_8$ so that $|G| = 2^4$ and then all maximal subgroups distinct from M are isomorphic to E_8 and $M \cong C_4 \times C_2$ or D_8. By noting that the number of abelian maximal subgroups of G is equal 0, 1 or 3, it follows that all maximal subgroups of G are abelian and so $d(G) = 2$ and G is minimal nonabelian. Maximal subgroups of G are E_8, E_8 and $C_4 \times C_2$. But then G is generated by two involutions and then $G \cong D_8$, a contradiction.

(i2) Suppose that G has a maximal subgroup $H_1 \cong C_2 \times Q_8$ or $H_1 \cong Q_8 * C_4$ with $Z(Q_8 * C_4) = Z(Q_8)$. Then $|G| = 2^5$ and all maximal subgroups are isomorphic to $C_2 \times Q_8$ or $Q_8 * C_4$, except one denoted with M.

Suppose for a moment that $|M'| > 2$. Then Theorem 145.8 implies that G' is abelian of type $(4, 2)$. But $|G| = 2^5$ and so a result of O. Taussky (Proposition 1.6) implies that G is of maximal class. It follows that G is metacyclic, a contradiction (noting that H_1 is nonmetacyclic). Hence we have $|M'| \leq 2$. If $|G'| = 8$, then again the result of Taussky implies that G is metacyclic, a contradiction. Hence we have $|G'| \leq 4$.

(i2a) First assume that G' is noncyclic, i.e., $G' \cong E_4$. Let H_i be any maximal subgroup of G distinct from M. If $H_i \cong Q_8 * C_4$ with $Q_8 \cap C_4 = Z(Q_8)$, then Q_8 is characteristic in H_i and so we have $Q_8 \trianglelefteq G$. But $|G/Q_8| = 4$ and so $G' \leq Q_8$ implying that G' is cyclic, a contradiction. It follows that all H_i with $H_i \neq M$ are isomorphic to $Q_8 \times C_2$ and so they are Dedekindian. If M is not Dedekindian, then Theorem 276.1 implies that $G' \cong C_4$, a contradiction. Hence M is also Dedekindian. Since $G' \cong E_4$, it follows that G is non-Dedekindian and so G is minimal non-Dedekindian but not minimal nonabelian. As a consequence of Lemma A we get $G \cong Q_{16}$ and so $G' \cong C_4$, a contradiction. We have proved that G' is cyclic and so $G' \cong C_2$ or $G' \cong C_4$. By Theorem 137.7, if $G' \cong C_4$, then $d(G) = 2$.

(i2b) Suppose that $d(G) = 2$ and $G' \cong C_4$. Let H_1, H_2, M be maximal subgroups of G, where M is not minimal nonmetacyclic.

If M (of order 2^4) is not minimal nonabelian, then each minimal nonabelian subgroup in G is isomorphic to D_8 or Q_8 and Q_8 is a subgroup of G. Then we may use Theorem 90.1, where case (a) of that theorem is not possible since in that case all minimal nonabelian subgroups of G are isomorphic to D_8. Also case (c) of that theorem is not possible since in that case we would have $G' \cong C_2$. Hence we must be in case (b) of that theorem implying that $G = HZ(G)$, where H is of maximal class and $\mho_1(Z(G)) \leq Z(H)$. Since $G' \cong C_4$, we have $|H| = 2^4$ and so $|G : H| = 2$. But then $d(G) = 3$, a contradiction. We have proved that M is minimal nonabelian.

If $H_1 \cong H_2 \cong Q_8 \times C_2$, then all minimal nonabelian subgroups of G, except one, are isomorphic to Q_8 and then Theorem 258.2 implies that G has an abelian subgroup X of index 2. But in that case G would have two distinct maximal subgroups which are not minimal nonmetacyclic, a contradiction. It follows that G has a maximal subgroup $H_1 = Q * Z$, where $Q \cong Q_8$, $Z \cong C_4$ and $Q \cap Z = Z(Q)$.

Since all elements in $H_1 - (Q \cup Z)$ are involutions and $Z = Z(H_1)$, it follows that Q is characteristic in H_1 and so $Q \trianglelefteq G$ so that $G' < Q$. Set $A = G'Z$ so that A is a maximal normal abelian subgroup of G. Here we note that G has no abelian subgroup of index 2 since M (being minimal nonabelian) is a unique maximal subgroup of G which is not minimal nonmetacyclic. Set $Y = C_G(G')$, where $Y \cap H_1 = A$ and $H_1 Y = G$. It follows that $G/A \cong E_4$ and so $A = \Phi(G)$ is abelian of type $(4, 2)$. Set $A_0 = \Omega_1(A) \cong E_4$ so that $A_0 \trianglelefteq G$ and $V = C_G(A_0)$ is a maximal subgroup of G and we have

$$V = A \cup (G - (H_1 \cup Y)).$$

Since $\Phi(G) = \mho_1(G) = A$, there is $x \in G$ such that $x^2 \in A - A_0$ implying that $o(x) = 8$. Then $x \in M$ and so M is minimal nonabelian of order 2^4 and exponent 8 and therefore $M \cong M_{2^4}$. Since $Z(M) \cong C_4$, we get $M = Y = C_G(G')$ and $V = C_G(A_0) \cong Q_8 \times C_2$. We have $G' = Z(M)$ and $G' < \langle x \rangle$ so that $\langle x \rangle \trianglelefteq G$. Assume for a moment that $G/\langle x \rangle \cong C_4$. Since $\langle x \rangle$ is a maximal normal abelian subgroup of G, $G/\langle x \rangle \cong C_4$ must act faithfully on $\langle x \rangle$, a contradiction. It follows that $G/\langle x \rangle \cong E_4$ and so $\Phi(G) = \langle x \rangle$, contrary to the fact that $\Phi(G)$ is abelian of type $(4, 2)$. This is a final contradiction and so:

(i2c) We have $G' \cong C_2$.

Since $Q \cong Q_8$ is a subgroup of G, we have $Q \trianglelefteq G$ and so $G = Q * C$, where $C = C_G(Q)$, $Q \cap C = Z(Q)$ and $|C| = 2^3$. Let Q_1, Q_2, Q_3 be three maximal subgroups of Q, where $Q_1 \cong Q_2 \cong Q_3 \cong C_4$. Then $Q_i * C$, $i = 1, 2, 3$, are three pairwise distinct maximal subgroups of G. It follows that C is nonabelian and so $C \cong Q_8$ or D_8 and therefore G is extraspecial of order 2^5 and $d(G) = 4$.

If $C \cong Q_8$, then $G = D_1 * D_2$ for some suitable $D_1 \cong D_2 \cong D_8$. But then G has two distinct maximal subgroups isomorphic to $C_2 \times D_8$, a contradiction. Hence we have $G \cong Q_8 * D_8$ and then each maximal subgroup of G is isomorphic to $Q_8 * C_4$ or $Q_8 \times C_2$, a contradiction.

(i3) Finally, assume that G has a maximal subgroup which is minimal nonmetacyclic of order 2^5. Then $|G| = 2^6$ and all maximal subgroups H_i, $i = 1, 2, \dots$, of G distinct from M are isomorphic to G^* from Theorem 69.1 (c).

Suppose, by way of contradiction, that G has an abelian subgroup A of index 2. Then all minimal nonabelian subgroups of G are isomorphic to

$$M_2(2, 2) = \langle a, b \mid a^4 = b^4 = 1,\ a^b = a^{-1} \rangle.$$

We may use Theorem 57.3. First suppose that we have part (a) of that theorem. If $v \in G - A$, then $o(v) = 4$ and $E = C_A(v)$ is elementary abelian of order ≥ 4. Let H be a maximal subgroup of G containing $E\langle v \rangle$ so that $H \cong G^*$ (from Theorem 69.1 (c)). But $E\langle v \rangle$ is a maximal abelian subgroup in H so that $E \cong E_4$ and $A \cap H \cong C_4 \times C_4$ (a unique abelian maximal subgroup in H). It follows that $A \cap H = \Omega_2(A)$ and so A is abelian of type $(8, 4)$. Let X be any maximal subgroup of G. If $X = A$, then $\mho_1(X)$ is abelian of type $(4, 2)$ containing E. If $X \neq A$, then $X \cong G^*$ and so $\mho_1(X) = E$. Hence $\mho_1(G) = \mho_1(X)$ and so $\Phi(G) = \mho_1(G)$ is of order 2^3 so that $d(G) = 3$. On the other hand, each maximal subgroup X of G distinct from A is isomorphic to G^* and so must contain $\Omega_2(A) \cong C_4 \times C_4$ and so $d(G) = 2$, a contradiction (where we note that for each $v \in G - A$, $C_A(v) = E \cong E_4$).

Also, in all cases of part (b) of Theorem 57.3 we get a contradiction. Indeed, in case (b1) of Theorem 57.3, we have $G = K \times V$ with $K \cong M_2(2,2)$ and $V \cong E_4$ and so G possesses three maximal subgroups isomorphic to $M_2(2,2) \times C_2$ containing K, a contradiction. In case (b2) of Theorem 57.3, $G = K \times V$, where $K \cong G^*$ and $|V| = 2$ and so all seven maximal subgroups of G containing V are non-isomorphic to G^*, a contradiction. In case (b3) of Theorem 57.3, $G = K$ is a special group of order 2^6 with $Z(K) \cong E_4$ all of whose maximal subgroups are isomorphic to G^*, a contradiction. In case (b4) of Theorem 57.3, $G = K$ is a splitting extension of $B = B_1 \times B_2$ with a cyclic subgroup $\langle b \rangle$ of order 4, where $B_1 \cong B_2 \cong C_4$ and b inverts each element of B. Then $(B_1\langle b \rangle) \times \Omega_1(B_2)$ and $(B_2\langle b \rangle) \times \Omega_1(B_1)$ are two distinct maximal subgroups of G which are both isomorphic to $M_2(2,2) \times C_2$, a contradiction. We have proved that G has no abelian subgroup of index 2.

Let K_1 be a unique abelian maximal subgroup of H_1 so that $K_1 \cong C_4 \times C_4$. Since K_1 is characteristic in H_1, we have $K_1 \trianglelefteq G$ and so K_1 is a maximal normal abelian subgroup of G. Since $G' \le K_1$, it follows that G is metabelian. Set

$$U = \Omega_1(K_1) = \Omega_1(H_1) = Z(H_1) = H_1' = \mho_1(H_1).$$

Let $H_1, H_2, \dots, H_r$ $(r \ge 2)$ be all maximal subgroups of G which are isomorphic to $H_1 \cong G^*$ so that $H_i > U$ and $U = Z(H_i)$ for all $i = 1, 2, \dots, r$. It follows that $U \le Z(G)$. On the other hand, $Z(G) < K_1$ since K_1 is a maximal normal abelian subgroup in G. Hence $U = Z(G) = \Omega_1(H_i)$, $i = 1, 2, \dots, r$.

Assume for a moment that $G/K_1 \cong C_4$. Then for each $g \in G - H_1$, $g^2 \in H_1 - K_1$ and so $o(g) = 8$. But then all elements $g \in G - H_1$ are of order 8 and so all $g \in G - H_1$ are contained in M and then $M = G$, a contradiction. We have proved that $G/K_1 \cong E_4$.

Let G_i/K_1, $i = 1, 2, 3$, be three subgroups of order 2 in G/K_1. Then we may assume that $G_1 \cong G_2 \cong G^*$. It follows $\mho_1(G_1) = U = \Omega_1(K_1)$, $\mho_1(G_2) = U$ and $\mho_1(G_3) < K_1$ with $\mho_1(G_3) \ge U$. Hence $\mho_1(G) = \Phi(G) < K_1$. It follows that there is a maximal subgroup G_4 of G which does not contain K_1. We have either $G_3 \cong G^*$ and then all three subgroups G_1, G_2, G_3 containing K_1 are isomorphic to G^* or $G_3 = M$. In the second case, $G_4 \cong G^*$ and G_4 does not contain K_1 so that if $K \cong C_4 \times C_4$ is a unique abelian maximal subgroup of G_4, then $G/K \cong E_4$ and if H_i/K, $i = 1, 2, 3$, are three subgroups of order 2 in G/K, then $H_i \cong G^*$, $i = 1, 2, 3$. We have proved that in any case G has a maximal normal abelian subgroup $K \cong C_4 \times C_4$ with $G/K \cong E_4$ so that if H_i/K, $i = 1, 2, 3$, are three subgroups of order 2 in G/K, then all three maximal subgroups H_1, H_2, H_3 are isomorphic to G^*.

Set $U = \Omega_1(K) \cong E_4$ so that $\Omega_1(G) = U$, $\exp(G) = 4$ and $U = Z(G)$. By Theorem 57.3, G has a minimal nonabelian subgroup L which is not isomorphic to $M_2(2,2)$. Hence L is not contained in any of H_1, H_2, H_3 and so L covers G/K. We have

$$L \cap K = Z(L) = \Phi(L) = \mho_1(L) \le Z(G)$$

implying $L \cong Q_8$. It follows that each minimal nonabelian subgroup of G is isomorphic to Q_8 or $M_2(2,2)$ and so we may use Theorem 92.6 to get a final contradiction. Indeed, since $\exp(G) = 4$, we must have part (b) of Theorem 92.6. Also, $\Omega_1(G) = U \cong E_4$ implies that we cannot have parts (b1) or (b2) of that theorem. Our basic hypothesis excludes case (b3) of that theorem. Again, the fact that $\Omega_1(G) \cong E_4$ excludes case (b4) of that theorem. Case (b5) of that theorem, where $G \cong Q_8 \times Q_8$, also contradicts our basic assumption. Since G has no abelian subgroup of index 2, case (b6) of Theorem 92.6 cannot occur. Finally, case (b7) of that theorem contradicts our basic hypothesis since all three maximal subgroups of G containing $C \cong M_2(2,2)$ are not isomorphic to G^*.

We have proved that we must have $p > 2$.

(ii) Suppose $p > 2$. Then each minimal nonmetacyclic subgroup is either a group of order p^3 and exponent p or a 3-group of order 3^4 and class 3 given in Theorem 69.1 (b).

(ii1) Assume that G has a maximal subgroup H_1 which is of order p^3 and exponent p. Then $|G| = p^4$ and $\exp(M) = p^2$ and so M is either abelian of type (p^2, p) or $M \cong M_{p^3}$. Also, $G' \leq H_1$ is elementary abelian, $M \cap H_1 \cong E_{p^2}$ and $d(M) = 2$ so that $d(G) \leq 3$. Since G has at least p maximal subgroups of exponent p, it would follow that if G is regular, then G is of exponent p, a contradiction. Hence G is irregular. But $|G| = p^4$ and so $\mathrm{cl}(G) \leq 3$ and Theorem 7.1 (b) gives $\mathrm{cl}(G) = 3$ and so $p = 3$. We have obtained a group stated in our theorem.

(ii2) Assume that G has a maximal subgroup H_1 which is a 3-group of order 3^4 and class 3 given in Theorem 69.1 (b). Then $|G| = 3^5$ and each maximal subgroup of G, except M, is isomorphic to H_1. Set $U = H_1'$ so that $U = \Omega_1(H_1) \cong E_9$ and $U \trianglelefteq G$ with $U \leq \Phi(G)$.

We have $|G : C_G(U)| = 3$ and so obviously $C_G(U) = M$. Since $U \leq Z(M)$ and $|M| = 3^4$, we get $|M'| \leq 3$ and so we have obtained groups appearing in our theorem. Indeed, since H_1 is irregular, so is G. Our theorem is proved. □

§ 281 Nonabelian p-groups in which any two distinct minimal nonabelian subgroups have a cyclic intersection

We solve here Problem 3402 by proving the following result.

Theorem 281.1. *Let G be a nonabelian p-group which is not minimal nonabelian but any two distinct minimal nonabelian subgroups have a cyclic intersection. Then $p = 2$ and we have one of the following possibilities:*
(a) *G is a 2-group of maximal class and order > 8,*
(b) *$G \cong Q_{2^n} \times V$, where $n \geq 3$, $\exp(V) \leq 2$ and $|G| > 8$,*
(c) *$G = \langle a, c \mid a^8 = c^4 = [a, c^2] = 1,\ [a, c] = a^2c^2,\ [a^2, c] = a^4\rangle$, where $|G| = 2^5$ with $\exp(G) = 8$, $G' = \langle a^2c^2\rangle \cong C_4$, $\Omega_1(G) = Z(G) = \langle a^4, c^2\rangle \cong E_4$ and $\langle a, c^2\rangle \cong C_8 \times C_2$ is a unique abelian maximal subgroup in G.*
Conversely, all the above groups satisfy the assumptions of the theorem.

Proof. Let G be a nonabelian p-group which is not minimal nonabelian but any two distinct minimal nonabelian subgroups have a cyclic intersection.

Suppose that G has an abelian subgroup U of order p^2 and type (p, p) such that $H = N_G(U) \neq C_G(U) = K$. Then $|H : K| = p$. For any $h \in H - K$, $U\langle h\rangle$ is minimal nonabelian. If $U\langle h\rangle$ does not contain all elements in $H - K$, then there is $g \in H - K$ such that $g \notin U\langle h\rangle$. But $U\langle g\rangle$ is minimal nonabelian distinct from $U\langle h\rangle$ and $U\langle h\rangle \cap U\langle g\rangle \geq U$, a contradiction. Hence $U\langle h\rangle$ must contain all elements in $H - K$ so that $U\langle h\rangle = H$ is minimal nonabelian. By our assumption, $H < G$. If U is characteristic in H, then $N_G(U) > H$, a contradiction. Hence U is not characteristic in H.

Let $H < L \leq G$ with $|L : H| = p$ and let r be any element in $L - H$ so that $U^r \neq U$ and $U^r \leq H$. It follows that $|\Omega_1(H)| \geq p^3$ and so (by Lemma 65.1) we have the following possibilities: $H \cong D_8$ or $H \cong S(p^3)$ (a nonabelian group of order p^3 and exponent $p > 2$) or $\Omega_1(H) \cong E_{p^3}$.

If $p = 2$ and $H \cong D_8$, then $C_G(H) < H$ and Proposition 10.17 gives that G is a 2-group of maximal class and these groups appear in part (a) of our theorem.

If $H \cong S(p^3)$, then consider a minimal nonabelian subgroup X in L which is distinct from H. In that case $X \cap H$ must be cyclic of order $\geq p^2$, a contradiction. It follows that H is a nonmetacyclic minimal nonabelian subgroup with $E = \Omega_1(H) \cong E_{p^3}$.

Let Y be any minimal nonabelian subgroup of L distinct from H. Then $Y \cap H$ is cyclic and let $y \in Y - H$ and assume for a moment that $o(y) = p$. Then y does not centralize E since $U < E$ and $y \notin K$. By Lemma 57.1, there is a minimal nonabelian subgroup S in $E\langle y\rangle$ containing y so that $|S \cap E| \geq p^2$, a contradiction. It follows that $o(y) \geq p^2$. Since Y has only one subgroup of order p, it follows $p = 2$ and $Y \cong Q_8$. Hence all minimal nonabelian subgroups of L, except H, are isomorphic to Q_8 and note that $\Omega_1(H) \cong E_{p^3}$. But by Theorem 258.2 (given also in § 280), H must be metacyclic,

https://doi.org/10.1515/9783110533149-025

a contradiction. We shall assume in what follows that G has an abelian subgroup U of order p^2 and type (p, p) such that $N_G(U) \neq C_G(U)$. In particular, G has no subgroup isomorphic to D_8.

Let M be any minimal nonabelian subgroup in G. Let $M < X \leq G$ with $|X : M| = p$. Since X is nonabelian and so X is generated by its minimal nonabelian subgroups, it follows that there are minimal nonabelian subgroups in X which cover X/M. Let L be any minimal nonabelian subgroup in X distinct from M (and so covering X/M). By our hypothesis, $M \cap L$ is cyclic and so L has a cyclic subgroup $M \cap L$ of order $\geq p^2$ and index p. It follows that $L \cong Q_8$ or $L \cong M_{p^n}$, $n \geq 3$, where in case $p = 2$ we have $n \geq 4$. If $L \ncong Q_8$, then L possesses an abelian subgroup U of type (p, p) such that $N_L(U) \neq C_L(U)$. But this was excluded in the previous paragraph. It follows that $p = 2$ and $L \cong Q_8$. If $M \ncong Q_8$, then all minimal nonabelian subgroups of X, except one, are isomorphic to Q_8. By Theorem 258.2 (given also in § 280), we get $M \cong M_2(2, 2)$ (since $M \ncong D_8$). We have proved that $p = 2$ and each minimal nonabelian subgroup in G is isomorphic to Q_8 or $M_2(2, 2)$ and so we may use Theorem 92.6 (given also in § 280).

In case (a) of Theorem 92.6, G is of exponent ≥ 8, G has a unique abelian maximal subgroup A, $|G'| > 2$ and all elements in $G - A$ are of order 4. Also we have $\Omega_1(A) = \Omega_1(G) \leq Z(G)$ and if $x \in G - A$, then x inverts each element in $A/\Omega_1(A)$. Each minimal nonabelian subgroup X of G is generated by its elements in $X - A$ of order 4. If all minimal nonabelian subgroups of G are isomorphic to Q_8, then by Corollary A.17.3, $G \cong Q_{2^n} \times V$, where $n \geq 3$ and $\exp(V) \leq 2$ and these are the groups stated in part (b) of our theorem. Hence we may assume that G has a minimal nonabelian subgroup $M \cong M_2(2, 2)$, where $Z(M) \leq Z(G)$. If there is an involution $i \in A - M$, then $\langle i, M\rangle = \langle i\rangle \times M$ and then there is $M_1 \leq \langle i, M\rangle$ with $M_1 \neq M$ and $M_1 \cong M_2(2, 2)$, where $M \cap M_1$ is noncyclic, a contradiction. Hence $\Omega_1(A) = \Omega_1(G) = Z(M) \cong E_4$. If there is another subgroup $M_2 \cong M_2(2, 2)$, $M_2 \neq M$, then $M_2 \cap M \geq Z(M) \cong E_4$, a contradiction. Hence G has exactly one minimal nonabelian subgroup $M \cong M_2(2, 2)$ and all other minimal nonabelian subgroups are isomorphic to Q_8. Since D_8 is not a subgroup in G, Theorem 258.2 implies that G is of order 2^5 and exponent 8 and is isomorphic to the group given in Theorem 258.2 (b) and this is the group stated in part (c) of our theorem.

If in case (b1) of Theorem 92.6, $K \cong Q_8$, then $G \cong Q_8 \times V$ with $\exp(V) = 2$ and these are the groups from part (b) of our theorem. If $K \cong M_2(2, 2)$, then $G = K \times V$ with $\exp(V) = 2$; then there are two distinct subgroups L_1, L_2 in G with $L_1 \cong L_2 \cong M_2(2, 2)$ and $L_1 \cap L_2 \geq Z(L_1) = Z(L_2) \cong E_4$, a contradiction.

In cases (b2) and (b3) of Theorem 92.6 we get a contradiction because in both cases there are two distinct subgroups $H_1 \neq H_2$ in K isomorphic to $M_2(2, 2)$ but $Z(H_1) = Z(H_2) \cong E_4$.

In case (b4) of Theorem 92.6 we have $K = Q \times C$, where $Q = \langle a, b\rangle \cong Q_8$ with $a^2 = b^2 = z$ and $\langle z\rangle = Z(Q)$ and $C = \langle c\rangle \cong C_4$ with $c^2 = d$. Then

$$\langle a, bc\rangle \cong \langle ac, b\rangle \cong M_2(2, 2) \quad \text{with } \langle a, bc\rangle \neq \langle ac, b\rangle$$

and $\langle a, bc\rangle \geq \langle z, d\rangle$, $\langle ac, b\rangle \geq \langle z, d\rangle$, a contradiction.

In case (b5) of Theorem 92.6 we get a contradiction because $Q_8 \times Q_8$ contains a subgroup $Q_8 \times C_4$ from (b4).

In case (b6) of Theorem 92.6, G has an abelian subgroup B of exponent 4 and an element $v \in G - B$ of order 4 which inverts each element in B. If $B = \langle a \rangle \times V$, where $o(a) = 4$ and $\exp(V) \leq 2$, then in case $v^2 = a^2$ we have $G \cong Q_8 \times V$, $\exp(V) = 2$, and in case $v^2 \neq a^2$, we have that $M_2(2,2) \times C_2$ is a subgroup of G, a contradiction (as above). Now suppose that $\langle a \rangle \times \langle b \rangle \cong C_4 \times C_4$ is a subgroup of B. If $v^2 \in \langle a^2, b^2 \rangle$, then we may assume that $v^2 = a^2$ and then $H_1 = \langle v, b \rangle \cong M_2(2,2)$, $H_2 = \langle v, ab \rangle \cong M_2(2,2)$ with $H_1 \neq H_2$ and $H_1 \cap H_2 \geq \langle a^2, b^2 \rangle \cong E_4$, a contradiction. If $v^2 \notin \langle a^2, b^2 \rangle$, then $M_2(2,2) \times C_2$ is contained in G, a contradiction (as above).

In case (b7) of Theorem 92.6, $K = Q * C$ is the central product of

$$Q = \langle a, b \rangle \cong Q_8 \quad \text{and} \quad C = \langle c, d \mid c^4 = d^4 = 1,\ c^d = c^{-1} \rangle \cong M_2(2,2)$$

with $Q \cap C = \langle c^2 d^2 \rangle = Z(Q)$, where K is special of order 2^6 with $Z(K) = \Omega_1(K) \cong E_4$. Here we must have $V = \{1\}$ since $C_2 \times M_2(2,2)$ cannot be a subgroup in G and so $G = K$. Consider $\langle ac, b \rangle \cong M_2(2,2)$, where $\langle ac, b \rangle \cap \langle c, d \rangle = \langle c^2, d^2 \rangle \cong E_4$, a final contradiction. Our theorem is proved. □

§ 282 p-groups with large normal closures of nonnormal subgroups

If $H < G$ is nonnormal in a p-group G, then $|G : H^G| \geq p$, where H^G is the normal closure of H in G, i.e., the least G-invariant subgroup of G containing H. Therefore, it is natural to consider the situation described in the following definition.

Definition 1. A non-Dedekindian p-group G such that $|G : H^G| = p$ for any nonnormal $H < G$ is said to be an LC-group (large closure).

Such p-groups were classified in Theorem 62.1 by the second author. In this section we offer a proof based on other ideas. This proof is also elementary.

We assume that G is an LC-group of order $> p^3$. It is easy to check that any 2-group of maximal class is an LC-group. If a 2-group G is not of maximal class, then, by Proposition 1.6, $|G : G'| > 2^2$.

In this section we classify p-groups that are LC-groups. It appears that if an LC-group G has order $> p^3$, it is metacyclic and, if $|G| > p^4$, then $p = 2$ and, provided G is not of maximal class, then exactly two of its epimorphic images of order $\frac{1}{2}|G|$ are of maximal class.

The main result of this section is the following theorem, which will follow from a number of lemmas.

Theorem 282.1. *If a p-group G is an* LC*-group, then one of the following holds:*
(a) $|G| = p^3$, $G \ncong \mathrm{Q}_8$.
(b) $G = \mathrm{M}_p(2, 2)$, *the nonabelian metacyclic group of order p^4 and exponent p^2.*
(c) *G is a 2-group of maximal class.*
(d) $p = 2$, $|G| = 2^{n+2} > 2^4$, *G is metacyclic,* $\mathrm{Z}(G) = \Omega_1(G) \cong \mathrm{E}_4$, $G/\mathrm{Z}(G) \cong \mathrm{D}_{2^n}$, G/G' *is abelian of type* $(4, 2)$, $G' \cong \mathrm{C}_{2^{n-1}}$, $\Phi(G)$ *is abelian of type* $(2^{n-1}, 2)$.

In what follows G is an LC-group of order $> p^3$ which is not a 2-group of maximal class.

Lemma 282.2. *The following statements hold:*
(a) *If* $\mathrm{E}_{p^2} \cong N \lhd G$ *(such an N exists, by Lemma 1.4), then* $N \leq \mathrm{Z}(G)$. *Moreover, if* $N \lhd G$ *is elementary abelian of index* > 2 *in G, then* $N \leq \mathrm{Z}(G)$.
(b) $\mathrm{d}(G) = 2$. *Moreover, G/G' possesses a cyclic subgroup of index p.*

Proof. (a) Indeed, if $K < N$ is of order p, then $K^G \leq N$, $|G : K^G| \geq |G : N| = \frac{1}{p^2}|G| > p$ so that $K \lhd G$ and we conclude that $N \leq \mathrm{Z}(G)$. Now the last assertion is obvious.

(b) Let $C < G$ be nonnormal. Then $C^G \leq CG' < G$ has index p so that G/G' has a cyclic subgroup CG'/G' of index p. In particular, as $G' \leq \Phi(G)$, we get $\mathrm{d}(G) = 2$. □

Lemma 282.3. *If* $|G| = p^4$. *then* $G \cong \mathrm{M}_p(2, 2)$.

Proof. By Lemma 1.4, there is in G a normal subgroup $R \cong \mathrm{E}_{p^2}$ since G is not a 2-group of maximal class. By Lemma 282.2 (a), $R = \mathrm{Z}(G)$ since $|G : R| > p$. One has $G' < R$

https://doi.org/10.1515/9783110533149-026

(otherwise, G is minimal nonabelian with the derived subgroup of order $> p$, contrary to Lemma 65.1). By Lemma 282.2 (b), $d(G) = 2$. Then G is minimal nonabelian. As all nonnormal subgroups of G have order $p^2 > p$ and G has no cyclic subgroup of index p, it follows easily from Lemma 65.1 that $G \cong M_p(2, 2)$. □

Lemma 282.4. *If $|G'| = p$, then $G \cong M_p(2, 2)$.*

Proof. As the (abelian) group G/G' has a cyclic subgroup of index p (Lemma 282.2 (a)), it follows that G/G' has two distinct cyclic subgroups U/G' and V/G' of index p. The subgroups U and V are distinct abelian of index p in G, and we conclude that $U \cap V = Z(G)$. As $d(G) = 2$, it follows that G is minimal nonabelian. If $G = M_p(m, n, 1)$ and $L < G$ is nonnormal of order p^m, then $L^G = L \times G'$ has index p in G so $n = 1$. Similarly, $n = 1$; then $|G| = p^3 < p^4$, contrary to the assumption. Thus, G is metacyclic. Let

$$G = \langle a, b \mid o(a) = p^m,\ o(b) = p^n,\ a^b = a^{p+1}\rangle, \quad A = \langle a\rangle, \quad B = \langle b\rangle.$$

As $B^G = BG'$, we get $m = 2$. Assume that $n > 2$ and set $X = \langle b^{n-2}a\rangle$. Then X is nonnormal in G and $X^G = XG'$ has index $p^{n-1} > p$ in G, a contradiction. Thus, $n = 2$ so that $G = M_p(2, 2)$. □

One may assume, by virtue of Lemmas 282.3 and 282.4, that $|G| > p^4$ and $|G'| > p$. Note that a non-Dedekindian epimorphic images of LC-groups are LC-groups. As G/G' has a cyclic subgroup of index p and $\exp(G/G') > p$, it follows that, if $p = 2$, the quotient group G/L, where L is G-invariant of index p in G' cannot be Dedekindian (Theorem 1.20).

Lemma 282.5. *If $|G'| = p^2$, then $p = 2$ and G is metacyclic, $G/\Omega_1(G') \cong M_2(2, 2)$.*

Proof. Let $L \triangleleft G$ be a subgroup of order p in G'; then, since G/L is non-Dedekindian (this follows from $d(G/L) = 2$ and $|G/L| > p^3$), we get $G/L \cong M_p(2, 2)$ (Lemma 282.4) hence $|G| = p^5$. Assume that $G' \cong E_{p^2}$; then $G' \le Z(G)$ (Lemma 282.2 (a)). As $|G'| = p^2$, G is not minimal nonabelian. By Lemma 282.2 (b), G/G' is abelian of type (p^2, p) so, by Lemma 65.2 (a), G is minimal nonabelian hence $|G'| = p$ (Lemma 65.2), a contradiction. Thus, $G' \cong C_{p^2}$. Let $R \cong E_{p^2}$ be a normal subgroup of G (R exists, by Lemma 1.4); then $R \ne G'$. By Lemma 282.2 (a), $R = Z(G)$ and, since the nonabelian group G/R is metacyclic of order p^3, we get $p = 2$ and $G/R \cong D_8$ (otherwise, G is minimal nonabelian so $|G'| = 2$). Let $C_4 \cong C/R < G/R$. Then $C \in \Gamma_1$ is abelian. If $\Omega_1(C) \cong E_8$, then $\Omega_1(C) \le Z(G)$ (see the proof of Lemma 282.2 (a)) since $|G : \Omega_1(C)| = 4 > 2$ so G is minimal nonabelian since $d(G) = 2$, and this is a contradiction. Thus, C is abelian of type $(8, 2)$. It follows that $G/\mho_2(G) \cong M_2(2, 2)$ and hence, by Supplement to Corollary 36.6, G is metacyclic. If $K < R$ is of order 2 and $K \nleq G'$ (K exists since $R \cong E_4$ and G' is cyclic), then $|(G/K)/(G/K)'| = 4$ so that G/K is of maximal class (Proposition 1.6). It follows that $G/R \cong D_8$ (note that $R = \Omega_1(G)$ – see Proposition 10.17). □

Next we assume that $|G| > 2^5$ (see Lemma 282.5). As $G/\mho_2(G) \cong M_2(2, 2)$ is metacyclic, so is G (Supplement to Corollary 36.6). Next, G/G' is abelian of type $(4, 2)$.

Lemma 282.6. *The group G is as in Theorem 282.1* (c).

Proof. If $T \lhd G$ is of index 2^5, then the group G/T is as in Lemma 282.5. It follows that $G/\mho_2(G) \cong M_2(2,2)$ is metacyclic so is G (Supplement to Corollary 36.6); see remark preceding the lemma. By Lemma 282.4 and Proposition 10.17, $R = \Omega_1(G) \cong E_4$ and, by Lemma 282.2 (a), $R \leq Z(G)$. If $K < R$ is of order 2 and $K \nleq G'$, then $|G/K : ((G/K)'| = 4$, so that G/K is of maximal class of order, say 2^{n+1} (Proposition 1.6). It follows that $G/R \cong D_{2^n}$. Thus, if G is an LC-group of order 2^{n+2}, then $G' \cong C_{2^{n-1}}$, G/G' is abelian of type $(4,2)$.

It is easy to prove that in fact the group of Theorem 282.1 is an LC-group. It suffices to show that if $H < G$ is nonnormal cyclic, then $|G : H^G| = 2$. Since $\Omega_1(G) \leq Z(G)$, we get $|H| > 2$. Then $H \cap Z(G) = K$ is of order 2. If $K \nleq G'$, then H/K is a nonnormal cyclic in the 2-group G/K of maximal class, and its normal closure has index 2 in G/K. If $K = \Omega_1(G')$, the similar assertion follows by induction. □

In fact, Lemma 282.6 completes the proof of Theorem 282.1.

Now we consider a dual, in some sense, situation in case of groups of exponent p.

Exercise 1. Let G be a nonabelian group of exponent p such that $E \cap Z(G) > \{1\}$ for any subgroup $E \cong E_{p^2}$. Study the structure of G provided $x \in G - Z(G)$ implies $C_G(x) = \langle x \rangle \times Z(G)$.

Exercise 2. Let G be a noncyclic p-group of order $> p^3$. Let for any $N \lhd G$ of index p^3 the quotient group G/N be metacyclic. If $p > 2$, then G is metacyclic. For $p = 2$, G may be nonmetacyclic.

Solution. One has $G/\mho_1(G) \cong E_{p^2}$. If $p > 2$, then G is metacyclic (Corollary 36.8). If $p = 2$, then the nonmetacyclic group $M_2(2,1,1)$ satisfies the hypothesis. □

Problem 1. Classify the non-Dedekindian p-groups G such that $|G : H^G| \leq p^2$ for all nonnormal $H < G$.

Problem 2. If G is as above, describe the structure of the quotient group $G/Z(G)$.

Problem 3. Study the p-groups such that $|H^G : H| \leq p$ for all $H \leq G$.

Problem 4. Let G be a nonabelian group of exponent p such that $|H/H_G| = p$ for any nonnormal $H < G$. Study the structure of $G/Z(G)$.

Problem 5. Study the nonabelian p-groups all of whose minimal nonabelian subgroups are normal.

Problem 6. Study the nonabelian p-groups G such that $|G : S^G| \leq p$ for any minimal nonabelian $S < G$.

Problem 7. Suppose that a nonabelian p-group G of order $> p^3$ is generated by subgroups isomorphic to E_{p^2} and the normal closure of any such nonnormal subgroup has index p in G. Describe the structure of G.

§ 283 Nonabelian p-groups with many cyclic centralizers

We solve here Problem 3567 by proving the following result.

Theorem 283.1. *Let G be a nonabelian p-group such that for each $H \in \Gamma_1$ and $x \in G - H$, the centralizer $C_H(x)$ is cyclic. Then one of the following holds:*
(a) *G is a 2-group of maximal class.*
(b) *$G \cong S(p^3)$, $p > 2$.*
(c) *$G \cong M_{p^n}$.*
Conversely, all the above groups satisfy the assumptions of our theorem.

Proof. Let G be a nonabelian p-group such that for each maximal subgroup H of G and each $x \in G - H$, we have $C_H(x)$ is cyclic.

If G has no normal subgroup isomorphic to E_{p^2}, then by Lemma 1.4, G is a 2-group of maximal class.

From now on, we assume that G has a normal subgroup $U \cong E_{p^2}$. Let M be any maximal subgroup of G containing U. By our hypothesis, each $x \in G - M$ acts nontrivially on U and so we must have $M = C_G(U)$. In particular, M is a unique maximal subgroup of G containing U and so $d(G/U) = 1$ and therefore G/U is cyclic. Let $1 \neq z \in U \cap Z(G)$ so that the group $G/\langle z \rangle$ is abelian as an extension of $U/\langle z \rangle$ of order p by cyclic group G/U, and therefore $G' = \langle z \rangle \cong C_p$.

Let N be a maximal subgroup of G which does not contain U. Such a maximal subgroup exists since G is nonabelian and so $d(G) > 1$. Then N covers G/U and $N \cap U = \langle z \rangle = G'$ so that N is abelian. Moreover, U centralizes M and therefore $M = U(M \cap N)$ is abelian. Let $u \in U - \langle z \rangle$ so that u centralizes $M \cap N$ and therefore our hypothesis implies that $M \cap N$ is cyclic. If $M \cap N = \langle z \rangle$, then $|G| = p^3$ and in that case either $G \cong D_8$ or $G \cong S(p^3)$ or $G \cong M_{p^3}$, $p > 2$.

If $M \cap N > \langle z \rangle$, then N is cyclic of order $\geq p^3$ and index p and so $G \cong M_{p^n}$, $n \geq 4$. Our theorem is proved. □

Problem 1. Classify the p-groups G such that $C_G(x)$ is cyclic for all (i) $x \in G - \Phi(G)$, (ii) $x \in G - \Omega_1(G)$.

Problem 2. Study the p-groups G such that $C_G(H) \leq H$ for all nonabelian $H < G$ nonincident with a fixed maximal subgroup M of G.

https://doi.org/10.1515/9783110533149-027

§ 284 Nonabelian p-groups, $p > 2$, of exponent $> p^2$ all of whose minimal nonabelian subgroups are of order p^3

We initiate a study of Problem 3925 by proving the following result.

Theorem 284.1. *Let G be a nonabelian p-group, $p > 2$, of exponent $> p^2$ all of whose minimal nonabelian subgroups are of order p^3. Then G has an abelian subgroup A of index p, where all elements in the set $G - A$ are of order $\leq p^2$.*

Proof. Let G be a nonabelian p-group, $p > 2$, of exponent $> p^2$ all of whose minimal nonabelian subgroups are of order p^3.

Let A be a maximal normal abelian subgroup in G so that $A < G$. By Lemma 57.1, for any $x \in G - A$, there is $a \in A$ so that $\langle x, a\rangle$ is minimal nonabelian. This implies that $1 \neq \mathrm{o}(x) \leq p^2$.

Suppose that $\exp(G/A) > p$. Then there is $A < B \leq G$ such that $B/A \cong \mathrm{C}_{p^2}$. Let $g \in B - A$ be such that the subgroup $\langle g\rangle$ covers B/A. Then $\mathrm{o}(g) = p^2$ (by the previous paragraph, all elements of the set $G - A$ have orders $\leq p^2$) and so $\langle g\rangle \cap A = \{1\}$. By Lemma 57.1, there is $y \in A$ so that $\langle g, y\rangle$ is minimal nonabelian and so $\langle g, y\rangle \cong \mathrm{M}_{p^3}$. But then $\langle g, y\rangle \cap A = \langle y\rangle \cong \mathrm{C}_p$ and so $\langle y\rangle \trianglelefteq \langle g, y\rangle$ and $\langle g, y\rangle = \langle g\rangle \times \langle y\rangle$ is abelian, which is a contradiction. We have proved that $\exp(G/A) = p$; therefore, for each $g \in G - A$, we have $g^p \in \Omega_1(A)$.

Assume that $|G : A| > p$. Let $A < H \leq G$, where $H/A \cong \mathrm{E}_{p^2}$ so that H is metabelian. But $\exp(A) \geq p^3$ (since $\exp(G) \geq p^3$ and all elements of the set $G - A$ have orders $\leq p^2$) so that $\mathrm{H}_p(H/\Omega_1(A)) = A/\Omega_1(A) \neq \{1\}$. We can apply a result of Hogan–Kappe [HogK] for the metabelian group $H/\Omega_1(A)$ stating that a nontrivial Hughes subgroup in a metabelian p-group is of index at most p. This gives here a contradiction and so we have proved that $|G : A| = p$. □

Problem 1. Study the nonabelian p-groups of exponent p^2, $p > 2$, all of whose minimal nonabelian subgroups have order p^3.

Note that the nonabelian 2-groups all of whose minimal nonabelian subgroups have order 2^3 are classified in § 90.

Problem 2. Study the nonabelian p-groups all of whose minimal nonabelian subgroups have exponent $\leq p^2$.

https://doi.org/10.1515/9783110533149-028

§ 285 A generalization of Lemma 57.1

The second author has proved the following result.

Lemma 57.1. *Let A be a maximal normal abelian subgroup of a nonabelian p-group G. Then for any $x \in G - A$ there is $a \in A$ such that the subgroup $\langle a, x\rangle$ is minimal nonabelian.*

The importance of this result is explained by the role of minimal nonabelian subgroups in p-group theory. In particular, a nonabelian p-group is generated by its minimal nonabelian subgroups. Recall Problem 860 asking to classify the nonabelian p-groups covering by minimal nonabelian subgroups.
Below we prove the following generalization of Lemma 57.1.

Theorem 285.1. *Let A be a normal abelian subgroup of a nonabelian p-group G such that $A \nleq \mathrm{Z}(G)$. Then for any $x \in G - \mathrm{C}_G(A)$ there is $a \in A$ such that the subgroup $\langle a, x\rangle$ is minimal nonabelian.*

Note that in the theorem we do not assume that A is maximal abelian subgroup, i.e., the strong inclusion $A < \mathrm{C}_G(A)$ is possible.

Proof. We proceed by induction on $|G|$. Take $x \in G - \mathrm{C}_G(A)$ (such an x exists, by hypothesis). As the subgroup $H = \langle x\rangle A$ is nonabelian, by the choice of x, and A is a noncentral normal abelian subgroup of H, one may assume that $G = H$ (indeed, if $H < G$, the result follows by induction applied to the pair $A < H$). Set $X = \langle x\rangle$ and let B be the least nonabelian subgroup of G containing X. One has $G = XA$ hence $B = X(B \cap A)$, by the modular law, and $B \cap A$ is a noncentral normal abelian subgroup of B (otherwise, since X is cyclic, the subgroup B will be abelian). If $B < G$, then, by induction, there is $a \in B \cap A$ such that the subgroup $\langle a, x\rangle$ is minimal nonabelian (note that Lemma 57.1 is not applicable here), and we are done. Thus, $B = G$. Let $X \leq D < G$, where the subgroup D is maximal in G. Then $D = X(D \cap A)$, by the modular law. As D and A are abelian (D is abelian, in view of minimality of $G = B$ since $X < D < B = G$) and, as $G = DA = X(D \cap A)A = XA$ (recall that $D \in \Gamma_1$ and $A \nleq D$), the intersection $D \cap A \leq \mathrm{Z}(G)$. Set $\overline{G} = G/(D \cap A)$. Then $\overline{G} = \overline{D} \times \overline{A}$, where $|\overline{A}| = |\overline{G} : \overline{D}| = p$ (as $D \in \Gamma_1$, the intersection $D \cap A$ is maximal in A) and so the subgroup $\overline{D} = \overline{X}$ is cyclic of index p in $\overline{G}$. Thus, $\overline{G}$ is abelian of type $(|\overline{X}|, p)$. There are in $\overline{G}$ two distinct cyclic subgroups $\overline{U}$ and $\overline{V}$ of index p; therefore their inverse images U and V are distinct abelian of index p in G. It follows that $U \cap V = \mathrm{Z}(G)$ and $|G : \mathrm{Z}(G)| = p^2$ (clearly, $\mathrm{Z}(G) = D \cap A$). Let $a \in A - \mathrm{C}_G(x)$. We claim that the two-generator nonabelian subgroup $S = \langle a, x\rangle$ is minimal nonabelian. Indeed, the intersection $S \cap \mathrm{Z}(G) \leq \mathrm{Z}(S)$ has index p^2 in S, by the product formula (note that $S\mathrm{Z}(G) = G$), so that $S \cap \mathrm{Z}(G) = \mathrm{Z}(S)$. It follows from $|S : \mathrm{Z}(S)| = p^2$ and $\mathrm{d}(S) = 2$ that $\mathrm{Z}(S) = \Phi(S)$, and we conclude that all maximal subgroups of S are abelian. Thus, S is minimal nonabelian, completing the proof. □

Lemma 57.1 follows from Theorem 285.1.

https://doi.org/10.1515/9783110533149-029

Corollary 285.2. *Suppose that G is a nonabelian 2-group such that, whenever $S \le G$ is minimal nonabelian, then $\Omega_1(S) \le Z(S)$. Then $\Omega_1(G) \le Z(G)$.*

Proof. Assume that there are in G two non-commuting involutions x, y. Then the dihedral subgroup $\langle x, y \rangle$ contains a subgroup $D \cong D_8$. Since $D = \Omega_1(D) \nleq Z(D)$, we get a contradiction since D is minimal nonabelian. It follows that $E = \Omega_1(G)$ is a normal elementary abelian subgroup of G. Assume that $E \nleq Z(G)$. Since $\langle G - E \rangle = G$, there is $x \in G - E$ that does not centralize E. By Theorem 285.1, there is $a \in E$ such that the subgroup $S = \langle a, x \rangle$ is minimal nonabelian. Since the involution $a \nleq Z(S)$, we get a contradiction. Thus, $\Omega_1(G) = E \le Z(G)$. □

Problem 1. Let G be a p-group such that $G = \langle x \rangle \cdot H$ (semidirect product), where the element x normalizes but not centralizes $H < G$. Is it true that $x \in S \le G$, where S is minimal nonabelian?

Problem 2. Study the nonabelian p-groups G such that the set $G - Z(G)$ is covered by minimal nonabelian subgroups. (This is a generalization of Problem 860.)

§ 286 Groups ofexponent p with many normal subgroups

If all subgroups of order p^2 in a nonabelian group G of exponent p are normal, then $G \cong \mathrm{S}(p^3)$. Indeed, assume that $|G| > p^3$. Let $S \le G$ be minimal nonabelian; then $S \cong \mathrm{S}(p^3)$. As all maximal subgroups of S are G-invariant, it follows that S is normal in G. By Theorem 10.28, G/S is elementary abelian so that $G' < S$ ($<$, by Lemma 1.4). Assume that $S < G$. As G is not of maximal class (Lemma 9.1), it follows that $\mathrm{C}_G(S) \nleq S$. Let $S < H \le S\mathrm{C}_G(S)$, where H is of order p^4. Then $H = S \times X$, where $|X| = p$. As all subgroups of of order p^2 in H are G-invariant, it is abelian, a contradiction. Thus, $|G| = p^3$.

1°. In this subsection we describe the nonabelian groups of exponent p all of whose subgroups isomorphic to $\mathrm{S}(p^3)$ are normal. As all nonabelian groups of order p^4 and exponent p satisfy the condition, in what follows we assume that $|G| > p^4$. We also describe the nonabelian p-groups of exponent p all of whose nonabelian subgroups of order p^4 are normal.

Proposition 286.1. *Let G be a nonabelian group of exponent p and order $> p^4$. If all its subgroups isomorphic to $\mathrm{S}(p^3)$ are G-invariant, then $|G'| = p$.*

Proof. Let $\mathrm{S}(p^3) \cong S < G$; then $S \triangleleft G$. By Theorem 10.28, all subgroups of G containing S are G-invariant so $G' \le S$. In fact, by Burnside, $G' < S$ (see Lemma 1.4). Set $D = \mathrm{C}_G(S)$. If $D \nleq S$, then G contains a subgroup isomorphic to $S \times C$, where $|C| = p$. In that case, the intersection of all subgroups isomorphic to $\mathrm{S}(p^3)$ in $S \times C$ has order p, and we are done since all those subgroups contain G', by the above. Now let $\mathrm{C}_G(S) < S$. Then G is of maximal class (Proposition 10.17) and order p^4 (if $|G| > p^4$, then G contains a nonnormal subgroup isomorphic to $\mathrm{S}(p^3)$, by Theorem 9.6 (f), contrary to the hypothesis). □

It follows easily from the proposition that if G is nonabelian group of exponent p and order p^4 with $|G'| > p$, then G is of maximal class.

Exercise 1. If a nonabelian group G of exponent p is not covered by subgroups isomorphic to E_{p^3}, then G is of maximal class. In particular, if that G is not generated by such subgroups, then it is of maximal class.

Solution. If $x \in G$ is not contained in a subgroup isomorphic to E_{p^3} and $x \in L \cong \mathrm{E}_{p^2}$, then $\mathrm{C}_G(L) = L$, and the result follows from Proposition 1.8. □

Applying the result of Exercise 1 to G_1, the fundamental subgroup of G of order $> p^3$, we see that G_1 is covered by subgroups isomorphic to E_{p^3}.

Exercise 2. Let G be a group of exponent p and order $> p^3$. If a normal $E \cong \mathrm{E}_{p^2}$ is not contained in a subgroup isomorphic to $\mathrm{S}(p^3)$, then $E \le \mathrm{Z}(G)$.

https://doi.org/10.1515/9783110533149-030

2°. Now assume that any nonabelian subgroup of order p^4 in a nonabelian group G of order $> p^5$ and exponent p is normal. Assume, in addition, that $|G'| > p$. Then, by Proposition 286.1, there is in G a nonnormal subgroup $S \cong \mathrm{S}(p^3)$. Let $S < T < G$, where $|T| = p^4$; then $T \triangleleft G$. Take $x \in G - T$ and assume that x normalizes S; then $S = T \cap \langle x, S\rangle \triangleleft G$, a contradiction (Proposition 286.1). It follows that $\mathrm{N}_G(S) = T$. One has $\mathrm{e}_3(T) \le p^2$ which implies $|G| = p^6$ (by hypothesis, $|G| > p^5$) and hence $\mathrm{e}_3(T) = p^2$. In that case, $T = S \times C$, where $|C| = p$. It follows from the above the following:

Proposition 286.2. *Let G be a nonabelian group of exponent p and order $> p^6$. If all nonabelian subgroups of order p^4 are G-invariant, then $|G'| \le p^2$.*

Let G be a nonabelian group of exponent p and order $> p^7$. Assume that all nonabelian subgroups of order p^5 are G-invariant and $|G'| > p$. Let $L \triangleleft G$ be of order p, $L < G'$. Then all nonabelian subgroups of order p^4 are normal in G/L so, by Proposition 286.2, $|G'/L| \le p^2$ hence $|G'| \le p^3$.

Problem 1. Study the nonabelian groups of exponent p all of whose subgroups isomorphic to $\mathrm{S}(p^3)$ are nonnormal. (Any group of maximal class of order p^5 and exponent p satisfies this condition.)

Problem 2. Study the p-groups G covered by maximal normal abelian subgroups. Consider the case $\exp(G) = p$ is detail.

Problem 3. Study the nonabelian groups of exponent p all of whose nonabelian subgroups of order p^5 are normal.

Problem 4. Is it true that if a p-group G contains a regular subgroup of index p^2, then it contains a normal regular subgroup of the same index? (Compare with Exercise 1.6.)

Problem 5. Let G be a nonabelian group of exponent p. Suppose that for any $E < G$ of order p^2 one has $E \cap \mathrm{Z}(G) > \{1|$. Study the structure of $G/\mathrm{Z}(G)$.

§ 287 p-groups in which the intersection of any two nonincident subgroups is normal

As it is known, the structure of a finite p-group depends on intersections of some pairs of their nonincident subgroups (there is in the book a great number of results justifying this assertion). For example, if the intersection of any two distinct maximal subgroups of a nonabelian p-group G of order $> p^3$ is cyclic, then G has a cyclic subgroup of index p.

In this section we study the p-groups in which the intersection of any two nonincident subgroups is normal. In the case $p > 2$ we classify those nonmetacyclic groups up to isomorphism. For the case $p = 2$ we obtain some useful information. Such groups we call NI-groups (normal intersection). For $p = 2$, a non-Dedekindian NI-group has no subgroup isomorphic to E_8; such groups are described in § 50. In conclusion we state a number of related problems.

In the following theorem the NI-groups for $p > 2$ are classified. Of course, all Dedekindian p-groups and groups of order p^3 are NI-groups.

Theorem 287.1. *If an* NI-*group G is a nonabelian p-group, $p > 2$, of order $> p^3$, then G has no subgroups isomorphic to E_{p^3} and one of the following holds:*

(a) *G is a p-group of maximal class and order p^4 without subgroup isomorphic to E_{p^3}.*

(b) *G is metacyclic.*

(c) *$G = E * C$ (central product), where $\Omega_1(G) = E \cong S(p^3)$, C is cyclic of order $> p$.*

In what follows, G is a nonabelian NI-group of order $> p^3$, $p > 2$.

Proof. The theorem is a consequence of a number of lemmas.

Lemma 287.2. *The* NI-*group G has no subgroup isomorphic to E_{p^3}.*

Proof. Assume that there is in G a subgroup $E \cong E_{p^3}$; then $E < G$. In that case, any subgroup of order p in E is the intersection of two its distinct subgroups of order p^2 and therefore, being G-invariant, belongs to $Z(G)$. It follows that $E \le Z(G)$. Let $C < G$ be nonnormal cyclic. Then the abelian subgroup $CE = C \times E_1$, where $E_1 \le E$ is of order $\ge p^2$. If U, V are distinct subgroups of order p in E_1, then $C = (C \times U) \cap (C \times V) \lhd G$, a contradiction. Thus, E does not exist. □

Any group of exponent p and order $> p^3$ has a subgroup isomorphic to E_{p^3}. Therefore, if an NI-group G is nonabelian, then it has no subgroup of order p^4 and exponent p. It follows that if G is a nonabelian NI-group of exponent p, then $G \cong S(p^3)$.

As G has no subgroup isomorphic to E_{p^3}, it is one of the groups of Theorem 13.7.

Lemma 287.3. *If a nonabelian metacyclic* NI-*group $G = BC$, where $C \lhd G$ and B are cyclic and $B \cap C = \{1\}$, and, in addition, if $p = 2$, G has no nonabelian sections of order* 8, *then G is minimal nonabelian.*

https://doi.org/10.1515/9783110533149-031

Proof. In view of Theorem 1.2, we have $|B|(= |G : C|) > p$ (otherwise, $G \cong M_p(p^n)$ in which the unique nonnormal subgroup is not an intersection of two nonincident subgroups). Let $C < M \in \Gamma_1$. Then $M = (M \cap B)C$, by the modular law. By hypothesis, $M \cap B \triangleleft G$ so that $[C, M \cap B] = \{1\}$. In that case,

$$C_G(M \cap B) \geq BC = G \implies M \cap B \leq Z(G).$$

By the product formula, $|B : (M \cap B)| = p$. Set $\overline{G} = G/(M \cap B)$. By Theorem 1.2, $\overline{G}$ is either abelian of type $(|C|, p)$ or it is isomorphic to $M_{p|C|}$. In both cases, $\overline{G}$ has two distinct cyclic subgroups $\overline{U}$ and $\overline{V}$ of index p. Their inverse images U, V are abelian of index p in G. It follows that $U \cap V = Z(G)$. As $d(G) = 2$ and $G/Z(G) \cong E_{p^2}$, we conclude that G is minimal nonabelian. □

It is easy to check that a metacyclic minimal nonabelian p-group, $p > 2$, is an NI-group (this follows from the fact that all noncyclic subgroups of G are normal and $\Phi(L) \leq Z(G)$ for any cyclic $L < G$). Next we assume that our NI-group G is nonmetacyclic.

In view of Lemma 287.2 and Theorem 13.7, it remains to consider the following two cases.

(1) Let an NI-group G be a p-group of maximal class and order $\geq p^4$. By Lemma 287.2, G has no subgroup isomorphic to E_{p^3}. It follows that $\exp(G) > p$. By Theorems 9.5 and 9.6 and Lemmas 287.2 and 1.4, $|\Omega_1(\Phi(G))| < p^3$. Therefore, if $p > 3$, then $|G| = p^4$. Now let $p = 3$. If $|G| > 3^4$, then, working in the metacyclic fundamental subgroup G_1 of G, we prove that $Z(G)$ contains a subgroup isomorphic to E_{3^2}, a contradiction. If $|G| = 3^4$, then any 3-group of maximal class that has no subgroup isomorphic to E_{3^3} is an NI-group (note that G_1, the fundamental subgroup of such a G, is abelian of type $(3^2, 3)$).

(2) Now let $G = EL$, where $\Omega_1(G) = E \cong S(p^3)$ and L is cyclic of order $> p$. Assume that $|L| > p^2$. Set $L_1 = C_L(E)$; then $|L : L_1| \leq p$. In that case all maximal subgroups of E are G-invariant (they are intersections of two nonincident subgroups). It follows that G/E' is abelian of type $(\frac{1}{p}|L|, p, p)$ so that, by Lemma 4.3, $G = E * L_0$ (the central product), where L_0 is cyclic. It remains to consider the case when $|G| = p^4$. It is easily seen, that such G is either of maximal class or $Z(G) \cong C_{p^2}$ and hence G is an NI-group.

Now Theorem 287.1 follows. □

Next we consider the case $p = 2$ shortly. As in Lemma 287.2, G has no subgroup isomorphic to E_8. It is also true that if $H < G$ is nonnormal, then $|H : H_G| = 2$.

Lemma 287.4. *Let H be a nonnormal subgroup of an (arbitrary) p-group G. Then there exists $N \triangleleft G$ such that $|H/(H \cap N)| = p$.*

Proof. Let $N \triangleleft G$ be as large as possible such that $H \nleq N$. Assume that $|HN/N| = p^{k+1}$, where $k > 0$. Let $M/N \triangleleft G/N$ be of order p^k. Then $HN \nleq M$ since

$$|HN| = p^{k+1}|N| > p^k|N| = |M|,$$

and so $H \nleq M$, contrary to a choice of N since $M > N$. Thus, $k = 0$ so that $|HN : N| = p$.

But

$$p = |HN : N| = \frac{|H||N|}{H \cap N||N|} = |H : (H \cap N)|,$$

and we are done. □

Lemma 287.5. *Let H be a nonnormal subgroup of an* NI*-group G. Then $|H : H_G| = p$.*

Proof. Let $N \lhd G$ be as large as possible such that $H \not\leq N$. By Lemma 287.4, we have $|H/(H \cap N)| = p$. As $H \cap N \lhd G$, we get $H \cap N = H_G$. Thus, $|H : H_G| = p$. □

B. Wilkens [Wil7] showed that if G is a 2-group of Lemma 287.5, then G has a normal abelian subgroup of index 4.

Proposition 287.6. *If an* NI*-group G of order $2^n > 2^3$ is neither metacyclic nor Dedekindian, then G has a normal subgroup $N \in \{C_2, E_4\}$ such that G/N is Dedekindian.*

Proof. The group G has no subgroup isomorphic to E_8 (Lemma 287.2). As G is nonmetacyclic, there is in G a minimal nonmetacyclic subgroup M. By Theorem 66.1, a nonabelian M is one of the following groups:
(i) $Q_8 * C_4 \cong D_8 * C_4$ of order 2^4,
(ii) the special group of order 2^5 with $d(M) = 3$ (its defined relations are presented in Theorem 66.1),
(iii) the group $Q_8 \times C_2$.
The subgroup M', being an intersection of two nonincident subgroups of M, is G-invariant; also $|M'| \leq 4$ (with equality only in case (ii)). The quotient group G/M' is an NI-group with subgroup $M/M' \cong E_8$. It follows that G/M' is Dedekindian (Lemma 287.2). Assume that $|M'| = 2$; then $|G'| \leq 4$. If $|G'| = 4$, then, by Theorem 1.23, there is in G' a normal subgroup L of index 2 such that G/L is non-Dedekindian. As $L \neq M'$, it follows that $G' \cong E_4$ and $G' \leq Z(G)$. Now let $|M'| = 4$; then $|G'| \leq 8$. Let again L be a G-invariant of index 2 in G' such that G/L is non-Dedekindian. Then $L \neq M'$. We offer the readers to consider this situation in detail. □

Exercise 1. Suppose that a nonabelian metacyclic p-group $G = BC$, where $C \lhd G$ and B are cyclic and $B \cap C = \{1\}$ has no nonabelian section of order 8. Is it true that if any two nonincident abelian subgroups of G have G-invariant intersection, then G is minimal nonabelian?

Exercise 2. Suppose that a nonabelian metacyclic p-group G is such as in Exercise 1. If the intersection of any two its distinct subgroups with cyclic subgroups of index p is G-invariant, then G is minimal nonabelian.

Solution. One may assume that $|B| > p$. Let $B < D < G$, where $|D : B| = p$. In that case, $D \cap C = \Omega_1(C) \lhd G$ has order p so that D is abelian of type $(|B|, p)$. The subgroup D has two distinct cyclic subgroups B and B_1 of index p and their intersection $E = B \cap B_1 \lhd G$ has order $\frac{1}{p}|B|$. Then $C_G(E) \geq BC = G$ so that $E \leq Z(G)$. In that case G/E is either abelian of type $(p, |C|)$ or isomorphic to $M_{p|C|}$. Then G/E has two distinct cyclic subgroups

$U/E = CE/E \cong C$ and V/E of index p. The subgroups U, V are distinct abelian of index p in G which implies that $U \cap V = Z(G)$ has index p^2 in G. It follows that G is minimal nonabelian. □

Exercise 3. If a 2-group G is a non-Dedekindian NI-group with cyclic center, then one of the following holds:
(a) G is a group of maximal class,
(b) $C_G(R)$ is abelian with cyclic subgroup of index 2,
(c) $C_G(R) \cong Q_8 \times C_2$,
where in (b) and (c) we assume that G is not of maximal class; then there is in G a normal subgroup $R \cong E_4$, and we have $|G : C_G(R)| = 2$.

Hint. Clearly, 2-groups of maximal class are NI-groups. Next we assume that G is not of maximal class. As $Z(G)$ is cyclic, any abelian subgroup of G has a cyclic subgroup of index p. Then there is in G a normal subgroup $R \cong E_4$ (Lemma 1.4). If $C_G(R)$ is abelian, it has a cyclic subgroup of index 2. Now assume that $C_G(R)$ is nonabelian. Moreover, any abelian subgroup of G has a cyclic subgroup of index p. As G has no subgroup isomorphic to E_8, we have $\Omega_1(C_G(R)) = R$. Let $S \leq C_G(R)$ be minimal nonabelian. Then S is metacyclic (here we use Lemma 65.1). In that case, $S \cong Q_8$ (here we use Lemma 65.1 again). In that case, by Corollary A.17.3, $C_G(R) = Q \times C$, where Q is generalized quaternion and $|C| = 2$ ($|C| = 2$, by Lemma 287.2). It is easily seen that Q has no nonnormal subgroup of order 4, and we conclude that $Q \cong Q_8$. □

Exercise 4. Assume that an NI-group G of order $2^n > 2^3$ contains a subgroup $D \cong D_8$ and suppose that G is not of maximal class. Then $|G'| = 2$.

Solution. Let $A < D$ be of order 4. Then $C = C_G(A) > A$ (Proposition 1.8) so that $A = C \cap D \lhd G$. Thus, all maximal subgroups of D are G-invariant so is D. Consideration of $\Omega_1(Z(G))D$ shows that $\Omega_1(Z(G)) = Z(D)$ has order 2 (Lemma 287.2) hence $Z(G)$ is cyclic. By Proposition 10.17, $C_G(D) \nleq D$. Let $D < U \leq C_G(D)D$ with $|U : D| = 2$. Then $U = D * C$, where $C \cong C_4$ (Lemma 287.2). One has $U/D' = U/U' \cong E_8$. It follows that the quotient group G/D' is Dedekindian. Assuming that $|G'| > 2$, we get $|G'| = 4$ (Theorem 1.20). In that case, by Proposition 1.23, there is in G' a G-invariant subgroup K of index 2 such that G/K is not Dedekindian. Then $E_4 \cong G' = D' \times K \leq Z(G)$, contrary to the hypothesis. Thus, G/D' is abelian so that $G' = D'$ is of order 2. □

Exercise 5. Assume that an NI-group G of order $2^n > 2^3$ contains a subgroup $D \cong Q_8$ and suppose that G is not of maximal class. Is it true that $|G'| = 2$?

Hint. Mimic the solution of Exercise 4 and apply part (a) of Theorem 1.17 and Lemma 287.2. □

Exercise 6. Assume that an NI-group G of order $2^n > 2^3$ contains a subgroup $D \cong M_{2^n}$. Find $|G'|$.

Exercise 7. Let H be a noncyclic subgroup of an NI-group G. Then $H' \lhd G$.

Exercise 8. If an NI-group G of order $2^n > 2^3$ is nonmetacyclic, then $|G'| \le 8$.

Hint. Take in G a minimal nonmetacyclic subgroup H. By Theorem 69.1. $\mathrm{d}(H) = 3$ and $|H'| \le 4$. By Exercise 7, $H' \lhd G$. It follows that G/H' is a Dedekindian NI-group since $\mathrm{d}(H/H') = 3$ (Lemma 287.2). □

Problem 1. Study the p-groups G such that $H \cap N \lhd G$ for any nonincident $H < G$ and $N \lhd G$.

Problem 2. Study the p-groups of order p^{n+1} containing $\le n + 2p + 1$ nontrivial normal subgroups.

Problem 3. Classify the nonabelian p-groups G in which any two nonincident abelian subgroups of G have G-invariant intersection.

Problem 4. Study the p-groups G such that $H \cap N \lhd G$ for any two distinct $H, N < G$ of distinct orders.

Problem 5. Study the irregular p-groups G in which the intersection of any two distinct maximal absolutely regular subgroups is G-invariant.

Problem 6. Study the irregular p-groups G in which the intersection of any two distinct maximal abelian subgroups is G-invariant.

Problem 7. Classify the p-groups that are lattice isomorphic to a group from Theorem 287.1 (c).

Problem 8. Classify the 2-groups of Exercises 3 and 4 containing a proper nonabelian subgroup of order 8. (The group $G = \mathrm{D}_8 * \mathrm{C}_4$ satisfies the condition.)

§ 288 Nonabelian p-groups in which for every minimal nonabelian $M < G$ and $x \in G - M$, we have $C_M(x) = Z(M)$

We solve here Problem 2686 by proving the following result.

Theorem 288.1 (Janko). *Let G be a nonabelian p-group in which for every minimal nonabelian $M < G$ and $x \in G - M$, we have $C_M(x) = Z(M)$. Then for each maximal abelian subgroup A of G, $|A : Z(G)| = p$, $\mho_1(G) \leq Z(G)$ and, if $p = 2$, then G is of class 2.*

Proof. Let G be a nonabelian p-group in which for every minimal nonabelian subgroup $M < G$ and $x \in G - M$, we have $C_M(x) = Z(M)$. Let A be any maximal abelian subgroup in G and let $A < B \leq G$, where $|B : A| = p$. Let M be a minimal nonabelian subgroup in a nonabelian group B. Then $B = AM$. If $M \cap A < A$, then consider an element $x \in A - M$. But then x centralizes the maximal subgroup $M \cap A > Z(M)$ of M, contrary to our basic assumption. It follows that $M = B$ is minimal nonabelian. In the case under consideration, A is maximal in M.

It follows that $Z(M) = \Phi(M) < A$ and $|A : Z(M)| = p$. By our basic assumption, each $y \in G - M$ centralizes $Z(M)$ and so $Z(M) = Z(G)$. We have proved that for each maximal abelian subgroup A of G, $Z(G) < A$ and $|A : Z(G)| = p$ and so each maximal abelian subgroup of G is of order $p|Z(G)|$. It follows that $\exp(G/Z(G)) = p$ so that $\mho_1(G) \leq Z(G)$.

Let $g \in G - Z(G)$ so that the subgroup $Z(G)\langle g\rangle$ is abelian. It follows that $g^p \in Z(G)$ and the subgroup $Z(G)\langle g\rangle$ is maximal abelian. As we have noted, $\mho_1(G) \leq Z(G)$ and $\exp(G/Z(G)) = p$ and all maximal abelian subgroups of G partition G modulo $Z(G)$. If $p = 2$, the $G/Z(G)$ is abelian so $\mathrm{cl}(G) = 2$. Our theorem is proved. □

It follows from the proof that the group G of Theorem 288.1 is covered by minimal nonabelian subgroups (such groups are subject of Problem 860).

Exercise 1. If a group G of exponent p satisfies the hypothesis of Theorem 288.1, then $G \cong S(p^3)$.

Hint. Let $A < G$ be maximal abelian; then $A < S \leq G$, where S is minimal nonabelian of order p^3. It follows that $|A| = p^2$. As G has no abelian subgroup of order $> p^2$, we get $|G| = p^3$ so $G \cong S(p^3)$. □

Exercise 2. Let a p-group G be as in the conclusion of Theorem 288.1. Is it true that G satisfies the hypothesis of that theorem.

Problem. Let a p-group G be an $\mathcal{A}_n$-group, $n > 2$. Study the structure of G if for any $\mathcal{A}_2$-subgroup S and $x \in G - S$ we have (i) $C_S(x) = \Phi(S)$, (ii) $C_S(x) = Z(S)$.

https://doi.org/10.1515/9783110533149-032

§ 289 Non-Dedekindian p-groups all of whose maximal nonnormal subgroups are conjugate

In this section we study the non-Dedekindian p-groups in which maximal nonnormal subgroups are conjugate. A more general question is stated in Problem 1, below.

We solve here Problem 2 from Appendix 109 by proving the following result.

Theorem 289.1 (Janko). *Let G be a non-Dedekindian p-group all of whose maximal nonnormal subgroups are conjugate and let M be one of them. Then* $\mathrm{cl}(G) = 2$ *and* $|M^G : M| = p$. *Also,* G/M^G, $\mathrm{N}_G(M)/M$ *and* G' *are all cyclic.*

In the proof of this theorem we shall use very heavily the following known result.

Theorem 231.1. *Let G be a non-Dedekindian p-group and let G_0 be the subgroup generated by all nonnormal subgroups of G, where we assume that $G_0 < G$. Then* $\mathrm{cl}(G) = 2$ *and both G/G_0 and G' are cyclic.*

Proof of Theorem 289.1. Let G be a non-Dedekindian p-group all of whose maximal nonnormal subgroups are conjugate. Let M be a fixed maximal nonnormal subgroup in G. Then $|G : M| > p$ and let G_0 be such that $M < G_0 < G$ and $|G_0 : M| = p$. Then $G_0 \trianglelefteq G$ because M was a maximal nonnormal subgroup in G. By hypothesis, G_0 is generated by all maximal nonnormal subgroups of G.

Let $x \in G - G_0$ and assume for a moment that $\langle x \rangle$ is not normal in G. In that case let X be a maximal nonnormal subgroup of G containing $\langle x \rangle$. But then X cannot be conjugate to M in G since all conjugates of M lie in G_0. We have proved that $\langle x \rangle \trianglelefteq G$ for all $x \in G - G_0$. Then each subgroup of G which is not contained in G_0 is normal in G. Hence all nonnormal subgroups in G are contained in $G_0 < G$. Let $g \in G$ be such that $M^g \neq M$ so that $\langle M, M^g \rangle = G_0 = M^G$ and M^g is nonnormal in G. It follows that G_0 is the subgroup generated by all nonnormal subgroups of G and $G_0 < G$. By Theorem 231.1, G/G_0 is cyclic, G is of class 2 and the derived subgroup G' is cyclic.

Set $K = \mathrm{N}_G(M)$ so that $K < G$ and G_0/M is a unique subgroup of order p in K/M. Indeed, if G_1/M is another subgroup of order p in K/M, then $G_1 \trianglelefteq G$, by the first paragraph of the proof, and $G_0 \cap G_1 = M$ so that $M \trianglelefteq G$, a contradiction. It follows that either K/M is cyclic or $p = 2$ and K/M is a generalized quaternion group. But K/G_0 as a subgroup of the cyclic group G/G_0 is cyclic and so the second case is not possible. It follows that K/M is cyclic and our theorem is proved. □

Problem 1. Study the non-Dedekindian p-groups all of whose maximal nonnormal subgroups have the same order.

Problem 2. Study the irregular p-groups all of whose nonnormal maximal abelian subgroups (i) are conjugate, (ii) have the same order. (Note that if $p > 2$, then there is in any irregular p-group a nonnormal maximal abelian subgroup.)

https://doi.org/10.1515/9783110533149-033

§ 290 Non-Dedekindian p-groups G with a noncyclic proper subgroup H such that each subgroup which is nonincident with H is normal in G

We solve here Problem 2491 by proving the following result.

Theorem 290.1. *Let G be a non-Dedekindian p-group with a noncyclic proper subgroup H such that each subgroup in G which is nonincident with H is normal in G. Then $H \trianglelefteq G$ and G/H is cyclic. Also,* $\mathrm{cl}(G) = 2$ *and the derived subgroup G' is cyclic.*

In the proof of this theorem we shall use Theorem 231.1 as stated in § 289.

Proof. Let G be a non-Dedekindian p-group with a noncyclic proper subgroup H such that each subgroup in G which is nonincident with H is normal in G. If H is not normal in G, then there is $g \in G$ such that $H^g \neq H$. But then H^g is nonincident with H and so $H^g \trianglelefteq G$, a contradiction. Hence $H \trianglelefteq G$.

Let $x \in G - H$. Then $\langle x \rangle$ is nonincident with H since H is noncyclic, and so $\langle x \rangle \trianglelefteq G$. Let X be any subgroup in G which is not contained in H. Then $\langle X - H \rangle = X$ and so $X \trianglelefteq G$. Hence all nonnormal subgroups in G are contained in H and so the subgroup G_0 generated by all nonnormal subgroups in G is contained in H and therefore $G_0 < G$.

We may use Theorem 231.1 implying that G/G_0 is cyclic and so G/H is cyclic. Also, G is of class 2 and G' is cyclic. The theorem is proved. □

Exercise 1. Study the non-Dedekindian p-groups containing a maximal cyclic subgroup H such that all subgroups of G nonincident with H are normal (quasinormal).

Problem 1. Study the nonmodular p-groups G containing a subgroup H such that all subgroups of G nonincident with H are quasinormal.

Problem 2. Study the p-groups all of whose subgroups of order $> p$ are normal (quasinormal). Moreover, study the nonabelian p-groups all of whose minimal nonabelian subgroups are normal (quasinormal).

Problem 3. Study the p-groups all of whose maximal abelian subgroups quasinormal. Is it true that if $p > 2$, then G is regular?

https://doi.org/10.1515/9783110533149-034

§ 291 Nonabelian p-groups which are generated by a fixed maximal cyclic subgroup and any minimal nonabelian subgroup

We solve here Problem 3 from § 182 by proving the following result.

Theorem 291.1 (Janko). *Let G be a nonabelian p-group which is not minimal nonabelian but is generated with a fixed maximal cyclic subgroup $Z = \langle g \rangle$ and any minimal nonabelian subgroup. Then G has an abelian subgroup H of index p and $\mathrm{d}(G) \le 3$. If $\mathrm{d}(G) = 3$, then $G = M * X$, where $M \ne G$ is minimal nonabelian and X is cyclic. If $\mathrm{d}(G) = 2$, then $g \in H - \Phi(G)$.*

Proof. Let G be a nonabelian p-group which is not minimal nonabelian but is generated by a fixed maximal cyclic subgroup $Z = \langle g \rangle$ and any minimal nonabelian subgroup. Let H be a maximal subgroup of G containing Z. Since H has no minimal nonabelian subgroup, it is abelian. By hypothesis, we have $\mathrm{d}(G) \le 3$.

(i) First assume $\mathrm{d}(G) = 3$. Then $g \notin \Phi(G)$ since $g \notin H \in \Gamma_1$, and $g^p \in \Phi(G)$ and so $G/Z\Phi(G) \cong \mathrm{E}_{p^2}$ and all $p + 1$ maximal subgroups of G containing $Z\Phi(G)$ are abelian (see the previous paragraph) implying that $Z\Phi(G) = \mathrm{Z}(G)$ hence $G/\mathrm{Z}(G) \cong \mathrm{E}_{p^2}$ which implies $|G'| = p$ (Lemma 1.1).

Let M be a minimal nonabelian subgroup in G; then M covers $G/\mathrm{Z}(G)$, by hypothesis, and so we have $G = M\mathrm{Z}(G)$ hence $M \trianglelefteq G$. It follows that $G = MZ$ (here we use the hypothesis) so that $\mathrm{Z}(G)/(M \cap \mathrm{Z}(G))$ is cyclic. Since $Z \le \mathrm{Z}(G)$, we get $G = M * Z$ (central product). Here M is minimal nonabelian and Z is cyclic.

(ii) Assume now that $\mathrm{d}(G) = 2$. Then $g \in H - \Phi(G)$ (indeed, $G = \langle g, S \rangle > S$, where $S < G$ is minimal nonabelian) and each minimal nonabelian subgroup in G covers G/H. The theorem is proved. □

https://doi.org/10.1515/9783110533149-035

§ 292 Nonabelian p-groups generated by any two non-conjugate minimal nonabelian subgroups

Any nonabelian p-group G is generated by minimal nonabelian subgroups (Theorem 10.28). In this section we describe the p-groups generated by any two non-conjugate minimal nonabelian subgroups.

We initiate here a study of Problem 3 from § 212 by proving the following result.

Theorem 292.1 (Janko). *Let G be a nonabelian p-group which is generated by any two non-conjugate minimal nonabelian subgroups. Then each maximal normal abelian subgroup has index $\leq p^2$ in G and so G is metabelian. Also, each minimal nonabelian subgroup M in G is contained in a unique maximal subgroup of G coinciding with the normal closure M^G.*

Note that all A_2-groups satisfy the assumption of this theorem.

Proof. Let A be a maximal normal abelian subgroup in G and let $A < B \trianglelefteq G$ with $|B : A| = p$. Assume that $B < G$. Let $B < C \trianglelefteq G$ with $|C : B| = p$, $b \in B - A$ and $c \in C - B$. By Lemma 57.1, there are elements $a_1, a_2 \in A$ such that the subgroups $M_1 = \langle b, a_1\rangle$ and $M_2 = \langle c, a_2\rangle$ are minimal nonabelian. Since M_1 and M_2 are non-conjugate in G, we get $\langle M_1, M_2\rangle = G$ implying $C = G$ (indeed, $M_1, M_2 \leq C$) and so in this case $|G : A| = p^2$. We have proved that whenever A is a maximal normal abelian subgroup in G, then $|G : A| \leq p^2$ and so the group G is metabelian.

Let $M < G$ be a minimal abelian subgroup. Let X be any maximal subgroup of G containing M. Then $M^G \leq X$. If $M^G \neq X$, then there is a minimal nonabelian subgroup $N < G$ contained in X such that $N \nleq M^G$ because X is nonabelian and so is generated by its minimal nonabelian subgroups (Theorem 10.28). Since N is not conjugate to M, our hypothesis implies that $\langle M, N\rangle = G$, contrary to $\langle M, N\rangle \leq X < G$. Hence $M^G = X$ and so each minimal nonabelian subgroup M in G is contained in a unique maximal subgroup of G which is equal to the normal closure M^G. □

Problem 1. Let a p-group G be an $\mathcal{A}_n$-group, $n > 1$. Describe the structure of G provided it is generated by any two minimal nonabelian subgroups of distinct orders.

Problem 2. Describe the p-groups G provided $G\langle A, S\rangle$, where $A < S$ is an arbitrary maximal abelian subgroup and $S \leq G$ is an arbitrary maximal nonabelian subgroup.

Problem 3. Study the irregular p-groups generated by any two distinct maximal regular subgroups.

https://doi.org/10.1515/9783110533149-036

§ 293 Exercises

In this section a number of exercises is presented; some of them present well-known results.

Exercise 1 (Modular law). For a group G, let $A, B, C < G$ be such that $G = BC$ and $B \leq A$. Then $A \cap BC = B(A \cap C)$.

Solution. Clearly, $B(A \cap C) \subseteq A \cap (BC)$. Let $a \in A \cap (BC)$; then $a = bc$, where $b \in B$ and $c \in C$. In that case, $c = b^{-1}a \in A \cap C$ hence $a = bc \in B(A \cap C$, and we conclude that $A \cap (BC) \subseteq B(A \cap C)$. The required equality is established. □

Below we use the following obvious fact. If G is a p-group and $A < G$ is characteristic in $\mathrm{N}_G(A)$, then $A \lhd G$ (otherwise, if $\mathrm{N}_G(A)$ is a subgroup of index p in $T \leq G$, then $A \lhd T$ implies $\mathrm{N}_G(A) \geq T$, a contradiction).

Exercise 2. Suppose that all minimal nonabelian subgroups of a nonabelian p-group G have orders $> p^3$. Then the subgroup $\Omega_1(G)$ is elementary abelian.

Solution. Let $E < G$ be elementary abelian of maximal possible order. Write $N = \mathrm{N}_G(E)$. Assume that $E < \Omega_1(N)$. Then there is $x \in \Omega_1(N) - E$ of order p. Set $H = \langle x, E\rangle$. By the maximal choice of E, the subgroup H is nonabelian. Therefore, by Lemma 57.1, there is $a \in E$ such that the subgroup $M = \langle a, x\rangle$ is minimal nonabelian. It follows from $\Omega_1(M) = M$ that $|M| = p^3$ (Lemma 65.1), contrary to the hypothesis. Thus, $E = \Omega_1(N)$ is characteristic in N, and we conclude that $N = G$ (see the paragraph preceding the exercise) and hence $\Omega_1(G) = \Omega_1(N) = E$ is elementary abelian. □

Exercise 3. If for any $\mathcal{A}_1$-subgroup S of a p-group G, $p > 2$, the subgroup $\Omega_1(S)$ is elementary abelian, then $\Omega_1(G)$ is elementary abelian.

Exercise 4. Suppose that all $\mathcal{A}_2$-subgroups of an $\mathcal{A}_n$-group G, $n > 1$, $p > 2$, are metacyclic. Then the subgroup $\Omega_1(G)$ is elementary abelian.

Solution. Let $E < G$ be elementary abelian of maximal order and set $N = \mathrm{N}_G(E)$. Assume that $\Omega_1(N) > E$. Take $x \in N - E$ of order p and set $H = \langle x, E\rangle$. Then the subgroup H is nonabelian, by a choice of E. Therefore, by Lemma 57.1, there is $a \in E$ such that $S = \langle a, x\rangle$ is minimal nonabelian. It follows from $\Omega_1(S) = S$ that $S \cong \mathrm{S}(p^3)$ (Lemma 65.1). Let $S < T \leq G$, where $|T : S| = p$. Then T is a nonmetacyclic $\mathcal{A}_2$-subgroup of order p^4, contrary to the hypothesis. Thus, the subgroup $\Omega_1(N) = E$ is characteristic in N, and this implies that $N = G$ and $\Omega_1(G) = E$. □

Exercise 5. Classify the p-groups G of exponent $> p$ such that, whenever $A, B < G$ are cyclic with $A \cap B = \{1\}$, then $[A, B] = \{1\}$.

Solution (Janko). Any minimal nonabelian subgroup of G must be isomorphic to Q_8 (Lemma 65.1). By Corollary A.17.3, $G = \mathrm{Q}_{2^m} \times E$, where $\exp(E) \leq 2$. □

https://doi.org/10.1515/9783110533149-037

Proposition 293.1 (Mann, see commentary to Problem 115). *Let G be a nonabelian p-group of exponent $> p$. If all minimal nonabelian subgroups of G are generated by elements of order p, then G has an abelian subgroup, say A, of index p and $A = \mathrm{H}_p(G)$.*

Proof. We proceed by induction on $|G|$. All nonabelian subgroups of G are generated by elements of order p (Theorem 10.28) and all minimal nonabelian subgroups of G have order p^3 (Lemma 65.1).

If $p = 2$, then by Theorems 10.28 and 10.33, the Hughes subgroup $\mathrm{H}_2(G)$ has index 2 in G and is abelian. Next we assume that $p > 2$.

Suppose that there is an abelian $A \in \Gamma_1$. Take $x \in G - A$. By Lemma 57.1, there is $a \in A$ such that $S = \langle a, x\rangle$ is minimal nonabelian. As $S \cong \mathrm{S}(p^3)$, we get $\mathrm{o}(x) = p$, and we conclude that $A = \mathrm{H}_p(G)$.

It remains to prove that G has an abelian subgroup of index p. If this is false, then there is a nonabelian $H \in \Gamma_1$ of exponent $> p$ since $\exp(G) > p$. By induction on $|G|$, there is in H an abelian subgroup of index p. As the number of abelian subgroups of index p in $H \lhd G$ is congruent to 1 (mod p) (Exercise 1.6 (a)), it follows that there is in H a G-invariant abelian subgroup A of index p; then $|G : A| = p^2$ so that G is metabelian. Let $A < F \in \Gamma_1$; then F is nonabelian. By the previous paragraph, all elements in the set $F - A$ have order p. Assume that G/A is cyclic; then $G = \langle x\rangle A$ for some element $x \in G - A$ of order $\geq p^2$. By Lemma 57.1, there is $a \in A$ such that $S = \langle a, x\rangle$ is minimal nonabelian. Then $S \cong \mathrm{S}(p^3)$, by hypothesis, so $\mathrm{o}(x) = p < p^2$, a contradiction. Thus, $G/A \cong \mathrm{E}_{p^2}$. As F is an arbitrary member of the set Γ_1 containing A and all elements of the set $F - A$ have order p, it follows that all elements of the set $G - A$ have order p. We conclude that $A = \mathrm{H}_p(G)$, which contradicts to [HogK] since G is metabelian. Thus, there is in the set Γ_1 an abelian member $A = \mathrm{H}_p(G)$. □

Exercise 6. Let G be a p-group of maximal class with abelian subgroup of index p. Study the p-groups H such that $\mathrm{cd}(H) = \mathrm{cd}(G)$ and $\mathrm{k}(H) = \mathrm{k}(G)$. It is true that H is also of maximal class?

Hint. By Ito's theorem on degrees (see Introduction, Theorem 17), one has

$$\mathrm{cd}(G) = \{1, p\}.$$

Isaacs and Passman [IsP1] have classified the p-groups G satisfying $\mathrm{cd}(G) = \{1, p\}$. □

Exercise 7. Classify the nonabelian 2-groups G such that $\alpha_1(G) - \alpha_1(\Phi(G)) \leq 4$. State a similar problem for $p > 2$.

Hint. See § 76. □

Exercise 8. Is it true that each nonabelian p-group G contains a minimal nonabelian subgroup S such that $S \cap \Phi(G) = \Phi(S)$?

Answer. The group D_{2^n}, $n > 3$, shows that this is not true. □

Exercise 9. If G is a p-group of order $> p^p$ and $\mathrm{Z}_{p+1}(G)$ is of maximal class, so is G.

Hint. Since $|Z_{p+1}(G)| \geq p^{p+1}$, one may assume that $|G| > p^{p+1}$ (see Theorem 9.5). By Theorem 9.5, the group G is irregular. Assume that G is not of maximal class. By Theorem 12.1 (a), there is in G a normal subgroup R of order p^p and exponent p and, clearly, $R \leq Z_{p+1}(G)$ hence $|Z_{p+1}(G)| = p^{p+1}$ (see Theorem 9.6). Note that $d(Z_{p+1}(G)) = 2$. By Theorem 13.5,

$$e_p(Z_{p+1}(G)) = e_p(G) \equiv 1 \pmod{p} \implies \exp(Z_{p+1}(G)) = p,$$

contrary to Theorem 9.5. Thus, G is of maximal class. □

The following assertion is due to Isaacs.

Exercise 10. Every p-group P is an epimorphic image of a p-group G such that $|Z(G)| = p$. Also, P is a subgroup of a p-group G such that $|Z(G)| = p$.

Hint. Let $G = P \wr C$, where $|C| = p$, be the standard wreath product of order $|C|^{|P|}|P|$. Clearly, G contains a subgroup isomorphic to P. Let B be the base of G; then $G/B \cong P$. It remains to note that $|Z(G)| = p$. □

Exercise 11. Let G be a 2-group of maximal class. Classify the 2-groups H such that $cd(H) = cd(G)$ and $k(H) = k(G)$.

Solution. Let $|G| = 2^n$, $|H| = 2^m$, $|H : H'| = 2^s$. One has

$$cd(H) = cd(G) = \{1, 2\}, \quad |G : G'| = 4, \quad k(H) = k(G) = 2^{n-2} + 3,$$

$$k(H) = 2^s + \frac{2^m - 2^s}{2^2} = 2^{m-2} + 3 \cdot 2^{s-2} = k(G) = 2^{n-2} + 3.$$

It follows from

$$2^{n-2} - 2^{m-2} = 3(2^{s-2} - 1) \implies m = n, \ s = 2$$

that H is a 2-group of maximal class of order $|G|$ (Proposition 1.6). However, we do not assert that $H \cong G$. □

We suggest to state a similar result for $p > 2$.

Exercise 12. Find the number of maximal cyclic subgroups in a minimal nonabelian p-group.

Exercise 13. Let G be a nonabelian p-group of exponent p. If for any $x \in G - Z(G)$ one has $|G : \langle x \rangle^G| = p$, then $G \cong S(p^3)$. (This result is a partial case of Janko's Theorem 62.1. See also § 282.)

Exercise 14. Let G be a noncyclic p-group of order $p^n > p^2$. Then the following assertions are equivalent:

(a) If $N > \{1\}$ is a nontrivial normal subgroup of G, it has only one G-invariant subgroup of index p.
(b) G is of maximal class.

Solution. By Lemma 9.1, (b) implies (a). Let us prove the reverse implication. One may assume that $|G| > p^3$. Note that all noncyclic epimorphic images of G satisfy (a). Applying (a) to $N = \Omega_1(Z(G))$, we see that $\Omega_1(Z(G))$ is cyclic, and this implies that $Z(G)$ is cyclic. Assume that $C \leq Z(G)$ is cyclic of order p^2. Then, by Lemma 1.4, there is in G a normal subgroup $R \cong E_{p^2}$. One has $R \cap C = L \cong C_p$ and $CR/L = Z(G/L) \cong E_{p^2}$, a contradiction. Thus, $|Z(G)| = p$. As a noncyclic $G/Z(G)$ satisfies (a), it is, by induction on $|G|$, of maximal class. Then G is also of maximal class since $|Z(G)| = p$. □

Exercise 15. Classify the nonnilpotent groups with exactly two conjugate classes of nonnormal subgroups.

Hint. Take in G a minimal nonnilpotent subgroup S and describe its structure and embedding in G. □

Exercise 16. Classify the nonabelian p-groups G which are $\mathcal{A}_n$-groups, $n > 2$, such that, whenever a nonabelian $H \leq G$ is of order $p^k \geq p^3$, then $\alpha_1(H) = p^{k-3}$.

Solution. Let $H \leq G$ be minimal nonabelian and let $|H| = p^k$. Then

$$1 = \alpha_1(H) = p^{k-3} \implies k = 3$$

so that all minimal nonabelian subgroups of G have order p^3. Let $H < T \leq G$, where $|T : H| = p$. Then $\alpha_1(T) = p^{4-1} = p$, and we see that T is of maximal class. By Exercise 10.10, G is of maximal class. Let G_1 be the fundamental subgroup of G and assume that G_1 is nonabelian. Then $|G_1| > p^3$ (see § 9) and, as above, G_1 is of maximal class, a contradiction. Thus, G is of maximal class with abelian subgroup of index p. □

Exercise 17. Study the p-groups G such that, whenever $H \in \Gamma_1$ and $h \in H - Z(G)$, there is $a \in G - H$ such that the subgroup $\langle x, h\rangle$ is minimal nonabelian.

Exercise 18. Is it true that a p-group G of class 2 and exponent p is special if it has no nontrivial direct factor?

Solution. Obviously, in the case under consideration, $Z(G) = \Phi(G)$. Since $\Phi(G) = G'$ (this holds for every group of exponent p), the group G is special. □

Exercise 19. Classify the minimal nonmetacyclic p-groups, $p > 2$.

Hint. If $|G| > p^3$, then G has no proper subgroup of order p^3 and exponent p. Therefore $\Omega_1(G) \cong E_{p^2}$. Use Theorem 13.7. □

Exercise 20. Let G be a p-group such that $c_1(G) = 1+p+p^2$. If $p > 2$, then $|\Omega_1(G)| = p^3$. (For $p = 2$, as the group $G = Q_8 * C_4$ of order 2^4 shows, a similar assertion is not true.)

Solution. By Lemma 1.4, G has a normal subgroup $R \cong E_{p^2}$. Take $x \in G - R$ of order p; then the subgroup $K = \langle x, R\rangle$ has order p^3 and exponent p and so

$$c_1(K) = 1 + p + p^2 = c_1(G),$$

and we conclude that $K = \Omega_1(G)$. □

If G is a nonabelian p-group, $p > 2$, and $\Phi(G)$ is cyclic, then $\Phi(G) \leq Z(G)$. (In that case, G is regular, by Theorem 7.1 (b).) Indeed, let $\Phi(G) \leq C < G$, where C is maximal cyclic; then $G = C\Omega_1(G)$ (to prove this, we use Theorem 1.2). As $\Omega_1(G)$ is generated by G-invariant abelian subgroups of type (p, p) containing $\Omega_1(\Phi(G))$ (check!), we get $\Omega_1(G) \leq C_G(\Phi(G))$. Therefore, $C_G(\Phi(G)) \geq C\Omega_1(G) = G$, and we are done.

Exercise 21. Suppose that a nonabelian p-group G is covered by $p + 1$ maximal abelian subgroups. Then $|G : Z(G)| = p^2$ and $G = SZ(G)$, where $S \leq G$ is minimal nonabelian.

Hint. By § 116, all these $p + 1$ maximal abelian subgroups have index p in G so their intersection coincides with $Z(G)$; then $|G : Z(G)| = p^2$. If $S \leq G$ is minimal nonabelian, then $|S : (S \cap Z(G))| = p^2$ and the assertion follows, by the product formula. □

Exercise 22. Study the nonmetacyclic p-groups G covered by $p + 1$ maximal metacyclic subgroups.

Hint. Take in G a minimal nonmetacyclic subgroup. □

Exercise 23. Study the irregular p-groups G covered by $p + 1$ maximal regular subgroups.

Exercise 24. Describe the representation groups of the minimal nonmetacyclic group of order 2^5. (See § 66.) Do this for the remaining minimal nonmetacyclic 2-groups.

Exercise 25. If all maximal regular subgroups of an irregular p-group G have class ≤ 2, then $p < 5$.

Solution. Let $M \leq G$ be minimal irregular and let $A < M$ be of index p. We get $\mathrm{cl}(A) \leq 2$ since A is a subgroup of a maximal regular subgroup of G. Thus, all maximal subgroups of M are of class ≤ 2. So, by Fitting's lemma, $\mathrm{cl}(M) \leq 4$. By Theorem 7.1 (b), $p < 5$. □

Exercise 26. If A is a maximal regular subgroup of an irregular p-group G, then $|A/\mho_1(A)| \geq p^{p-1}$.

Hint. Let $A < B \leq G$, where $|B : A| = p$; then B is irregular. By Theorem 9.8 (a), one has that $|B/\mho_1(B)| \geq p^p$ implies $|A/\mho_1(B) \geq p^{p-1}$. □

Exercise 27. If G is an irregular p-group, then $|(G/Z(G) : \mho_1(G/Z(G))| \geq p^p$.

Hint. By Remark 7.2, the quotient group $G/Z(G)$ is not absolutely regular. Use Theorem 9.8 (a). □

Exercise 28. A nonabelian group of order p^4 is not covered by nonabelian subgroups of order p^3.

Hint. The group G has an abelian subgroup A of index p. If A is a unique such subgroup, then $d(G) = 2$, and the result is clear. Otherwise, $|Z(G)| = p^2$, and $Z(G)$ is not covered by nonabelian subgroups of order p^3 since their centers coincide with $G' < Z(G)$ (here we use Lemma 1.1). □

Exercise 29. Compute the number of subgroups in 2-groups of maximal class.

Hint. The group $G = \mathrm{D}_8$ contains $1+c_1(G)+s_2(G)+1 = 1+5+3+1 = 10$ subgroups. The group $H = \mathrm{Q}_8$ contains $1 + 1 + 3 + 1 = 6$ subgroups. Use Hall's enumeration principle and induction. □

Exercise 30. If G is an irregular p-group, then $|\Omega_1(G')| \geq p^{p-1}$.

Hint. Use Theorems 9.8 (c) and 7.2 (d). □

Exercise 31. If all proper subgroups of a p-group G, $|G| > p^4$, $p \neq 3$, are two-generator, then either G is metacyclic or minimal nonmetacyclic of order 2^5.

Solution. By hypothesis, $\mathrm{d}(G) \leq 3$. The assertion holds if $p = 2$. Indeed, every maximal subgroup of G is metacyclic (Theorem 36.6), and the result follows from Theorem 66.1. Next we assume that $p > 3$. By hypothesis, G has no subgroup isomorphic to E_{p^3}. Now the result follows from Theorem 13.7.[1] □

Exercise 32. Does there exist a p-group G such that $\mathrm{Aut}(G) \cong \mathrm{E}_{p^3}$?

Exercise 33. Find the number of involutions and cyclic subgroups of order 4 in an extraspecial group of order 2^{2n+1}.

Hint. Use the Frobenius–Schur formula for the number $t(G)$ of solutions of the equation $x^2 = 1$ in a 2-group G. By that formula, the required number is either $2^{2n} - 2^n$ or $2^{2n} + 2^n$. Using this, we find $c_2(G) = \frac{|G|-t(G)}{2}$. □

Exercise 34. Let C be a nonnormal cyclic subgroup of a p-group G. If $|G : C^G| = p$, then G/G' has a cyclic subgroup CG'/G' of index p so $\mathrm{d}(G) = 2$.

Exercise 35. Let C be a nonnormal cyclic subgroup of a p-group G. If G/C^G is cyclic, then $\mathrm{d}(G) = 2$.

Solution. As $C^G \leq CG'$, the group G/CG' is cyclic. Next, $CG'/G' \cong C/(C \cap G')$ is also cyclic hence G/G' is metacyclic. The result follows since $G' \leq \Phi(G)$. □

Exercise 36. Study the p-groups G such that all subgroups of $\Phi(G)$ are G-invariant.

Hint. Prove that $\Phi(G)$ has no minimal nonabelian subgroup. □

Exercise 37. Classify the nonabelian groups G of exponent p in which the normalizer of any nonnormal subgroup of order p^2 coincides with its centralizer.

Hint. Let $S \leq G$ be minimal nonabelian. Then all maximal subgroups of S are G-invariant so is S. By Theorem 10.28, all nonabelian subgroups of G are normal. It follows that $\Phi(G)$ has no minimal nonabelian subgroup so it is abelian. □

---

1 If G is a 3-group of maximal class and order $> 3^4$, then all its subgroups are two-generator.

Exercise 38. Let the minimal nonabelian group $G = \mathrm{M}_p(m, n)$ be metacyclic. Show that any representation group Γ of G is metacyclic.

Solution. Let $\mathrm{M}(G)$ be the Schur multiplier of G. As the nonabelian p-group $\Gamma/\mathrm{M}(G) \cong G$ is metacyclic and $|\Gamma' : \mathrm{M}(G)| = p$, it follows that Γ is also metacyclic (Theorem 36.1; see also Theorem 47.2). □

Exercise 39. Let Q be a normal subgroup of a p-group G such that $G/Q \cong \mathrm{E}_{p^2}$, and let $M_1/Q, \dots, M_{p+1}/Q$ be all subgroups of order p in G/Q. Assume that $\mathrm{c}_1(M_i) = p + 1$ for all i. Then one of the following holds:
(a) $\mathrm{c}_1(Q) = 1$ and $\mathrm{c}_1(G) = 1 + p + p^2$.
(b) $\mathrm{c}_1(Q) = p + 1 = \mathrm{c}_1(G)$.

Solution. One has by Hall's enumeration principle,

$$\mathrm{c}_1(G) = \sum_{i=1}^{p+1} \mathrm{c}_1(M_i) - p\mathrm{c}_1(Q) = (p+1)(p+1) - p\mathrm{c}_1(Q).$$

It follows from the displayed formula that $\mathrm{c}_1(Q) \in \{1, p+1\}$. If $\mathrm{c}_1(Q) = 1$, then one has $\mathrm{c}_1(G) = 1 + p + p^2$. If $\mathrm{c}_1(Q) = p + 1$, then $\mathrm{c}_1(Q) = p + 1$. □

Exercise 40. Let Q be a normal subgroup of a p-group G such that $G/Q \cong \mathrm{E}_{p^2}$, and let $M_1/Q, \dots, M_{p+1}/Q$ be all subgroups of order p in G/Q. Let $n > 1$ and assume that $\mathrm{c}_n(M_i) = p$ for all i. Use Hall's enumeration principle to find $\mathrm{c}_n(G)$.

Exercise 41. Classify the minimal nonabelian p-groups G of exponent $> p$ all of whose cyclic subgroups of order $> p$ are G-invariant.

Answer. The groups $G \in \{\mathrm{Q}_8, \mathrm{D}_8, \mathrm{M}_{p^n}\}$ are metacyclic. □

Exercise 42. Let G be a prime power $\mathcal{A}_n$-group, $n > 1$, and suppose that $\mathrm{C}_G(A) < A$ for any $\mathcal{A}_1$-subgroup $A < G$ of minimal order. Study the structure of G.

Exercise 43. Is it true that if all minimal nonabelian subgroups of a nonabelian 2-group G are isomorphic to Q_8, then $\Omega_1(G) \le \mathrm{Z}(G)$?

Hint. Use Corollary A.17.3. □

Exercise 44. A 2-group G is abelian if and only if all its metacyclic sections are abelian.

Solution. Assume that G is nonabelian. Then G contains a minimal nonabelian subgroup S. By hypothesis, S is nonmetacyclic. By Lemma 65.1,

$$S = \langle a, b \mid \mathrm{o}(a) = 2^m,\ \mathrm{o}(b) = 2^n,\ [a, b] = c,\ \mathrm{o}(c) = 2,\ [a, c] = [b, c] = 1\rangle \cong \mathrm{M}_2(m, n.1).$$

In that case, $S/\langle a^2, b^2\rangle \cong \mathrm{D}_8$ is nonabelian metacyclic, a contradiction. Thus, S does not exist so G is abelian. □

Exercise 45. A p-group G, $p > 2$, is abelian if and only if all its sections of order p^3 and all metacyclic sections are abelian.

Exercise 46. Classify the noncyclic p-groups that are generated by any two nonincident cyclic subgroups such that the product of their orders is maximal possible.

Solution. Let $C < G$ be cyclic of maximal order and let $C < L \leq G$, where $|L : C| = p$. Assume that L has two distinct cyclic subgroups of index p. Then $L = G$ and G is one of the following groups: (i) abelian of type $(|C|, p)$, (ii) isomorphic to $\mathrm{M}_{p|C|}$, (iii) isomorphic to Q_8. Now assume that L has only one cyclic subgroup of index p. Then L is a 2-group of maximal class (Theorem 1.2). In that case, by Exercise 10.10, G is a 2-group of maximal class, and any such G satisfies the hypothesis. □

Exercise 47. Classify the p-groups G having an upper central series

$$\{1\} < H_1 < \cdots < H_n = G$$

such that each H_i is isolated in H_{i+1}, $i = 1, \ldots, n-1$. Is it true that $\exp(G) = p$?

Exercise 48. Is it true that a p-group G is abelian if and only if $\mathrm{k}(G) = \mathrm{k}(G/N)\mathrm{k}(N)$ for each $N \lhd G$?

Exercise 49. Suppose that a nonabelian p-group G contains an abelian subgroup A of index p. Prove that, whenever $H \leq G$ is nonabelian, then $\mathrm{Z}(H) < A$.

Solution. The subgroup $H \cap A$, being a maximal abelian in the nonabelian group H, contains $\mathrm{Z}(H)$. □

Exercise 50. Let H be a nonabelian subgroup of a p-group G and let $L < H$ be maximal abelian. If L is not isolated in H and $L \leq A < G$, where A is abelian, then A is not isolated in G.

Hint. Use Exercise A.101.7. □

Exercise 51. Let G be a minimal nonnilpotent group with nonabelian derived subgroup G'. Prove that any representation group Γ of G is minimal nonnilpotent.

Hint. By Schur, if G' is a p-subgroup, then the Schur multiplier $\mathrm{M}(G)$ of G is a p-group. Now, $\mathrm{M}(G) \leq \Gamma' \cap \mathrm{Z}(\Gamma) \leq \Phi(\Gamma)$. It follows that any maximal subgroup, say H, of Γ contains $\mathrm{M}(G)$. As $H/\mathrm{M}(G)$, a proper subgroup of $\Gamma/\mathrm{M}(G) \cong G$, is nilpotent and $\mathrm{M}(G) \leq \mathrm{Z}(H)$, it follows that H is nilpotent. □

Exercise 52. If G is a p-group of order $> p^{p+1}$ such that $|G : G'| = p^2$ and $\overline{G} = G/\mathrm{K}_{p+1}(G)$ has an abelian subgroup of index p, then G is of maximal class.

Hint. The quotient group $\overline{G}$ is of maximal class, by Lemma 1.1 and induction, so is G, by Theorem 9.7. □

Exercise 53 (see Theorem 71.8). Let G be a two-generator p-group, let $H \in \Gamma_1$ and $G' > H'$. If $x^p \in H'$ for all $x \in G - H$, then G/H' is of maximal class.

Solution. Set $\overline{G} = G/H'$; then $\overline{G}$ is nonabelian with the abelian subgroup $\overline{H}$ of index p and $\Omega_1(\overline{G}) \geq \langle \overline{G} - \overline{H} \rangle = \overline{G}$ hence $\Phi(\overline{G}) = \overline{G}'$ so that $|\overline{G} : \overline{G}'| = p^2$. By Lemma 1.1, we have $|Z(\overline{G})| = p$. By induction, either $|\overline{G}/Z(\overline{G})| = p^2$ or $\overline{G}/Z(\overline{G})$ is of maximal class. In both cases $\overline{G}$ is of maximal class. □

Exercise 54. Let G be a nonabelian p-group of exponent p and order $> p^5$. Suppose that all nonabelian subgroups of G of order p^4 are G-invariant. Then $|G'| \leq p^2$.

Hint. Let $L < G'$ be G-invariant of order p. Then all nonabelian subgroups of G/L of order p^3 are normal. It is easy to prove that $|(G/L)'| \leq p$ so that $|G'| \leq p^2$. □

Exercise 55. A group G is nilpotent if and only if $H^G < G$ for all $H < G$.

Hint. If K contains the normalizer of a Sylow subgroup of G, then $N_G(K) = K$ and, if G is nonnilpotent, then $K^G = G$ (here we use Frattini's argument). □

Exercise 56. Classify the noncyclic p-groups G such that $L_1 \cap L_2 = \{1\}$ for any two nonincident cyclic $L_1, L_2 < G$.

Solution. Assume that $\exp(G) > p$ and let $A < G$ be maximal cyclic of order $> p$. Let $A < B \leq G$, where $|B : A| = p$. By Theorem 1.2, $p = 2$ and $B \cong D_{2|A|}$ is of maximal class. By Exercise 10.10, G is of maximal class, and it is easy to see that G is dihedral. □

Exercise 57. If G is a non-elementary abelian 2-group, then $\Phi(G)$ cannot be a maximal cyclic subgroup of G.

Solution. Assume that $|\Phi(G)| = 2$. Any cyclic subgroup of G of order 4 contains $\Phi(G)$, and we are done in this case. Now let $|\Phi(G)| > 2$ and let $\Phi(G)$ be cyclic. Let $L < \Phi(G)$ be of index 2. Then $\Phi(G/L) = \Phi(G)/L$ is a subgroup of index 2 in a cyclic subgroup C/L of a group G/L, by the above, since $\exp(G/L) > 2$. As $L = \Phi(\Phi(G)) \leq \Phi(C)$ and C/L is cyclic, the subgroup C is also cyclic and $C > \Phi(G)$. □

Exercise 58. Let G be a nonabelian 2-group and suppose that any subgroup of G containing $\Phi(G)$ as a subgroup of index 4, has no nonabelian subgroups of order 8. If $\Phi(G)$ is cyclic, then $\Phi(G) \leq Z(G)$, $|G'| = 2$.

Solution. Let $\Phi(G) < L$, where $L < G$ is maximal cyclic (L exists, by Exercise 57; moreover, $|L : \Phi(G)| = 2$). Let $L < M \leq G$, where $|M : L| = 2$. Then M is either abelian or isomorphic to $M_{2|L|}$, by assumption and Theorem 1.2. It follows that $\Phi(G) \leq Z(M)$. Since such subgroups M generate G, we get $C_G(\Phi(G)) = G$ hence $\Phi(G) \leq Z(G)$. Thus, $\mathrm{cl}(G) = 2$. Therefore, if $x, y \in G$, then one has, since $x^2 \in \Phi(G) \leq Z(G)$,

$$[x, y]^2 = [x^2, y] = 1.$$

Since G', being a subgroup of $\Phi(G)$, is cyclic, we get $|G'| = 2$. □

Exercise 59 (P. Hall; see [Gor1, Theorem 5.4.9]). Let all characteristic abelian subgroups of a nonabelian 2-group G be cyclic. Describe the structure of G.

Exercise 60. Let the subgroup M of maximal class and order $> 2^3$ has index 2 in a 2-group G. If G is not split over M, then G is of maximal class.

Hint. One has $c_1(G) = c_1(M) \equiv 1 \pmod 4$ so G is a group of maximal class (Theorem 1.17 (a)). Obviously, the pair (G, M) is either $(SD_{2^{n+1}}, D_{2^n})$ or $(Q_{2^{n+1}}, Q_{2^n})$. □

Exercise 61. Let G be a p-group. If $\Phi(G)$ is two-generator, it is metacyclic.[2]

Exercise 62. Let $\nu(G)$ be the sum of orders of cyclic subgroups of a p-group G and let $\sigma_1(G)$ be the sum of divisors of $|G|$. Prove that $\nu(G) = \sigma_1(G)$.

Solution. Let $|G| = p^n$. One may assume that G is noncyclic. Then there is $T \lhd G$ such that $G/T \cong E_{p^2}$. We proceed by induction on $|G|$. By induction, $\nu(H) = \sigma_1(H)$ for $H < G$ so that $\nu(H)$ depends only on $|H|$. Let H_i/T, $i = 1, \dots, p+1$, be all subgroups of index p in G/T. Let $|G| = p^n$. Then, by induction on n,

$$\nu(H_1) = \cdots = \nu(H_{p+1}) = \sigma_1(H_1) = \cdots = \sigma_1(H_{i+1}) = 1 + p + \cdots + p^{n-1}.$$

It is easy to check that

$$\nu(G) = \sum_{i=1}^{p+1} \nu(H_i) - p\nu(T).$$

Indeed, if $C < G$ is cyclic, then each summand $|C|$ appears on the right hand-side of the above display only one time. Therefore, by induction,

$$\begin{aligned} \nu(G) &= (p+1)\sigma_1(H_1) - p\sigma_1(T) \\ &= (p+1)(1 + p + \cdots + p^{n-1}) - p(1 + p + \cdots + p^{n-2}) \\ &= 1 + p + \cdots + p^{n-1} + p^n = \sigma_1(G). \end{aligned}$$ □

Exercise 63. Let G be a non-Dedekindian p-group such that for each nonnormal cyclic $L < G$ the normal closure $L^G \in \Gamma_1$ is of maximal class. Is it true that $p = 2$ and G is of maximal class?

Exercise 64. Let a p'-automorphism α of a p-group P induce the identity on the quotient group $P/\Phi(P)$. Then $\alpha = \mathrm{id}_P$.

Solution. Let $G = \langle\alpha\rangle \cdot P$ be a natural semidirect product and set $H = \langle\alpha\rangle \cdot \Phi(P)$. Then $H \lhd G$, by hypothesis, and, by Frattini's argument,

$$G = N_G(\langle\alpha\rangle)H = N_G(\langle\alpha\rangle)(\langle\alpha\rangle \cdot \Phi(P)) = N_G(\langle\alpha\rangle)\Phi(P) = N_G(\langle\alpha\rangle),$$

since $\Phi(P) \le \Phi(G)$. It follows that α centralizes P so that $\alpha = \mathrm{id}_P$. □

Exercise 65. Let p, q be distinct primes and b the order of $p \pmod q$. Let ϕ be an automorphism of order q of a p-group G. Then there is in G a ϕ-invariant elementary abelian or special subgroup E such that $p^b \le E| \le p^{2b+[b/2]}$. In particular, if $b = 1$, then the natural semidirect product of $\langle\phi\rangle$ and G is supersolvable.

2 This is a partial case of Theorem 44.12. See also Appendix 119.

Hint. Use Frobenius' normal p-complement theorem and the following addition [Gol] to Theorem A.22.1: If S be a p-closed minimal nonnilpotent $\{q, p\}$-subgroup with $\phi \in S$ and $S' \in \mathrm{Syl}_p(S)$ is nonabelian, then $Z(S')$ is a ϕ-invariant elementary abelian subgroup of order $\leq p^{[b/2]}$. The subgroup S' is either elementary abelian or special. The second assertion follows by induction on $|G|$ (see also [BZ, Exercise 3.19]). □

Exercise 66. Check the correctness of the following assertion: Given an abelian p-group A, each p-group P is an epimorphic image of a p-group with center isomorphic to A.

Exercise 67. Let G be a nonabelian p-group containing an abelian subgroup of index p. Is it true that if $k(G)$ is maximal possible, then $|G : Z(G)| = p^2$?

Exercise 68. Let G be a nonabelian p-group containing an abelian subgroup of index p. Is it true that if $k(G)$ is minimal possible, then $|Z(G)| = p$ (in that case, G is of maximal class)?

Exercise 69. Let A be an abelian p-group. Then there is a p-group G such that $A \cong Z(G)$ has index p^2 in G.

Solution. If $A = C_{p^n}$, then one can take $G = M_{p^{n+2}}$. If A is abelian of type (p^m, p^n), then one can take $G = M_p(m+1, n+1)$. If $A = U \times V$, where U is abelian of type (p^m, p^n) and V is a nonidentity abelian p-group, one can take $G = M_p(m+1, n+1) \times V$. In all cases, $|G : Z(G)| = p^2$. □

Exercise 70. Let A be an abelian p-group. Does there exist a p-group G such that $A \cong Z(G)$ has index $p^n > p^2$ in G?

Exercise 71. Let $p > 2$ and $G = \Sigma_{p^n} \in \mathrm{Syl}_p(S_{p^n})$. Is it true that G has only one maximal normal abelian subgroup?

Hint. We may assume that $n > 2$. One has $G = \Sigma_{p^{n-1}} \wr C_p$ (C_p is a passive factor). Let B be the base of G. Check that B is the unique maximal abelian normal subgroup of G. If this is true, then G has no normal cyclic subgroup of order p^2 (indeed, $\exp(B) = p$ and $|B| > p$). □

Exercise 72. Let a 2-group $G = H \wr C_4$ be of order $4^{|H|} \cdot 4$. Is it true that the base subgroup of this wreath product is the unique maximal normal abelian subgroup of G?

About critical subgroups see § 269.

Exercise 73. Study the p-groups G containing a critical subgroup H of index p.

Exercise 74. Study the p-groups containing a minimal nonabelian critical subgroup.

Exercise 75. Study the p-groups G with critical subgroup G' $(\Phi(G), \mho_1(G))$.

Exercise 76. Let $A < B < G$, where $|B : A| = p$. Study the structure of a p-group G provided A, B are critical in G.

Exercise 77. A nonabelian two-generator critical p-group is minimal nonabelian.

Solution. If $x, y \in G$, then

$$[x, y]^p = [x^p, y] = 1$$

so that G' is elementary abelian. Now the result follows from Lemma 65.2 (a). □

Exercise 78. Study the nonabelian groups of order $> p^4$ and exponent p all of whose nonabelian subgroups of order p^3 are nonnormal.

Exercise 79. Classify the non-Dedekindian p-groups all of whose nonnormal subgroups are conjugate (have equal order).

Exercise 80. Study the 2-groups G with a nonabelian $H \in \Gamma_1$ such that any minimal nonabelian subgroup of G not contained in H is (i) isomorphic to D_8, (ii) of order 8, (iii) isomorphic to $M_2(2, 2)$.

Exercise 81. Let G be a 3-group of maximal class with nonabelian fundamental subgroup G_1. Study the 3-groups H such that $\mathrm{cd}(H) = \mathrm{cd}(G)$ and $\mathrm{k}(H) = \mathrm{k}(G)$.

Exercise 82. Let H be a normal subgroup of a p-group G such that $\Phi(H) \le Z(H) \ge [G, H]$ and $C_G(H) = Z(H)$. Is it true that $K = C_{\mathrm{Aut}(G)}(H)$ is a p-group?

Exercise 83. Let H be a normal subgroup of a p-group G, $\chi \in \mathrm{Irr}(G)$. Then $\chi(1)$ divides $|G : H|\psi(1)$ for some $\psi \in \mathrm{Irr}(H)$. In particular, if H is abelian, then $\chi(1) \mid |G : H|$.

Hint. Let $\psi \in \mathrm{Irr}(\chi_H)$. Consider the inertia group of ψ in G. □

Exercise 84. Classify the irregular p-groups G without regular subgroups of order p^{p+1}.

Solution. Let $A < B \le G$, where $|A| = p^p$ and $|B : A| = p$. Then B is irregular of order p^{p+1} so of maximal class, By Exercise 10.10, G is of maximal class. Since the fundamental subgroup G_1 is regular, its order is $\le p^p$. Then

$$|G| = p|G_1| \le p^{p+1}$$

so $|G| = p^{p+1}$ since G is irregular. □

Exercise 85. Let ϕ be an automorphism of order q of a p-group G, $q \ne p$ is prime. Let $\mathcal{M}$ be the set of all proper ϕ-invariant elementary abelian and special subgroups of G. If ϕ act trivially on all members of the set $\mathcal{M}$, then $\phi = \mathrm{id}_G$.

Hint. Use Theorem A.22.1. □

Exercise 86. Let $G = X \times Y$ be a finite group and let $A \le G$. Is it true that

$$A = (A \cap X) \times (A \cap Y)$$

provided $\mathrm{GCD}(|X|, |Y|) = 1$?

Exercise 87. Classify the irregular p-groups all of whose regular subgroups are absolutely regular.

Solution. As G has no subgroup of order p^p and exponent p, it is of maximal class with $|\Omega_1(G)| = p^{p-1}$ (here we use Theorem 12.1 (a)). □

Exercise 88. Prove that a p-group G of order $> p^4$ and exponent p possesses two subgroups A, B of order p^2 such that $A \cap B = \{1\}$.

Solution. Let $R \lhd G$ be of order p^2 and $C = C_G(R)$. Let $A/R < C/R$ be G-invariant of order p; then $A \cong E_{p^3}$. Take $L < R$ of order p; then $A = L \times B$, where $|B| = p^2$. Let $K < C$ be a subgroup of order p such that $K \nleq A$. Set $D = L \times K$. Then B and D are of order p^2 and $B \cap D = \{1\}$. □

Exercise 89 ([FT, Lemma 8.5]). Let G be a group of odd order and let p be the smallest prime divisor of $|G|$. If a Sylow p-subgroup of G has no subgroup isomorphic to E_{p^3}, then G is p-nilpotent.

Solution. By Theorem A.22.1, G has no minimal nonnilpotent subgroup whose derived subgroup is a p-subgroup (here we also use Theorem 13.7). Then, by Theorem A.52.2, G is p-nilpotent. □

Exercise 90. Study the p-groups all of whose minimal nonabelian subgroups are critical.

Exercise 91. Check correctness of the following assertion: If H is a quasinormal subgroup of index p^2 in a p-group G, then $H \lhd G$.

Answer. This assertion is not correct as the group $G = M_{p^3}$ shows. □

Exercise 92. Let A be a proper normal subgroup of a p-group $G = \Omega_1(G)$. Describe the structure of G provided for any $x \in G - A$ of order p the subgroup $\langle x, A\rangle$ is of maximal class. (Compare with Exercise 10.10.) Consider in detail the following case: $A \cong E_{p^p}$ (in this case, $\langle x, A\rangle \cong \Sigma_{p^2}$, by Exercise 9.13).

Exercise 93. Suppose that H is a proper nonabelian normal subgroup of a p-group G and let $c(G/H)$ be the number of nonidentity cyclic subgroups of G/H. Is it true that then $\alpha_1(G) \geq c(G/H)$?

Hint. Use Theorem 10.28 and Lemma 57.1. □

Exercise 94. Using Theorem 13.2 (b), study the p-groups G, $p > 2$, of order $> p^{p+1}$ such that $\Omega_2^{\#}(G)$ is of maximal class.

Exercise 95. Study the nonabelian metacyclic p-groups G such that, whenever $A, B < G$ are nonincident, then $A \cap B \lhd G$. (See § 287.)

Exercise 96. Let G be a minimal nonabelian p-group generated by elements of order p^n. If $\exp(G) > p^n$, then $G \cong D_8$.

Solution. If $|G| = p^3$, then $G \cong D_8$. Next assume that $|G| > p^3$. As G is irregular (Theorem 7.2 (b)), we get $p = 2$. By Lemma 65.1, we have $\exp(\Omega_1(G)) = 2$. Since the abelian group $G/\Omega_1(G)$ is generated by elements of order 2^{n-1}, by hypothesis, we get $\exp(G) = 2^n$, a contradiction. □

Exercise 97. Suppose that a p-group G, $p > 2$, has a maximal elementary abelian subgroup E of order p^2. Then G has no a subgroup of order p^{p+1} and exponent p permutable with E.

Solution. Assume that $R \leq G$ is of order p^{p+1} and exponent p with $RE = ER$. Set $H = ER$. Then $C_H(E) = E$ so that $H = ER$ is of maximal class with subgroup R of order p^{p+1} and exponent p, contrary to Theorems 9.5 and 9.6. Thus, R does not exist. □

Exercise 97 is a partial case of the main result of [GM] in the case when the subgroup R is abelian.

Exercise 98 (Existence of Sylow subgroups). If p^s divides the order of a group G, then there is in G a subgroup of order p^s.

Solution. We proceed by induction on $|G|$ assuming that G is a minimal counterexample. If $G = \langle x \rangle$ is cyclic of order $p^s t$, then $\langle x^t \rangle \leq G$ is of order p^3, contrary to the assumption. Next we assume that G is noncyclic. If $Z(G)$ has a subgroup X of order p, then, by induction, G/X has a subgroup Y/X of order p^{s-1}. Then $Y \leq G$ is of order p^s, contrary to the assumption. Next we assume that $Z(G)$ has no subgroup of order p. A noncyclic abelian G has two distinct maximal subgroups M, N. It follows from $G = MN$ that one of subgroups M, N, say M, has order divisible by p. By induction, M contains a subgroup of order p, and we are done, by the above. Now let G be nonabelian and $Z(G)$ has no subgroup of order p. The above argument shows that p does not divide $|Z(G)|$. Let $z_1, \dots, z_k$ be distinct representatives of all noncentral G-classes. Then

$$|G| = |Z(G)| + \sum_{i=1}^{k} |G : C_G(z_i)|.$$

Since p does not divide $|Z(G)|$, it follows that $p \nmid |G : C_G(x_i)|$ for some i. Then p^s divides $|C_G(x_i)|$. Since $C_G(x_i) < G$, there is in $C_G(i)$ a subgroup of order p^s, by induction. □

Exercise 99. Does there exist a p-group G of maximal class having two distinct nontrivial critical subgroups?

Exercise 100. Let G_0 be a metacyclic p-group, $p > 2$. Classify the p-groups G such that (i) $s_2(G) = s_2(G_0)$, (ii) $c_2(G) = c_2(G_0)$.

Hint. (i) Assume that G is nonmetacyclic and $|G| > p^3$. In that case G has no subgroup isomorphic to E_{p^3} so one can use Theorem 13.7.

(ii) In the case $c_2(G)$, one can use also Theorem 13.2 (b) and Theorem 12.1 (a). □

Exercise 101. Classify the nonabelian p-groups G such that $|H : H'| = p^2$ for all nonabelian two-generator $H < G$.

Hint. All minimal nonabelian subgroups have order p^3. □

Exercise 102. Prove that representation groups of absolutely regular p-groups are regular.

Hint. Use Mann's Remark 7.2. □

Exercise 103. Is it true that if G is a special 2-group, then one has either $\Omega_1(G) = G$ or $\Omega_2^\#(G) = G$?

Hint. One has $\Omega_2^\#(G) = \mathrm{H}_2(G)$. Assume that $\mathrm{H}_2(G) < G$; then $|G : \mathrm{H}_2(G)| = 2$, $\mathrm{H}_2(G)$ is abelian and all elements in $G - \mathrm{H}_2(G)$ are involutions. Therefore, $\Omega_1(G) = G$. □

Exercise 104. Classify the p-groups G such that $|H : \mathrm{Z}(H)| \le p^2$ for all $H < G$.

Solution. By Lemma 1.1, we have $|H'| \le p$. Now the classification of such p-groups G follows from Theorem 137.7. □

Exercise 105. Study the p-groups G such that $|G/\mathrm{Z}(G)| = p^3$.

Hint. Obviously, we have $G/\mathrm{Z}(G) \in \{\mathrm{D}_8, \mathrm{E}_{p^3}, \mathrm{S}(p^3)\}$ and G is not minimal nonabelian. Let $\overline{G} = G/\mathrm{Z}(G) \cong \mathrm{D}_8$. If $\overline{A} < \overline{G}$ is cyclic of index 2, then $A \in \Gamma_1$ is abelian,

$$|G'| = \frac{1}{2}|G : \mathrm{Z}(G)| = 4$$

and $A/\mathrm{Z}(G) \cong G'$ so that $G' \cong \mathrm{C}_4$ (Lemma 1.1). If $\mathrm{Z}(G) < B \in \Gamma_1 - \{A\}$, then $|B'| \le 2$. □

Exercise 106. Describe the group $\mathrm{Aut}(G)$, where G is a p-group of maximal class with an abelian subgroup of index $p > 2$. (For $p = 2$, see Theorem 34.8.).

Exercise 107. Describe the groups $\mathrm{Aut}(G)$ and $\mathrm{Aut}(\mathrm{Aut}(G))$, where $G \cong \mathrm{M}_p(2.2)$.

Exercise 108. Classify the non-Dedekindian p-groups G all of whose nonnormal abelian subgroups are cyclic.

Exercise 109. Describe the group $\mathrm{Aut}(G)$, where G is a representation group of the group $\mathrm{M}_p(2, 2)$.

Exercise 110. Let G be a p-group, $p > 2$, and let α be a p'-automorphism of G. If α fixes all elements of G of order p, then $\alpha = \mathrm{id}_G$. (This is not true for $p = 2$.)

Solution. Let $W = \langle\alpha\rangle \cdot G$ be the natural semidirect product. As W has no minimal nonnilpotent subgroup (Theorem A.22.1), it is nilpotent hence $\alpha = \mathrm{id}_G$. □

Exercise 111. Let G be a nonabelian p-group, $\theta \in \mathrm{Irr}_1(G)$. Then $\theta(1)^2$ divides the number $\sum_{\chi\in\mathrm{Irr}(G),\,\chi(1)<\theta(1)} \chi(1)^2$.

Hint. This follows from $|G| = \sum_{\chi\in\mathrm{Irr}(G)} \chi(1)^2$. □

Exercise 112. The following statements hold:

(i) (B. M. Schein, personal communication) Let $\delta(G)$ be the minimal degree of a faithful representation of a (finite) group G by permutations. Classify the groups G such that $\delta(G) = |G|$.

(ii) Find $\delta(\mathrm{M}_p(m,n,1))$, $\delta(\mathrm{M}_p(m,n))$.
(iii) Find $\delta(G)$, where G is minimal nonmetacyclic p-group.
(iv) Find $\delta(A)$, where A is an abelian p-group of given type.

Hint. (i) Let $H_1, \dots, H_k < G$ be such that $\bigcap_{i=1}^k (H_i)_G = \{1\}$ and the number

$$\mathcal{N} = \sum_{i=1}^{k} |G/H_i|$$

is minimal possible. Then $\delta(G) = \mathcal{N}$. For example,

$$\delta(\mathrm{D}_{2^n} \times \mathrm{C}_2) = 2^{n-1} + 2 = \delta(\mathrm{SD}_{2^n} \times \mathrm{C}_2).$$

If $A, B < G$ are nonincident of coprime orders, then $\delta(G) \le |G/A| + |G/B| < |G|$, a contradiction. It follows that G is a p-group for some prime p. The same argument shows that if $p > 2$, then G has only one subgroup of order p so G is cyclic (Proposition 1.3). Now let G be noncyclic; then $p = 2$. In that case, G does not contain a subgroup isomorphic to E_8. If $\exp(G) = 2$, then $G \cong \mathrm{E}_4$. Now let $\exp(G) > 2$. If $\mathrm{c}_2(G) = 1$, then G is dihedral and $\delta(G) = \frac{1}{2}|G| < G|$. There are no two nonincident subgroups of G of orders 2 and 4, Therefore, $\mathrm{c}_1(G) = 1$ so that G is a generalized quaternion group. □

Exercise 113. Let G be a p-group, $p^n > 2$ and among maximal abelian normal subgroups of G of exponent $\le p^n$ choose A with maximal $\mathrm{d}(A)$. Then $\Omega_n(\mathrm{C}_G(\Omega_n(A)) = \Omega_n(A)$.

Hint. Let $x \in \mathrm{C}_G(\Omega_n(A)) - A$ be of order $\le p^n$. We have $B = \langle x, \Omega_n(A)\rangle > A$ is abelian and $\mathrm{d}(B) = \mathrm{d}(A) + 1$. By Theorem 10.1, there is $C \lhd G$ of exponent $\le p^n$ containing $\Omega_n(A)$ as a subgroup of index p, a contradiction. □

Exercise 114 (P. Hall). If G is a solvable group, $a_1, \dots, a_n \in G$ have pairwise coprime orders and $a_1 \dots a_n = 1$, then $a_1 = \dots = a_n = 1$.

Solution. We proceed by induction on $|G|$. Set $\overline{G} = G/G'$; then $\overline{a}_1 \cdots = \overline{a}_n = \overline{1}$. As $\overline{G}$ is abelian, we get $\overline{a}_1 = \dots = \overline{a}_n = \overline{1}$ so that $a_1, \dots, a_n \in G'$. As $|G'| < |G|$, we get, by induction, $a_1 = \dots = a_n = 1$. □

This is also true for the nonsolvable groups, but the proof depends on Thompson's theorem on groups all of whose local subgroups are solvable [Tho3].

Exercise 115 ([FT, (3.10)]). Let $P \in \mathrm{Syl}_p(G)$ and let $A \le P$ be a maximal normal abelian subgroup of P. Then $\mathrm{C}_G(A) = A \times B$, where B is a p'-subgroup of G.

Solution. One may assume that $P < G$. In view of Burnside's theorem on normal p-complement, one may assume that P is nonabelian; then $A < P$. Set $N = \mathrm{N}_G(A)$. Since $P, \mathrm{C}_G(A) \le N$, one may assume, without loss of generality, that $N = G$; then $A \lhd G$ and hence $U = \mathrm{C}_G(A)$ is normal in G. It follows that $U \cap P \lhd P$. The subgroup $(U \cap P)A = A$ so that $U \cap P = A$. Thus, $A \in \mathrm{Syl}_p(\mathrm{C}_G(A))$ is contained in the center of $\mathrm{C}_G(A)$. By Burnside's normal p-complement theorem, A is a direct factor of $\mathrm{C}_G(A)$: $\mathrm{C}_G(A) = A \times B$. □

Exercise 116. Is the following assertion true? Let $P \in \mathrm{Syl}_p(G)$ and let $A \lhd P$ be such that $\mathrm{C}_P(A) \leq A$. Then $\mathrm{C}_G(A) = \mathrm{Z}(A) \times B$, where $B < G$ is a p'-subgroup.

Exercise 117. Prove that there does not exist a nonabelian p-group containing exactly $p + 2$ pairwise distinct maximal abelian subgroups.

Solution. Assume that $\mathcal{A}(G) = \{A_1, \ldots, A_{p+2}\}$ is the set of pairwise distinct maximal abelian subgroups of a p-group G, $|A_1| \geq \cdots \geq |A_{p+2}|$. Then $|G : A_1| = p$ (otherwise, $|\mathcal{A}(G)| > p^2 > p + 1$). In that case $A_1 \cap A_i = \mathrm{Z}(G)$ for all $i > 1$, so that $|A_i| = p|\mathrm{Z}(G)|$ for $i > 1$. Also $|G : A_2| > p$ (otherwise, $|G : \mathrm{Z}(G)| = p^2$ and then $|\mathcal{A}(G)| = p + 1 < p + 2$). Set $\overline{G} = G/\mathrm{Z}(G)$. Then

$$\overline{G} = \bigcup_{i=1}^{p+2} \overline{A}_i$$

is a partition. Let $|\overline{G}| = p^n (\geq p^3)$. Then

$$|\overline{G}| = |\overline{A}_1| + |\overline{A}_2^{\#}| + \cdots + |\overline{A}_{p+2}^{\#}| = p^{n-1} + (p+1)(p-1) = p^{n-1} + p^2 - 1 < p^n,$$

a contradiction. □

Exercise 118 (Janko). Show that there does not exist a nonabelian p-group G such that $\mathrm{Z}(\mathrm{C}_G(x)) \leq \Phi(\mathrm{C}_G(x))$ for any $x \in G - \mathrm{Z}(G)$.

Solution. Write $\mathrm{C}_G(x) = M_x$ for $x \in G - \mathrm{Z}(G)$. Then $x \in \mathrm{Z}(M_x) \leq \Phi(M_x) \leq \Phi(G)$. Hence, $G = \langle G - \mathrm{Z}(G) \rangle \leq \Phi(G)$, a contradiction. □

In particular, there does not exist a nonabelian p-group G such that, for any $x \in G - \mathrm{Z}(G)$ the centralizer $\mathrm{C}_G(x)$ is special or minimal nonabelian.

Exercise 119. Let G be a p-group of maximal class with abelian subgroup of index p. Is it true that if L and G are lattice isomorphic, then L is also of maximal class?

Exercise 120 ([FT, Lemma 8.9]). Let $E \cong \mathrm{E}_{p^3}$ be a subgroup of a p-group G, $p > 2$. If $A \cong \mathrm{E}_{p^2}$ is G-invariant, then A is contained in a G-invariant subgroup isomorphic to E_{p^3}.

Solution. As $\mathrm{C}_{AE}(A)$ is G-invariant elementary abelian of order $> |A|$, we are done, by Theorem 10.1. □

Exercise 121 ([Tho3, Lemma 5.23]). Let A be a nonabelian G-invariant subgroup of a p-group G and let $B < A$ be G-invariant abelian. If F is a maximal G-invariant abelian subgroup of A containing B, then $\mathrm{C}_A(F) = F$.

Exercise 122. Suppose that all subgroups of order p^4 of a p-group G, $|G| > p^4$, are isomorphic. Describe the structure of $H < G$, where $|H| = p^4$, and G. (This was solved by the second author; see § 304.)

Exercise 123 ([Tho3, Lemma 5.24]). Suppose that a p-solvable group G has no subgroup isomorphic to E_{p^3}, $p > 2$. Then each p-chief factor of G has order p or p^2.

Hint. One may assume that $O_{p'}(G) = \{1\}$. Then Theorem 13.7 yields the structure of $P \in \mathrm{Syl}_p(G)$. Let us consider all p-groups from conclusion of Theorem 13.7. If P is either metacyclic or 3-group of maximal class without subgroup isomorphic to E_{3^3}, then any its subgroup is two-generator so G has no p-chief factor of order $> p^2$. We suggest the reader to consider the remaining cases when either $R \cong \Sigma_9$ of $P = \Omega_1(P)C$, where $\Omega_1(P) \cong S(p^3)$ and the subgroup C is cyclic, taking in mind that $C_G(U) \le U$, where $U = O_p(G)$. □

Exercise 124. Let A be a cyclic subgroup of a p-group $G > \{1\}$ and $|A|^2 \ge |G|$. Then $A_G > \{1\}$. What will be if $A \cong Q_{2^n}$ and $|G| \le 2^{2n}$?

Solution. Assume that $A_G = \{1\}$. Then there is an element $x \in G$ such that $A^x \cap A = \{1\}$ (here we use the cyclicity of the subgroup A). In that case, we have $AA^x = G$, contrary to Lemma A.28.8. (The group $G \cong \Sigma_9 \in \mathrm{Syl}_3(S_9)$) contains a subgroup $A \cong E_9$ such that $A_G = \{1\}$ and $|A|^2 = |G|$.) □

Exercise 125. Find a maximal abelian subgroup of minimal order in Σ_{p^n}.

Exercise 126. Find $|\mathrm{Aut}(\Sigma_{p^2})|$, $|\mathrm{Aut}(E_{p^2} \wr C_p|$, $|\mathrm{Aut}(C_{p^2} \wr C_p)|$, $|\mathrm{Aut}(C_p \wr C_{p^2})|$ and $|\mathrm{Aut}(C_p \wr E_{p^2})|$.

Exercise 127. Classify the p-groups G of order $> p^{p+1}$ all of whose maximal subgroups, except one, are of maximal class.

Solution. If $d(G) = 2$, then G is of maximal class (Theorem 12.12 (a)). The equality $d(G) = 3$ is impossible, by Theorem 13.6. □

Exercise 128. Prove that $\exp(A \wr C_p)) > \exp(A)$.

Hint. Let $L \le A$ be cyclic with $|L| = \exp(A)$. Find $\exp(L \wr C_p)$. □

Exercise 129. Classify the nonabelian p-groups G all of whose maximal cyclic subgroups are self-centralizing.

Solution. Obviously, $Z(G)$ is cyclic. Assume that G is not generalized quaternion. Then G contains an abelian subgroup $R = Z_1 \times Z_2$ of type (p, p) (Lemma 1.4 and Theorem 1.2). Let $Z_i \le L_i < G$, where L_i is a maximal cyclic subgroup of G, $i = 1, 2$. Then

$$\{1\} < Z(G) < C_G(L_i) = L_i, \quad i = 1, 2,$$

a contradiction since $L_1 \cap L_2 \le Z_1 \cap Z_2 = \{1\}$. Thus, G is a generalized quaternion group. □

Exercise 130. Does there exist a p-group G such that $N_G(C)$ is minimal nonabelian for any nonnormal maximal cyclic $C < G$?

Exercise 131. Suppose that A is an abelian subgroup of index p in a nonabelian p-group G. Then all $\ne A$ maximal abelian subgroups of G have order $p|Z(G)|$. If $x \in G - A$, then $|C_G(x) : Z(G)| = p$ so $C_G(x)$ is abelian.

Solution. Let $B \neq A$ be a maximal abelian subgroup of G. Then $AB = G$ and so $A \cap B = Z(G)$, and we conclude that $|B| = p|Z(G)|$, by the product formula. Let $x \in G - A$; then $C_G(x) = \langle x, Z(G)\rangle$ is a maximal abelian subgroup of G of order $p|Z(G)|$. □

Exercise 132. Let a p-group G be such that, whenever $x \in H - Z(G)$, where $H \in \Gamma_1$, then $C_G(x) \leq H$. Prove that $Z(S) \leq Z(G)$ for all minimal nonabelian $S \leq G$ not contained in H.

Hint. Suppose that $S < G$ is an $\mathcal{A}_1$-subgroup not contained in H (see Theorem 10.28). Clearly, $Z(S) \leq H \cap S$ since $H \cap S$ is maximal subgroup of S. Assume that there is $x \in Z(S) - Z(G)$. Then $C_G(x) \geq S$. Since $S \nleq H$, we get a contradiction. Thus, $Z(S) \leq Z(G)$, i.e., the centers of all $\mathcal{A}_1$-subgroups S of G are contained in $Z(G)$. □

Exercise 133. Classify the nonabelian groups G of exponent p that are not generated by subgroups isomorphic to E_{p^3}.

Hint. Let $H < G$ be generated by all subgroups isomorphic to E_{p^3} of G and let $x \in G - H$. Then $C_H(x)$ has no subgroup of order p^2. Hence $C_{\langle x,H\rangle}(x) \cong E_{p^2}$ so that the subgroup $\langle x, H\rangle$ is of maximal class (Proposition 1.8). Thus, any subgroup of G containing H as a subgroup of index p is of maximal class. By Exercise 10.10, G is of maximal class hence $|G| \leq p^p$, by Theorem 9.5. (Note that any nonabelian group G of exponent p is generated by nonabelian subgroups of order p^3, by Theorem 10.28. The group Σ_{p^2} (of exponent p^2), $p > 2$, is not generated by subgroups isomorphic to E_{p^3}.) □

Suppose that H, a proper subgroup of a p-group G, $p > 2$, is generated by all subgroups isomorphic to E_{p^3} from G and, in addition, all elements of the set $G - H$ have order p. If $x \in G - H$, then $C_H(x)$ has no abelian subgroup of type (p, p) so it is cyclic. Let $R < H$ be a G-invariant abelian subgroup of type (p, p). Then we have $C_G(R) \leq H$ (otherwise, if $x \in C_G(R) - H$, then $\langle R, x\rangle \cong E_{p^3}$ is not contained in H, a contradiction). As $|G : C_G(R)| = p$, we get $H \in \Gamma_1$.

Exercise 134. Prove that any nonabelian p-group G of exponent p and order $> p^k > p^3$ is generated by nonabelian subgroups of order p^k.

Solution. Let $S < H < G$, where S is nonabelian of order p^3 and $|H| = p^k$. Therefore, the subgroup D generated by all nonabelian subgroups of order p^k contains all minimal nonabelian subgroups of G so $D = G$. Now the result follows from Theorem 10.28. □

Exercise 135. Is it true that a p-group of maximal class of order p^5 and exponent $p \geq 5$ is not covered by its minimal nonabelian subgroups?

Exercise 136. Study the p-groups G such that the normal closure of any nonnormal abelian subgroup is of maximal class.

Exercise 137. Classify the p-groups in which the intersection of all subgroups of index p^2 is equal to $\{1\}$.

Exercise 138. A group G of order p^4 with derived subgroup of order p^2 is of maximal class.

Solution. If $|Z(G)| = p^2$, then G has an abelian maximal subgroup and one has $|G'| = p$ (Lemma 1.1), a contradiction. Thus, $|Z(G)| = p$ so that $Z(G) < G'$. As $G/Z(G)$ is nonabelian (use Theorem 10.17), we get $cl(G) = 3$ hence G is of maximal class. □

Exercise 139. Let G be a p-group of order p^5 with $|G'| = p$. Is it possible that $|Z(G)| = p^2$?

Solution. Suppose that $|Z(G)| = p^2$. Let $A < G$ be minimal nonabelian (G is not minimal nonabelian since $|G'Z(G) = p^3 > p^2$). Then $G = AC_G(A)$ (Lemma 4.3). If $|A| = p^4$, then $C_G(A) = Z(G)$ has order p^3, a contradiction. If $|A| = p^3$, then $C_G(A)$ is nonabelian and therefore $G = A * C_G(A)$ is extraspecial, $|Z(G)| = p < p^2$, a contradiction. □

Exercise 140. Let G be a p-group of order p^7 with $|G'| = p$. Describe all possibilities for $|Z(G)|$.

Exercise 141. Let G be a p-group of order p^n with $|G'| = p$.
(i) Is it possible that $|Z(G)| = p^{n-3}$?
(ii) One has $|G : Z(G)| = p^{2k}$ for some k. (Use Lemma 4.2.)

Exercise 142. Let $H > \{1\}$ be the intersection of all minimal nonabelian subgroups of a nonabelian p-group G. If $M < G$ does not contain H, then M is abelian.

Exercise 143 (Theorem 9.11). If G is a p-group, $p > 2$, such that $G/\mho_1(G) \cong E_{p^2}$, then it is metacyclic.

Solution. We proceed by induction on $|G|$. By Theorem 9.8 (a), G is absolutely regular. It follows from Theorem 7.2 (d) that $|\Omega_1(G)| = p^2$ and $|\Omega_2(G)| \le p^4$. Assume that G is nonmetacyclic. Then it contains a minimal nonmetacyclic subgroup H. It follows that $\exp(H) > p$. By Theorem 69.1, H is a 3-group of maximal class and order 3^4, a contradiction since H is irregular (see Theorem 9.5). Thus, G is metacyclic. □

In particular, if G is a p-group, $p > 2$, and $G/\mho_2(G)$ is metacyclic so is G, by Supplement to Corollary 36.6. (The last assertion is also true for $p = 2$.)

Exercise 144. Is it true that if a two-generator irregular p-group G, $p > 2$, of order $> p^{p+1}$ contains an absolutely regular subgroup H of index p and $H/\Omega_1(H)$ is noncyclic, then G is of maximal class?

Solution. Assume that G is not of maximal class. By Theorem 12.1 (b), $G = H\Omega_1(G)$; then $|\Omega_1(G)| = p|\Omega_1(H)|$ and $H \cap \Omega_1(G) = \Omega_1(H)$. In that case,

$$G/\Omega_1(H) = (H/\Omega_1(H)) \times (\Omega_1(G)/\Omega_1(H))$$

so that $d(G/\Omega_1(H)) > 2$, a contradiction. Thus, G is of maximal class. □

Exercise 145. Let G be a group of order p^p and exponent p. Find the minimal value of $s_2(G)$.

Exercise 146. Prove that if a p-group G is nonabelian and all elements of the set $G - A$ have order p for any maximal abelian subgroup $A < G$, then $\exp(G) = p$.

Hint. All maximal abelian subgroups of G are isolated. Apply Theorem 252.1. □

Exercise 147. Is it true that if a p-group G of exponent $> p$ is nonabelian and all elements of the set $G - S$ have order p for any minimal nonabelian subgroup $S < G$, then there is in G an abelian subgroup of index p?

Hint. All minimal nonabelian subgroups of G are isolated. Apply Theorem 251.1. By that theorem, G has an abelian subgroup of index p. Treat this situation in detail using Exercise A.101.7. □

Exercise 148. Prove that if a nonmetacyclic p-group G is nonabelian and all elements of the set $G - M$ have order p for any maximal metacyclic $M < G$, then $\exp(G) = p$.

Solution. All maximal metacyclic subgroups of G are isolated so $\exp(G) = p$. □

Exercise 149. If G is a group of order p^n and $\mathrm{s}_{n-2}(G) \le p + 1$, then G is metacyclic.

Solution. Clearly, $\mathrm{d}(G) = 2$. If $H \in \Gamma_1$, then $\mathrm{d}(H) \le 2$. Thus, G and all its maximal subgroups are two-generator. Now the result follows from Corollary 36.6. □

Exercise 150. Suppose that G is a p-group such that $H < G$ implies $\mathrm{d}(H) < \mathrm{d}(G)$ for all $H < G$. Then $G' = \Phi(G)$.

Hint. One has $\mathrm{d}(G) = \mathrm{d}(G/G')$. Then

$$\mathrm{d}(H) = \mathrm{d}(G),$$

where H is the inverse image of $\Omega_1(G/G')$ in G, and this implies that $H = G$, i.e., $\exp(G/G') = p$, and we conclude that $G' = \Phi(G)$. □

Exercise 151. Suppose that G is a p-group such that $G/\mathrm{K}_3(G) \cong \mathrm{M}_{p^n}$. Is it true that $\mathrm{K}_3(G) = \{1\}$?

Solution. Note that $\mathrm{K}_3(G) \le \Phi(G)$ so that $\mathrm{d}(G) = 2$. Assume that $\mathrm{K}_3(G) > \{1\}$. Let $R < \mathrm{K}_3(G)$ be G-invariant of index p. To obtain a contradiction, one may assume that $|R| = p$. The group G/R contains two distinct cyclic subgroups U/R and V/R of index p; then $U, V \in \Gamma_1$ are distinct abelian and $U \cap V = \mathrm{Z}(G)$ has index p^2 in G. Since

$$\mathrm{d}(G) = \mathrm{d}(G/R) = 2,$$

the group G is minimal nonabelian. Since $|G'| = p^2$, this contradicts Lemma 65.1. Thus, $\mathrm{K}_3(G) = \{1\}$. □

Exercise 152. If a two-generator p-group G of order $> p^3$, $p > 2$, has a cyclic Frattini subgroup, then G is metacyclic.

Hint. All members of the set Γ_1 are metacyclic so G is either metacyclic or minimal nonmetacyclic. The second case is impossible in view of main results of §§ 66, 69. □

Exercise 153. Study the 2-groups G all of whose subgroups of order 2^5 are minimal nonmetacyclic.

Hint. All minimal nonabelian subgroups of G are isomorphic to $M_2(2, 2)$. □

Exercise 154. All maximal subgroups of an abelian p-group G are isomorphic if and only if G is homocyclic.

Hint. Assume that G is noncyclic and $\exp(G) > p$. Then the minimal number of generators of each member of the set Γ_1 is equal to $d(G)$. It follows that $\Omega_1(G) \leq \Phi(G)$. Consider the quotient group $G/\Omega_1(G)$ and use induction. (Otherwise, use the basic theorem on abelian p-groups.) □

Exercise 155. All epimorphic images of order $\frac{1}{p}|G|$ of an abelian p-group G are isomorphic if and only if G is homocyclic.

Hint. Assume that G is not homocyclic. Let $G = Z_1 \times \cdots \times Z_k$, all Z_i are cyclic and $|Z_1| > |Z_n|$. Then $G/\Omega_1(Z_1) \not\cong G/\Omega_1(Z_n)$. □

Exercise 156. A p-group G is absolutely regular if and only if $G/K_p(G)$ is.

Hint. One has $|G/\mho_1(G)| < p^p$. Use Theorem 9.8 (a). □

Exercise 157. Classify the p-groups G such that the quotient group $\overline{G} = G/K_{p+1}(G)$ has no normal subgroup of order p^p and exponent p.

Hint. By Theorem 12.1 (a), $\overline{G}$ is either absolutely regular or of maximal class. If $\overline{G}$ is absolutely regular so is G (Theorem 9.8 (a)). If $\overline{G}$ is of maximal class so is G, by Theorem 9.7. In that case, $|G| = p^{p+1}$ and $|\Omega_1(G)| = p^{p-1}$. □

Exercise 158. Let $G = F \wr H$ be a wreath product of p-groups F and H, where the active factor F is transitive. Prove that $d(G) = d(F) + d(H)$. In particular, $d(\Sigma_{p^n}) = n$.

The following few exercises allow us to repeat some known material (they do not present new results).

Exercise 159. If A, B are p-groups, then

$$\Phi(A \times B) = \Phi(A) \times \Phi(B), \quad d(G) = d(A) + d(B).$$

Hint. Recall that $\Phi(G) = \mho_1(G)G'$. Therefore, we have $\Phi(A) \times \Phi(B) \leq \Phi(G)$. However, $(A \times B)/(\Phi(A) \times \Phi(B))$ is elementary abelian. □

Exercise 160. The order of a minimal nonabelian subgroup H of a special p-group is $\leq p^5$.

Hint. One has $|H \cap Z(G)| \leq |\Omega_1(H)| \leq p^3$ and $|H/(H \cap Z(G))| = p^2$. □

Exercise 161. Prove that all minimal nonabelian subgroups of an extraspecial p-group G have order p^3.

Hint. Let $S < G$ be minimal nonabelian. Clearly, $G' = H'$. As H/G' is elementary abelian, we get $S/G' \cong E_{p^2}$, and we are done. □

Exercise 162. Study the special p-groups with abelian subgroup A of index p.

Hint. By Lemma 1.1, $|G| = p|G'||Z(G)| = p|G'|^2$. □

Exercise 163. Let a p-group G be not of maximal class and $H < G$ of order $\geq p^p$. Find all possible numbers of subgroups in G of order $p|H|$ which are of maximal class.

Exercise 164. Let A be a maximal normal abelian subgroup of a nonabelian p-group G. Then the following conditions are equivalent:
(a) A is contained in the union of minimal nonabelian subgroups of G.
(b) G is covered by minimal nonabelian subgroups.

Hint. By Lemma 57.1, the set $G - A$ is covered by minimal nonabelian subgroups of G. □

Exercise 165. Let A be a maximal abelian subgroup of a nonabelian group G of exponent p. If all subgroups of G containing A as a subgroup of index p have class > 2, then $A \triangleleft G$.

Solution. Assume that the subgroup A is nonnormal in G. Let $K = A_G$; then $K > \{1\}$ since $Z(G) < A$, and $K < A$, by assumption. By Theorem 10.1, there is in G a normal abelian subgroup B containing K as a subgroup of index p (note that $p > 2$); then $A \cap B = K$ (by the choice of K, we get $B \nleq A$). Write $H = AB$. By Fitting's lemma (Introduction, Theorem 21), we have $\mathrm{cl}(H) = 2$ since $A, B \triangleleft H$ are abelian and $|H : A| = p$, a contradiction. □

Exercise 166. State an analog of Exercise 165 for p-groups of exponent $> p$.

Exercise 167. Study the p-groups of exponent p containing only one normal subgroup of index p^2.

Exercise 168. If a p-group G has exactly p nonabelian maximal subgroups, then one has $\mathrm{d}(G) = 2$.

Solution. By Lemma 1.6, $|\Gamma_1| \leq p + (p + 1) < p^2 + p + 1$ so that $\mathrm{d}(G) = 2$. □

Exercise 169. Study the nonabelian p-groups G containing a maximal subgroup M such that the set $G - M$ is the union of $p(p - 1)$ conjugate G-classes. (Example: G is a 2-group of maximal class with cyclic $M \in \Gamma_1$.)

Exercise 170. If all $\mathcal{A}_2$-subgroups of a p-group G are of maximal class, then G is also of maximal class.

Solution. Let $S < G$ be minimal nonabelian of minimal order and $S < H$, where $|H : S| = p$. Then H is an $\mathcal{A}_2$-subgroup so it is of maximal class. It follows from Exercise 10.10 that G is also of maximal class. □

Exercise 171. A p-group G is Dedekindian if all its subgroups of order $\geq |G'|$ are normal.

Hint. Assume that G is non-Dedekindian. Then, by Proposition 1.23, there is in G' an G-invariant subgroup L of index p such that G/L is non-Dedekindian. Let $H/L < G/L$ be nonnormal; then $|H| \geq |G'|$, a contradiction. □

Let a two-generator p-group G be of order $> p^4$ and let $M \in \Gamma_1$ be of maximal class. Then M_1, the fundamental subgroup of M, is characteristic in M so G-invariant. As

$$|G : M_1| = p^2 = |G : \Phi(G)|,$$

it follows that

$$M_1 = \Phi(G).$$

Indeed, G is of maximal class so $\Phi(G)$ is its unique normal subgroup of index p^2.

Exercise 172. Prove that if G is a two-generator p-group of exponent $> p^2$, then $\Omega_k^\#(G)$ is not of maximal class for $k > 2$.

Hint. We use induction on $|G|$. Let $a \in G$ be of order $> p^2$ and let $a \in H \in \Gamma_1$. Assume that H is of maximal class. Then G is of maximal class (Theorem 12.12 (a)). By Theorem 13.19 (a), $a \in H_1$, where H_1 is the fundamental subgroup of H. By the paragraph, preceding the exercise, $H_1 = \Phi(G) < G_1$. Thus, $\Omega_k^\#(G) \leq G_1$, so that $\Omega_k^\#(G)$ is absolutely regular of exponent $> p^2$. As $\Omega_k^\#(G)$ is of maximal class, its order is $> p^p$ so it is irregular (Theorem 9.5). This is a contradiction. □

Exercise 173. Study the p-groups all of whose noncyclic abelian subgroups are normal.

Hint. If G is non-Dedekindian, it has no subgroup isomorphic to E_{p^3}. □

Exercise 174. At least $p + 1$ maximal subgroups of a nonabelian p-group G contain together all maximal abelian subgroups of G.

Solution. Maximal abelian subgroups cover G. Denote by $\mathcal{M}$ the set of maximal subgroups from the previous sentence. By Remark 116.1, at least $p + 1$ subgroups may cover G so that $|\mathcal{M}| \geq p + 1$. □

Since the centralizers of noncentral elements of G cover a nonabelian p-group G, it follows that the number of such centralizers is at least $p + 1$. Similarly, the number of those members of the set Γ_1 that contain together all maximal cyclic subgroups of G is at least $p + 1$.

Exercise 175. Study the p-groups covered by $p + 1$ minimal nonabelian subgroups.

Hint. Let $G = \bigcup_{i=1}^{p+1} A_i$, where all A_i are minimal nonabelian; then $A_i \in \Gamma_1$ for all i (Remark 116.1). Therefore, if $d(G) = 2$, then G is an $\mathcal{A}_2$-group. Let $d(G) > 2$ and assume that $H \in \Gamma_1$ is neither abelian nor minimal nonabelian. We have $H = \bigcup_{i=1}^{p+1}(H \cap A_i)$, where all $H \cap A_i$ are abelian. It follows that $|H : Z(H)| = p^2$. □

Exercise 176. If the number μ of nonabelian maximal subgroups of a p-group G is $> p + 1$, then $\mu \geq p^2$.

Hint. One has $d(G) > 2$. Use Exercise 1.6. □

Exercise 177. Does there exist a noncyclic abelian p-group of exponent $> p > 2$ all of whose maximal subgroups are pairwise non-isomorphic?

Exercise 178. Let a 2-group G contain a subgroup M of maximal class and index 2.
(i) Then G is either of maximal class or split over M.
(ii) (Mann) If $M \cong \mathrm{SD}_{2^n}$, then G is split over M.

Solution. (i) Assume that G is not split over M. Then $c_1(G) = c_1(M) \equiv 1 \pmod 4$ so G is of maximal class (Theorem 1.17 (a)). In the case under consideration, we have $G \cong \mathrm{SD}_{2^{n+1}}$ and either $M \cong \mathrm{D}_{2^n}$ or else $G \cong \mathrm{Q}_{2^n}$, $n > 3$.

(ii) By Theorem 1.2, G is not a group of maximal class. Then there is in G a normal subgroup $R \cong \mathrm{E}_4$ (Lemma 1.4). As $R \not\leq M$ (indeed, $|M| > 2^3$), there is an involution $a \in R - M$. In that case, $G = \langle x \rangle \cdot M$ (a semidirect product), as asserted. □

Exercise 179. Let G be a p-group of order $> p^4$. Is it true that G possesses a subgroup of order p^4 which is either abelian or minimal nonabelian?

Answer. "No" as an extraspecial group of order p^5 shows. □

Exercise 180 ([Bla7, Lemma 2]). Suppose that a minimal nonabelian p-group $G \not\cong \mathrm{Q}_8$. Let $R(G)$, the intersection of all nonnormal subgroups of G, be $> \{1\}$. Then $p = 2$ and $G \cong \mathrm{M}_2(2, 2)$.

Exercise 181. Classify the abelian p-groups G such that $\exp(\mathrm{Aut}(G)) = p$.

Exercise 182. As we have noted, a nonabelian two-generator critical p-group is minimal nonabelian. Describe the nonabelian p-groups all of whose nonabelian two-generator subgroups are critical.

Exercise 183. Study the irregular p-groups of exponent $p^e > p$ all of whose regular subgroups of exponent p^e are absolutely regular.

Exercise 184. Classify the nonmodular p-groups all of whose subgroups of index $\geq p^3$ are quasinormal.

Hint. If our group G is nonmodular, it contains a non-quasinormal subgroup H of index p^2. As all maximal subgroups of H are quasinormal, H is not generated by them hence it is cyclic. The p-groups with cyclic subgroup of index p^2 are classified in [Nin] (see also § 84). □

Exercise 185. (i) Describe the subgroups of $\mathrm{Aut}(\mathrm{M}_{p^n})$.
(ii) Study the p-groups containing a critical subgroup isomorphic to M_{p^n}.
(iii) Study the p-groups containing a critical subgroup isomorphic to $\mathrm{M}_p(2, 2)$.

Exercise 186. Let G be a nonabelian p-group of order $> p^3$, $p > 2$. If all abelian subgroups of G of order p^3 are isomorphic, then one and only one of the following holds:
(a) $\exp(G) = p$.
(b) G is a 3-group of maximal class and order $\leq 3^5$.
(c) $G = \mathrm{S}(p^3) * C$ (central product), where C is cyclic of order p^2.
(d) G is a minimal nonabelian metacyclic group of order p^4 and exponent p^2.

Hint. Assume that $\exp(G) > p$. Obviously, $\exp(G) = p^2$ and all abelian subgroups of order p^3 in G are of type (p^2, p). □

Exercise 187. Let $G > \{1\}$ be a solvable group and let A be a maximal characteristic abelian subgroup of G. Set $C = \mathrm{C}_G(A)$ and assume that $C > A$. Let $H/A = \mathrm{Sc}(C/A)$ be the socle of C/A. Then H is characteristic in G. Prove that $\mathrm{C}_G(H) = \mathrm{Z}(H) = A$.

Solution. It suffices to prove the last equalities in the case when H is nonabelian. One has $\mathrm{Z}(H) = A$, by the choice of A. Write $L = \mathrm{C}_G(H)$; then L is characteristic in G. One has

$$L \leq \mathrm{C}_G(A) = C.$$

It suffices to prove that $L = A$. Assume that this is false. Then $\mathrm{Sc}(L/A)$ is a nonidentity characteristic subgroup of G/A. In that case, since $L \cap H = A$, we get

$$LH/A = (L/A) \times (H/A) \leq \mathrm{Sc}(C/A) = H/A,$$

which is a contradiction. □

Exercise 188. Let G be a minimal nonmetacyclic p-group. Study the p-groups H such that $\mathrm{cd}(H) = \mathrm{cd}(G)$ and $\mathrm{k}(H) = \mathrm{k}(G)$. Consider in detail the case when $|G| = 2^5$.

Exercise 189. Let G be a nonabelian p-group of exponent $> p > 2$ all of whose minimal nonabelian subgroups are of exponent p. Then $A = \mathrm{H}_p(G)$ is abelian of index p in G.

Solution. By the footnote to Problem 115 in Volume 1, G has an abelian subgroup A of index p. By Lemma 57.1, for any $x \in G - A$ there is $a \in A$ such that $S = \langle a, x\rangle$ is minimal nonabelian. By hypothesis, $S \cong \mathrm{S}(p^3)$ is of exponent p so that $\mathrm{o}(x) = p$. It follows that all elements of the set $G - A$ have order p, and we conclude that $A = \mathrm{H}_p(G)$ since $\exp(G) > p$. □

Exercise 190. Study the p-groups all of whose noncyclic metacyclic subgroups are normal.

Exercise 191. A group G of exponent p^e is abelian if and only if all its elements of order p^e are contained in its center.

Solution. If $x \in G$ is of order p^e and $y \in G$, then

$$\Omega_e^{\#}(\langle x, y\rangle) = \langle x, y\rangle \leq \mathrm{Z}(G),$$

and we are done. □

Exercise 192. Let $H \in \Gamma_1$ be an isolated subgroup of a p-group G of maximal class and order $> p^{p+1}$. Then $H = G_1 = \mathrm{H}_p(G)$.

Hint. Assume that H is irregular. The fundamental subgroup G_1 is absolutely regular and generated by its cyclic subgroups of order $> p$. Therefore, there is a cyclic $C < G_1$ of order $> p$ that is not contained in H. Then $C \cap H > \{1\}$ so H is not isolated. Thus, we must have $H = G_1$. As all elements of the set $G - G_1$ have order p, we get $G_1 = \mathrm{H}_p(G)$. □

Exercise 193. Classify the irregular p-groups G all of whose maximal subgroups are either absolutely regular or of exponent p.

Hint. One has $\exp(G) > p$ (Theorem 7.2 (b)). Let $H \in \Gamma_1$ be of exponent $> p$; then H is absolutely regular. If G is of maximal class, then $|G| = p^{p+1}$ since there in Γ_1 a member of exponent p (Theorems 9.5 and 9.6). Next assume that G is not of maximal class. Then, by Theorem 12.1 (b), $G = H\Omega_1(G)$, where $\Omega_1(G)$ is of order p^p and exponent p. If $\Omega_1(G) < T \in \Gamma_1$, then T is neither absolutely regular not of exponent p, a contradiction. Thus, in the case under consideration we also have $|G| = p^{p+1}$. □

Exercise 194. If $p > 3$ and H is a normal metacyclic subgroup of an irregular p-group G of maximal class, $|G| > p^4$, then $|H| < p^3$.

Hint. Clearly, we have $H < G_1$. However, any G-invariant subgroup of order p^3 has exponent p. □

Exercise 195. Let W be an abelian subgroup of type $(4, 4)$ of a 2-group G and let $C = \mathrm{C}_G(W)$. If all abelian subgroups of C containing W are two-generator, then $\Omega_2(C) = W$ and C is metacyclic.

Solution. If an involution $a \in C$, then $A = \langle a, W\rangle$ is abelian. It follows from $\mathrm{d}(A) = 2$ that $a \in W$, and we conclude that $\Omega_1(C) = \Omega_1(W)$. Let $b \in C$ be of order 4 and set $B = \langle b, W\rangle$. The subgroup B is abelian so that $\mathrm{d}(B) = 2$. It follows that $b \in W$ so that $\Omega_2(C) = W$. Assume that C is nonmetacyclic. Then there is in C a minimal nonmetacyclic subgroup M. Since $\exp(M) \le 4$ (Theorem 66.1), it follows that $M \le W$, a contradiction. Thus, M does not exist so that C is metacyclic.[3] □

Exercise 196. Study the p-groups of exponent $> p^2$ all of whose noncyclic abelian subgroups of order p^4 are isomorphic. (See Exercise 186.)

Exercise 197. Is it true that if a p-group G contains two distinct maximal characteristic abelian subgroups, then it has two distinct proper critical subgroups?

Exercise 198. Let G be a representation group of the abelian group $\mathrm{C}_{p^2} \times \mathrm{C}_{p^2}$. Is it true that $\mathrm{Z}(G) \cong \mathrm{C}_{p^2} \cong G'$?

3 A similar result also holds for p-groups, $p > 2$.

Exercise 199. A subgroup H of a p-group G is said to be quasi-critical if it is characteristic in G, $[G, H] \leq Z(H)$ and $C_G(H) = Z(H)$. Present a quasi-critical p-group which is not critical. Is it true that the representation group of a homocyclic p-group of rank > 1 is quasi-critical? Classify the abelian p-groups all of whose representation groups are quasi-critical.

The critical subgroups are more important than quasi-critical ones since the latter class has a more complicated structure.

Exercise 200. Find all primary $\mathcal{A}_2$-groups that are critical.

Exercise 201. Describe the p-groups with isolated center (derived subgroup, Frattini subgroup). (The derived subgroup of a nonmetacyclic minimal nonabelian p-group is isolated, by Lemma 65.1.)

Exercise 202. (a) Is it true that the representation group of an elementary abelian 2-group is critical?
(b) Find a necessary condition for the representation groups of a noncyclic abelian p-group A to be quasi-critical (see Exercise 199).

Exercise 203. Find all primary $\mathcal{A}_2$-groups containing a nontrivial isolated subgroup.

Exercise 204. Study the nonabelian critical p-groups containing a characteristic subgroup of index p.

Exercise 205. Classify the nonabelian p-groups in which the intersection of any two nonincident nonnormal subgroups is cyclic.

Exercise 206. Classify the modular $\mathcal{A}_2$-groups.

Exercise 207. Study the group of those automorphisms of a p-group G that fix all G-invariant subgroups.

Exercise 208. Describe the characteristic subgroups in primary $\mathcal{A}_n$-groups, $n = 1, 2$.

Exercise 209. Let $G = \Sigma_{p^2} \in \mathrm{Syl}_p(S_{p^2})$. Find the numbers of minimal nonabelian subgroups in G.

Exercise 210 ([FT, Lemma 8.3]). The following statements hold:
(a) Let G be a p-group, $p > 2$, and let $A \triangleleft G$ be maximal abelian with maximal $d(A)$. Then $\Omega_1(C_G(\Omega_1(A)) = \Omega_1(A)$.
(b) Let G be a p-group and let $A \triangleleft G$ be maximal abelian of exponent $p^n > 2$ with maximal $d(A)$. Then $\Omega_n(C_G(\Omega_n(A)) = \Omega_n(A)$.

Solution. (a) Assume that $x \in \Omega_1(C_G(\Omega_1(A)) - \Omega_1(A))$ has order p. Set $E_0 = \Omega_1(A) \times \langle x \rangle$. By Theorem 10.1, there is in G a G-invariant elementary abelian subgroup E_1 such that $E < E_1$ and $|E_1 : E| = p$. This is a contradiction since $d(E_1) = d(A) + 1 > d(A)$. □

The result of the following exercise is a partial case of Theorem 10.4.

Exercise 211. If a p-group G, $p > 2$, contains a subgroup $E \cong \mathrm{E}_{p^3}$, then it contains a normal subgroup isomorphic to E.

Solution. By Lemma 1.4, there is in G a normal subgroup $R \cong \mathrm{E}_{p^2}$. Write $H = RE$. Then $\mathrm{C}_H(R) > R$ so that R is contained in subgroup $E_1 \cong \mathrm{E}_{p^3}$ of H. Now the result follows from Theorem 10.1. (Compare this solution with the proof of the corresponding result in [GLS, Lemma 10.17]. A similar result does not hold for $p = 2$.) □

Exercise 212. Let G be a group of exponent p. If there is in G a subgroup $E \cong \mathrm{E}_{p^4}$, then there is in G a normal subgroup isomorphic to E.

Solution. Assume that this is false. In that case $p > 2$, and $|G : E| \geq p^3$, by Exercise 1.7; then $|G| \geq p^7$. By Exercise 211, there is in G a normal subgroup $R \cong \mathrm{E}_{p^3}$. By Theorem 10.1,

$$\mathrm{C}_G(R) = R.$$

Then G/R, being a subgroup of $\mathrm{Aut}(R)$, has order $\leq p^3$. In that case,

$$|G| \leq |R||\mathrm{Aut}(R)|_p \leq p^6 < p^7,$$

a contradiction. (This is a partial case of Theorem 10.5.) □

Exercise 213. Suppose that a p-group G contains an abelian subgroup A of type (p^2, p). Is it true that G contains a normal subgroup isomorphic to A?

Hint. The group $G = \Sigma_{p^2} \times \mathrm{C}_p$, $p > 2$, has an abelian subgroup A of type (p^2, p) but has no normal subgroup isomorphic to A. Find a similar example for $p = 2$. □

Exercise 214. Is it true that the following assertions for a nonabelian p-group G are equivalent?

(a) If $A \leq G$ is minimal nonabelian, then all maximal subgroups of A are maximal abelian in G.

(b) If $B \leq G$ is nonabelian, then all maximal abelian subgroups of B are maximal abelian in G.

Exercise 215. Let W be an abelian subgroup of type $(4, 2)$ of a 2-group G and let $C = \mathrm{C}_G(W)$. Is the following assertion true? If all abelian subgroups of C containing W are two-generator, then either $W < W_1 \leq C$, where W_1 is abelian of type $(4, 4)$ or $\Omega_2(C) = W$ and C is either abelian of type (p^k, p) or isomorphic to M_{p^n}.

Exercise 216. Let G be a p-group of order $> p^3$, $p > 2$, and let $a_n(G)$ be the number of abelian subgroups of order p^n in G. Is it true that $a_3(G) \equiv 1 \pmod p$?

Exercise 217. Let G be an extraspecial p-group of order $> p^3$ and exponent p^2, $p > 2$. Then:

(a) $G = A_1 * \cdots * A_m$, where A_i are minimal nonabelian of order p^3, $A_i \cap A_j = \mathrm{Z}(G)$.

(b) $|G| = p^{2m+1}$.

(c) The subgroups A_i can be ordered so that $A_1 \cong \cdots \cong A_{m-1} \cong \mathrm{S}(p^3)$, $A_m \cong \mathrm{M}_{p^3}$.

(d) $G = B_1B_2$, where B_1, B_2 are maximal abelian subgroups of order p^{m+1} such that $B_1 \cap B_2 = Z(G)$, $\exp(B_1) = p$, B_2 has exactly $m - 1$ invariants that equal p and one invariant that equals p^2.

(e) The group G has exactly $p - 1$ nonlinear irreducible characters, say, $\chi_1, \dots, \chi_{p-1}$, and all of them have degree p^m. One has $\chi_i = \mu_i^G = \nu_i^G$, where $\mu_i \in \mathrm{Lin}(B_1)$ and $\nu_i \in \mathrm{Lin}(B_2)$ for $i = 1, \dots, p-1$. Next, $\chi_i(1)\chi_i = \lambda_1^G$, where $\lambda_i = \mathrm{Lin}^*(Z(G))$, for all $i \leq p - 1$.

(f) All maximal abelian subgroups of G have order p^{m+1}.

Hint. Let $S < G$ be minimal nonabelian; then $S' = G'$ and $\exp(S/G')$, being a subgroup of rank 2 of the elementary abelian p-group G/G', has order p^2. It follows that all minimal nonabelian subgroups of G have order p^3.

(a) This follows from the previous paragraph and Lemma 4.3.

(b) This follows from (a) and the product formula.

(c) As $\exp(G) = p^2$, one may assume that $\exp(A_m) = p^2$; then $A_m \cong \mathrm{M}_{p^3}$. Assume that $A_i \cong \mathrm{M}_{p^3}$ for some $i < m$. Set $H = A_i * A_m$; then H is extraspecial of order p^5. By Theorem 7.1 (b), Theorem 7.2 (d) and Lemma 1.1, $\Omega_1(H)$ is nonabelian of order p^4 and exponent p. Let $S_i < \Omega_1(H)$ be minimal nonabelian. Then $S_i \cong \mathrm{S}(p^3)$. By Lemma 4.3, $H = S_i * S_m$, where $S_m \cong \mathrm{M}_{p^3}$. Replace in the decomposition of G the factor $A_i * S_m$ by $S_i * A_m$. Making this for all $i < m$, if necessary, we prove (c).

(d) For $i < m$, set $A_i = R_{i1}R_{i2}$, where $R_{i1} \cong R_{i2} \cong \mathrm{E}_{p^2}$, and $A_m = \Omega_1(A_m)C$, where $C \cong \mathrm{C}_{p^2}$. Next, set

$$B_1 = R_{11} * \cdots * R_{m-1,1} * \Omega_1(A_m), \quad B_2 = R_{12} * \cdots * R_{m-1,2} * C.$$

Then

$$B_1 \cong \mathrm{E}_{p^{m+1}}, \quad B_2 \cong \mathrm{E}_{p^{m-1}} \times \mathrm{C}_{p^2}, \quad B_1 \cap B_2 = Z(G)$$

are abelian, and hence, by the product formula, $G = B_1B_2$, completing the proof of (d). Note that $B_1, B_2 \triangleleft G$ since $G' = B_1 \cap B_2$.

(e) The class number $\mathrm{k}(G)$ of G equals $p^{2m} + (p - 1)$. Therefore,

$$|\mathrm{Irr}_1(G)| = |\mathrm{Irr}(G)| - |\mathrm{Lin}(G)| = p - 1,$$

and let the degrees of nonlinear irreducible characters of G are p^{a_i}, $i = 1, \dots, p - 1$. In that case,

$$p^{2m+1} = |G| = p^{2m} + \sum_{i=1}^{p-1} p^{2a_i}. \tag{1}$$

By Ito's theorem on degrees (see Theorem 17 from Introduction) and (c), $a_i \leq m$ for all i. Therefore, it follows from (1) that $a_i = m$ for all i so that $\mathrm{cd}(G) = \{1, p^m\}$. Therefore, B_1, B_2 are maximal abelian subgroups of G, by the above mentioned Ito's theorems.

(f) Let $U < G$ be maximal abelian; then $U \triangleleft G$ since $G' = Z(G) < U$. By (e), we have $|U| \leq p^{m+1}$. Let $|U \cap A_i| = p^2$ for $i \leq k$ and $U \cap A_i = G'$ for $i > k$. Then

$$U = (U \cap A_1) * \cdots * (U \cap A_k).$$

If $k < m$, then some maximal subgroup L of A_{k+1}, $k < m$, centralizes $U \cap A_i$ for $i \leq k$ so $U * L > U$ is abelian, a contradiction. Thus, $k = m$ so that $|U| = p^{m+1}$. □

Let $\mathrm{ES}(2, 2n+1)$ be an extraspecial group of order 2^{2n+1}. Then it follows from the Frobenius–Schur formula on the number of involutions that (note that $\mathrm{Irr}_1(G) = \{\chi\}$ and the Frobenius–Schur indicator $\nu_2(\chi) \in \{-1.1\}$)

$$c_1(\mathrm{ES}(2, 2n+1)) \in \{2^{2n} - 2^n - 1, 2^{2n} + 2^n - 1\}.$$

Exercise 218. Let G be a p-group of class 2 and $m = p$ if $p > 2$ and $m = 4$ if $p = 2$. If $\exp(G/Z(G)) = p$, then $(xy)^m = x^m y^m$ for all $x, y \in G$.

Solution. It follows from Exercise 13 (a) in Introduction and induction on i that

$$(xy)^i = x^i y^i [y, x]^{i(i-1)/2} = x^i y^i [y, x^{i(i-1)/2}] \quad (i > 0). \tag{2}$$

Let $p > 2$ and $i = p$; then $[y, x^{p(p-1)/2}] = 1$, and (2) yields

$$(xy)^p = x^p y^p$$

since $x^{p(p-1)/2} \in Z(G)$. Now let $p = 2$; then $m = 4$. By (2),

$$(xy)^4 = x^4 y^4 [y, x]^6 = x^4 y^4 [y, x^6] = x^4 y^4 \tag{3}$$

since $x^6 \in Z(G)$. □

Exercise 219. Let A be a p'-group of automorphisms of an abelian p-group G. If G is not a direct product of two nontrivial A-invariant subgroups, then G is homocyclic.

Hint. Apply Theorem 6.1. □

Exercise 220. Study the p-groups G such that, whenever A is abelian and $|A|^2 < |G|$, then $A \lhd G$.

Exercise 221. Let G be an extraspecial 2-group of order $> 2^3$. Then $\Omega_1(G) = G = \Omega_2^{\#}(G)$. (See § 245.)

Exercise 222. Let a p-group G be an $\mathcal{A}_2$-group with $|G'| = p$. Then $G = AZ(G)$ and $|G : A| = p$, where $A < G$ is an $\mathcal{A}_1$-subgroup.

Solution. By Lemma 65.2 (a), $d(G) = 3$. By Lemma 4.3, $G = AZ(G)$, where A is minimal nonabelian. Clearly, $|G : A| = p$. □

Exercise 223. A nonabelian 2-group $G = \Omega_1(G)$ is generated by subgroups isomorphic to D_8.

Solution. Let H be generated by all subgroups of G that are isomorphic to D_8; then H is nonabelian. Indeed, if $x, y \in G$ are non-commuting involutions, then $\langle x, y\rangle$ contains a subgroup isomorphic to D_8. Assume that $H < G$. Take an involution $x \in G - H$. Assume that there is in H an involution y such that $xy \neq yx$. Then $S = \langle x, y\rangle$ is dihedral and $S \nleq H$. Since S is generated by subgroups isomorphic to D_8, we get $S \leq H$, which is a contradiction. Thus, x centralizes all involutions in $H = \Omega_1(H)$. Take in H a dihedral subgroup F and set $L = F \times \langle x \rangle$; then $L \nleq H$. As L is generated by subgroups isomorphic to D_8, we get $L \leq H$, which is a contradiction. Thus, $H = G$. □

Exercise 224. Let $p > 2$. Is it true that a nonabelian p-group $G = \Omega_1(G)$ is generated by subgroups isomorphic to $S(p^3)$?

Exercise 225. A p-group $G = \Omega_1(G)$ if and only if there is $x \in G - H$ of order p for any $H \in \Gamma_1$.

Hint. Assume that $\Omega_1(G) = M < G$. If $M \leq H \in \Gamma_1$, then the set $G - H$ has no elements of order p, a contradiction. □

Exercise 226. Let $\mathcal{AS}$ be the set of maximal abelian and minimal nonabelian subgroups of a nonabelian p-group G. Study the structure of G if all members of the set $\mathcal{AS}$ have the same order.

Hint. Let $E < G$ be elementary abelian of maximal order and $N = N_G(E)$. Assume that $x \in N - E$ is of order p and set $H = \langle x, E \rangle$. Then H is nonabelian. By Lemma 57.1, there is $a \in E$ such that $S = \langle a, x \rangle$ is minimal nonabelian. Then $S \cong S(p^3)$ or D_8 so all members of the set $\mathcal{AS}$ have order p^3. If x does not exist, then $E = \Omega_1(N)$ so that $N = G$ hence $\Omega_1(G)$ is elementary abelian. □

Exercise 227. Let G be a p-group. Prove that $[K_i(G), K_j(G)] \leq K_{i+j}(G)$.

Hint. Use induction on j and the Three Subgroups Lemma. □

Exercise 228. Let H be a critical p-group and let P be a Sylow p-subgroup of the holomorph of H. Find sufficient condition for H to be critical in P. Consider the cases when H is (i) abelian, (ii) extraspecial, (iii) minimal nonabelian.

Exercise 229. Given an abelian p-group G, find the minimal numbers of generators of a Sylow p-subgroups of $\mathrm{Aut}(G)$ and the holomorph of G.

Exercise 230. Classify the extraspecial p-groups G in which G' is the unique nontrivial characteristic subgroup.

The following exercise is in fact an open problem.

Exercise 231. Suppose that all maximal abelian subgroups of a nonabelian p-group G are normal. Is it true that the class (derived length) of G is bounded?

Exercise 232. Let H be a proper normal subgroup of order p^n of a p-group G. Assume that there is in H only one G-invariant subgroup of order p^i for $i = 1, \dots, n-1$. Describe the structure of H.

Exercise 233. It is known that any extraspecial p-group G of order $> p^3$ and exponent p^2, $p > 2$, has at least three nontrivial characteristic subgroups of different orders: G', $\Omega_1(G)$ of order $\frac{1}{p}|G|$ and $Z(\Omega_1(G)) \cong E_{p^2}$. Classify those G that have exactly three nontrivial characteristic subgroups.

Exercise 234. Each irregular p-group G contains at least $p+1$ pairwise non-conjugate maximal regular subgroups.

Hint. Maximal regular subgroups cover G. Use § 116. □

Exercise 235. Does there exist a p-group admitting an irredundant covering by p^2 its subgroups of maximal class and index p?

Exercise 236. Find the derived subgroup of the holomorph $\mathcal{H}$ of the abelian group A of type $(4, 2)$.

Hint. The subgroups A and $\mathcal{H}'$ are nonincident since $\mathrm{Aut}(A) \cong \mathrm{D}_8$ and $\mathcal{H}$ is not of maximal class. □

Exercise 237. If $H_1, \dots, H_{p+1} < G$ cover G, then all $H_i \in \Gamma_1$.

Hint. See Remark 116.1. □

Exercise 238. Given n, present an abelian p-group containing all types of abelian p-groups of order $\leq p^n$. Find the minimal order $p^{\mathrm{o}(n)}$ of such a group. Let $\mathcal{U}_n$ be the set of such groups of order $p^{\mathrm{o}(n)}$ and let $U_n \in \mathcal{U}_n$. Is it true that the type of U_n is determined uniquely? Let V_n be a nonabelian p-group of minimal order, which we denote $p^{w(n)}$, that contains all types of abelian p-groups of orders $\leq p^n$; let $\mathcal{V}_n$ be the set of all such groups of order $p^{w(n)}$. Compare the numbers $\mathrm{o}(n)$ and $w(n)$. Describe the set of all groups of $\mathcal{U}_n$ and $\mathcal{V}_n$. Find $\mathrm{o}(n)$ and $w(n)$ for $n \leq 7$. Is it true that $\mathrm{o}(7) = 16 = w(7)$ and $\mathrm{o}(8) = 20 = w(8)$?

Discussion. Obviously, the abelian group A_n of type $(p, p^2, \dots, p^n)$ contains all abelian p-groups of orders $\leq p^n$. Any abelian p-group satisfying the condition is a subgroup of A_n. The abelian group of type (p, p, p^3), a proper subgroup of A_3, also satisfies the condition. Clearly, $\mathrm{o}(3) = p^5$. Next (check!),

$$\mathrm{o}(4) = 8, \quad \mathrm{o}(5) = 10, \quad \mathrm{o}(6) = 14, \quad \mathrm{o}(7) = 16$$

and

$$w(2) = 3 = \mathrm{o}(2), \quad w(3) = 5 = \mathrm{o}(3).$$

One has $w(n) = \mathrm{o}(n)$ for $n \leq 7$. The nonabelian group

$$\mathrm{M}_p(1, 6, 1) \times \mathrm{C}_{p^3} \times \mathrm{C}_{p^2} \times \mathrm{C}_p$$

of order p^{14} contain all abelian p-groups of order $\leq p^6$. □

Sketch of the solution. Let us consider the following partition:

$$\begin{aligned} M = {} & 100 + 50 + 33 + 25 + 20 + 16 + 14 + 12 + 11 + 10 \\ & + 9 + 8 + 2 \cdot 7 + 2 \cdot 6 + 4 \cdot 5 + 5 \cdot 4 + 8 \cdot 3 + 17 \cdot 2 + 50 \cdot 1. \end{aligned}$$

Thus, $\mathrm{o}(100) = 482$. The number 9 stands on eleventh position since there is a partition containing $11 = [\frac{100}{9}]$ parts equal 9. The part 3 stands on $33 = [\frac{100}{3}]$-th position since there is a partition containing 33 parts equal 3. This explains how to obtain $o(n)$ for an arbitrary n. □

If the abelian p-group possesses the desired property, then $|\Omega_1(A)| = p^n$ since $E_{p^n} < A$. Similarly,

$$|\Omega_2(A)| = p^{n+[n/2]}, \quad |\Omega_3(A) = p^{n+[n/2]+[n/3]},$$

and so on. These considerations led M. Roitman to the following result:

Theorem. *The abelian p-group of type $(p^{k_1}, p^{k_2}, \dots, p^{k_n})$, where $k_i = [\frac{n}{i}]$ for all $i \le n$, is the unique abelian p-group of minimal order containing all abelian p-groups of orders $\le p^n$. Thus, $o(n) = p^N$, where $N = \sum_{i=1}^{n} [\frac{n}{i}]$.*

Exercise 239. Find the minimal order of a p-group G containing all types of (i) minimal nonabelian subgroups of order $\le p^4$, (ii) subgroups of maximal class and order $\le p^4$, (iii) two-generator subgroups of order $\le p^4$.

Exercise 240. Prove that a p-group in which any two subgroups of equal order are permutable is modular.

Solution. Assume that G is nonmodular. Then it has a section $M/N \in \{D_8, S(p^3)\}$. As D_8 and $S(p^3)$ does not satisfy the condition, we get a contradiction. Thus, G is modular. □

Exercise 241. Let G be a p-group with he irregular subgroup $\Phi(G) = \Omega_1(\Phi(G))$ (then $|\Phi(G)| > p^{p+1}$). Is it true that $\Phi(G)$ contains a G-invariant subgroup of order p^{p+1} and exponent p?

Answer. Yes. Indeed, by Lemma 1.4, $\Phi(G)$ is not of maximal class hence it contains a G-invariant subgroup R of order p^p and exponent p (Theorem 13.5). Let $x \in \Phi(G) - R$ be of order p; by Lemma 1.4, $|R \cap Z(\Phi(G))| > p$. Then the subgroup $\langle x, R\rangle$ is regular so exponent p. □

Exercise 242. Does there exist a p-group G containing a non-isolated subgroup H such that, whenever $H < L < G$ and $|L : H| = p$, then all elements of the set $L - H$ have order p?

Exercise 243 (Theorem 12.12 (a)). Let a p-group G be not of maximal class and let $M \in \Gamma_1$ be of maximal class. Then $d(G) = 3$.

Solution. Assume that $d(G) = 2$. Let $F < M$ be of index p. Then, by Remark 10.5, $F \lhd G$. Therefore, $G/\Phi(M)$ is abelian of type (p^2, p). Let $L < \Phi(M)$ be G-invariant of index p. Then G/L is minimal nonabelian (Lemma 65.2 (a)). It follows that M/L is abelian of order p^3 so M is not of maximal class, contrary to the hypothesis. □

Exercise 244. Suppose that G is a p-group of order $> p^6$. If all normal abelian subgroups of G of order p^4 are two-generator, then G has no normal subgroup isomorphic to E_{p^3}. (For such groups see Theorem 13.7 and § 50.)

Solution. Assume that there is in G a normal subgroup $E \cong \mathrm{E}_{p^3}$. Then $\mathrm{C}_G(E) > E$ since $|\mathrm{Aut}(E)|_p = p^3$. Let A/E be a G/E-invariant subgroup of order p in $\mathrm{C}_G(E)/E$. Then A is G-invariant abelian of order p^4, $E < A$ hence $\mathrm{d}(A) > 2$, contrary to the hypothesis. □

Exercise 245. Let $H < G$ be a proper nonidentity Hall subgroup of a group G and let, whenever a cyclic $C < G$ is such that $C \cap H > \{1\}$, then $C \leq H$ (such an H is said to be isolated in G).

(i) If, in addition, $H < G$ is normal, then G is a Frobenius group with kernel H.

(ii) If $H < \mathrm{N}_G(H)$, then H is nilpotent, by Thompson's theorem on groups admitting a fixed-point-free automorphism of prime order.

Exercise 246. Let a 2-group G be such that G/G' is abelian of type $(2, 2^2, \dots, 2^n)$. Prove that $\mathrm{Aut}(G)$ is a 2-group. The same is true provided G/G' is abelian of type $(2^{a_1}, 2^{a_2}, \dots, 2^{a_n})$, where $a_1 < a_2 < \cdots < a_n$. Is the following assertion true? If a subgroup $M \leq \Phi(G)$ is characteristic in G and $\mathrm{Aut}(G/M)$ is a 2-group, so is $\mathrm{Aut}(G)$.

Solution of correctness of the last assertion. Assume that $\phi \in \mathrm{Aut}(G)$ is of odd order. Then ϕ induces the identity automorphism on G/M. Let $A = \langle \phi \rangle$ and let $W = A \cdot G$ be the natural semidirect product of A and G. Then AM is normal in W. By Frattini's lemma,

$$W = \mathrm{N}_W(A)AM = \mathrm{N}_W(A)M.$$

Since $M \leq \Phi(G) \leq \Phi(W)$, we get $W = \mathrm{N}_W(A) = A \times G$, and we are done. □

Exercise 247. Describe the minimal nonabelian 2-groups G such that $\mathrm{Aut}(G)$ is a 2-group. (See [Mal6] and [Kur].)

Exercise 248. Describe the group $\mathrm{Aut}(G)$, where G is minimal nonmetacyclic group of order 2^5. Is it true that $3 \mid |\mathrm{Aut}(G)|$?

Exercise 249. Let $H \cong \mathrm{M}_2(2, 2)$ be a subgroup of index 2 in a 2-group G. Describe the case when G is not split over H.

Exercise 250. Let G be a p-group of exponent $> p > 2$. If all cyclic subgroups of G of order $> p$ are normal, then G is regular.

Solution. It suffices to prove that $H = \langle x, y \rangle$ is regular for all $x, y \in G$. If one of the elements x, y has order $> p$, then H, being metacyclic, is regular. Now let $\mathrm{o}(x) = \mathrm{o}(y) = p$. Take $z \in G$ of order $> p$. Then $K = \langle x, z \rangle$ is regular and G-invariant, by Theorem 1.2. Then $\Omega_1(K) \cong \mathrm{E}_{p^2}$ is G-invariant. Similarly, $L = \Omega_1(\langle y, z \rangle) \cong \mathrm{E}_{p^2}$ is G-invariant. In that case $F = \Omega_1(K)\Omega_1(L)$ is of class $\leq 2 < p$ (Fitting's lemma) is regular (Theorem 7.1 (b)) so $H \leq F$ is regular. □

Exercise 251. Find $t = |\mathrm{Aut}(G)|$, where

$$G = \langle a, b \mid \mathrm{o}(a) = 2^2,\ \mathrm{o}(b) = 2^n,\ n > 2,\ a^b = 2^3 \rangle$$

is metacyclic of order 2^{n+2}.

Hint. Find the number of bases $\{u, v\}$ of the group G such that $o(v) = 4$, $o(u) = 2^n$ and $\langle v\rangle \lhd G$. □

Exercise 252. Find the irregular p-groups G of maximal class such that $c_k(G) = c_k(R)$ $(k = 1, 2, \dots)$ for some regular p-group R.

Hint. Use counting theorems for p-groups of maximal class from § 12 and Theorem 13.2 to check that the following assertions are correct: $p > 2$, $|G| = p^{p+1}$ and $|\Omega_1(G)| \in \{p^{p-1}, p^p\}$. □

Exercise 253. Study the p-groups G of exponent $> p > 2$, all of whose nonabelian maximal subgroups have exponent p (it is assumed that G is neither abelian nor minimal nonabelian).

Hint. One has $\exp(G) = p^2$ since there is in the set Γ_1 a nonabelian member. As there are in Γ_1 two distinct nonabelian members (Exercise 1.6), we get $\Omega_1(G) = G$ hence G is irregular (Theorem 7.2 (b)) so that $\mathrm{cl}(G) \ge p$ (Theorem 7.1 (b)). As G is minimal irregular, we get $d(G) = 2$. Note that G has exactly one abelian subgroup, say A, of index p since $\mathrm{cl}(G) \ge p > 2$. By hypothesis, $\exp(A) = p^2$. Let $H \in \Gamma_1$ be nonabelian. Then $A \cap H = \Phi(G)$ is a maximal subgroup of A of exponent p. Therefore, $\mho_1(A) \lhd G$ has order p. All maximal subgroups of $G/\mho_1(A)$ have exponent p so that $\exp(G/\mho_1(A)) = p$, i.e., $\mho_1(G) = \mho_1(A)$. As $\exp(A/\mho_1(A)) = p$, it follows that $\mathrm{cl}(G/\mho_1(A)) \le p-1$ (Exercise 387 (a)) so that $\mathrm{cl}(G) = p$ since G is irregular (Theorem 7.1 (b)). □

Exercise 254. Given n, if all minimal nonabelian subgroups of a nonabelian p-group G have exponent $> p^n > 2$, then $\Omega_n(G)$ is abelian.

Solution. Let $A < G$ be a G-invariant abelian subgroup of exponent $\le p^n$ of maximal order. We claim that $A = \Omega_n(G)$. Assume that there is in $G - A$ an element x of order $\le p^n$. Set $H = \langle x, A\rangle$. By Theorem 10.1, the subgroup H is nonabelian (otherwise, there is an abelian G-invariant subgroup $> A$ of exponent $\le p^n$). Then, by Lemma 57.1, there is $a \in A$ such that the subgroup $S = \langle a, x\rangle$ is minimal nonabelian. Since $\Omega_n(S) = S$ and $p^n > 2$, it follows from Exercise 96 that $\exp(S) \le p^n$, contrary to the hypothesis. □

Exercise 255. If a regular p-group G contains an isolated maximal subgroup H, then $\exp(G) = p$.

Hint. One has $G = \langle G - H\rangle \le \Omega_1(G)$. Therefore, $\Omega_1(G) = G$ implies $\exp(G) = p$ (Theorem 7.2 (b)). □

Exercise 256. Classify the nonmetacyclic p-groups G all of whose proper epimorphic images are metacyclic.

Solution. If $d(G) > 2$, then $\Phi(G) = \{1\}$ so that $G \cong E_{p^3}$. Now let $d(G) = 2$. Assume that $p > 2$. Then $|G/\mho_1(G)| \ge p^3$, and we conclude that $|G| = p^3$. Now let $p = 2$. Then $\overline{G} = G/\mho_2(G)$ is nonmetacyclic (Supplement 1 to Corollary 36.6) so that $\mho_2(G) = \{1\}$

and G is nonabelian. Assume that $|G'| > 2$. If $L \le G' \cap Z(G)$ is of order 2, then G/L is metacyclic so is G, by Theorem 36.1, a contradiction. Thus, $|G'| = 2$ and so G is minimal nonabelian (Lemma 65.2 (a)). In that case, as easily seen, G is nonmetacyclic of order 2^4, i.e., $G \cong M_2(2, 1, 1)$. □

Remark 1 ([Man2]). We will prove that all minimal nonabelian subgroups of a nonabelian group G of exponent p are normal if and only if either $|G'| = p$ or G is of maximal class of order p^4. Clearly, $p > 2$. Assume that $|G| > p^4$. Let $S < G$ be minimal nonabelian; then $S \lhd G$ is of order p^3. In that case, all subgroups of G, containing S, are G-invariant (Theorem 10.28) hence $S \ge G' = \Phi(G)$ (Theorem 1.20). By Lemma 1.4, G' is abelian hence $G' < S$. Assume that $|G'| = p^2$. In that case G' is contained in all minimal nonabelian subgroups of G, $G' \nleq Z(G)$ and we conclude that $C = C_G(G') \in \Gamma_1$. As C has no minimal nonabelian subgroup since $G' \le Z(C)$, it follows that C is abelian. By Lemma 1.1, one has $|G : Z(G)| = p|G'| = p^3$; then

$$|Z(G)| = \frac{|G|}{p^3} \ge p^2.$$

Therefore, there is $x \in Z(G) - S$. Set $H = S \times \langle x \rangle$. Then $|\Phi(H)| = p < p^2 = |G'|$ hence among of p^2 nonabelian subgroups of order p^3 in H there is one not containing G' (indeed, $H/G' \cong E_{p^2}$ contains exactly $p + 1 < p^2$ subgroups of order p), a contradiction. Thus, if $|G| > p^4$, then $|G'| = p$. A similar argument shows that the same conclusion holds provided all minimal nonabelian subgroups of a nonabelian p-group G (of exponent $> p$) have order p^3 and all such subgroups are G-invariant.

Let a group G be of order $> p^3$ and exponent p and suppose that all subgroups of G of order p^2 are normal. Let $H \le G$ be of order p^4 and let $E \cong E_{p^3}$ be a subgroup of H. As all subgroups of E of order p as intersections of subgroups of order p^2 are H-invariant, it follows that $E \le Z(H)$ so that H is abelian. Thus, all subgroups of G of order p^4 are abelian so G has no minimal nonabelian subgroup and this implies that G is abelian.

Remark 2. Let G be a nonabelian group of exponent p and order $> p^3$. We claim that, provided $N_G(L) = C_G(L)$ for any nonnormal $L < G$ of order p^2, then $G = S \times E$, where S is extraspecial and E is abelian. By the preceding paragraph, G has a nonnormal subgroup of order p^2. Let $S < G$ be minimal nonabelian. Then all maximal subgroups of S do not lie in the centers of their normalizers so they are G-invariant, and we conclude that $S \lhd G$. By Remark 1, either $|G'| = p$ or G is of maximal class and order p^4. However, by Exercise 9.1 (b), the group G is not of maximal class. Thus, $|G'| = p$. By Lemma 4.3, $G = E \times A$, where E is extraspecial and A is abelian. Let $L < G$ of order p^2 be nonnormal and $y \in G - L$ normalizes but not centralizes L. Then $G' = S' \le L$, a contradiction. Thus, $N_G(L) = C_G(L)$ hence G satisfies the hypothesis.

Exercise 257. Classify the p-groups G such that G' is the unique minimal normal subgroup of G. Is it true that G is extraspecial?

Answer. “No”. Indeed, the group $G = M * N$, where $M \cong M_{p^4} \cong N$ and $|G| = p^7$, is a counterexample. □

Exercise 258. The number $\mathrm{cs}(G)$ of chief series of a p-group G is $\equiv 1 \pmod p$. Study the p-groups G such that $\mathrm{cs}(G) = 1 + kp$, $k \le 3$.

Hint. If $L_1, \dots, L_k$ are all G-invariant subgroups of order p, then

$$\mathrm{cs}(G) = \sum_{i=1}^{k} \mathrm{cs}(G/L_i).$$

Use induction. □

Exercise 259. The number $\mathrm{ch}(G)$ of maximal chains of a p-group G is $\equiv 1 \pmod p$.

Hint. One has $\mathrm{ch}(G) = \sum_{H \in \Gamma_1} \mathrm{ch}(H)$. By induction, if $H \in \Gamma_1$, then $\mathrm{ch}(H) \equiv 1 \pmod p$. Since $|\Gamma_1| \equiv 1 \pmod p$, we get $\mathrm{ch}(G) \equiv |\Gamma_1| \equiv 1 \pmod p$. □

Exercise 260. Find $\mathrm{ch}(G)$, where G (i) has a cyclic subgroup of index p, (ii) G is a minimal nonabelian p-group, (iii) is a primary $\mathcal{A}_2$-group.

Hint. (i) If $G = D_{2^n}$, then $\mathrm{ch}(G) = 2^n - 1$. If $G = Q_{2^n}$, then $\mathrm{ch}(G) = 2^{n-1} - 1$. If $G = SD_{2^n}$, then $\mathrm{ch}(G) = 3 \cdot 2^{n-2} - 1$. If $G = M_{p^n}$ or $C_{p^{n-1}} \times C_p$, then $\mathrm{ch}(G) = 1 + (n-1)p$. To prove the statement, use induction. □

Exercise 261. If all $\mathcal{A}_2$-subgroups of a p-group G are of maximal class, then G is of maximal class. Is it true that G has an abelian subgroup of index p?

Hint. Let $S < G$ be minimal nonabelian subgroup of minimal order. If $S < H \le G$, where $|H : S| = p$, then H is an $\mathcal{A}_2$-subgroup so it is of maximal class. By Exercise 10.10, G is of maximal class. If G is a 3-group of maximal class and order $> 3^4$ with nonabelian fundamental subgroup, then all its $\mathcal{A}_2$-subgroups are of maximal class so the answer on the question is negative. □

Exercise 262. If it follows from $A \cap B = \{1\}$ that $[A, B] = \{1\}$ for any maximal cyclic subgroups A, B of a p-group G, then G is abelian.

Solution. Assume that G is nonabelian. Let $S \le G$ be minimal nonabelian. Then $S \not\cong Q_8$. In that case, there are maximal cyclic subgroup $U, V < S$ such that $U \cap V = \{1\}$ but $[U, V] > \{1\}$. If $U \le A < G$, $V \le B < G$, where A, B are maximal cyclic, then $A \cap B = \{1\}$ and $[A, B] \ge [U, V] > \{1\}$, a contradiction. Thus, S does not exist so G is abelian. □

Exercise 263. Compute the number of maximal chains in a extraspecial p-group of order p^{2n+1}.

Exercise 264. Compute the number of maximal chains in minimal nonmetacyclic group of order 2^5. Do the same for all other minimal nonmetacyclic 2-groups.

Exercise 265. Find the numbers of chief series in minimal nonabelian p-groups.

Exercise 266. Find the numbers of maximal chains in primary $\mathcal{A}_2$-groups.

Exercise 267. Find the number of maximal chains in an abelian p-group of given type. Consider in detail the homocyclic p-groups.

Exercise 268. Classify the p-groups G such that, whenever $H < G$ is nonabelian and $h \in H - \mathrm{Z}(G)$, then $\mathrm{C}_G(h) < H$.

Exercise 269. Prove that all maximal subgroups of the regular wreath product $W = R \wr C$, where R is an irregular passive factor and $|C| = p$, are irregular.

Hint. The Frattini subgroup $\Phi(W)$ contains a subgroup isomorphic to R. □

Exercise 270. Let G be a p-group, $\mathrm{d}(G) = 3$. Is it true that if $\alpha_1(G) < 1 + p + p^2$, then G has a normal abelian subgroup of index p^2?

Hint. See § 76. □

Exercise 271. Suppose that $p > 2$ and a p-group G possesses p pairwise distinct maximal subgroups containing all minimal nonabelian subgroups of G. Study the structure of G. (Example: G is a p-group of maximal class and order $> p^3$ with an abelian subgroup of index p.)

Exercise 272. If $A, B < G$ are distinct maximal subgroups of a nonabelian p-group G such that any minimal nonabelian subgroup of G is contained either in A or in B, then $p = 2$ and A, B, C are all members of the set Γ_1 containing $A \cap B$, where C is abelian.

Solution. Set $D = A \cap B$ and let C/D be such that $C \in \Gamma_1 - \{A, B\}$. Assume that C is nonabelian. In that case, there is in C a minimal nonabelian subgroup S not contained in D (Theorem 10.28). Since $S \nleq A$, $S \nleq B$, we get a contradiction. Thus, C is abelian hence $p - 1$ members of the set Γ_1 containing D are abelian. If $p > 2$, then $D = \mathrm{Z}(G)$ so that A, B are abelian, contrary to the hypothesis. Thus, $p = 2$. □

Exercise 273 (Janko). Let A be a maximal normal abelian subgroup of a nonabelian p-group G. If any other maximal abelian subgroup of G has the cyclic intersection with A, then either A is cyclic or $|G : A| = p$.

Hint. Assume that A is noncyclic and $|G : A| > p$. Then there is in A a G-invariant subgroup $R \cong \mathrm{E}_{p^2}$. Take $x \in \mathrm{C}_G(R) - A$ and let $B < G$ be a maximal abelian subgroup containing $\langle x, R \rangle$. Then $B \neq A$ and $R \leq A \cap B$ so $A \cap B$ is noncyclic, a contradiction. □

Exercise 274. Suppose that a p'-group A acts on a p-group P and centralizes a normal abelian subgroup Q of P such that $\mathrm{C}_P(Q) = Q$. Then A centralizes P.[4]

4 This is a partial case of [GLS, Lemma 11.8].

Hint. Let W be the natural semidirect product of A and Q and set $W = AP$. Let us consider $C_G(Q) = A \times Q = W$ (note that W has no minimal nonnilpotent subgroup). As $W \lhd G$, we get $A \lhd G$, and the result follows. □

Exercise 275. Let G be a nonabelian p-group and let $H \lhd G$ be of minimal index such that G/H is nonabelian. Then $Z(G/H)$ is cyclic.

Exercise 276. Suppose that a p-group G has a maximal elementary abelian subgroup $E \cong E_{p^2}$. If G has a normal elementary abelian subgroup $F \cong E_{p^n}$, $n \geq p > 2$, then $EF \cong \Sigma_{p^2}$.

Solution. Set $H = EF$. Then $C_H(E) = E$, so that H is of maximal class (Proposition 1.8). By Exercise 9.13, $n = p$ so that $H =\cong \Sigma_{p^2}$. □

The group $P = E_{p^n}$ admits an automorphism of order $p^n - 1$, acting irreducibly on P ([Gor1, Corollary 6.4]). Indeed, let us consider P as the additive group of the Galois field $GF(p^n)$. Let the mapping $\phi : P \to P$ be defined by $\phi : a \mapsto a\xi$, where $a \in P$ and ξ is a generator of the (cyclic) multiplicative group of the field $GF(p^n)$. Then ϕ is an automorphism of P of order $o(\xi) = p^n - 1$.

Exercise 277. Study the p-groups $G = \Omega_1(G)$, $p > 2$, all of whose subgroups of order p^p and exponent p are abelian.

Solution. If G is regular, then $G = \Omega_1(G)$ is elementary abelian since $\exp(G) = p$ and G has no minimal nonabelian subgroup (Lemma 65.1). Now assume that G is irregular. Let $E < G$ be of order p^p and exponent p; then E is elementary abelian. By Theorem 13.5, one may assume that $E \lhd G$. Let $E \leq E_1$, where $E_1 \lhd G$ is maximal elementary abelian in G; then $H = \langle x, E_1 \rangle$ is nonabelian for some $x \in G - E_1$ of order p (Theorem 10.1). By Lemma 57.1, there is $y \in E_1$ such that $S = \langle x, y \rangle$ is minimal nonabelian; then $S \cong S(p^3)$. If $S \leq M < H$, where $|M| = p^p$ and $M = \langle x \rangle (M \cap E_1)$, then M is nonabelian of order p^p and exponent p, contrary to the hypothesis. □

Exercise 278. Let A be the abelian group of type (p^2, p, p). Study a Sylow p-subgroup of $\mathrm{Aut}(A)$.

Exercise 279. Let A be the abelian group of type (p^n, p), $n > 2$. Study Sylow p-subgroups of $\mathrm{Aut}(A)$ and the holomorph of A.

Exercise 280. If G is a p-group, its width equals $\log_p(|G/\mho_1(G)|)$. We offer readers to study, for $k \leq p - 1$, the p-groups of width $> k$, all of whose maximal subgroups of width k are isolated. See Theorem A.118.4.

Exercise 281 ([Gor1, Exercise 5.14]). Construct a 3-group G of maximal class and order 3^5.

Hint. The fundamental subgroup G_1 of G is either abelian of type $(9, 9)$ or the metacyclic group $M_3(2, 2)$ of order 3^4 and exponent 9 (see Theorem 9.6). □

Exercise 282. Construct a 3-group of maximal class and order 3^{2n+1}, $n > 2$, with homocyclic fundamental subgroup.

Exercise 283. Construct a 3-group of maximal class and order 3^{2n+1}, $n > 2$, whose fundamental subgroup is (metacyclic) minimal nonabelian.

Exercise 284. Construct a p-group of maximal class and order p^n, n arbitrary, with abelian fundamental subgroup.

Exercise 285. Construct a p-group of maximal class and order p^n with isolated fundamental subgroup.

Exercise 286. Let $G = B \cdot A$, where $B = \langle b \rangle \cong C_{p^n}$ and $A \cong E_{p^{p^n}}$, $A \lhd G$ and $A \cap B = \{1\}$. Suppose that B cyclically permutes the elements of a basis of A. Prove that G/G' is abelian of type (p^n, p) and $Z(G) \cong C_p$.

Solution. Clearly, $C_G(A) = A$. Let $\{a_1, a_2, \dots, a_k\}$ $(k = p^n)$ be a basis of A such that $a_i^b = a_{i+1}$, $i < k$, $a_k^b = a_1$. Then $[a_i, b] = a_i^{-1} a_{i+1}$, $i < k$. These $p^n - 1$ commutators are linearly independent and so $|G'| = p^{p^n - 1}$ since $G' < A$. As $|A : G'| = p$, the quotient group G/G' is abelian of type (p^n, p). Let

$$c = \prod_{i=1}^{k} a_i^{b^{s_i}} \in Z(G).$$

It follows from $c^b = c$ that $s_1 = \dots = s_k$ so that $Z(G) = \langle a_1 \dots a_k \rangle$ has order p. Note that $|G'||Z(G)| = |A|$, and this agrees with [Isa1, Lemma 12.12]. It is possible to show that $\mathrm{cl}(G) = p^n$. Clearly, $G = C_{p^n} \wr C_p$ with active factor C_{p^n}. □

Exercise 287. Study the p-groups of maximal class admitting a fixed-point-free automorphism.

Hint. Observe that the order of a fixed-point-free automorphism of G divides the number $|Z(G)^\#| = p - 1$. Therefore, a p-group of maximal class, $p \le 3$, does not admit a fixed-point-free automorphism. □

Exercise 288. Does there exist a 2-group G such that $\mathrm{Aut}(G) \cong M_2(2, 2)$?

Exercise 289. Study the noncyclic p-groups that are not covered by subgroups of index p^2.

Hint. There is $H \in \Gamma_1$ that is not covered by its subgroups of index p. It follows that H is cyclic. Use Theorem 1.2. □

Exercise 290. Study the irregular p-groups all of whose proper irregular subgroups are of maximal class.

Hint. Let $R < G$ be maximal regular and let $R < H \le G$, where $|H : R| = p$. Assume that $H < G$. Then H is of maximal class and so, by Exercise 10.10, G is of maximal class. If G is not of maximal class, it is minimal irregular. □

Exercise 291. Suppose that a p-group G contains a normal subgroups $E \cong \mathrm{E}_{p^3}$ and $A \cong \mathrm{E}_{p^5}$, $p > 2$. Then $E < E_1 \lhd G$, where $E_1 \cong \mathrm{E}_{p^4}$.

Hint. Use Theorem 10.1. □

Exercise 292. Classify the non-Dedekindian metacyclic p-groups G all of whose noncyclic subgroups are normal.

Hint. Prove that $|G'| \le p^2$. If G has no normal abelian subgroup of type (p, p), it is isomorphic to Q_{16}. Now let $R \cong \mathrm{E}_{p^2}$ be G-invariant. Then G/R is Dedekindian. If $p > 2$, then $|G'| = p$ and G is minimal nonabelian (Lemma 65.2 (a)). Let $|G'| > p$; then $p = 2$ and $|G'| = 4$. Indeed, let $R \cong \mathrm{E}_4$ be G-invariant. Then G/R is Dedekindian. Then there is $S/R \lhd G/R$ of order 2 such that G/S is abelian. As a cyclic G' is a subgroup in a noncyclic S, we get $|G'| = 4$. □

Exercise 293. (i) Find the nontrivial isolated subgroups in a minimal nonabelian p-group.
(ii) Classify the minimal nonabelian p-groups without nontrivial isolated subgroups.

Hint. (ii) Such groups are metacyclic. □

Exercise 294 ([GLS, Lemma 10.15]). Let P be a p-group, $p > 2$. Let $Q \lhd P$ contain a proper P-invariant subgroup $E \cong \mathrm{E}_{p^k}$. If $E < X \le Q$, where $X \cong \mathrm{E}_{p^{k+1}}$, then there is in Q a P-invariant subgroup isomorphic to X.

Solution. Let $\mathcal{M}$ be the set of subgroups isomorphic to X in Q containing E. By Theorem 10.1, one has $|\mathcal{M}| \equiv 1 \pmod{p}$. Considering the action of P on $\mathcal{M}$, we find a P-invariant $X_0 \in \mathcal{M}$. □

Exercise 295. Let P be a p-group. Let $Q \lhd P$ contain a P-invariant abelian subgroup A of exponent $p^n > 2$. If $A < X \le Q$, where X is abelian of exponent p^n and order $p|A|$, then there is in Q a P-invariant abelian subgroup T of order $p|A|$ and exponent p^n. (This is a generalization of Exercise 294.)

Exercise 296. Describe the characteristic subgroups in Σ_{p^2}.

Exercise 297. Show that the generalized quaternion group is the unique nonabelian prime-power group covered by self-centralizing cyclic subgroups.

Hint. Such a group has only one subgroup of order p. □

Exercise 298. Classify the two-generator p-groups containing exactly $1 + p + p^2$ subgroups of index p^2.

Exercise 299. Describe the 2-groups containing a maximal subgroup isomorphic to M_{2^n}. (See § 74.)

Exercise 300. Let G be minimal nonabelian p-group. Describe the group of those automorphisms of G that act trivially on (i) $G/\Phi(G)$, (ii) $\Omega_1(G)$, (iii) $\Phi(G)$.

Exercise 301. Study the non-Dedekindian p-groups G such that $C \cap Z(G) = \{1\}$ for any nonnormal abelian $C < G$.

Hint. Suppose that an abelian $C < G$ is nonnormal. Assume that distinct $L_1, L_2 < Z(G)$ are of order p. Then $C \ngtr L_i$, $i = 1, 2$, and $C \times L_1$ and $C \times L_2$ are distinct G-invariant abelian subgroups (otherwise, we have $C \cap (L_1 \times L_2) > \{1\}$, by the product formula). Then $C = (C \times L_1) \cap (C \times L_2) \lhd G$, a contradiction. Thus $Z(G)$ is cyclic so G is a monolith. Assume that a cyclic $C < G$ is nonnormal of order $> p$. Then $C \times L \lhd G$, where $L = \Omega_1(Z(G))$. In that case, $\Omega_1(C) \lhd G$, a contradiction. Thus, all nonnormal cyclic subgroups of G have order p. Such a group G are described in § 63. □

Exercise 302. Suppose that a nonabelian p-group G of order p^m contains an abelian subgroup A_1 of index p and $|Z(G)| = p^z$. Let X be a maximal set of pairwise noncommuting elements of G. Is it true that $|X| = p^{m-1-z} + 1$?

Solution. Let $X = \{x_1, \dots, x_k\}$ and let $\{A_1, \dots, A_t\}$ be the set of all pairwise distinct maximal abelian subgroups of G. Then $A_1 \cap A_i = Z(G)$ and $|A_i| = p|Z(G)|$ for all $i > 1$. Set $B_i = A_i - Z(G)$ for all i; then $|B_i| = (p-1)p^z$ for $i > 1$ and $B_1 = p^{m-1} - p^z$, where $|Z(G)| = p^z$. One has $B_i \cap B_j = \emptyset$ for $i \neq j$. If $1 < j$ and $x_1 \in B_1$, $x_j \in B_j$, then $x_1x_j \neq x_jx_1$ since $C_G(x_1) = A_1$ does not contain x_j. Similarly, $x_ix_j \neq x_jx_i$ for distinct $i > 1$ and $j > 1$. As

$$G - Z(G) = \bigcup_{i=1}^{m} B_i$$

is a partition and each B_i contains exactly one element, say x_i, of the set X, we get $k = t$ and

$$p^m - p^z = |G - Z(G)| \implies p^m - p^z = (p^{m-1} - p^z) + (t-1)(p-1)p^z.$$

Therefore,

$$|X| = k = t = 1 + \frac{p^{m-1}(p-1)}{p^z(p-1)} = 1 + p^{m-1-z}.$$

In particular, if G is of maximal class, then $z = 1$ hence $|X| = p^{m-2} + 1$. If $G \cong M_{p^m}$, then $z = m - 2$ and $|X| = 1 + p$. □

Exercise 303. Find the number of maximal abelian subgroups in the extraspecial group G of order p^{2m+1}.

Exercise 304. Let G be an extraspecial group of order p^{2m+1} and exponent p and $x \in G - Z(G)$. Find the number of minimal nonabelian subgroups of G containing x.

Solution. All minimal nonabelian subgroups of G have order p^3 and contain $G' = Z(G)$. One has $|G : C_G(x)| = p$. All containing x subgroups of $C_G(x)$ of order p^3 are abelian since $\langle G', x\rangle \le Z(C_G(x))$. If $y \in G - C_G(x)$, then $K = \langle x, y\rangle$ is minimal nonabelian and $|K \cap C_G(x)| = p^2$. The number of elements in $K - C_G(x)$ is equal to $p^3 - p^2$. Therefore, the required number is equal to

$$\frac{|G - C_G(x)|}{p^3 - p^2} = \frac{p^{2m}(p-1)}{p^2(p-1)} = p^{2m-2}.$$ □

Exercise 305 ([GLS, Lemma 10.35]). Assume that $|P| = p^k \geq p^3$ and $P = Q\langle x\rangle$ with $Q \in \Gamma_1$ and $|C_P(x)| = p^2$. Let $z \in Z(P)^\#$. Then there exists $a \in Q$ with

$$z = [a, x, x, \dots, x] \quad (k - 2 \text{ times}).$$

Solution. By Proposition 1.8, P is of maximal class so $|P'| = |\Phi(P)| = p^{k-2}$. Now

$$P = Q\langle x\rangle = QC_P(x) \implies |Q : C_Q(x)| = |P : C_P(x)| = p^{k-2}.$$

This is the number of commutators $[a, x] = x^{-a}x$ for $a \in Q$, so each element of P' has this form. If $k = 3$, this completes the solution since $P' = \langle z\rangle$. In general, let $P_1 = P'\langle x\rangle$; then $P' < P_1 \in \Gamma_1$ and P_1 is of maximal class. Moreover, $Z(P) < P' < P_1$. By induction to $|P_1|$ and $|P'|$ in place of $|P|$ and $|Q|$, respectively, we get $[b, x, \dots, x] = z\,(k - 3 \text{ times})$ for some $b \in P'$, and since $b = [a, x]$ with $a \in Q$, the required conclusion holds. □

Exercise 306. If p is the minimal prime divisor of the order of a supersolvable group G, then it is p-nilpotent.

Solution. Assume that this is false. Then there is in G a minimal nonnilpotent subgroup $S = Q \cdot P$, where $S' = P \in \mathrm{Syl}_p(S)$. As all subgroups of G are supersolvable but S is not (Theorem A.22.1), we get a contradiction. □

Exercise 307. Is it true that a p-group G all of whose two elements of equal order are commute is abelian?

Hint. Assume that this is false and G is a counterexample of minimal order. By induction, we can show that G is minimal nonabelian. In that case, $|G| > p^3$ so that $\exp(G) = p^e > p$. As $\Omega_e^\#(G) = G$, there are in G two non-commuting elements of order p^e, a contradiction. □

Exercise 308. (a) Classify the minimal nonabelian p-groups G such that, whenever $H < G$ is noncyclic and $|H|^2 < |G|$, then $H \trianglelefteq G$.
(b) Check the following assertion: Let G be a p-group of order $> p^4$, $p > 2$, such that, whenever $H < G$ is noncyclic and $|H|^2 < |G|$, then $H \trianglelefteq G$. Then $\Omega_1(G)$ is abelian.

Exercise 309. Let M be a subgroup of maximal class and index $p > 2$ in a p-group G, $|G| > p^{p+2}$. Is it true that if G is not split over M, then it is of maximal class?

Hint. Use Theorems 13.2 (a) and 13.5. □

Exercise 310. Describe a Sylow 3-subgroup of the group $\mathrm{Aut}(\Sigma_9)$.

Exercise 311. Classify the p-groups G such that some $H \in \Gamma_1$ contains no maximal cyclic subgroup of G.

Solution (Janko). If C is a maximal cyclic subgroup of H, then, not being a maximal cyclic subgroup of G, we get $C = \mho_1(C_1)$ for some maximal cyclic subgroup C_1 of G since $\exp(G/\Phi(G)) = p$. It follows that $\mho_1(G) \geq H$. Thus, $|G : \mho_1(G)| \leq |G : H| = p$, and we conclude that G is cyclic. □

Exercise 312. Let a p-group G be irregular of order $> p^{p+1}$ and such that $|\Omega_1(G)| = p^p$. Then one of the following holds:
(a) $\Omega_1(G) \leq \Phi(G)$.
(b) All members of the set Γ_1 not containing $\Omega_1(G)$ are absolutely regular.
(c) All members of the set Γ_1 not containing $\Omega_1(G)$ are of maximal class.

Hint. Use Theorems 12.1 (b), 12.12 (b) and 9.8 (a). □

Exercise 313. Classify the irregular p-groups of order $> p^{p+2}$ all of whose maximal subgroups are either absolutely regular or of maximal class.

Hint. Assume that G is not of maximal class. Then there is in G a normal subgroup R of order p^p and exponent p (Theorem 12.1 (a)). If H/R is maximal in G/R, then H is neither absolutely regular not of maximal class (Theorem 9.6 (c)). Thus, G is of maximal class. (See Theorem 9.6.) □

Exercise 314 (Janko). Let G be a nonabelian p-group such that, whenever $A < G$ is not contained in $\Phi(G)$, then $A \lhd G$. Then $p = 2$ and G is Dedekindian.

Solution. Suppose that G is non-Dedekindian. Then the subgroup G_0 generated by all non-G-invariant subgroups of G is contained in $\Phi(G)$. By Theorem 231.1, the quotient group G/G_0 is cyclic. Then G is cyclic, a contradiction. Thus, G is Dedekindian so $p = 2$ since G is nonabelian. □

Exercise 315. Study the two-generator 2-groups G such that $\mathrm{Aut}(G)$ is a 2-group.

Exercise 316. Study the noncyclic p-groups G of order $> p^3$ in which the set of all normal subgroups of index $> p$ is a chain.

Exercise 317. Classify the nonabelian p-groups G all of whose maximal abelian subgroups have index p in G.

Answer. One has $G = S\mathrm{Z}(G)$, where S is minimal nonabelian. Indeed, $|G : \mathrm{Z}(G)| = p^2$. If $S \leq G$ is minimal nonabelian, then

$$|S : \mathrm{Z}(S)| = |S : (S \cap \mathrm{Z}(G))| = p^2$$

so $G = S\mathrm{Z}(G)$, by the product formula. □

Exercise 318. Study the irregular p-groups G all whose maximal regular subgroups have index p. (Example: $G = R \times C$, where R is minimal irregular and $|C| = p$.)

Exercise 319. Study the non-absolutely regular p-groups all of whose maximal absolutely regular subgroups have index p.

Hint. For the solution, see Appendix 110. □

Exercise 320. Let G be an abelian group of type (p^m, p^n). Describe the subgroup of $\mathrm{Aut}(G)$ consisting of all automorphisms of G acting trivially on (i) $\Omega_1(G)$, (ii) $G/\mho_1(G)$.

Exercise 321. Let p be a prime divisor of the order of a group G and $P \in \mathrm{Syl}_p(G)$.
(i) If $p > 2$ and $\Omega_1(P) \le Z(G)$, then G is p-nilpotent.
(ii) If $p = 2$ and $\Omega_2(P) \le Z(G)$, then G is 2-nilpotent.

Hint. This follows from Frobenius' normal p-complement theorem (Theorem A.52.2) and Theorem A.22.1. Indeed, in both cases, G has no p-closed minimal nonnilpotent subgroup of order dividing by p. □

Exercise 322. Classify the 3-groups all of whose $\mathcal{A}_2$-subgroups are of maximal class.

Hint. Any 3-group of maximal class which is an $\mathcal{A}_2$-group is of order 3^4. Use Exercise 10.10 to show that G is of maximal class. Any 3-group of maximal class and order $> 3^3$ satisfies the condition. □

Exercise 323. Suppose that G is a nonmetacyclic p-group of order $> p^3$. All maximal metacyclic subgroups have index p in G if an only if G is either minimal nonmetacyclic or a group from Theorem 13.7 (c).

Solution. The minimal nonmetacyclic p-groups satisfy the hypothesis. Let $p = 2$ and $M \le G$ be minimal nonmetacyclic. If $K_i < M$ is maximal in M and $K_i \le L_i$, where L_i is metacyclic of index 2 in G, $i = 1, 2$, then $L_1 \ne L_2$ and so

$$\mathrm{d}(G) \ge \mathrm{d}(M) = 3 \text{ and } \mathrm{d}(G) \le \mathrm{d}(L_1) = 3 \implies \mathrm{d}(G) = 3,$$

and we conclude that all members of the set Γ_1 are metacyclic, hence G is minimal nonmetacyclic. Now assume that $p > 2$ and G is not minimal nonmetacyclic. Let $H \in \Gamma_1$ be metacyclic; then H is absolutely regular since $|H/\mho_1(H)| \le p^2 < p^p$ (Theorem 9.8 (a)). Let G be not a 3-group of maximal class. Then, by Theorem 9.8 (a), $G = H\Omega_1(G)$, where $\Omega_1(G)$ is of order p^3 and exponent p. Assume that $\Omega_1(G) \cong \mathrm{E}_{p^3}$. As for the case $p = 2$, we conclude that G is minimal nonmetacyclic, contrary to the assumption. Now let $\Omega_1(G) \cong \mathrm{S}(p^3)$. Then G is a group from Theorem 13.7 (c) or a 3-group of maximal class such that all members of the set Γ_1 not containing $\{\Omega_1(G)\}$ are metacyclic. Any group of Theorem 13.7 (c) satisfies the hypothesis. If G is a 3-group of maximal class, then it has no subgroup isomorphic to $\mathrm{S}(3^3)$. In that case, if $M < G$ is of order 3^3 and $M \nleq G_1$ (the fundamental subgroup of G), then M is a maximal metacyclic subgroup of G so that $M \in \Gamma_1$. But then G is minimal nonmetacyclic, contrary to the assumption.[5] □

Remark 3. Suppose that G is a group such that 4 divides $|G|$. If all 2-elements of G of order ≤ 4 are contained in $Z(G)$, then G is 2-nilpotent. Assume that this is false. Then there is in G a 2-closed minimal nonnilpotent subgroup S of even order. Let $P_1 \in \mathrm{Syl}_2(S)$. Since $\exp(P_1) \le 4$, we get $P_1 \le Z(G)$ hence S is nilpotent, a contradiction. Thus, S does not exist. Therefore, by Frobenius' normal p-complement theorem (Theorem A.52.2), our group G is p-nilpotent (i.e., it has a normal p-complement.

5 For additional details and a more general result, see Appendix 110.

Exercise 324. The following statements hold:
(a) Any two-generator group of order p^4 and exponent p is of maximal class.
(b) Any nonabelian two-generator group of order p^4 is either minimal nonabelian or of maximal class.
(c) Let G be a nonabelian group of exponent p. If all nonabelian subgroups of G are two-generator, then G is of maximal class with an abelian subgroup of index p.

Hint. (c) It suffices to consider the case when $|G| > p^4$. Use Proposition 10.17 and Exercise 10.10. □

Exercise 325. Let $n > 1$ and suppose that P is a noncyclic p-group such that $c_n(P) = 1$. Give a proof independent of Theorem 1.17 (b) that $p = 2$, G is dihedral if $n = 2$ and G is a 2-group of maximal class if $n > 2$.

Solution. One may assume that P is not a 2-group of maximal class. Then there is in P a normal subgroup $R \cong E_{p^2}$. Let $C_0 < P$ be unique cyclic of order p^n; then $C_0 \triangleleft P$. Set $H = RC_0$. If $C_0 \cap R = \{1\}$, then H, the abelian subgroup of type (p^n, p, p), satisfies $c_n(H) = p^2 > 1$, a contradiction. If $|H \cap C_0| = p$ and H is not a 2-group of maximal class, then $c_n(H) = p > 1$, a contradiction. Now let H be a 2-group of maximal class; then $H \cong D_8$ and $n = 2$. As P is not of maximal class, then $C_P(H) \nleq H$ (Proposition 10.17). Then $c_2(HC_P(H)) > 1$, a contradiction. Thus, G is a 2-group of maximal class. Now all remaining assertions are obvious. □

Let $\epsilon_k(G)$ be the number of elementary abelian subgroups of order p^k in a p-group G. By Theorem 1.17 (b), if G is neither cyclic nor a 2-group of maximal class, then $\epsilon_2(G) \equiv 1 \pmod p$.

Exercise 326. Let E be an extraspecial group of order 2^{1+2m}, $m \geq 3$. Is $\epsilon_3(E)$ odd?

Exercise 327 (Proposition 1.8). If a p-group G contains a noncentral element x such that $|C_G(x)| = p^2$, then G is of maximal class.

Solution (compare with [Bla12, proof of Theorem 2]). One may assume that $|G| > p^3$. If $C_G(x) < M \leq G$, where $|M : C_G(x)| = p$, then $M = N_G(C_G(x))$ is of maximal class (and order p^3), and the result follows from Exercise 10.10. □

Exercise 328 (Proposition 10.17). If a p-group G contains a proper nonabelian subgroup B of order p^3 such that $C_G(B) < B$, then G is of maximal class.

Hint. Let $B < H \leq G$, where $|H : B| = p$. Then $|Z(H)| = p$ and $H/Z(H)$ is nonabelian (check!). In that case, H is of maximal class, and the result follows from Exercise 10.10. To simplify checking, use Lemmas 65.2 (a) and 4.3. (By Theorem 9.6 (f), a p-group X of maximal class possesses a nonabelian subgroup of order p^3 containing its centralizer in X.) □

Exercise 329 (Remark 10.5). Suppose that a p-group G contains a subgroup L such that $N = N_G(L)$ is of maximal class. Then G is also of maximal class.

Solution. One may assume that $N < G$; then L is not characteristic in N. It follows that $|N : L| = p$. One has $Z(G) < N$ so that $|Z(G)| = p$, $Z(G) < L$ and $C_G(N) = Z(G)$. If $|N| = p^3$, then G is of maximal class (Exercise 328). Let $|N| > p^3$. In that case, $G/Z(G)$ is of maximal class, by induction, and the assertion follows since $|Z(G)| = p$. □

Exercise 330. Let G be a p-group.
(a) If $|G| > p^3$ and $|Z_3(G)| = p^3$, then $Z_3(G)$ is abelian.
(b) For $|G| > p^4$ and $K = Z_4(G)$ if order p^4, study the structure of K.

Solution. (a) Set $Z = Z_3(G)$. Assume that Z is nonabelian. Let $R < Z$ be G-invariant of index p. Put $C = C_G(R)$; then $|G : C| = p$ and $Z \cap C = R$. But C contains an abelian G-invariant subgroup, say $A > R$, of order p^3. As $A \leq Z_3(G) = Z$, we get a contradiction. Thus, Z is abelian. □

Exercise 331. Let $n > 3$ and let a p-group G of order $> p^n$ be such that $|Z_n(G)| = p^n$. Then $Z_n(G)$ is not of maximal class.

Solution. Assume that $Z = Z_n(G)$ is of maximal class. As $Z(G) < Z_n(G)$, we get $|Z(G)| = p$. Next we proceed by induction on n. Write $\overline{G} = G/Z(G)$. Then $Z_{n-1}(\overline{G})$ (of order $p^{n-1} > p^2$) is not of maximal class, by Exercise 330 (a) or by induction, and we are done. (For a related result, see Appendix 117.) □

Exercise 332. Let a p-group G contain two elementary abelian subgroups E and E_0 of orders p^3 and p^5, respectively, and let $E \triangleleft G$. If $p > 2$, then E is contained in G-invariant elementary abelian subgroup E_1 of order p^4.

Hint. Consider $C_{EE_0}(E)$ and take into account that a Sylow p-subgroup of $\mathrm{Aut}(E)$ is nonabelian of order p^3. Use Theorems 10.1 and 10.5. □

Exercise 333 ([GLS, Lemma 11.19]). Assume that $X \cong C_{p^n-1}$, $G = (X, P)$ is a Frobenius group whose kernel is a p-group P. Then every chief p-factor of G is has order $\geq p^n$.

Hint. Take in G a minimal normal subgroup R and, considering the Frobenius group (X, R), prove that $R \cong E_{p^n}$. □

Exercise 334. Classify the non-absolutely regular p-groups all of whose proper epimorphic images are absolutely regular.

Hint. Use Theorem 9.8 (a). Then G is of order p^p and exponent p. □

Exercise 335. Let an elementary abelian q-group Q act on a p-group P. If $p \equiv 1 \pmod q$, then G, the natural semidirect product Q and P, is supersolvable.

Solution. Working by induction on $|P|$, it suffices to prove that there is in G a normal subgroup of order p (indeed, any proper epimorphic image of G satisfies the hypothesis). One may assume that Q acts faithfully on P. By setting $\Omega = \Omega_1(Z(P))$, it suffices to find in $H = Q \cdot \Omega$, the natural semidirect product of Q and Ω, a Q-invariant subgroup of order p. Therefore, one may assume, without loss of generality, that $H = G$. First assume that $|Q| = q$. Let $S \leq G$ be minimal nonnilpotent. Then S' is G-invariant of order p (Theorem A.22.1). Next assume that $|Q| > q$. In that case, there is in Q an element x such that $G > C_G(x) > Q$ since G is not a Frobenius group. Then $P_1 \in \mathrm{Syl}_p(C_G(x))$ is Q-invariant. By induction, there is in P_1 a Q-invariant subgroup L of order p. Then $L \lhd G$, and we are done. □

Exercise 336. Let a 2-group P be given by $P = D * A$, where D is a 2-group of maximal class, $A = C_4$ and $D \cap A = Z(D)$. Show that P has no subgroup isomorphic to E_8.

Solution. Assume that $E \cong E_8$ is a subgroup of P. Then the subgroup $H = EA$ is abelian of type $(4, 2, 2)$ so $|D| > 2^3$. In that case $H \cap D$ is maximal in H, by the product formula. A maximal subgroup of H is either isomorphic to E_8 or abelian of type $(4, 2)$. Since D has no such subgroups, we get a contradiction. □

Exercise 337. Let a 2-group P be given by $P = D * Q$, where $D \cong D_{2^m}$, $Q \cong Q_{2^n}$ and $D \cap Q = Z(D)$. Show that P has no subgroup isomorphic to E_8.

It follows that in the group from Exercise 337 all abelian subgroups have ranks < 3.

Exercise 338. Classify the nonabelian critical groups G of exponent p.[6]

Exercise 339. Study the p-groups G with absolutely regular Frattini subgroup and such that $|\Omega_1(G)| = p^p$.

Hint. See Appendix 8. □

Exercise 340. Prove that a p-group G, $p > 2$, has order p^{p+1} provided the subgroup $H = \Omega_2^{\#}(G) = \langle x \in G \mid o(x) = p^2 \rangle$ is irregular of order p^{p+1}.

Hint. The subgroup H is of maximal class (Theorem 7.1 (b)). We claim that $H = \Omega_p(G)$. Indeed, assume that there is in G a cyclic subgroup C of order p^3. Then $H \cap C \cong C_{p^2}$ so that, by the product formula, H is maximal in $F = CH$ of order p^{p+2}. By Theorems 9.5 and 9.6 (e), any p-group of maximal class and order p^{p+2}, $p > 2$, has exponent p^2. Therefore, F is not of maximal class. Then, by Theorem 12.12 (b), $F/K_p(F)$ is of order p^{p+1} and exponent p so that $\exp(F) = p^2 < p^3 = \exp(C)$, a contradiction. Thus, C does not exist, and we conclude that $H = \Omega_p(G)$. To obtain a contradiction, it suffices to assume that $|G| = p^{p+2}$. If G is of maximal class, then, by Theorem 9.6 (e), $H \geq G_1$, where the fundamental subgroup G_1 of G is absolutely regular of order $p^{p+1} = |H|$, which implies $H = G_1$, a contradiction since H is irregular. Thus, G is not of maximal class. Then, by Theorem 13.6, the number of subgroups of maximal class and order p^{p+1} in G is equal to p^2. By hypothesis, H contains more than one maximal subgroups of exponent p^2 and these subgroups are maximal absolutely regular in G. Let A be one of such subgroups. As G is not of maximal class, there is, by Exercise 10.10,

6 It is easy to show that $G = S \times E$, where S is special and E is abelian. However, the special p-groups of exponent p are critical.

a subgroup $T \le G$ such that $A < T$ is of index p and T is not of maximal class. Then T is regular (Theorem 7.1 (b)) of exponent p^2. This is a contradiction since $\Omega_2^\#(T) = T \neq H$ and $|T| = |H|$. Thus, $G = H$ is of maximal class and order p^{p+1}. □

Remark 4. Let G be a p-group. If $k > 2$, then $\Omega_k^\#(G)$ is not of maximal class, by Theorem 13.19 (a). If $\Omega_2^\#(G)$ is of maximal class and order $> p^{p+1}$, then G is of maximal class (see §§ 12–13).

We say that an element x is half-central in a p-group P provided $|P : C_P(x)| \le p$. Recall that $r(G)$ is the maximum of ranks of abelian subgroups of a p-group G.

Exercise 341 ([GLS, Lemma 10.20]). Let P be a p-group.
(i) If x is half-central in P, then $r(C_P(x)) = r(P)$.
(ii) If $U \cong E_{p^2}$ is normal in P, then $r(C_P(U)) = r(P)$.

Solution. One may assume that P is nonabelian.

(i) Let $A < P$ be elementary abelian with $r(A) = r(P)$ and $B = A \cap C_P(x)$. One may assume that $x \notin A$. Then $r(B) \ge r(A) - 1$ and

$$r(C_P(x)) = r(B \times \langle x \rangle) = r(A) = r(P).$$

(ii) As $|P : C_P(U)| \le p$, all elements of U are half-central, and therefore the equality follows from (i). □

Exercise 342. A subgroup H is *half-normal* in a p-group G provided $|G : N_G(H)| \le p$. Prove that a non-Dedekindian p-group G possesses a nonnormal half-normal subgroup.

Solution. By Theorem 1.23, there is a G-invariant $L < G'$ of index p such that G/L is non-Dedekindian. Therefore, it suffices to show that G/L has a nonnormal half-normal subgroup. One may assume, without loss of generality, that $L = \{1\}$; then $|G'| = p$. As G is non-Dedekindian, it contains a nonnormal cyclic subgroup H. The subgroup $M = H \times G'$ is G-invariant and contains exactly p cyclic subgroups of order $|H|$ not containing G'. It follows that there are at most p subgroups conjugate with H in G, and we conclude that

$$\{1\} < |G : N_G(H)| \le p.$$

Thus, $|G : N_G(H)| = p$, by a choice of H, so that H is nonnormal half-normal subgroup in G. □

It follows from the solution of Exercise 342 that a non-Dedekindian p-group G contains a nonnormal half-normal subgroup H such that H/H_G is cyclic. About p-groups all of whose subgroups are half-normal, see § 122. It appears that $|G : Z(G)| \le 2^4$ if $p = 2$ (Janko) and $|G : Z(G)| \le p^3$ if $p > 2$ (J. Bohanon).

Exercise 343. Study the half-normal subgroups of special p-groups.

Exercise 344. Let G be a p-group, $p \neq 3$. If the mapping $\phi : x \mapsto x^3$ is an endomorphism of G, then G is abelian.

Solution. By hypothesis, $(xy)^3 = x^3y^3$. Assuming that G is nonabelian, take in G a minimal nonabelian subgroup S. Take $x, y \in S$ such that $[x, y] \neq 1$. As $\mathrm{cl}(S) = 2$, one has

$$x^3y^3 = (xy)^3 = x^3y^3[y, x]^{3(3-1)/2} = (xy)^3[y, x]^3 \implies [y, x]^3 = 1 \implies [y, x] = 1,$$

which is a contradiction. Thus, S does not exist so G is abelian.[7] □

Exercise 345. Let p, n be such that p does not divide $\frac{1}{2}n(n-1)$. If the mapping $\phi : x \mapsto x^n$ is an endomorphism of a p-group G, then G is abelian.

Hint. Mimic the solution of Exercise 344. □

Exercise 346 ([GLS6, Lemma 2.13]). Let x be an involution in a 2-group P. Then $|P : C_P(x)| \leq 2$ if and only if $|\langle x\rangle^P| \leq 4$.

Solution. Let $U = \langle x\rangle^P$. If $|P : C_P(x)| \leq 2$, then the subgroup $U \leq C_P(x)$ is generated by at most two commuting involutions constituting a P-class so $|U| \leq 4$. Conversely, if $|U| \leq 4$, then $|P : C_P(U)| \leq 2$ so $|P : C_P(x)| \leq 2$. □

Exercise 347. Let x be an element of order $p > 2$ in a p-group P. If $|P : C_P(x)| \leq p$, then $\langle x\rangle^P$ is elementary abelian of order $\leq p^p$.

Exercise 348 ([GLS6, Lemma 2.16]). Let a 2-group P be of the form $P = \langle a\rangle T$, where $|P : T| = 2$ and $T = R \times R^a$ for some $R < T$. Then there is an involution in $P - T$ and $P = R \wr C_2$.

Hint. We proceed by induction on $|P|$. If $|R| = 2$, then $P \cong D_8$, and the result is obvious. Now let $|R| > 2$. Let $M < R$ be maximal in R. Then $Q = M \times M^a \lhd P$. By induction, $T_1 = \langle a\rangle Q$ is such that the set $T_1 - (M \times M^a)$ contains an involution x. Now the last assertion follows. □

Exercise 349. Let a p-group P be of the form $P = T\langle a\rangle$, where $p > 2$, $|P : T| = p$ and $T = R \times R^a \times \cdots \times R^{a^{p-1}}$ for some $R < T$. Then there is an element of order p in $P - T$ and $P = R \wr C_p$.

Hint. If $|R| = p$, then $T \cong E_{p^p}$ and $P \cong \Sigma_{p^2}$ (Proposition 1.8 and Exercise 9.13). Therefore, there is in $P - T$ an element of order p (indeed, Σ_{p^2} contains two distinct subgroups of order p^p and exponent p). Now let $|R| > p$. If $S < R$ is of index p, then $U = S \times S^a \times \cdots \times S^{a^{p-1}} \lhd G$. By induction, the set $U\langle a\rangle - (S \times S^a \times \cdots \times S^{a^{p-1}})$ contains an element of order p. Now the last assertion is obvious. □

Exercise 350 (Janko). Let G be a nonabelian p-group such that, whenever $A < G$ is not contained in $\Phi(G)$, then $A \lhd G$. Then $p = 2$ and G is Dedekindian.

---

7 It is known the following result. If a finite $3'$-group G admits an endomorphism $x \mapsto x^3$, then it is abelian. It suffices to prove this considering minimal nonnilpotent groups.

Solution. Suppose that G is non-Dedekindian. Then the subgroup G_0 generated by all non-G-invariant subgroups of G is contained in $\Phi(G)$. By Theorem 231.1, the quotient group G/G_0 is cyclic. Then G is cyclic, a contradiction. Thus, G is Dedekindian so $p = 2$ since G is nonabelian. □

Exercise 351. Let G be a nonabelian p-group and p^a a minimal order of a G-invariant subgroup not contained in $Z(G)$. Then there is $x \in G - Z(G)$ such that $|G : C_G(x)| \le p^{a-1}$.

Solution. Let $M \lhd G$ be of minimal order p^a not contained in $Z(G)$. Then

$$|M : (M \cap Z(G))| = p.$$

Take $x \in M - Z(G)$. One has

$$|M - Z(G)| = (p-1)p^{a-1}.$$

All elements which are G-conjugate with x are contained in the set $M - Z(G)$, and this implies that $|G : C_G(x)| \le p^{a-1}$. □

Exercise 352. Let G be a nonabelian p-group with cyclic center. Then there is in the set $G - Z(G)$ a half-central element x, i.e., $|G : C_G(x)| = p$.

Solution. If $|Z(G)| = p$ and $M/Z(G)$ is normal of order p, then $x \in M - Z(G)$ satisfies $|G : C_G(x)| = p$. Now let $|Z(G)| > p$. Then G is not of maximal class so it contains a normal subgroup $R \cong E_{p^2}$. As $R \not\le Z(G)$, one can take $x \in R - Z(G)$. □

Exercise 353. (i) Is it true that the subgroup $\mho_1(G)$ is not isolated in a p-group G of exponent $> p$?

(ii) Study the p-groups whose Frattini subgroup is isolated.

Exercise 354. Study the nonabelian p-groups G such that, whenever $S \le G$ is minimal nonabelian, then all subgroups of S are half-normal in G.

Exercise 355. Study the nonabelian p-groups G such that the covering of G by all maximal abelian subgroups is irredundant. (Example: G has an abelian subgroup of index p.) Consider in detail the groups G for which $|A| = p|Z(G)|$ for all maximal abelian subgroups A of G.

Exercise 356. Study the nonmetacyclic p-groups G such that the covering of G by all maximal metacyclic subgroups is irredundant.

Exercise 357. Let $E \cong E_{p^2}$ be nonnormal maximal elementary abelian subgroup of a p-group G and let $H \le G$ be of exponent p and of order $> p^2$. If $E \le N_G(H)$, then the subgroup EH is of maximal class.

Solution. As $C_{EH}(E) = E$, so the subgroup EH is of maximal class (Proposition 1.8). □

Exercise 358. Let $H \cong \Sigma_{p^2}$ be a subgroup of a p-group G and suppose that, whenever $A < H$ is absolutely regular of index p, then A is maximal absolutely regular subgroup of G. Study the structure of the group G.

Exercise 359. If a p-group G is covered by any $p+1$ pairwise distinct minimal nonabelian subgroups, then it is an $\mathcal{A}_2$-group.

Hint. All minimal nonabelian subgroups of G are maximal, and this implies that G is an $\mathcal{A}_2$-group. □

Exercise 360. Let G be a non-Dedekindian p-group all of whose nonnormal cyclic subgroups are self-centralizing. Then $G \cong Q_{2^n}$, $n > 3$ is a generalized quaternion group.

Solution. One has $\Omega_1(G) \le Z(G)$. If $C < G$ is nonnormal cyclic, then $Z(G) < C$ so that $|\Omega_1(G)| = p$, and Proposition 1.3 implies the result. □

Exercise 361. Is it true that the group of exponent $p^e > p$ is abelian if any two elements of different orders commute?

Answer. "No". The minimal nonabelian group $M_p(2,2)$ satisfies the hypothesis. □

Exercise 362. If a p-group G, $p > 2$, has a nonmetacyclic minimal nonabelian subgroup S but has no normal subgroup isomorphic to E_{p^3}, then $S \cong S(p^3)$.

Hint. Use Lemma 65.1 and Theorem 10.4. □

Exercise 363. (i) Describe the p-groups containing a cyclic subgroup of index p^2 but having no normal cyclic subgroup of index p^2.
(ii) Moreover, describe the p-groups containing an absolutely regular subgroup of index p^2 but containing no normal absolutely regular subgroup of index p^2.

Hint. To solve (i), use [Nin]. □

Exercise 364. Study the p-groups G, $p > 3$, of order $> p^p$ all of whose subgroups of order p^p are two-generator.

Hint. Let $M \le G$ be of order p^{p+1}. By hypothesis, all maximal subgroups of M are two-generator. Set $\overline{M} = M/\mho_1(M)$. By Theorem 5.8,

$$|\overline{M}| \le p^3 < p^{p-1}.$$

It follows that M is absolutely regular (Theorem 9.8 (a)). This implies that G has no subgroup of order p^{p-1} and exponent p. As a consequence of Theorem 12.1 (b), G is absolutely regular. □

Exercise 365. Let ϕ be an automorphism of prime order $q > 2$ of a 2-group G of exponent > 2. Is it true that there is $x \in G$ of order > 2 such that $\phi(x) \ne x$?

Exercise 366. Let G be a 3-group of exponent > 3. Suppose that $\phi \in \mathrm{Aut}(G)$ is of prime order $q > 3$. Is it true that there is $x \in G$ of order > 3 such that $\phi(x) \ne x$?

Exercise 367. Let G be a p-group of exponent $> p$ and $|G/\mho_1(G)| = p^w$. Suppose that a prime $q \ne p$ does not divide $p^i - 1$ for $i = 1, \dots, w$. Is it true that q does not divide $|\mathrm{Aut}(G)|$?

Exercise 368 ([HH, Lemma 1.2.5]). Let G be a p-solvable group with $O_{p'}(G) = \{1\}$. Write $P = O_p(G)$ and $\overline{G} = G/\Phi(P)$. Then $C_{\overline{G}}(\overline{P}) = \overline{P}$.

Hint. Using the Frattini argument, we prove that $O_{p'}(\overline{G}) = \{\overline{1}\}$. Therefore, one may assume that $\Phi(P) = \{1\}$; then $O_p(G) = P$ and P is elementary abelian. It easily follows from this that $C_G(P) = P$. □

Exercise 369. If a p-group G, $p > 2$, is such that $Z_3(G)$ is metacyclic, then G is either metacyclic or a 3-group of maximal class.

Hint. Use Theorem 13.7. (One can obtain an essential information on the case $p = 2$ using § 50). □

Exercise 370. (i) Let a p-group G be regular of exponent p^e. If $|G/\mho_1(G)| = p^k$, then $|G| \le p^{ek}$.
(ii) Let a Q_8- and E_8-free 2-group G of exponent 2^e, $e > 2$, be such that $|\Omega_i(G)| = 2^{2i}$ for $i = 1, 2, 3$. Study the structure of G.

Hint. (ii) Use the classification of minimal nonmetacyclic 2-groups given in § 66. □

Exercise 371. Using Lemma 1.1 and Theorem 1.2, prove that if G has only one subgroup of order p, it is either cyclic or a generalized quaternion group.

Solution. We use induction on $|G|$. If G is abelian, it is cyclic, by a basic theorem on abelian p-groups. Now let G be nonabelian. Then G has no subgroup of type (p, p). By Lemma 1.4, it has a cyclic subgroup of index p so the result follows, by Theorem 1.2. □

Exercise 372. Using Theorem 1.2, prove that if p-group G of order $> p^n$ has only one subgroup of order $p^n > p$, then it is cyclic.

Exercise 373. If all subgroups isomorphic to E_{p^2} are conjugate in a p-group G, $p > 2$, then $\Omega_1(G) \cong E_{p^2}$, and then G is either metacyclic of a 3-group of maximal class.

Hint. The number of conjugates to any subgroup is a power of p. We have $e_2(G) = 1$ (Lemma 1.4) so that $\Omega_1(G) \cong E_{p^2}$. Therefore, G is either metacyclic or a 3-group of maximal class. □

Exercise 374. A regular 2-group G is abelian.

Solution. Assume that G is nonabelian. Then it contains a minimal nonabelian subgroup $S = \langle x, y\rangle$. As S is regular and $|S'| = 2$, one has, for some $z \in S'$, the following chain of implications:

$$(xy)^2 = x^2y^2z^2 = x^2y^2 \implies yx = xy \implies S \text{ is abelian},$$

a contradiction. □

Exercise 375. Let G be a p-group such that $K = \Omega_2(G)$ ($K = \Omega_2^\#(G)$) is irregular of maximal class. Is it true that G is also of maximal class?

Exercise 376. Study the p-groups G such that $G/\mathrm{K}_3(G)$ is minimal nonabelian metacyclic of order p^4 and exponent p^2.

Exercise 377. If G is a p-group of order $> p^{p+1}$ such that $G/\mathrm{K}_{p+1}(G)$ is of maximal class, so is G.

Solution. Clearly, $K = \mathrm{K}_{p+1}(G)$ is the unique normal subgroup of index p^{p+1} in G. Since $\mathrm{K}_{p+1}(G) < G'$, we get $|G : G'| = p^2$. Let H/K be absolutely regular of index p in G/K (see Theorem 9.5). Assume that H is not absolutely regular. Then $H/\mho_1(H)$ is of order $\geq p^p$ and exponent p (Theorem 9.8 (a)). It follows that $\mho_1(H) \leq K$ since $|G : \mho_1(H)| \geq p^{p+1}$ and $\mho_1(H) \lhd G$. Assume that $|H : \mho_1(H)| \leq p^p$. Let $\mho_1(H) \leq U \leq H$, where $U \lhd G$ and $|H : U| = p^p$. Then we must have $U = K$ since $|U| = |K|$, by the said in the first sentence. It follows that $\exp(H/K) = p$, contrary to a choice of H. Thus, the subgroup H is absolutely regular. If G is not of maximal class, then $G = H\Omega_1(G)$, where $|\Omega_1(G)| = p^p$ (Theorem 12.1 (b)). In that case, $L = H \cap \Omega_1(G)$ is G-invariant of index p in $\Omega_1(G)$ and G/L is not of maximal class (it is important that $|G| > p^{p+1}$); then G/K is not of maximal class since $|G/L : (G/L)'| > p^2 = |G : G'|$, a contradiction. Thus, G is of maximal class. □

Exercise 378. Study the 2-groups G admitting an involutory automorphism ϕ such that the group $\mathrm{C}_G(\phi)$ has only one involution.

Hint. See §§ 48–49. □

Exercise 379. If a nonabelian p-group G contains $< p + 1$ nonabelian subgroups of index p, then $\mathrm{d}(G) = 2$.

Hint. Assume that $\mathrm{d}(G) > 2$; then G is not minimal nonabelian. The set Γ_1 has at most $p + 1$ abelian members (Exercise 1.6 (a)) so

$$|\Gamma_1| \leq p + (p + 1) < 1 + p + p^2,$$

a contradiction. □

Exercise 380. For a nonabelian p-group G with $\mathrm{d}(G) = 3$ find the maximal possible number of subgroups of index p^2.

Exercise 381. Suppose that a p-group G, $p > 2$, has no normal subgroup isomorphic to E_{p^n}. Is it true that the ranks of subgroups of G are bounded?

Exercise 382. Classify the groups G of order $> p^3$ and exponent p that are not covered by subgroups isomorphic to E_{p^3}.

Solution. If $x \in G$ is not contained in a subgroup isomorphic to E_{p^3} and $x \in U < G$, where $U \cong \mathrm{E}_{p^2}$, then $\mathrm{C}_G(U) = U$. By Proposition 1.8 and Theorem 9.5, G is of maximal class and order $\leq p^p$. Assume that $|G| > p^3$. Then G_1, the fundamental subgroup of G, is covered by subgroups isomorphic to $\mathrm{S}(p^3)$. Such a group G satisfies the hypothesis (see Theorem 9.6 (f)). □

Exercise 383. Classify the groups G of order $> p^3$ that are not covered by abelian subgroups of order p^3.

Hint. Suppose that $x \in G$ is not contained in abelian subgroup of order p^3. Let $x \in A < G$, where A is of order p^2. Then $C_G(A) = A$, so G is of maximal class. Apply the above argument to G_1, the fundamental subgroup of G (see Exercise 382). □

Exercise 384. If a group G of order $> p^4$ and exponent p is not covered by nonabelian subgroups of order p^4, then $C_G(x)$ is abelian for all $x \in G - Z(G)$.

Solution. Suppose that $x \in G - Z(G)$ is not contained in a nonabelian subgroup of order p^4; then x is not contained in a nonabelian subgroup of order p^3. If $C_G(x)$ is non-abelian, it contains a minimal nonabelian subgroup L. As $|L| = p^3$ (Lemma 65.1), $x \notin L$. Then x is contained in the nonabelian subgroup $\langle x\rangle \times L$ of order p^4, a contradiction. Thus, $C_G(x)$ is abelian for all $x \in G - Z(G)$. □

Exercise 385. Let S be a proper nonabelian subgroup of a p-group G. Suppose that, whenever $A < S$ is maximal abelian and $B < G$ is maximal abelian containing A, then $B \in \Gamma_1$. Prove that $G = SZ(G)$ and all $p + 1$ maximal abelian subgroups are maximal in S.

Hint. Since S has at least $p + 1$ distinct maximal abelian subgroups, the set Γ_1 has exactly $p + 1$ distinct abelian members (see Exercise 1.6). Therefore, S has exactly $p + 1$ maximal abelian subgroups (indeed, if distinct $U, V < S$ are distinct maximal abelian and $U \le A \in \Gamma_1$ and $V \le B \in \Gamma_1$ are distinct abelian, then $A \neq B$ since the subgroup $\langle U, V\rangle$ is nonabelian). It follows that $|G : Z(G)| = p^2$ and $G = SZ(G)$.[8] □

Exercise 386. Let $S \cong S(p^3)$ be a proper normal subgroup of a p-group G and suppose that all maximal subgroups of G not containing S are metacyclic. Study the structure of G.

Hint. By Lemma 1.4, $S \nleq \Phi(G)$. □

Exercise 387. The following statements hold:

(a) Suppose that a group G of exponent p contains an abelian subgroup A of index p. Prove that $\mathrm{cl}(G) \le p - 1$.

(b) Suppose that the Frattini subgroup of a group G of exponent p is abelian. Prove that $\mathrm{cl}(G) \le (p - 1)\mathrm{d}(G)$.

(c) If a p-group G contains an elementary abelian subgroup E of index p, then $\mathrm{cl}(G) \le p$.

Solution. (a) Assume that $|Z(G)| = p$. Then $|G : G'| = p^2$ (Lemma 1.1). One may assume that $|G| > p^3$. In that case, $|Z(G/Z(G))| = p$, by Lemma 1.1, so, by induction, $G/Z(G)$ is of maximal class. Then G is also of maximal class since $|Z(G)| = p$. Now the result follows,

8 If $\mathrm{d}(S) = 2$, then S is minimal nonabelian.

by Theorem 9.5. Now let $|Z(G)| > p$ and let distinct $L_1, L_2 < Z(G)$ be of order p. Then, by induction, $cl(G/L_i)) \le p - 1$, $i = 1, 2$, hence $cl(G) \le p - 1$ since G is isomorphic to a subgroup of $(G/L_1) \times (G/L_2)$.

(b) Write $\overline{G} = G/\Phi(G)$ and let $\overline{G} = \overline{H}_1 \times \dots \overline{H}_d$, where $d = d(G)$ and $|\overline{H}_i| = p$ for $i = 1, \dots, d$. By (a), $cl(H_i) \le p - 1$ for all i, and therefore, by Fitting's lemma,

$$cl(G) \le \sum_{i=1}^{d} cl(H_i) \le (p-1)d.$$

(c) One may assume that $G' > \{1\}$. If $|Z(G)| = p$, then G is of maximal class, and by Theorem 9.5 and Exercise 9.13, G is of order $\le p^{p+1}$ which implies that $cl(G) \le p$. Now let $|Z(G)| > p$. Arguing as in (b) and using induction, one obtains that $cl(G) \le p$. □

The following stronger result holds [Hup1, Satz III.10.10]: If a regular p-group G has an abelian subgroup of index p, then $cl(G) \le p - 1$. It follows that if the Frattini subgroup of a regular p-group G is abelian, then $cl(G) \le (p-1) \cdot d(G)$.

Exercise 388. Suppose that the Frattini subgroup of a p-group G is elementary abelian. Prove that $cl(G) \le p \cdot d(G)$.

Hint. Use Exercise 387 (c) and Fitting's lemma. □

Exercise 389. Prove that, provided a p-group G, $p > 2$, contains a subgroup of order p^{11} and exponent p, then it contains a normal subgroup isomorphic to E_{p^5}.

Hint. We know that $|Aut(E_{p^n})|_p = p^{n(n-1)/2}$. Using this fact, Theorems 10.5 and 10.1 and the main result of [Kon1], prove that our group G contains a normal subgroup isomorphic to E_{p^5}. □

Exercise 390. Is it true that, provided a p-group G contains a subgroup of exponent p and order $p^{1+n(n-1)/2}$, then it contains a subgroup isomorphic to E_{p^n}?

Exercise 391. Study the p-groups with minimal nonabelian Frattini subgroup.

Hint. By Theorem 44.12, the Frattini subgroup of our group is metacyclic. □

Exercise 392. Let $P = M_p(2, 2)$ be a Sylow 2-subgroup of a group G, $p > 2$. Is it true that G is 2-nilpotent?

Exercise 393. Let $P \cong D_{2^n}$ be a Sylow subgroup of a group G. Prove that the centralizer $C_G(x)$ is 2-nilpotent for any involution $x \in G$.

Hint. Use Frobenius' theorem on normal p-complement (Theorem A.52.2) and Theorem A.22.1. □

Exercise 394. Let $P \cong SD_{2^n}$ be a Sylow subgroup of a group G and L a noncentral subgroup of order 2 in P. Prove that the centralizer $C_G(L)$ is 2-nilpotent.

Exercise 395. Let a 2-group P of maximal class be a Sylow subgroup of a group G and let L be a subgroup of order 4 in P. Prove that the centralizer $C_G(L)$ is 2-nilpotent.

Exercise 396. Let the minimal nonmetacyclic group P of order 2^5 be a Sylow subgroup of a group G and let $L < P$ be of order 2. Prove that the centralizer $C_G(L)$ is 2-nilpotent.

Hint. Use Frobenius' theorem on normal p-complement (Theorem A.52.2) and Theorem A.22.1. □

Exercise 397. Let a metacyclic minimal nonabelian 2-group P of order > 8 be a Sylow 2-subgroup of a group G and let $L < P$ be of order 2. Is it true that the centralizer $C_G(L)$ is 2-nilpotent?

Exercise 398. Suppose that the Frattini subgroup of a nonabelian two-generator p-group G is cyclic. Describe the structure of G.

Solution. Any maximal subgroup of G has a cyclic subgroup of index p so metacyclic. If G is metacyclic, it follows from $\Phi(G) = \mho_1(G)$ that G has a cyclic subgroup of index p. Now let G be nonmetacyclic. Then it is two-generator minimal nonmetacyclic so $G \in \{S(p^3), D\}$, where D is a group of maximal class and order 3^4 with $|\Omega_1(D)| = 3^2$ (§§ 66, 69). But $\Phi(D)$ is noncyclic. □

Exercise 399. Let $p > 2$ and let M be an absolutely regular subgroup of index p of a p-group G. If G does not split over M, then G is either absolutely regular or of maximal class.

Hint. Assume that G is neither absolutely regular nor of maximal class. Then, by Theorem 12.1 (a), there is in G a normal subgroup R of order p^p and exponent p. As $R \nleq M$, there is $x \in R - M$ (of order p). As $G = \langle x \rangle \cdot M$, a semidirect product, we get a contradiction. □

A p-group G is said to be L_n-group if $\Omega_1(G)$ is of order p^n and exponent p and $G/\Omega_1(G)$ is cyclic of order $> p$. By Theorem 17.4 (c), all proper subgroups of an L_p-group G are regular so, if G is irregular, it is two-generator.

Exercise 400. If G is an L_p-group of exponent $p^e > p^2$, then $|G/\mho_1(G)| = p^p$ and $\mho_1(G) \cong C_{p^{e-1}}$.

Solution. One has $|G| = p^{p+(e-1)}$, $G = CR$, where $C \cong C_{p^e}$ and $R = \Omega_1(G)$. It follows that $|G/\mho_1(G)| = p^p$. Indeed, this is true if G is regular (Theorem 7.2 (d)). If G is irregular, then $|G/\mho_1(G)| \geq p^p$ (Theorem 9.8 (a)). However, if $|G/\mho_1(G)| > p^p$, then $\exp(G) < p^e$, a contradiction. □

Exercise 401. If all maximal abelian subgroups of a nonabelian p-group G, $p > 2$, are normal, then G is regular.

Hint. Mimic the solution of Exercise 250 (but in this case the solution is simpler). □

Exercise 402. Suppose that a nonabelian p-group $G = \Omega_1(G)$, $p > 2$, and $H < G$ is generated by all subgroups isomorphic to $S(p^3)$ from G. If $E \triangleleft G$ is an elementary abelian subgroup of maximal order, then the set $H \cup E$ contains all elements of order p from G.

Solution. Assume that $x \in G - (H \cup E)$ is of order p. Then the subgroup $M = \langle x, E\rangle$ is nonabelian, by Theorem 10.1. By Lemma 57.1, there is $a \in E$ such that the subgroup $S = \langle a, x\rangle$ is minimal nonabelian. It follows from $\Omega_1(S) = S$ that $S \cong \mathrm{S}(p^3)$. As $S \nleq H$, we get a contradiction, Thus, x does not exist. □

If $G = \Sigma_9 \in \mathrm{Syl}_3(\mathrm{S}_9)$, then $\Omega_1(G) = G$ and the set Γ_1 has exactly two members of exponent 3.

Exercise 403. Let H be a proper subgroup of maximal class of a p-group G. If all subgroups of G containing H as a subgroup of index p are two-generator, then G is of maximal class.

Hint. Use Theorem 12.12 (a) and Exercise 10.10. □

Exercise 404. Let a p-group $G = A \times B$ and let $H \leq A$ and $F \leq B$ be characteristic. Is it true that $H \times F$ is characteristic in G?

Answer. "No". Indeed, let $A \cong B \cong \mathrm{C}_{p^n}$, $n > 2$. Let $A > H \cong \mathrm{C}_{p^2}$ and $B > F \cong \mathrm{C}_p$. Then $H \times F$ is not characteristic in G. □

Exercise 405. Let G be an irregular 3-group of maximal class and order $> 3^4$. Is it true that if a 3-group L and G are lattice isomorphic, then L is of maximal class?

Hint. One has $|L/\mho_1(L)| = |G/\mho_1(G)| = 3^3$ so L is not absolutely regular. Next,

$$\mathrm{e}_3(L) = \mathrm{e}_3(G) \not\equiv 1 \pmod 3$$

so L is irregular (Theorem 7.2 (d)). It follows that L is of maximal class (Theorem 13.5). We suggest to consider the case $|G| = 3^4$. □

Exercise 406. Let G be a p-group such that $\mathrm{c}_1(G) = 1 + p + \cdots + p^{p-1}$. Is it true that $|\Omega_1(G)| = p^p$?

Exercise 407. Classify the p-groups G of exponent p such that $\mathrm{s}_n(G) = 1 + p + 2p^2$ for some $n > 1$. (See Theorem 5.9.)

Exercise 408. Study the irregular p-groups G such that $\mathrm{c}_2(G) = p^{p-1}$. (See Theorem 13.2 (b).)

Exercise 409 ([Hig]). Classify the nonnilpotent solvable groups all of whose nonidentity elements have prime orders.

Hint. The Fitting subgroup $\mathrm{F}(G)$ is a p-group of exponent p for some prime p. Let $A/F(G)$ be the Fitting subgroup of $G/\mathrm{F}(G)$. Then A is a Frobenius group with the Frobenius kernel $\mathrm{F}(G)$ so that $|A/\mathrm{F}(G)| \cong \mathrm{C}_q$, a prime $q \neq p$. Assume that $A < G$ and let B/A be the Fitting subgroup of G/A. Then $B/\mathrm{F}(G)$ is a Frobenius group with kernel $A/\mathrm{F}(G)$ so $B/A \cong \mathrm{C}_r$, a prime $r \neq q$. If, in addition, $r \neq p$, then B is a Frobenius group so subgroup of B of order rq is cyclic, a contradiction (see [BZ, Chapter 10]). Thus, $r = p$. It follows that $p \mid q - 1$ so that $p < q$. Assume that $B < G$. If $B < C \leq G$, where $|C : B|$ is a prime, then $|C : B| = q$ (let us consider $G/\mathrm{F}(G)$). Then $q < p$, a contradiction. Thus, $B = G$, $G = D \cdot \mathrm{F}(G)$, $\mathrm{F}(G)$ is a group of exponent p, D is nonabelian of order pq, $p \mid q - 1$, $D' \cdot \mathrm{F}(G)$ is a Frobenius group.[9] □

Let $\mathrm{e}_n(G)$ be the number of subgroups isomorphic to E_{p^n} in a p-group G.

Exercise 410. If a p-group G is irregular, then $\mathrm{e}_{p-1}(G) \equiv 1 \pmod p$. Is it true that if $p > 2$ and $k < p - 1$, then $\mathrm{e}_k(G) \equiv 1 + p \pmod{p^2}$?

Hint. Use Hall's enumeration principle and Theorem 12.1 (a). □

Exercise 411. If all nonabelian subgroups of a special p-group G are special, then $|G| = p^3$.

Solution. Assume that $|G| > p^3$. Let $S \le G$ be minimal nonabelian. As S is special, it follows that $\mathrm{Z}(S) = S'$ is of order p hence $|S| = p^3$. Assume that $S < G$ and let $S < M \le G$, where $|M : S| = p$. Then the nonabelian subgroup M is not special (check!), a contradiction. □

Exercise 412. Suppose that a p-group G is neither cyclic nor a 2-group of maximal class. Then $\mathrm{e}_2(G) \equiv 1 \pmod p$.

Hint. One may assume that $|G| > p^3$. We proceed by induction on $|G|$. Since (Sylow)

$$\mathrm{e}_2(G) + \mathrm{c}_2(G) = \mathrm{s}_2(G) \equiv 1 \pmod p,$$

it suffices to show that $\mathrm{c}_2(G) \equiv 0 \pmod p$ (Theorem 1.17 (b)). By Lemma 1.4, there is in G a normal subgroup $R \cong \mathrm{E}_{p^2}$. If G/R is cyclic, then $\mathrm{e}_2(G) \equiv 1 \pmod p$, and we are done. Now let G/R be noncyclic. Let $R \le T \lhd G$ be such that $G/T \cong \mathrm{E}_{p^2}$, and let $H_1/T, \dots, H_{p+1}/T$ be all subgroups of order p in G/T. Then (check!)

$$\mathrm{c}_2(G) = \sum_{i=1}^{p+1} \mathrm{c}_2(H_i) - p\mathrm{c}_2(T).$$

If $|G| > p^4$, then all noncyclic subgroups H_i are not 2-groups of maximal class so $\mathrm{c}_2(G) \equiv 0 \pmod p$, by the displayed formula and induction, and we are done. In the case $|G| = p^4$ we suggest the reader to finish the solution.[10] □

Exercise 413. Let $p > 2$ and $n > 3$. Denote by $\mu_n(G)$ the number of metacyclic subgroups of order p^n in a p-group G. Prove that if a p-group G is neither metacyclic nor a 3-group of maximal class, then $\mu_n(G) \equiv 0 \pmod p$.

---

9 We suggest the reader to classify the nonnilpotent solvable groups all of whose elements have prime power orders, the main result of [Hig].

10 This is a partial case of Theorem 1.17 (b). We suggest to prove the last theorem for $\mathrm{c}_n(G)$, $n > 2$.

Hint. First consider the case when $|G| = p^{n+1}$ and apply Theorem 12.1 (a). When $|G| > p^{n+1}$, apply Hall's enumeration principle, induction and the result obtained in the previous sentence.[11] □

Exercise 414. Let $n > p$. Denote by $\rho_n(G)$ the number of absolutely regular subgroups of order p^n in a p-group G. Prove that if a p-group G is neither absolutely regular nor of maximal class, then $\rho_n(G) \equiv 0 \pmod{p}$.

Hint. Use Theorem 12.1 (a) and Hall's enumeration principle. □

We suggest the reader to consider in Exercise 414 the case $n = p$, using Theorem 13.5.

Exercise 415 (Theorem 13.6). Let $n \geq p + 1$. Denote by $\delta_n(G)$ the number of subgroups of maximal class and order p^n in a p-group G. Prove that if p-group G is not of maximal class, then $\delta_n(G) \equiv 0 \pmod{p^2}$.

Hint. If $|G| = p^{n+1}$, this follows from Theorem 12.12 (c). If $|G| > p^{p+1}$, use Hall's enumeration principle and some results of § 12. Use also Remark 10.5. □

Exercise 416. Let G be a p-group of order p^n and $\mathrm{d}(G) = d$. Then

$$|\mathrm{Aut}(G)| = |\mathrm{GL}(d, p)||\Phi(G)|^d$$

if and only if whenever B_1 and B_2 are two (ordered) minimal systems of generators of G, then there exists $\phi \in \mathrm{Aut}(G)$ such that $B_2 = B_1^{\phi}$.

Hint. The group $\mathrm{Aut}(G)$ acts in the fixed-point-free way on each its orbit on the set of ordered bases of G. The above presented number is the number of all bases of G. □

Exercise 417. Let $H \triangleleft G$ and suppose that H is the union of $\mathrm{k}_G(H)$ conjugate G-classes. Let G_1 be the fundamental subgroup of a p-group G of maximal class and order p^n. Is it true that, provided $n > p + 1$, then $\mathrm{k}(G) = \mathrm{k}_G(G_1) + p(p - 1)$?

Hint. Use Theorem 9.6. □

Exercise 418. Find $\mathrm{k}(G)$, where G is a 3-group of maximal class and order 3^n.

Hint. Take into account that the fundamental subgroup G_1 of G is either abelian or minimal nonabelian. □

Exercise 419. Let S be a minimal nonabelian group of order p^n. Is it true that a group G of order p^n such that $\mathrm{k}(G) = \mathrm{k}(S)$ is also minimal nonabelian?

Answer. "No". Indeed, $\mathrm{k}(\mathrm{M}_{16}) = 10 = \mathrm{k}(\mathrm{D}_8 \times \mathrm{C}_2)$. □

Exercise 420. Classify the nonabelian groups of order p^n with maximal possible $\mathrm{k}(G)$. Is it true that this will be if and only if $|G : \mathrm{Z}(G)| = p^2$?

11 If G is a minimal nonmetacyclic group of order 2^5, then $\mu_4(G) = 7 \equiv 1 \pmod{2}$.

Exercise 421. Classify the p-groups containing exactly p minimal nonabelian subgroups of order p^3.

Hint. Let $S < G$ be minimal nonabelian of order p^3 and let $S < D \leq G$, where $|D : S| = p$. Then $C_D(S) < S$ so that D is of maximal class. Also $C_G(S) < S$ hence, by Exercise 10.10. G is of maximal class. Assume that $|G| > p^4$. Note that one can assume that $S \nleq G_1$, where G_1 is the fundamental subgroup of G. One has $|G : N_G(S)| = p$, and we conclude that $N_G(S) = D$ so $|G| = p^5$. Let $M \neq D$ be nonabelian subgroup of index p in G. Then $\alpha_1(M) = p$ so D and M together contain at least $2p - 1$ minimal nonabelian subgroups, a contradiction. Thus, $G = D$. □

Exercise 422. Classify the nonnilpotent groups G all of whose nonnormal subgroups have the same order.

Hint. Let $S = PQ$ be minimal nonnilpotent subgroup of G, where $Q = S' \in \mathrm{Syl}_q(S)$ and a cyclic $P \in \mathrm{Syl}_p(S)$. Then $|Q| = q$ (Theorem A.22.1) and S is normal in G. In that case, $N_G(P) = P$ since $N_G(P)$ is not G-invariant so, by Frattini's lemma, $G = S$ is minimal nonabelian. (Note that then all nonnormal subgroups of G are conjugate.) □

Exercise 423. Is it true that if a group G of maximal class of order p^5 and exponent p contains a multiple of p^2 nonabelian subgroups of order p^3, then G has an abelian subgroup of index p.

Exercise 424. Suppose that A is an elementary abelian 2-group of automorphisms of a p-group G of exponent p and order $> p$, $p > 2$. Then there is in G a normal A-invariant subgroup of order p^2.

Hint. The group W, which is the natural semidirect product of A and G, is supersolvable (Exercise 335). □

Exercise 425 (see [HH, Lemma 1.2.3]). If G is a p-solvable group with $O_{p'}(G) = \{1\}$, then $C = C_G(O_p(G)) \leq O_p(G)$.

Solution. One has $C \lhd G$. Assume that $C \nleq O_p(G)$; then $O_p(G) < O_p(G)C$. Let $M/O_p(G) \leq O_p(G)C/O_p(G)$ be a minimal normal subgroup of $G/O_p(G)$. Then $M = O_p(G) \times L$, where $L \lhd G$ is a π'-subgroup. As $L > \{1\}$ is contained in $O_{p'}(G) = \{1\}$, we get a contradiction. (Compare with Exercise 368.) □

Exercise 426. Let $\mu_k(G)$ be the number of subgroups of maximal class and order p^k in a p-group G of exponent p (in that case, $k \leq p$, by Theorem 9.5). Find all possible residues $\mu_k(G) \pmod{p^2}$.

Exercise 427. Let G be a minimal nonabelian p-group. Study the p-groups H such that $\mathrm{cd}(H) = \mathrm{cd}(G)$ and $\mathrm{k}(H) = \mathrm{k}(G)$.

Solution. Let

$$|G| = p^n, \quad |H| = p^m, \quad |H : H'| = p^s.$$

Then

$$\mathrm{cd}(H) = \mathrm{cd}(G) = \{1, p\}$$

and

$$\mathrm{k}(G) = |\mathrm{Irr}(G)| = p^{n-1} + p^{n-2} - p^{n-3},$$
$$\mathrm{k}(H) = p^s + \frac{p^m - p^s}{p^2} = p^{m-2} + p^{s-2}(p^2 - 1).$$

Assume that $n \geq m$. It follows that

$$p^{n-2} + p^{n-3}(p^2 - 1) = p^{m-2} + p^{s-2}(p^2 - 1) \implies p^m(p^{n-m} - 1) = (p^{n-1} - p^s)(p^2 - 1). \quad (4)$$

It follows from (4) that $s = n - 1$ and $m = n$. Set $|\mathrm{Z}(H)| = p^z$. Then

$$\mathrm{k}(G) = p^{n-1} + p^{n-3}(p - 1) = \mathrm{k}(H) = p^z + \frac{p^n - p^z}{p} = p^{n-1} + p^{z-1}(p - 1). \quad (5)$$

Since H is nonabelian, it follows that $z = n - 2$, i.e., $|H : \mathrm{Z}(H)| = p^2$. In that case, $H = S\mathrm{Z}(H)$, where S is minimal nonabelian. We suggest the reader to check if all such H satisfy the hypothesis. □

Exercise 428. Check that the following conditions on a p-group G and its subgroup H are equivalent:
(i) All maximal cyclic subgroups of H are maximal cyclic in G.
(ii) H is isolated in G.

Exercise 429. All maximal cyclic subgroups of a nonabelian p-group G are isolated if and only if one of the following holds:
(i) $\exp(G) = p$.
(ii) G is a dihedral 2-group.

Solution. Assume that G has a maximal cyclic subgroup C of order $> p$. If $C < D \leq G$, where $|D : C| = p$, then D is dihedral (Theorem 1.2). By Exercise 10.10, G is a 2-group of maximal class, and it is clear now that G is dihedral. □

Exercise 430. Prove that all noncyclic abelian subgroups of the group Σ_{p^2} are elementary abelian.

Exercise 431. If a p-group G contains exactly p^2 nonabelian maximal subgroups, then $\mathrm{d}(G) = 3$ and $\mathrm{cl}(G) = 2$.

Exercise 432. Let G be a p-group of exponent $p^e > p$. Show that for each cyclic $L < G$ of order p^e the centralizer $\mathrm{C}_G(L)$ is not minimal nonabelian.

Exercise 433. Let D be a nonabelian p-group containing a characteristic subgroup L such that D/L is a nonabelian group with cyclic center. Then for any p-group G, the Frattini subgroup of G has no G-invariant subgroup isomorphic to D.

Hint. Use Lemma 1.4. □

Exercise 434. Any group G of order 2^6 is metabelian.

Hint. If $|G : G'| = 2^2$, then G is of maximal class so metabelian. Now let $|G : G'| > 2^2$; then $|G'| \leq 2^3$. If $|G'| = 2^3$, then G' is abelian, by Lemma 1.4 since G' is a G-invariant subgroup of $\Phi(G)$. If $|G'| < 2^3$, then G' is abelian. □

There exists a nonmetabelian group of order p^6 for $p > 2$, a Sylow p-subgroup of the holomorph of E_{p^3}.

Let us prove, by induction on n, that if a 2-group G has order 2^{3n}, then its derived length $\mathrm{dl}(G) \leq n$. This is true for $n \leq 2$, by Exercise 434. Let $n \geq 3$. One may assume that $|G'| > 2^3$. Let $R < G'$ be G-invariant of order 2^3; then R is abelian (Lemma 1.4). One has $|G/R| = 2^{3(n-1)}$. Therefore, by induction, $\mathrm{dl}(G/R) \leq n - 1$, and hence $\mathrm{dl}(G) \leq n$.

Now let $p > 2$. Similarly, one can prove that the derived length of a p-group G of order p^{3n-1} does not exceed n.

Essentially better estimates, which are due to Hall and Mann, one can be found in Appendix 6. By Theorem A.6.10, if a p-group G has the derived length 4, then $|G| \geq p^{13}$. By Theorem A.6.9, if a p-group G has the derived length k, then $|G| = p^n$, where $n \geq 2^{k-1} + 2(k-1) - 2$.

Exercise 435. Let E be a normal subgroup of exponent $p^n > 2$ of a p-group G such that the set $N_G(E) - E$ has no element of order $\leq p^n$ (i.e., $\Omega_n(N_G(E)) \leq E$). Is the following assertion true? If $\alpha \in \mathrm{Aut}(G)$ induces the identity automorphism on E, then $o(\alpha)$ is a power of p.

Exercise 436. Describe all possible minimal nonabelian subgroups of a p-group G containing exactly p distinct maximal abelian subgroups of exponent $> p$.

Exercise 437. Given $n > 1$, let G be a p-group such that $c_n(G) = p^{p-1}$. Describe the subgroup $\Omega_n^{\#}(G)$.

Exercise 438. Study the p-groups G such that, whenever $H \in \Gamma_1$, $\alpha_1(G) - \alpha_1(H) \leq p$. (See § 76.)

Exercise 439. Find a minimal order of a group G of exponent p whose minimal degree of a representation by permutations is $\frac{p+1}{p^3}|G|$.

Exercise 440. Let $\phi \in \mathrm{Aut}(P)$ be of prime order q, where P is a p-group, $p \neq q$, $d(P) = d$. Is it true that if ϕ acts irreducibly on $P/\Phi(P)$ and trivially on $\Phi(P)$, then d is the least positive integer such that $q \mid p^d - 1$?

Hint. Let $S \leq G$ be minimal nonnilpotent, where G is a natural semidirect product of $\langle\phi\rangle$ and P and $\phi \in S$. Use Theorem A.22.1. □

Exercise 441. Let ϕ be a fixed-point-free automorphism of a group G. Is it true that $o(\phi) \mid |G^{\#}|$?

Exercise 442. Let $G \in \{M_p(m,n,1), M_p(m,n)\}$. Study the p-groups L that are lattice isomorphic with G.

Exercise 443. Let G be an irregular p-group all of whose maximal subgroups of exponent p have order p^p and normal. Is it true that either $|\Omega_1(G)| = p^p$ or $|G| = p^{p+1}$? (See § 300.)

Exercise 444. Study the p-groups all of whose noncyclic maximal abelian subgroups are elementary abelian (homocyclic).

Exercise 445. Let $H < G$ be an abelian H_p-subgroup of a p-group G. Then one has $|G : H| = p$.

Hint. Use [HogK]. Another solution. Let $S \leq G$ be minimal nonabelian. It follows from $S = \langle S - H \rangle$ that $|S| = p^3$, and we conclude that $|G : H| = p$ (see Theorem 10.33 for $p = 2$ and Mann's commentary to Problem 115 for $p > 2$). □

Exercise 446. The following statements hold:

(i) Let N, a normal subgroup of a p-group G, has no G-invariant subgroups isomorphic to E_{p^2}. Then N is either cyclic or a 2-group of maximal class.

(ii) Let G be a nonabelian p-group. If $Z_2(G)$ is cyclic, then G is a 2-group of maximal class and order $> 2^3$.

(iii) Let N be a normal subgroup of a p-group G. If $N \cap Z_2(G)$ is cyclic, then N is either cyclic or a 2-group of maximal class.

(iv) Describe the p-groups G such that $Z_p(G)$ is absolutely regular.

Hint. Apply Lemma 1.4, Theorems 1.17 (b) and 13.5. □

Exercise 447. Study a p-group G, $p > 2$, such that $Z_3(G)$ is metacyclic.

Hint. Observe that there is no in G normal subgroup of order p^3 and exponent p. Use Theorem 13.7. □

Exercise 448. Study the nonabelian p-groups with exactly one non-isolated minimal nonabelian subgroup.

Exercise 449. Let a 2-group P of order 2^7 (of order 2^{10}) possesses an automorphism of order 31 (of order 127). Is it true that P is special provided $P' = Z(P)$?

Exercise 450. Classify the p-groups G containing a nonabelian subgroup H of order p^3 such that, whenever $H < M \leq G$ with $|M : H| = p$, then $\mathrm{cl}(M) = 3$.

Hint. Use Exercise 10.10. □

Exercise 451. Study the special groups of order 2^6 possessing an automorphism of order 7. (A Sylow 2-subgroup of the simple Suzuki group $Sz(2^3)$, which is special of order 2^6, admits an automorphism of order 7.)

Exercise 452. If G is a group and $\chi \in \mathrm{Irr}(G)$, then $\chi(1)^2 \leq |G : Z(G)|$.

Solution. Let $Z = \mathrm{Z}(G)$. For any $\chi \in \mathrm{Irr}(G)$ one has $\chi_Z = \chi(1)\mu$ for $\mu \in \mathrm{Lin}(G)$ so that $|\chi(z)| = \chi(1)$ for $z \in Z$. Thus,

$$|G| = \sum_{x \in G} |\chi(x)|^2 \geq \sum_{z \in Z} |\chi(z)|^2 = |Z|\chi(1)^2 \implies \chi(1)^2 \leq |G : Z|. \qquad \square$$

Exercise 453. If G is a p-group and $\exp(G/G') = p$, then $G'/\mathrm{K}_3(G)$ is elementary abelian.

Solution. Without loss of generality, one may assume that $\mathrm{K}_3(G) = \{1\}$; then $\mathrm{cl}(G) = 2$. If $x, y \in G$, then $[x, y]^p = [x, y^p] = 1$. As G' is abelian, it is elementary. $\square$

Exercise 454 ([FT, Lemma 8.7]). Let A be a p'-group of automorphisms of a p-group P with $[P, A] = P$ and $[A, \Phi(P)] = \{1\}$. Then $\Phi(P) \leq \mathrm{Z}(P)$.

Solution. One has

$$[A, \Phi(P), P] = \{1\}$$

and

$$[\Phi(P), P, A] \leq [\Phi(P), A] = \{1\}$$

so that, by the Three Subgroups Lemma,

$$[P, \Phi(P)] = [P, A, \Phi(P)] = \{1\} \implies [\Phi(P) \leq \mathrm{Z}(P). \qquad \square$$

Exercise 455. Suppose that a p-group of exponent $> p$ is not absolutely regular and suppose that $\tau(G) < p$, where $\tau(G)$ is the number of maximal absolutely regular subgroups of exponent $> p$ in G. Then G is of maximal class.

Solution. Let $\mathcal{A}(G)$ be the set of maximal absolutely regular subgroups of exponent $> p$ in G. By hypothesis, $1 < \tau(G) < p$. Let $H \in \mathcal{A}(G)$ and $H < K \leq G$, where $|K : H| = p$. If K has no normal subgroup of order p^p and exponent p, then it is of maximal class (Theorem 12.1 (a)). If this holds for any choice of K, then G is of maximal class (Exercise 10.10). It follows from Theorems 9.5 and 9.6 that $K = G$, i.e., $H \in \Gamma_1$. If $|G| > p^{p+1}$, then $|\tau(G)| = 1$. Now assume that for some choice of K there is in K a normal subgroup $R = \Omega_1(K)$ of order p^p and exponent p (Theorem 12.1 (a)); then $K = RH$ so that $R \nleq \Phi(K)$. Every maximal subgroup of K not containing R has exponent $> p$. Assume that $R \nleq L < K$, where $|K : L| =$. Then $|\Omega_1(L)| = |M \cap L| = p^{p-1}$ so that L is either absolutely regular or of maximal class. Since there are in K at least $p + 1$ such subgroups L, one may assume that L is irregular of maximal class. Then, by Theorem 12.12 (b), the quotient group $K/\mho_1(K)$ is of order p^{p+1} and exponent p. Such a group K has no absolutely regular subgroup of index p, a contradiction since H is absolutely regular subgroup of index p in K. This case is impossible. $\square$

Problem. Study the non-absolutely regular p-groups G of exponent $> p$ satisfying $\tau(G) = p$.

Exercise 456. Classify the nonabelian p-groups G such that $|S : (S \cap \Phi(G))| = p^2$ for any minimal nonabelian $S \leq G$.

Solution. Since $\Phi(G)$ has no minimal nonabelian subgroups, it is abelian. Let A with $\Phi(G) < A < G$, where $|A : \Phi(G)| = p$; then A has no minimal nonabelian subgroups, by hypothesis, so it is abelian. Since all such A generate G, it follows that $C_G(\Phi(G)) = G$ so that $\Phi(G) \le Z(G)$. If $d(G) = 2$, then G is minimal nonabelian. If $d(G) > 2$, then G also satisfies the hypothesis. □

Exercise 457. Let A be a maximal abelian subgroup of exponent ≤ 4 of a 2-group P and let ϕ be an automorphism of P of odd order which centralizes $\Omega_2(C_P(A))$. Is it true that $\phi = \mathrm{id}_P$?

Exercise 458. Does there exist a nonabelian p-group all of whose maximal abelian subgroups are characteristic?

Exercise 459. Find $\alpha_1(\Sigma_{p^2})$, that is, the number of minimal nonabelian subgroups in $\Sigma_{p^2} \in \mathrm{Syl}_p(S_{p^2})$.

Exercise 460. Let G be a p-group satisfying $|\mathcal{A}(G)| > p + 1$. Is it true that then

$$|\mathcal{A}(G)| \ge p^2 + 1?$$

Here $\mathcal{A}(G)$ is the set of all maximal abelian subgroups of G.

Exercise 461. If P is a Sylow p-subgroup of a p-solvable group G and $O_{p'}(G) = \{1\}$, then $Z(P) \le O_p(G)$.

Solution. By Exercise 425, $C_G(O_p(G)) \le O_p(G)$. It follows that

$$Z(P) \le C_G(O_p(G)) \le O_p(G),$$

and we are done. In particular, it is easy to prove by induction on $|P|$ that the p-length of G does not exceed $\mathrm{cl}(P)$. □

Exercise 462. Is it true that the p-group G, $p > 2$, all of whose sections of order p^p are abelian are regular?

Hint. Assume that G is minimal counterexample. Then G is minimal irregular p-group so $d(G) = 2$. By Theorem 9.8 (a), $|G/\mho_1(G)| \ge p^p$. If $H/\mho_1(G) \lhd G/\mho_1(G)$ is of index p^p, then G/H is elementary abelian of order p^p so that $d(G) \ge p > 2$, a contradiction. □

Exercise 463. Let k be a positive integer and let G be a group of order p^n, where $n > \frac{k(k+1)}{2}$. Is it true that G has a normal abelian subgroup of order p^{k+1}?

Exercise 464. Classify the p-groups G such that $\Omega_1(G)$ is minimal nonabelian.

Solution. By Lemma 65.1, $|\Omega_1(G)| = p^3$ so that $\Omega_1(G) \in \{D_8, S(p^3)\}$. One may assume that $|G| > p^3$. Let $p = 2$. Then G is of maximal class, by Theorem 1.17 (a), and now it is clear that $G \cong SD_{16}$. If $p > 2$, then G is a group of Theorem 13.7 (b)–(c). □

Exercise 465. Let G be a p-group and let A be a nonidentity p'-group of automorphisms of G. Let $e = 1$ if $p > 2$ and $e = 2$ if $p = 2$. Then A is faithful on $\Omega_e(G)$.

Hint. If $a \in A^\#$ centralizes $O_e(G)$, then the group $W = \langle a, G\rangle$ has no minimal nonnilpotent subgroup, by Theorem A.22.1. Therefore. the group G is a direct factor of W hence $a = \mathrm{id}_G$, a contradiction. □

Exercise 466. Let A be a p'-group of automorphisms of a nonabelian p-group G. If any minimal nonabelian subgroup S of G is A-invariant and A acts trivially on $S/\Phi(S)$, then $A = \{1\}$.

Hint. Use Theorem 10.28 and Exercise 465. □

Exercise 467. Suppose that G is a 2-group and $\alpha \in \mathrm{Aut}(G)$ of odd order acts trivially on $\Omega_2^\#(G) > \{1\}$. Is it true that α is the identity automorphism?

Hint. See Remark 3. □

Exercise 468. An irregular p-group G, $p > 3$, contains a maximal subgroup H with $\mathrm{d}(H) > 2$.

Hint. Use Theorems 9.8 (a) and 5.8 (b). □

Exercise 469 (Exercise 1.8 (a)). Let G be a nonmetacyclic minimal nonabelian p-group. Then G' is a maximal cyclic subgroup of G so it is isolated in G.

Hint. One may assume that $|G| > p^3$ (otherwise, $\exp(G) = p > 2$). Then $\Omega_1(G) \cong \mathrm{E}_{p^3}$ and $G' < \Omega_1(G)$. One has $H/G' = \Omega_1(G/G') \cong \mathrm{E}_{p^2}$. Obviously, $\Omega_1(G) \le H$ and, since $|H| = |\Omega_1(G)|$, we get $H = \Omega_1(G)$. Assume that $G' < C < G$, where C is cyclic of order p^2. Then $C \nleq H(\cong \mathrm{E}_{p^3})$. It follows that $CH/G' \cong \mathrm{E}_{p^3}$, a contradiction. Thus, G' is a maximal cyclic subgroup of G so it is isolated in G. □

Exercise 470. If G' is the unique minimal normal subgroup of a nonmetacyclic p-group G of order $> p^3$, then $\mathrm{d}(G) > 2$.

Hint. Use Lemmas 65.2 (a) and 65.1. □

Exercise 471. Let $S \cong \mathrm{S}(p^3)$ be a proper subgroup of a p-group G, $p > 3$. Assume that, whenever $L < S$ is maximal and M is a maximal metacyclic subgroup of G containing L, then $M \in \Gamma_1$. Prove that $G = SC$, where $S = \Omega_1(G)$ and $C < G$ is cyclic.

Solution. Obviously, $L < M$, $G = MS$, by the product formula. One has $\mathrm{N}_G(L) \ge SM = G$ so that $L \lhd G$, and we conclude that $S \lhd G$. If $\Omega_1(G) = S$, the result follows from Theorem 13.7. Now let $\Omega_1(G) > S$. Let $x \in G - S$ be of order p and set $H = \langle x, S\rangle$; then H is regular (Theorem 7.1 (b)). It follows that $\exp(H) = p$ (Theorem 7.2 (b)). However, $H \cap M$ is the nonmetacyclic subgroup of a metacyclic group M, a contradiction. □

Exercise 472. Classify the nonabelian p-groups all of whose nonabelian subgroups have cyclic centers.

Exercise 473. Study the 2-groups whose automorphism groups are 2-groups of maximal class.

Exercise 474 (see [BZ, § 11.3]). Let a group G of order g be a p-group with the class number $\mathrm{k}(G)$, $H \le G$,

$$\mathrm{T}(G) = \sum_{\chi \in \mathrm{Irr}(G)} \chi(1), \quad \mathrm{f}(G) = \frac{\mathrm{T}(G)}{g}, \quad \mathrm{mc}(G) = \frac{\mathrm{k}(G)}{g}, \quad |H| = h.$$

Then:

(i) $\mathrm{f}(G) \le \mathrm{f}(H)$, $\mathrm{mc}(G) \le \mathrm{mc}(H)$.

(ii) If G is minimal nonabelian p-group, then $\mathrm{mc}(G) = \frac{p^2+p-1}{p^3}$, $\mathrm{f}(G) = \frac{2p-1}{p^2}$.

(iii) If G is a nonabelian p-group of order p^n, then

$$\mathrm{mc}(G) \le \frac{p^2+p-1}{p^3} \implies \mathrm{k}(G) \le (p^2+p-1)p^{n-3}$$

and

$$\mathrm{f}(G) \le \frac{2p-1}{p^2} \implies \mathrm{T}(G) \le (2p-1)p^{n-2}.$$

In particular, G is abelian if at least one of the following conditions fulfilled:

$$\mathrm{mc}(G) > \frac{5}{8}, \quad \mathrm{f}(G) > \frac{3}{4}.$$

These results are best possible.

Solution. (i) One has

$$\mathrm{k}(G) = |\mathrm{Irr}(G)| \le \sum_{\phi \in \mathrm{Irr}(H)} |\mathrm{Irr}(\phi^G)| \le \mathrm{k}(H) \cdot \frac{g}{h} \implies \mathrm{mc}(G) \le \mathrm{mc}(H)$$

and

$$\mathrm{T}(G) = \sum_{\chi \in \mathrm{Irr}(G)} \chi(1) \le \sum_{\phi \in \mathrm{Irr}(H)} \phi^G(1) = \frac{g}{h} \sum_{\phi \in \mathrm{Irr}(H)} \phi(1) = \frac{g}{h}\mathrm{T}(H) \implies \mathrm{f}(G) \le \mathrm{f}(H).$$

(ii) If G is minimal nonabelian group of order p^n, then (see Lemma 65.1)

$$\begin{aligned}\mathrm{k}(G) &= |\mathrm{Z}(G)| + \frac{|G - \mathrm{Z}(G)|}{p} = p^{n-2} + \frac{p^n - p^{n-2}}{p} \\ &= p^{n-2} + p^{n-1} - p^{n-3} \implies \mathrm{mc}(G) = \frac{p^2+p-1}{p^3},\end{aligned}$$

and

$$\mathrm{T}(G) = \frac{g}{g'} + \frac{g - \frac{g}{g'}}{p^2} \cdot p = p^{n-1} + p^{n-1} - p^{n-2} = 2p^{n-1} - p^{n-2} \implies \mathrm{f}(G) = \frac{2p-1}{p^2}.$$

(iii) If $H \le G$ is minimal nonabelian, then (the last formula follows from (ii))

$$\mathrm{T}(G) = \sum_{\chi \in \mathrm{Irr}(G)} \chi(1) \le \sum_{\phi \in \mathrm{Irr}(H)} \phi^G(1) \le \frac{g}{h} \sum_{\phi \in \mathrm{Irr}(H)} \phi(1) \implies \mathrm{f}(G) \le \mathrm{f}(H) = \frac{2p-1}{p^2}. \quad \square$$

Exercise 475 ([BZ, Lemma 11.7]). For a group G one has $\mathrm{f}(G)^2 \leq \mathrm{mc}(G)$ with equality if and only if G is abelian.

Solution (A. Mann). Let $\mathrm{Irr}(G) = \{\chi_1, \dots, \chi_r\}$, $r = \mathrm{k}(G)$. Consider two r-dimensional vectors

$$a = (\chi_1(1), \dots, \chi_r(1)), \quad b = (1, \dots, 1).$$

By the Schwarz inequality, $(a \cdot b)^2 \leq \|a\|^2 \cdot \|b\|^2$ with equality if and only if $a = yb$ for some real y. Next,

$$a \cdot b = \mathrm{T}(G),\ \|a\|^2 = |G|,\ \|b\|^2 = \mathrm{k}(G) \implies \mathrm{T}(G)^2 \leq g \cdot \mathrm{k}(G) = g^2 \cdot \mathrm{mc}(G),$$

which is equivalent to $\mathrm{f}(G)^2 \leq \mathrm{mc}(G)$. If $\mathrm{f}(G)^2 = \mathrm{mc}(G)$, then $a = yb$ for some real y, and hence $\chi_1(1) = \dots = \chi_r(1)$ and G is abelian since $1 \in \mathrm{cd}(G)$. □

Exercise 476. Is it true that, provided G is a p-group with $\mathrm{f}(G) = \frac{1}{p}$, then $p = 2$?

Solution. Let

$$\overline{\delta}(G) = \{p^k \cdot 1, a_1 \cdot p^{c_1}, \dots, a_s \cdot p^{c_s}\}$$

be the degree vector of G (this means that $g/g' = p^k$ and there are in $\mathrm{Irr}_1(G)$ exactly a_i characters of degree p^{c_i} for all i). By [BZ, Proposition 11.6.2], due to Mann, $a_i = (p-1)a_i'$, where a_i are positive integers, $i = 1, \dots, s$. Let $0 < c_1 < \dots < c_s$. It follows from

$$|G| = p^m = p^k + (p-1)[a_1' p^{2c_1} + \dots + a_s' p^{2c_s}]$$

and

$$\mathrm{T}(G) = p^{m-1} = p^k + (p-1)[a_1' p^{c_1} + \dots + a_s' p^{c_s}]$$

that

$$|G| - \mathrm{T}(G) = p^{m-1}(p-1) = (p-1)[a_1' p^{c_1}(p^{c_1} - 1) + \dots + a_s' p^{c_s}(p^{c_s} - 1)],$$

and so

$$p^{m-1} = a_1' p^{c_1}(p^{c_1} - 1) + \dots + a_s' p^{c_s}(p^{c_s} - 1) \equiv 0 \pmod 2,$$

hence $p = 2$. □

Exercise 477. Study the p-groups G such that $\mathrm{f}(G) \geq \frac{1}{p+1}$.

Exercise 478. Let $H \leq G$. Then $\mathrm{mc}(H) = \mathrm{mc}(G)$ if and only if $\mathrm{f}(H) = \mathrm{f}(G)$. In that case, $G' = H'$ and all normal subgroups of H are normal in G.

Exercise 479. Study the p-groups G satisfying
(a) $\mathrm{f}(G) = \frac{2p-1}{p^2}$,
(b) $\mathrm{mc}(G) = \frac{p^2+p-1}{p^2}$.

Exercise 480. Suppose that F, H are distinct critical subgroups of a p-group G. Is it true that if $\mathrm{Z}(F) = \mathrm{Z}(H)$, then FH is critical in G?

Exercise 481. Study the p-groups all of whose maximal subgroups, except one, are minimal nonmetacyclic.

Exercise 482. Is it true that a p-group G is modular if any two its subgroups of equal order are permutable?

Hint. If G is nonmodular, then D_8 and $S(p^3)$ are sections of G and they do not satisfy the condition. □

Exercise 483. If each nontrivial normal subgroup of the two-generator nonabelian p-group G has only one G-invariant subgroup of index p, then G is of maximal class. The converse is also holds.

Hint. Note that the condition is inherited by epimorphic images, $Z(G)$ is cyclic. Use Lemma 1.4. □

Exercise 484. Classify the nonabelian Dedekindian p-groups.

Solution. If $S \leq G$ is minimal nonabelian, then $S \cong Q_8$ (Lemma 65.1). Therefore, by Corollary A.17.3, $G = S \times E$, where $\exp(E) \leq 2$. □

Exercise 485. Let $G > \{1\}$ be a finite group such that, whenever $a, b \in G - \Phi(G)$, then $\pi(o(a)) = \pi(o(b))$ (here $\pi(o(a))$ is the set of prime divisors of $o(a)$). Then G is a p-group for some prime p.

Solution. Assume that p, q are two distinct prime divisors of $|G/\Phi(G)|$ (it is easy to prove, using Frattini's lemma, that $\pi(G/\Phi(G)) = \pi(G)$). Then there are in $G - \Phi(G)$ a p-element x and q-element y, contrary to the hypothesis. Thus, $G/\Phi(G)$ is p-group for some prime p. In that case, G is also p-group. □

Exercise 486. Study the non-Dedekindian p-groups all of whose cyclic subgroups are TI-subgroups.

A p-group G is said to be *monotone* [Man11] if $H \leq K \leq G \Rightarrow d(H) \leq d(K)$. Abelian and metacyclic p-groups, obviously, are monotone. Also 3-groups of maximal class $\not\cong \Sigma_{3^2}$ are monotone. The irregular p-groups of maximal class, $p > 3$, are non-monotone (this follows from Theorems 9.8 (a) and 5.8 (a)). All extraspecial p-groups are monotone.

The following two results are easy consequences of Corollary 36.6.

Exercise 487. If a 2-group is monotone, then every its two-generator subgroup is metacyclic.

Exercise 488. If a p-group, $p > 2$, is monotone, then every its minimal nonabelian subgroup is either isomorphic to $S(p^3)$ or metacyclic.

Exercise 489. A p-group G is Dedekindian if and only if $|H|^2 \leq |G| \Rightarrow H \lhd G$.

Exercise 490. Prove that a minimal nonnilpotent group that is generated by minimal nonabelian subgroups is minimal nonabelian.

Exercise 491. If a metacyclic p-group G contains a proper normal subgroup H isomorphic to $M_2(p, p)$, then it possesses a cyclic subgroup of index 4.

Exercise 492. Classify the metacyclic p-groups all of whose noncyclic subgroups of equal order are isomorphic.

Exercise 493. If a metacyclic group G of exponent p^e contains a normal cyclic subgroup A of order p^e and G has no section isomorphic to Q_8, then the quotient group G/A is cyclic.

Solution. Assume that G is a counterexample. Then $e > 1$, $|G| > p^{e+1}$ and $|G| > p^4$ (check!). By Proposition 10.19, there is no in G a nonabelian subgroup of order p^3.

(i) Assume that $G/A \cong E_{p^2}$. Clearly, G is nonabelian. Then $A = \Phi(G) = \mho_1(G)$. Since A is cyclic, it is contained in a cyclic subgroup of order $p|A|$, a contradiction.

(ii) Assume that $|G : A| > p^2$. Then, by Lemma 1.4 and Theorem 1.2, the group G/A contains a subgroup $S/A \cong E_{p^2}$. By (i), S contains a cyclic subgroup L of index p, a contradiction since $|L| > |A|$. □

§ 294 p-groups, $p > 2$, whose Frattini subgroup is nonabelian metacyclic

In this section we present the proof of the following result.

Theorem 294.1 (G. L. Lange [Lan]). *Let $\Phi(G)$ be a nonabelian metacyclic Frattini subgroup of a p-group G, $p > 2$. Then there is in G a metacyclic subgroup T such that $\Phi(T) = \Phi(G)$.*

Let us prove that if G is a 3-group of maximal class, then $\Phi(G)$ is abelian (this fact we use in the proof of our theorem). Assume that this is false. Then $|G| > 3^4$. As G has a normal subgroup of type $(3, 3)$ (Lemma 1.4), it has no normal cyclic subgroup of order p^2 (Exercise 9.1 (b)). It follows that $|G_1'| = 3$. Next, $\mathrm{d}(\Phi(G)) = 2$ (Exercise 9.13). In that case, by Lemmas 36.6 and 65.2 (a), the subgroup G_1 is minimal nonabelian. As $\Phi(G) < G_1$, the subgroup $\Phi(G)$ is abelian, contrary to the assumption.

Proof of Theorem 294.1. Suppose that G is a counterexample of minimal order. Then G is nonmetacyclic. One has $\Phi_0 = \Omega_1(\Phi(G)) \cong \mathrm{E}_{p^2}$ (Lemma 1.4). Since $|G : \mathrm{C}_G(\Phi_0)| \le p$, we get $\Phi_0 \le \mathrm{Z}(\Phi(G))$. Since $\Phi(G)$ is nonabelian, it follows that the quotient $\Phi(G)/\Phi_0$ is noncyclic, and we conclude that $|\Phi(G)| \ge p^4$ and $\Phi_0 \le \Phi(\Phi(G))$. Let T be the largest subgroup of G containing $\Phi(G)$ and such that $\Omega_1(T) = \Phi_0 \le \mathrm{Z}(T)$ (this is legitimate since $|G : \mathrm{C}_G(\Phi_0)| \le p$). Then T is metacyclic (Theorem 12.1 (a), Theorem 9.8 (a) and Corollary 36.6) and $|T : \Phi(G)| \le p^2$. One has $T < G$ (note that G is a counterexample).

Let $T < H < G$, where $|H : T| = p$. Assume that H is a p-group of maximal class; then $p = 3$ (see § 9). Then $H < G$. Let $H < F \le G$, where $|F : H| \le 3$. Assume that F is a 3-group of maximal class. As $F/T \cong \mathrm{E}_{3^2}$, then $T = \Phi(F)$, a contradiction (see the paragraph preceding the theorem). Then there is in F a G-invariant subgroup $E_F \cong \mathrm{E}_{3^3}$. Let $G/H = F_1/H \times \cdots \times F_k/H$, where all F_i/H are of order 3. By Theorem 13.7, there is in F_i a normal subgroup $E_i \cong \mathrm{E}_i$. In view of Theorem 10.4, one may assume that all E_i are normal in G. Set $A = E_1 \dots E_k$; then $\Phi_0 \le \mathrm{Z}(A)$, $AH = G$ and $A \cap H = \Phi_0$. It follows that $\mathrm{cl}(A) \le 2$ so A is regular (Theorem 7.1 (b)) and we conclude that $A = \Omega_1(G)$ is of exponent p. In that case, $H \cap A = \Phi_0$ so that

$$\overline{G} = G/\Phi_0 = (H/\Phi_0) \times (A/\Phi_0) = \overline{H} \times \overline{A}.$$

It follows that

$$\Phi(\overline{H}) = \Phi(\overline{G}) \implies \Phi(H) = \Phi(G)$$

since $\Phi_0 \le \Phi(H)$. As H is of maximal class, its Frattini subgroup is nonabelian, a contradiction. Thus, $\Phi(G) < T$. Hence, any subgroup H of G containing T as a subgroup of index p is not of maximal class. By Theorems 13.7 and 10.4, H contains a G-invariant subgroup $E_H \cong \mathrm{E}_{p^3}$; then $H = TE_H$. Repeating a similar argument, we show that $G/\Phi_0 = T/\Phi_0 \times A/\Phi_0$, where A/Φ_0 is elementary abelian. As above, this implies $\Phi(T) = \Phi(G)$. As T is metacyclic, we are done. □

https://doi.org/10.1515/9783110533149-038

Recall that a p-group G is absolutely regular if $|G/\mho_1(G)| < p^p$. By Theorem 9.8 (a), absolutely regular p-groups are regular.

Problem. Let $\Phi(G)$ be the absolutely regular Frattini subgroup of a p-group G, $p > 2$. Suppose that $\exp(\Phi(G)) > p$. Find a condition implying $\Phi(G) = \Phi(T)$, where $T \leq G$ is either absolutely regular or a group of maximal class.

§ 295 Any irregular *p*-group contains a non-isolated maximal regular subgroup

Recall that a subgroup H of a p-group G is said to be *isolated* (in G) if

$$C \cap H > \{1\} \implies C \leq H \quad \text{for any cyclic } C \leq G.$$

If $H < G$ is isolated and $H < M < G$, then H is isolated in M. If, in addition, $|M : H| = p$, then all elements of the set $M - H$ have order p and hence $\Omega_1(M) = M$. In that case, therefore, if $\exp(H) > p$, then the subgroup M is irregular and $H = \mathrm{H}_p(G)$ is the Hughes subgroup of M. The subgroups $\{1\}$ and G are isolated in G.

Below we prove the following property of irregular p-groups.

Proposition 295.1. *An irregular p-group G contains a non-isolated maximal regular subgroup.*

Proof. Assume that all maximal regular subgroups of G are isolated in G. Let $p = 2$; then G is nonabelian and all regular subgroups of G are abelian. In that case, all maximal abelian subgroups are isolated in G. Therefore, by Theorem 252.1, $\exp(G) = 2$ so G is regular, a contradiction.

Now let $p > 2$. By Theorem 7.1 (b), $\exp(G) = p^e > p$. Let $L < G$ be cyclic of order p^e and let $L \leq R < G$, where R is a maximal regular subgroup of G. Then R is noncyclic (indeed, any p-group with cyclic subgroup of index $p > 2$ is regular, by Theorem 7.1 (c) so that $L < R$ and $|\Omega_1(R)| > p$, by Proposition 1.3). Let $R < M \leq G$, where $|M : R| = p$; then M is irregular, by a choice of R. As R is isolated in G so in M and $|M : R| = p$, all elements of the set $M - R$ have order p. As $\Omega_1(R)$ is characteristic in $R \lhd M$, we get $\Omega_1(R) \lhd M$. Let $x \in M - R$ and $X = \langle x \rangle$; then $|X| = p$. Set $H = X\Omega_1(R)$. Since $H - \Omega_1(R) \subset M - R$, all elements of the set $H - \Omega_1(R)$ have order p so that $\exp(H) = p$ hence the subgroup H is regular (Theorem 7.1 (b)). Let $H \leq S < G$, where S is a maximal regular subgroup of G. As $x \notin R$ and $x \in \langle x, R \rangle = H \leq S$, it follows that $R \neq S$. Being regular, R is generated by its cyclic subgroups of order $\exp(R) = p^e (> p)$ (in other words, $R = \Omega_e^{\#}(R)$). Therefore, there is in R a cyclic subgroup C of order $p^e > p$ not contained in S. But $\Omega_1(C) < \Omega_1(R)$ ($<$ since $\Omega_1(R)$ is noncyclic). Therefore,

$$\{1\} < C \cap \Omega_1(R) \leq C \cap H \leq C \cap S.$$

Since $C \nleq S$, the maximal regular subgroup S is not isolated in G. □

Alternate proof of Proposition 295.1. As above, assume that all maximal regular subgroups are isolated in G. Let $M \leq G$ be minimal irregular and let $H < M$ be maximal; then H is regular and $\mathrm{d}(M) = 2$. Let $H \leq R < G$, where R is maximal regular in G; then R is isolated in G hence the subgroup $H = M \cap R$ is isolated in M (Exercise A.101.7). Thus, all maximal subgroups of M are isolated. Assume that $G = M$. By Exercise A.101.7, $\Phi(G)$ is isolated in $H \in \Gamma_1$ and $|H : \Phi(G)| = p$ since $\mathrm{d}(G) = 2$. Then

https://doi.org/10.1515/9783110533149-039

we have $\Omega_1(H) \geq \langle H - \Phi(G)\rangle = H$ so $\exp(H) = p$ since H is regular (Theorem 7.2 (b)). Thus, all maximal subgroups of G have exponent p which implies that $\exp(G) = p$ and hence G is regular (Theorem 7.1 (b)), contrary to the assumption. Now let $M < G$. By the above, all maximal subgroups of M are isolated so, by the above, M is regular, a final contradiction. □

Exercise 1. Prove that the p-groups G all of whose maximal subgroups are isolated have exponent p.

Solution. By Proposition 295.1, the group G is regular. Let $H \in \Gamma_1$; then $|G : H| = p$ and $\Omega_1(G) = \langle G - H\rangle = G$ so $\exp(G) = p$ (Theorem 7.2 (b)). □

Problem 1. Study the p-groups without nontrivial isolated subgroups.

Problem 2. Study the irregular p-groups containing exactly one non-isolated maximal subgroup.

§ 296 Non-Dedekindian p-groups all of whose nonnormal maximal cyclic subgroups are C-equivalent

Two subgroups C_1, C_2 of a p-group G are said to be C-equivalent if $C_1^G = C_2^G$. It is asked, in Problem 4041 to classify the non-Dedekindian p-groups all of whose nonnormal maximal cyclic subgroups are C-equivalent. It is proved in Theorem 296.1 that all such groups are metacyclic. Note that in metacyclic group D_{16} not all nonnormal cyclic subgroups are C-equivalent.

Theorem 296.1. *Let G be a non-Dedekindian p-group all of whose nonnormal maximal cyclic subgroups are* C*-equivalent. Then G is metacyclic of class* 2.

Proof. Let G be a non-Dedekindian p-group all of whose nonnormal maximal cyclic subgroups are C-equivalent. Let Z be a maximal cyclic subgroup which is not normal in G and set $H = Z^G$ so that $H < G$.

We claim that $H = G_0$ is the subgroup generated by all nonnormal subgroups in G. Indeed, if there is a nonnormal subgroup X which is not contained in H, then there is $x \in X - H$ such that the cyclic subgroup $\langle x\rangle$ is not normal in G. Let L be a maximal cyclic subgroup in G which contains $\langle x\rangle$. Then L is also not normal in G but $L^G \neq H = Z^G$, contrary to our basic assumption.

We have proved that $G_0 = H = Z^G$ is the subgroup generated by all nonnormal subgroups of G. Moreover, if $L < G$ is an arbitrary nonnormal, then $L^G = G_0$. By Theorem 231.1, G is of class 2, $G/G_0 \neq \{1\}$ is cyclic and, if $g \in G - G_0$ is such that $\langle g\rangle$ covers G/G_0, then $\langle g\rangle \trianglelefteq G$. We claim that $\mathrm{d}(G) = 2$. Suppose that this is false so that $G/\Phi(G)$ is elementary abelian of order $\geq p^3$. Let $\langle y\rangle$ be a nonnormal maximal cyclic subgroup in G. As $\langle y\rangle^G = G_0$ and G/G_0 is cyclic, it follows that $y \not\leq \Phi(G)$. Then $y^p \in \Phi(G)$ and $\langle y\rangle\Phi(G) \trianglelefteq G$ so that

$$\langle y\rangle^G \leq \langle y\rangle\Phi(G)$$

and, by our basic assumption,

$$\langle y\rangle^G = H = G_0.$$

But, since $\mathrm{d}(G) > 2$, the group $G/(\langle y\rangle\Phi(G))$ is noncyclic, a contradiction. We have proved that $\mathrm{d}(G) = 2$.

If $g \in G - G_0$ is such that $\langle g\rangle$ covers the cyclic group G/G_0, then $g \in G - \Phi(G)$ (indeed, if $g \in \Phi(G)$, then $G = G_0\langle g\rangle \leq G_0\Phi(G) < G$, which is a contradiction) and $\langle g\rangle \trianglelefteq G$ since $\langle g\rangle \not\leq G_0$. Assume that the quotient group $G/\langle g\rangle$ is noncyclic. As $2 = \mathrm{d}(G) \geq \mathrm{d}(G/\langle g\rangle)$, we get $\mathrm{d}(G/\langle g\rangle) = 2$. It follows that $g \in \Phi(G)$, contrary to the choice of g. Thus, $G/\langle g\rangle$ is cyclic so that the group G is metacyclic, completing the proof. □

It is possible to define C-equivalence to any two subgroups: Two subgroups A, B of a p-group G are C-equivalent if $A^G = B^G$. A normal subgroups is self-C-equivalent.

https://doi.org/10.1515/9783110533149-040

In view of Theorem 10.28, if any two minimal nonabelian subgroups of a nonabelian p-group G are C-equivalent, then G itself is minimal nonabelian.

Problem 1. Study the non-Dedekindian p-groups with exactly two classes of C-equivalent nonnormal maximal abelian subgroups.

Problem 2. Study the p-groups with exactly p classes of C-equivalent nonnormal minimal nonabelian subgroups.

Problem 3. Study the nonmetacyclic p-groups all of whose minimal nonmetacyclic subgroups are C-equivalent.

§ 297 On 2-groups without elementary abelian subgroup of order 8

The 2-groups without elementary abelian subgroup of order 8 are described in detail in § 50.

To describe the orders of $2'$-automorphisms of a 2-group, we analyze in Proposition 297.1 the minimal nonnilpotent subgroups of its (partial) holomorph.

Proposition 297.1. *Suppose that a 2-group P has no subgroup isomorphic to E_8. If $q > 2$ is a prime divisor of $|\mathrm{Aut}(P)|$, then $q \in \{3, 5\}$.*

Proof. Assume that $|P| > 8$; otherwise, $P \in \{C_8, C_4 \times C_2, Q_8, D_8\}$, and we are done since

$$\mathrm{Aut}(C_8) \cong E_4, \quad \mathrm{Aut}(C_4 \times C_2) \cong D_8 \cong \mathrm{Aut}(D_8), \quad \mathrm{Aut}(Q_8) \cong S_4. \tag{1}$$

Suppose that $\phi \in \mathrm{Aut}(P)$ has a prime order $q > 2$. Let G be a natural semidirect product of $A = \langle \phi \rangle$ and P and let $S \leq G$ be minimal nonnilpotent such that $A < S$ (this is possible, by Sylow's theorem). Then $P_1 \in \mathrm{Syl}_2(S)$ is either elementary abelian or special, A acts irreducibly on $P_1/\Phi(P_1)$ (Theorem A.22.1). Taking in mind our aim, one may assume that $G = S$ is minimal nonnilpotent; then $P_1 = P$. If P is elementary abelian, then $P \cong E_4$, by hypothesis, and $q = 3$, by Sylow. Now let P be special. Then $|Z(P)| \leq 4$, by hypothesis.

Assume that $|Z(P)| = 2$; then P is extraspecial of order, say 2^{2m+1}, $m > 1$ (see (1)). The group P has a subgroup isomorphic to E_{2^m} and therefore $m = 2$. In that case, $P \cong Q_8 * D_8$ (note that the group $Q_8 * Q_8 \cong D_8 * D_8$ has a subgroup isomorphic to E_8). As A acts irreducibly on $P/\Phi(P)$, we get $q = |A| \mid (2^4 - 1)$ implying $q = 5$.

Now let P be special with noncyclic center; then $\Omega_1(P) = Z(P) \cong E_4$, by hypothesis, and this implies $|P| > 2^4$. By the Frobenius–Schur formula (see [BZ, § 4.6, formula (16)]), one has

$$4 = 1 + t(P) = \sum_{\chi \in \mathrm{Irr}_1(P)} \nu_2(\chi)\chi(1), \tag{2}$$

where $t(P)$ denotes the number of involutions in P, the Frobenius–Schur indicator $\nu_2(\chi) \in \{-1, 1\}$ if $\bar{\chi} = \chi$ and $\nu_2(\chi) = 0$ if $\bar{\chi} \neq \chi$. Let $|P/P'| = 2^{2m}$ (Theorem A.22.1 since, in the case under consideration, P/L is extraspecial for each $L < Z(P)$ of index 2; note that $|Z(P) : \ker(\chi)| = 2$ for each $\chi \in \mathrm{Irr}_1(P)$, where $\mathrm{Irr}_1(P)$ is the set of all nonlinear irreducible characters of P). If $\chi \in \mathrm{Irr}_1(P)$, then $P/\ker(\chi)$ is extraspecial of order 2^{1+2m} so that $\chi(1) = 2^m$. Therefore, by (2), since

$$\mathrm{Irr}_1(P/\ker(\chi)) = \{\chi\}, \quad |\mathrm{Irr}_1(G)| = 3, \quad c_1(Z(G)) = t(G) = 3, \quad |\mathrm{Lin}(G)| = |G/G'| = 2^{2m},$$

one has

$$4 = 1 + t(P) = 2^{2m} + 2^m \sum_{\chi \in \mathrm{Irr}_1(P)} \nu_2(\chi) = 2^m(2^m + \tau), \quad \text{where } \tau = \sum_{\chi \in \mathrm{Irr}_1(P)} \nu_2(\chi). \tag{3}$$

https://doi.org/10.1515/9783110533149-041

It follows from (3) that $m = 2$ and so $\tau = -3$, and we conclude that $\nu_2(\chi) = -1$ for each $\chi \in \mathrm{Irr}_1(P)$. Then we have $|P| = 2^{2+2m} = 2^6$. Applying to $G/\ker(\chi)$ ($\chi \in \mathrm{Irr}_1(G)$) the Frobenius–Schur formula on the number of involutions again, we get

$$t(G/\ker(\chi)) = |G/G'| - \chi(1) - 1 = 16 - 4 - 1 = 11$$

so that $G/\ker(\chi) \cong Q_8 * D_8$. As ϕ acts irreducibly on $P/Z(P) \cong E_{2^4}$, we get $o(\phi) = q = 5$ (Theorem A.22.1). □

There is in the Atlas [HS] a special group No. 187 (denote it by X) of order 2^6 and exponent 4, it has exactly three involutions generating $Z(X) \cong E_4$. The group X has automorphisms of orders 3 and 5 and possesses all properties described in the second part of the proof of Proposition 297.1; in particular, all three epimorphic images of X of order 2^5 are extraspecial isomorphic to $Q_8 * D_8$.

It follows that if P is as in Proposition 297.1, then $\mathrm{Aut}(P)$ is a $\{2, 3, 5\}$-group.

Problem. Prove an analog of Proposition 297.1 for 2-groups without normal elementary abelian subgroups of order 8.

§ 298 Non-Dedekindian p-groups all of whose subgroups of order $\le p^s$ ($s \ge 1$ fixed) are normal

We prove here a deep result of Passman [Pas, Theorem 1.6].

Theorem 298.1. *Let G be a non-Dedekindian p-group all of whose subgroups of order $\le p^s$ ($s \ge 1$ fixed) are normal in G. Then $\Omega_s(G) \le Z(G)$.*

Proof. Let G be a non-Dedekindian p-group all of whose subgroups of order $\le p^s$ (where $s \ge 1$ is fixed) are normal in G.

We show first by induction on s that if the quaternion group Q_8 is not involved in $\Omega_s(G)$, then $\Omega_s(G) \le Z(G)$. This is trivial for $s = 1$, so we assume $s \ge 2$. Let $x, y \in G$ with $o(x) \le p^s$, $x \ne 1$, and consider the subgroup $H = \langle x, y\rangle$. Since, by hypothesis, $\langle x\rangle \trianglelefteq G$, we have $\langle x\rangle \trianglelefteq H$ and also $J = \Omega_1(\langle x\rangle) \trianglelefteq G$ (note that $|J| = p$). Now in G/J all subgroups of order $\le p^{s-1}$ are normal and $\Omega_s(G)/J \ge \Omega_{s-1}(G/J)$. Thus, the group Q_8 is not involved in $\Omega_{s-1}(G/J)$ and so, by induction, we get here $\Omega_{s-1}(G/J) \le Z(G/J)$ implying $[x, y] \in J$ and hence $H' \le J$. Assume that the subgroup H is nonabelian; then $H' = J$ has order p so H is minimal nonabelian, by Lemma 65.2 (a). We consider two cases.

(i) Suppose first that $\langle y\rangle \ge J$. Let L be a nonnormal subgroup of H. Then L does not contain $H' = J$ and so $L \cap \langle y\rangle = \{1\}$. Thus

$$|L| \le |H : \langle y\rangle| \le o(x) \le p^s.$$

This is a contradiction since $|L| \le p^s$ implies that $L \trianglelefteq G$. Thus, the subgroup H is Dedekindian. If H is nonabelian, then Q_8 is contained in H. Since $s > 1$, the subgroup Q_8 is contained in $\Omega_s(G)$, contrary to the assumption. Thus, the subgroup H is abelian and so $[x, y] = 1$.

(ii) Now assume that $\langle y\rangle \not\ge J$ so that $\langle x\rangle \cap \langle y\rangle = \{1\}$. If $\langle y\rangle \trianglelefteq H$, then

$$[x, y] \in \langle x\rangle \cap \langle y\rangle = \{1\}$$

and so H is abelian. Now suppose that $\langle y\rangle$ is not normal in H so that H is nonabelian and $o(y) \ge p^{s+1}$, by hypothesis. Let $z \in \langle y\rangle$ have order p^s. By Lemma 65.2 (a), H is minimal nonabelian (indeed, we have $|H'| = p$ and $d(H) = 2$) and so $z \in Z(H)$ (since $z \in \Phi(\langle y\rangle \le \Phi(H) = Z(H))$. Set $w = xz$. Then w is of order p^s and $w^{p^{s-1}} = x^{p^{s-1}} z^{p^{s-1}}$ is not contained in J since $\langle x\rangle \cap \langle y\rangle = \{1\}$. But $\langle w\rangle \trianglelefteq H$, and so

$$[H, \langle w\rangle] \le \langle w\rangle \cap J = \{1\}.$$

Thus, $w \in Z(H)$ implying that $x = wz^{-1} \in Z(H)$ and so $H = \langle x, y\rangle$ is abelian. We have proved that $[x, y] = 1$ and so $\Omega_s(G) \le Z(G)$ since $\Omega_s(G)$ is generated by all elements x of order $\le p^s$ and y was an arbitrary element in G.

https://doi.org/10.1515/9783110533149-042

It remains to consider the case $p = 2$ and Q_8 is involved in $\Omega_s(G)$. We first observe by induction on t with $1 \leq t \leq s$ ($s \geq 2$) that $\Omega_t(G)$ is of exponent $\leq 2^t$. This is clear for $t = 1$. Let $t < s$ and suppose that $\Omega_t(G)$ is of exponent $\leq 2^t$. Now $\Omega_{t+1}(G)$ is generated by normal cyclic subgroups of order $\leq 2^{t+1}$ so $\Omega_{t+1}(G)/\Omega_t(G)$ is generated by normal and hence central subgroups of order 2. Thus, $\Omega_{t+1}(G)/\Omega_t(G)$ is elementary abelian and so

$$\exp(\Omega_{t+1}(G)) \leq \exp(\Omega_{t+1}(G)/\Omega_t(G)) \exp(\Omega_t(G)) \leq 2 \cdot 2^t = 2^{t+1}.$$

We have proved that $\exp(\Omega_s(G)) \leq 2^s$.

Let Y be any subgroup in $\Omega_s(G)$. Then Y is generated by its cyclic subgroups and since each of them has order $\leq \exp(\Omega_s(G)) \leq 2^s$, each one is normal in G. Then $Y \trianglelefteq G$ and so $Y \trianglelefteq \Omega_s(G)$. Therefore, $\Omega_s(G)$ is a Dedekindian group and since Q_8 is involved in $\Omega_s(G)$, it follows that $\Omega_s(G) = Q \times A$ is Hamiltonian, where $Q \cong Q_8$ and A is elementary abelian.

Finally, we show that $G = \Omega_s(G)$. If not, then clearly $s = 2$ and G has an element x of order 8. Then x^2 is an element of order 4 in $\Omega_s(G) = Q \times A$ so there exists an element $y \in Q \times A$ of order 4 with $[x^2, y] \neq 1$. Now, $o(y) = 4$ implying that $\langle y \rangle \trianglelefteq G$. But then acting with $\langle x \rangle$ of order 8 on $\langle y \rangle \cong C_4$, we see that x^2 centralizes $\langle y \rangle$, a contradiction. We have proved that $\Omega_s(G) = G$ and so G is Dedekindian. This is a final contradiction since G was supposed to be non-Dedekindian. The theorem is proved. □

§ 299 On p'-automorphisms of p-groups

Let, for a prime p, $\epsilon(p) = 1$ for $p > 2$ and $\epsilon(2) = 2$. Recall that if a p'-automorphism ϕ of a p-group P induces the identity automorphism on $\Omega_{\epsilon(p)}(P)$, then $\phi = \mathrm{id}_P$ (this follows easily from Frobenius' normal p-complement Theorem A.52.2 and Theorem A.22.1). We use this fact in the proof of the following theorem.

Theorem 299.1. *Suppose that a p-group P admits a p'-automorphism ϕ of a prime order q and $p \equiv 1 \pmod q$. Then there is in P a normal ϕ-invariant abelian subgroup A of exponent $\leq p^{\epsilon(p)}$ on which ϕ acts nontrivially.*

Proof. Assume, by way of contradiction, that ϕ centralizes all ϕ-invariant normal abelian subgroups of P of exponent $\leq p^{\epsilon}(p)$; then ϕ centralizes all ϕ-invariant abelian subgroups of P (Theorem A.52.2) and hence P is nonabelian (Theorems A.52.1 and A.22.1). Let H be a critical subgroup of P. By Exercise 269.2 (a), ϕ induces on H a nontrivial automorphism so H is nonabelian, by the above. Then ϕ acts nontrivially on $H/\Phi(H)$ (Burnside). The subgroup $\mathrm{Z}(H)$ is a ϕ-invariant normal (even characteristic) abelian subgroup of P. By the above, ϕ acts trivially on $\mathrm{Z}(H)$. Consider the action of ϕ on the elementary abelian p-group $H/\mathrm{Z}(H)$ (see the definition of a critical subgroup). As $[P, H] \leq \mathrm{Z}(H)$, all subgroups of $H/\mathrm{Z}(H)$ are P-invariant (indeed, in the case under consideration, $H/\mathrm{Z}(H)$ is a central subgroup of $P/\mathrm{Z}(H)$). As the natural semidirect product of groups $\langle\phi\rangle$ and $H/\mathrm{Z}(H)$ is supersolvable (Exercise 293.335), the group $H/\mathrm{Z}(H)$ is a direct product of ϕ-invariant subgroups $U_1/\mathrm{Z}(G)), \dots, U_d/\mathrm{Z}(G))$ of order p (Maschke's theorem). The abelian subgroups $U_1, \dots, U_d$ are ϕ-invariant and normal in P so ϕ centralizes them, by hypothesis. As $U_1 \dots U_d = H$, the automorphism ϕ centralizes H, a contradiction. Thus, there is in P a normal ϕ-invariant abelian subgroup A on which ϕ induces a nontrivial automorphism. Then ϕ induces a nontrivial automorphism in subgroup $\Omega_{\epsilon(p)}(A)$. □

In particular, if a p-group P of odd order admits an automorphism ϕ of order 2, there is in P a normal ϕ-invariant elementary abelian subgroup E of exponent p such that ϕ induces on E a nonidentity automorphism. Note that condition $p \equiv 1 \pmod q$ is essential in Theorem 299.1.

https://doi.org/10.1515/9783110533149-043

§ 300 On p-groups all of whose maximal subgroups of exponent p are normal and have order p^p

In this section we study the p-groups G satisfying the following properties:

(i) $\exp(G) > p$.

(ii) All maximal subgroups of exponent p in G are normal and have the same order p^p. The set of all such subgroups is denoted by $\mathcal{E}_p = \mathcal{E}_p(G)$.

All members of the set $\mathcal{E}_p$ are regular (Theorem 7.1 (b)).

In the proof of our main result we prove the following:

Theorem 300.1. *Suppose that a p-group G, $p > 2$, is such that $|\mathcal{E}_p| > 1$. Then the following hold:*

(a) *If distinct $R, S \in \mathcal{E}_p$, then RS is of maximal class and order p^{p+1}.*

(b) $\Phi = \bigcap_{R\in\mathcal{E}_p} R$ *is of order p^{p-1}, $\Phi \le \Phi(G)$.*

(c) *G has only one normal subgroup of order p^i and exponent p for $i < p$. In particular, G is a monolith.*

(d) *If $R \in \mathcal{E}_p$ and $x \in G - R$ is of order p, then the subgroup $H = \langle x, R\rangle$ is of maximal class and $H \lhd G$.*

(e) *If $K \lhd G$ is abelian of type (p, p), then $|\Omega_1(C_G(R))| = p^p$.*

(f) *Let distinct $R, S \in \mathcal{E}_p$ be as in* (a) *and $H = RS$. Take in p-group H of maximal class and order p^{p+1} (see* (a)*) an absolutely regular subgroup A of index p. Then A is a maximal absolutely regular subgroup of G.*

Proof. Let $R \in \mathcal{E}_p$ and take in the set $G - R$ an element x of order p. Set $H = \langle R, x\rangle$; then $\Omega_1(H) = H$, $\exp(H) > p$, by (ii), and $|H| = p^{p+1}$; therefore, H is irregular of maximal class (Theorems 7.2 (b) and 7.1 (b)). If $x \in R_1 < H$, where $|H : R_1| = p$, then $R_1 = \langle x\rangle(R_1 \cap R)$ is of exponent p so that $H = RR_1 \lhd G$, by (ii). Take $S \in \mathcal{E}_p - \{R\}$ and set $T = RS$. If $|T| = p^{p+1}$ and $y \in S - R$, then $T = \langle R, y\rangle$ is of maximal class and order p^{p+1}, by the first sentence of the proof since $o(y) = p$. Now assume that $|T| > p^{p+1}$. Let $R < U < T$, where $|U : R| = p$. By the modular law, $U = R(U \cap S)$. As $|U \cap S| = |U : R| = p$, one has $\Omega_1(U) = U$, and we conclude that U is of maximal class. Then, by Exercise 10.10, T is of maximal class so, by Theorem 9.6 (c), we get $|T| = p|R| = p^{p+1}$, contrary to the assumption. Thus, the product of any two distinct members of the set $\mathcal{E}_p$ is of maximal class and order p^{p+1} and their intersection, by the product formula, has order p^{p-1}.

If G is of maximal class, then $|G| = p^{p+1}$ (Theorem 9.6 (c)).

Next suppose that $|G| > p^{p+1}$. Let R, S and $T = RS$ be such as in the first paragraph of the proof (in that case, T is of maximal class and order p^{p+1}, by the above). As $1 < e_p(T) < p + 1$ (here $e_p(X)$ is the number of subgroups of order p^p and exponent p in a p-group X) and $e_p(G) \equiv 1 \pmod p$ (Theorem 13.5), there is in G a subgroup U of order p^p and exponent p such that $U \not\le T$. By the above, $|R \cap U| = p^{p-1}$, hence it follows that $|T \cap U| = p^{p-1}$, and we conclude that (recall that $d(T) = 2$) $U \cap T = U \cap R = \Phi(T)$

https://doi.org/10.1515/9783110533149-044

has index p^2 in T. Let W be a subgroup generated by all members of the set $\mathcal{E}_p$; then $|W| \geq p^{p+1}$ and $\Phi(W) = \Phi(T) \lhd G$ has order p^{p-1} and exponent p. Let K be a G-invariant subgroup of order p^2 in $\Phi(W)$ and set $C = C_G(K)$. Then $|G : C| = p$ since $K \nleq Z(T)$ (T is of maximal class). Since $|\mathcal{E}_p| > 1$, it follows from Theorem 12.1 (b) that C is not absolutely regular. Therefore, since C is not of maximal class, there is in C a G-invariant subgroup V of order p^p and exponent p, by Theorem 13.5. It is easily seen that it is possible to choose V so that $K < V$. If $z \in C - V$ is of order p, then $\langle V, z\rangle$ is of maximal class, by the above, a contradiction since the central subgroup $K \cong E_{p^2}$ is contained in $\langle V, z\rangle$. Thus, the set $C - V$ has no elements of order p so that $\Omega_1(C) = V$.

By the above, the subgroup $\Omega_1(Z(G))$ is contained in all members of the set $\mathcal{E}_p$ so it coincides with the center of some subgroup of maximal class (and order p^{p+1}. It follows from (a) that $Z(G)$ is cyclic. Let $L \neq K$ be G-invariant abelian subgroup of G of type (p, p). Then $R \cap L > \{1\}$ for any $R \in \mathcal{E}_p$. If $L \nleq R$, then RL is of maximal class and order $p^{p+1} > p^3$, contrary to Exercise 9.1 (b) (indeed, RL contains two distinct normal subgroups of order p^2). Thus, K is the unique normal subgroup isomorphic to E_{p^2} in G.

Assertion (c) is now obvious since any G-invariant subgroup of order $< p^p$ and exponent p is contained in Φ. Indeed, assume that there is in G a normal subgroup M of order $< p^p$ and exponent p such that $M \nleq R \in \mathcal{E}_p$. As above, MR is of maximal class and order p^{p+1}, contrary to Exercise 9.1 (b).

Let R, S, H, A be as in the statement of (f) and assume that $A < B < G$, where B is an absolutely regular subgroup of G such that $|B : A| = p$; then $|B| = p^{p+1}$. Set $T = BH$ (note that $H \lhd G$); then $|T| = p^{p+2}$, $|T : B| = p$, by the product formula, and T is not of maximal class (Theorem 9.6 (c)). By Theorem 12.12 (b), $T/K_p(T)$ is of order p^{p+1} and exponent p. Such a T cannot have an absolutely regular subgroup B of index p. Thus, the subgroup A of order p^p is a maximal absolutely regular subgroup of G. □

Suppose that a p-group G is as in Theorem 300.1 and A is as in Theorem 300.1 (f). Suppose, in addition, that G is not of maximal class; then $|G| > p^{p+1}$ since G is irregular. By Exercise 10.10, there exists in G a subgroup B containing A as a subgroup of index p such that B is not of maximal class. Then B is regular but not absolutely regular. Obviously, $A \lhd G$. As $B = A\Omega_1(B)$, where $\Omega_1(B)$ is of order p^p and exponent p, it follows that $B \lhd G$. All the above shows that $\Omega_1(G/A)$ is elementary abelian.

Note that $\Sigma_{p^2} \in \mathrm{Syl}_p(S_{p^2})$ satisfies the condition of the theorem.

Exercise 1. If the subgroup W in the third paragraph of the proof of Theorem 300.1 is two-generator, then $G = W$ is of maximal class and order p^{p+1}.

Hint. See the proof of Theorem 300.1. □

Exercise 2. Study the p-groups all of whose maximal subgroups of exponent p are quasinormal and have order p^p.

Exercise 3. Let A be maximal absolutely regular subgroup of an irregular p-group G. Then $|\Omega_1(A)| = p^{p-1}$.

Solution. Let $A < B \leq G$, where $|B : A| = p$. If B is regular, then

$$|\Omega_1(B)| = |B/\mho_1(B)| = p^p,$$

and the result is obvious. Now assume that B is irregular. Then the result follows from Theorem 9.8 (a). □

Problem 1. Does there exist a p-group G of exponent $> p$ all of whose maximal subgroups of exponent p have pairwise distinct orders?

Problem 2. Does there exist a nonabelian p-group all of whose maximal elementary abelian subgroups have pairwise distinct orders?

Problem 3. Study the p-groups of exponent $> p$ all of whose maximal subgroups of exponent p are normal and have order p^{p+1}.

§ 301 p-groups of exponent $> p$ containing $< p$ maximal abelian subgroups of exponent $> p$

A nonabelian p-group of exponent $> p$ contains a maximal abelian subgroup of exponent $> p$. Below we classify the nonabelian p-groups containing $< p$ maximal abelian subgroups of exponent $> p$. The presented proof shows the role of minimal nonabelian subgroups for solving this problem.

Theorem 301.1. *Let G be a nonabelian p-group G of exponent $> p$. If G has $< p$ pairwise distinct maximal abelian subgroups of exponent $> p$, then it has exactly one such subgroup, say A. In that case, $A = \mathrm{H}_p(G)$ is abelian of index p in G. In particular, if $p = 2$, then G is a generalized dihedral 2-group. All p-groups G with the abelian $\mathrm{H}_p(G) \in \Gamma_1$ satisfy the condition.*

Proof. Suppose that G has $< p$ pairwise distinct maximal abelian subgroups of exponent $> p$ (obviously, all those subgroups are G-invariant). Let $S \leq G$ be minimal nonabelian.

(i) Assume that $S = G$, i.e., G is minimal nonabelian, and $S \ncong \mathrm{D}_8$. The remaining minimal nonabelian groups of order p^3 and exponent $> p$ are Q_8 and M_{p^3} have $\geq p$ maximal abelian subgroups of exponent $> p$. In what follows we assume that $|G| > p^3$. Then G has at least p maximal (abelian) subgroups of exponent $> p$.

Indeed, suppose that G is metacyclic. Let $R = \Omega_1(G)$; then $R \cong \mathrm{E}_{p^2}$. As $|G| > p^3$, then all $p + 1 > p - 1$ maximal (abelian) subgroups of G have exponent $> p$. Now assume that G is nonmetacyclic. Then $\Omega_1(G) \cong \mathrm{E}_{p^3}$ (Lemma 65.1). If $|G| > p^4$, then all maximal (abelian) subgroups of G have exponent $> p$. Let $|G| = p^4$. Then all p members of the set $\Gamma_1 - \{\Omega_1(G)\}$ have exponent $> p$. Thus, if $G = S$ is minimal nonabelian, then $p = 2$ and $G \cong \mathrm{D}_8$ has only one maximal (abelian) subgroup of exponent > 2 which coincides with $\mathrm{H}_2(G)$.

(ii) Now let $S < G$ and $S \notin \{\mathrm{D}_8, \mathrm{S}(p^3)\}$. Then, by (i), there are in S pairwise distinct maximal (abelian) subgroups $B_1, \dots, B_p$. Let $B_i \leq A_i$, $i = 1, \dots, p$, where A_i is a maximal abelian subgroup of G for all $i \leq p$. In that case $A_1, \dots, A_p$ are pairwise distinct maximal abelian subgroups of exponent $> p$ of G (indeed, two distinct subgroups B_k, B_r generate S so cannot lie in an abelian subgroup simultaneously). This contradicts the hypothesis. Thus, if G has $< p$ maximal abelian subgroups of exponent $> p$, then any its minimal nonabelian subgroup is either isomorphic to D_8 or isomorphic to $\mathrm{S}(p^3)$. In the first case, by Theorem 10.32, G is a generalized dihedral group with maximal abelian subgroup $A = \mathrm{H}_2(G)$ of index 2. In the second case, by Mann's remark to Problem 115, G has an abelian subgroup $A = \mathrm{H}_p(G)$ of index p.

Conversely, let G have an abelian subgroup $A = \mathrm{H}_p(G)$ of index p and let $B \neq A$ be maximal abelian subgroup of G. Then $B - A \subseteq G - A$. Since all elements of the set $G - A$ have order p and $\langle B - A \rangle = B$, it follows that $B \in \{\mathrm{E}_{p^2}, \mathrm{E}_{p^3}\}$ has exponent p. We see

https://doi.org/10.1515/9783110533149-045

that, in the considered case, A is the unique maximal abelian subgroup of G having exponent $> p$. □

Problem 1. Classify the nonabelian p-groups containing exactly p (exactly $p + 1$) maximal abelian subgroups of exponent $> p$.

Problem 2. Classify the nonabelian p-groups of exponent $p^e > p$ containing $< p$ (exactly p) pairwise distinct maximal abelian subgroups of exponent p^e.

Problem 3. (i) Classify the minimal irregular p-groups containing $< p$ pairwise distinct maximal regular subgroups of exponent $> p$.

(ii) Classify the irregular p-groups with exactly one maximal regular subgroup of exponent $> p$.

§ 302 Alternate proof of Theorem 109.1

Below we offer alternate proof of the following:

Theorem 302.1 (Theorem 109.1). *Suppose that $p > 2$ and L is a metacyclic p-group of order $> p^3$ without cyclic subgroup of index p. An element $a \in \mathrm{Aut}(L)$ of order p does not centralize $\Omega_1(L)$ if and only if $p = 3$, the partial holomorph $G = \langle a\rangle \cdot L$ is a 3-group of maximal class and one of the following holds:*
(a) *L is abelian of type $(3^m, 3^n)$, $m > 1$, $n > 1$ and $|m - n| \le 1$.*
(b) *$L = \langle a, b \mid o(a) = 3^m,\ o(b) = 3^n,\ a^b = a^{1+3^{m-1}}\rangle \cong \mathrm{M}_3(m, n)$, $|m - n| \le 1$.*

Note that if G is a 3-group of maximal class and order $> 3^4$, then its fundamental subgroup is metacyclic and it is either abelian of type $(3^m, 3^n)$ or minimal nonabelian $\mathrm{M}_3(m, n)$ with $|m - n| \le 1$ in both cases. (See Theorem 9.6.)

For the following corollary of the theorem we offer an independent proof.

Proposition 302.2. *Let $P \cong \mathrm{C}_{p^m} \times \mathrm{C}_{p^n}$, $p > 3$, $m > 1$, $n > 1$. Suppose that $a \in \mathrm{Aut}(P)$ is of order p. Then the element a centralizes $\Omega_1(P)$.*

Proof. Let $G = \langle a\rangle \cdot P$ be the natural semidirect product with kernel P. Note that $|G| = p|P| = p^{m+n+1} \ge p^5$. One has $|G/\mho_1(G)| = p^3 < p^{p-1}$ as $\mho_1(G) = \mho_1(P)$ (indeed, the group $\overline{G} = G/\mho_1(G)$ is a split extension of the group $\overline{P} \cong \mathrm{E}_{p^2}$ by the group $\overline{\langle a\rangle}$ of order p), the group G is absolutely regular (Theorem 9.8 (a)) so regular. By Theorem 7.2 (d), $|\Omega_1(G)| = |\overline{G}| = p^3$; then $G/\Omega_1(G)$ is noncyclic. Indeed,

$$G/\Omega_1(G) = P\Omega_1(G)/\Omega_1(G) \cong P/(\Omega_1(G \cap P) = P/\Omega_1(P)$$

is abelian of type (p^{m-1}, p^{n-1}) and $m, n > 1$). Therefore, by Theorem 13.7 (see there part (c)), $E = \Omega_1(G) \cong \mathrm{E}_{p^3}$. As $a \in E$ and $E \cap P = \Omega_1(P)$, the automorphism a centralizes $\Omega_1(P)$. □

Condition $m, n > 1$ is essential in Proposition 302.2 (see Theorem 13.7 (c)).

Proof of Theorem 302.1. We use notation introduced in the statement. Let $G = \langle a\rangle L$ be a natural semidirect product of $\langle a\rangle$ with kernel L. One has

$$R = \Omega_1(L) \cong \mathrm{E}_{p^2}, \quad H = \langle a, R\rangle \cong \mathrm{S}_{p^3}, \quad L \cap H = R, \quad |G : \mathrm{C}_G(R)| = p$$

since a does not centralize R and $p > 2$ (this fact we have used above two times). Since a normalizes R but does not centralize R, by hypothesis, the subgroup $H = \langle R, a\rangle$ is of order p^3 and exponent p since $p > 2$; then $H \cong \mathrm{S}(p^3)$.

Assume that $H = \Omega_1(G)$. Then G has no normal subgroup isomorphic to E_{p^3} so G is a group of Theorem 13.7 (c) (note that 3-group of maximal class and order $> 3^4$ has no normal subgroup of order isomorphic to $\mathrm{S}(3^3)$, by Theorem 9.6 (c)). In particular, G/H is cyclic. However,

$$G/H = PH/H \cong P(P \cap H) = P/R$$

https://doi.org/10.1515/9783110533149-046

is noncyclic since $m, n > 1$, and this is a contradiction. Thus,

$$H < \Omega_1(G). \tag{1}$$

It follows from (1) that $\exp(\Omega_1(G)) > p$ (otherwise, $L \cap \Omega_1(G)$ is nonmetacyclic).

If G is a 3-group of maximal class, then either (a) or (b) hold (see Theorem 9.6).

Next we assume that G is not a 3-group of maximal class. As $p > 2$, (the metacyclic p-group) L is absolutely regular and therefore, by Theorem 12.1 (b), one has $G = L\Omega_1(G)$, where $\Omega_1(G)$ is of exponent p. As $L \cap \Omega_1(G) = \Omega_1(L) \cong \mathrm{E}_{p^2}$, we get $|\Omega_1(G)| = p^3 = |H|$, and this implies $H \cong \mathrm{S}(p^3)$, contrary to (1). Thus, G is a 3-group of maximal class, completing the proof. □

Compare the above proof with the proof of Theorem 109.1. The obtained result improves essentially the corresponding result in [MS] (see also [AS, Lemma A.1.30]) where L is abelian of type (p^2, p^2), $p > 3$ (moreover, that result follows from Proposition 302.2). Note that in Theorem 302.1, the metacyclic subgroup L need not necessarily be abelian, its structure, by the theorem, is restricted.

Problem 1. Suppose that L is a metacyclic 2-group without cyclic subgroup of index 2. Suppose that an element $a \in \mathrm{Aut}(L)$ of order 2 does not centralize $\Omega_2(L)$. Describe the structure of the holomorph $G = \langle a \rangle \cdot L$.

Problem 2. Study an action of an automorphism of order p on a homocyclic p-group of exponent $> p$.

§ 303 Nonabelian p-groups of order $> p^4$ all of whose subgroups of order p^4 are isomorphic

We solve here Problem 2 from Appendix 103 by proving the following result.

Theorem 303.1. *Let G be a nonabelian p-group of order $> p^4$ all of whose subgroups of order p^4 are isomorphic to a group V. Then $\exp(G) \le p^2$ and one of the following holds:*

(a) *If V is abelian, then V is of type (p^2, p, p),*

$$|G| \le p^7, \quad \Omega_1(G) = Z(G) \cong E_{p^3}$$

and each minimal nonabelian subgroup in G is nonmetacyclic of order p^5 and exponent p^2, i.e., $V \cong M_p(2,2,1)$).

(b) *If V is nonabelian, then $p > 2$ and $V \cong S(p^3) \times C_p$ or $V \cong M_{p^3} \times C_p$ or V is nonmetacyclic minimal nonabelian of order p^4 and exponent p^2, i.e., $V \cong M_p(2,1,1)$. Also, $|G| \le p^6$ and G has a self-centralizing normal subgroup $X \cong E_{p^3}$ and either $\exp(G) = p$ or $X = \Omega_1(G)$.*

Proof. Let G be a nonabelian p-group of order $> p^4$ all of whose subgroups of order p^4 are isomorphic. If G has no normal subgroup isomorphic to E_{p^2}, then by Lemma 1.4, G is a 2-group of maximal class and so G does not satisfy our hypothesis. It follows that G has a normal subgroup U isomorphic to E_{p^2}.

Suppose that G/U (of order $\ge p^3$) has no normal subgroup isomorphic to E_{p^2}. By Lemma 1.4, G/U is either cyclic or a 2-group of maximal class distinct from D_8. Suppose in addition that G/U has a cyclic subgroup $H/U \cong C_{p^3}$. Take $x \in H - U$ such that $\langle x\rangle$ covers H/U. If $\langle x\rangle \cap U \ne \{1\}$, then $\langle x\rangle \cong C_{p^4}$, a contradiction (since $U \cong E_{p^2}$ and so, as follows easily from the hypothesis, each subgroup of order p^4 is noncyclic). Hence $\langle x\rangle \cap U = \{1\}$ and so $\langle x\rangle \cong C_{p^3}$. Since $|H : C_H(U)| \le p$, we get $A = U \times \langle x^p\rangle \cong E_{p^2} \times C_{p^2}$ is of order p^4. Let $\langle x\rangle < H_1 < H$ with $|H_1 : \langle x\rangle| = p$ so that $|H_1| = p^4$ and $\exp(H_1) = p^3$. But

$$|A| = p^4 = |H_1| \text{ and } \exp(A) = p^2 \implies H \ncong H_1,$$

a contradiction. We have proved that G/U does not contain a subgroup isomorphic to C_{p^3}. By the above (since G/U is a 2-group of maximal class distinct from D_8 and without a cyclic subgroup of order 2^3), we must have $G/U \cong Q_8$ and so $|G| = 2^5$. Because $|G : C_G(U)| \le 2$, it follows that (noting that each maximal subgroup of G/U is isomorphic to C_4) G has an abelian maximal subgroup (of order 2^4). By our hypothesis, each maximal subgroup of G is abelian and so G is minimal nonabelian. In particular, $G' = \langle z\rangle \cong C_2$ and $z \notin U$ so that $U \times \langle z\rangle \cong E_8$ and $G/(U \times \langle z\rangle) \cong E_4$. It follows that $\exp(G) = 4$ and so G is a unique minimal nonabelian group of order 2^5 and exponent 4, and we conclude that $G \cong M_2(2,2,1)$. Since here each maximal subgroup of G is abelian of type $(4,2,2)$, it follows that this group satisfies our hypothesis.

From now on we assume that G/U has a normal subgroup $V/U \cong E_{p^2}$, where $V \trianglelefteq G$, $|V| = p^4$; then $\exp(V) \le p^2$ so that we have also proved that $\exp(G) \le p^2$.

https://doi.org/10.1515/9783110533149-047

(i) Suppose that V is abelian. Then each minimal nonabelian subgroup in G is of order $\geq p^5$. But $\exp(G) \leq p^2$ and so (by Lemma 65.1) each minimal nonabelian subgroup in G is of order $\leq p^5$. Hence each minimal nonabelian subgroup M in G is of order p^5 and exponent p^2 (it follows that $M \cong \mathrm{M}_p(2, 2, 1)$). Then M is the unique group all of whose maximal subgroups (of order p^4) are abelian of type (p^2, p, p). Hence V is abelian of type (p^2, p, p).

Let A be any maximal abelian subgroup in G. Let $A < X \leq G$ with $|X : A| = p$. Then X is nonabelian and since each minimal nonabelian subgroup is of order p^5, it follows that $|X| \geq p^5$ and so $|A| \geq p^4$. Each subgroup of order p^4 in A is isomorphic to V and so is of type (p^2, p, p). This obviously implies that $A \cong V$ is of order p^4. We have proved that each maximal abelian subgroup in G is of order p^4 and type (p^2, p, p).

Let A be a maximal normal abelian subgroup in G; then A is of order p^4 and type (p^2, p, p). Suppose for a moment that there is an element $g \in G - A$ of order p. By Lemma 57.1, there is $a \in A$ such that $\langle g, a\rangle$ is minimal nonabelian and so $\langle g, a\rangle \cong M$. By the structure of M, we have

$$\Phi(M) = \mathrm{Z}(M) = \Omega_1(M) \cong \mathrm{E}_{p^3}.$$

This gives a contradiction since $o(g) = p$ and $g \notin \Phi(\langle g, a\rangle)$. We have proved that all elements in $G - A$ are of order p^2 and $\exp(G/A) = p$ and $\Omega_1(G) = \Omega_1(A) \cong \mathrm{E}_{p^3}$.

Let x be any element in $G - A$ so that $x^p \in \Omega_1(A)$ and $A\langle x\rangle$ is nonabelian of order p^5. Hence $A\langle x\rangle \cong M$ is minimal nonabelian and so $\Omega_1(A) = \mathrm{Z}(A\langle x\rangle)$. In particular, all elements $x \in G - A$ centralize $\Omega_1(A)$ implying that

$$\Omega_1(A) = \mathrm{Z}(G) = \Omega_1(G).$$

Also we have

$$\mho_1(G) \leq \mathrm{Z}(G) = \Omega_1(A).$$

Now G stabilizes the chain $A > \Omega_1(A) > \{1\}$ and $\mathrm{C}_G(A) = A$ so that $|G/A| \leq p^3$ and therefore $|G| \leq p^7$.

(ii) Suppose that the subgroup V is nonabelian; then G has no abelian subgroup of order p^4. Let X be a maximal normal abelian subgroup in G containing U. Since $|G : \mathrm{C}_G(U)| \leq p$, we have $X > U$ and so $|X| = p^3$ and hence either $X \cong \mathrm{E}_{p^3}$ or $X \cong \mathrm{C}_{p^2} \times \mathrm{C}_p$. Since G/X acts faithfully on X, we have that G/X is isomorphic to a subgroup of $\mathrm{S}(p^3)$ in case $p > 2$ and G/X is isomorphic to a subgroup of D_8 in case $p = 2$ and in any case $|G| \leq p^6$ and $|G/X| \geq p^2$.

Since $\mathrm{C}_G(U) > X$ (because $|G/X| \geq p^2$), we have $\mathrm{Z}(V) \cong \mathrm{E}_{p^2}$. Also, $V/\mathrm{Z}(V) \cong \mathrm{E}_{p^2}$ and so all maximal subgroups of V containing $\mathrm{Z}(V)$ are abelian implying that $V' \cong \mathrm{C}_p$ and $V' \leq \mathrm{Z}(V)$. Also, $\Phi(V) \leq \mathrm{Z}(V)$. If $\Phi(V) = V'$, then $V \cong \mathrm{C}_p \times V_1$, where V_1 is nonabelian of order p^3. If $\Phi(V) = \mathrm{Z}(V)$, then V is minimal nonabelian of order p^4 and exponent p^2.

(ii1) Suppose that $p = 2$ so that $\exp(G) = 4$. If all minimal nonabelian subgroups of G are isomorphic to D_8 or if all minimal nonabelian subgroups of G are isomorphic to Q_8, then by Theorem 10.33 and Corollary A.17.3, G has an abelian subgroup of index 2 (and so of order $\geq 2^4$), a contradiction. Suppose that all minimal nonabelian subgroups of G are isomorphic to $M_2(2,2)$ (the metacyclic minimal nonabelian group of order 2^4 and exponent 4), then, by Theorem 57.3, we have a contradiction (because here in any case G would have an abelian subgroup of order 2^4). Finally, suppose that all minimal nonabelian subgroups in G are isomorphic to $M_2(2,1,1)$ (the nonmetacyclic minimal nonabelian group of order 2^4 and exponent 4). By Theorem 57.4, $\Omega_1(G)$ is a self-centralizing elementary abelian subgroup. In our case we have $X = \Omega_1(G) \cong E_8$. If $\exp(G/X) = 4$, then there are involutions in $G - X$, a contradiction. Hence $\exp(G/X) = 2$ and so G/X is elementary abelian. Since G/X is also a subgroup of D_8, we get $G/X \cong E_4$ and so $|G| = 2^5$. If $X \nleq \Phi(G)$, then there is a maximal subgroup H of G with $X \nleq H$, where $|H| = 2^4$ and $\Omega_1(H) = H \cap X \cong E_4$, a contradiction. Hence $X = \Phi(G)$ and, by Theorem 51.3, the structure of G is uniquely determined. In particular, there are involutions in $G - X$, contrary to $\Omega_1(G) = X$. We have proved that there are no groups in case $p = 2$.

(ii2) We have $p > 2$. First suppose that $V \cong C_p \times S(p^3)$ so that $\exp(G) = p$ and each subgroup of order p^4 is nonabelian and is isomorphic to $C_p \times S(p^3)$. Let X be a maximal normal abelian subgroup of G containing $U \cong E_{p^2}$ so that $X \cong E_{p^3}$. Since X is self-centralizing in G, we have that G/X is isomorphic to a subgroup of $S(p^3)$ and $|G/X| \geq p^2$.

Now suppose that $V \cong C_p \times M_{p^3}$ so that each minimal nonabelian subgroup in G is isomorphic to M_{p^3}. Let X be a maximal normal abelian subgroup of G containing $U \cong E_{p^2}$, where $U \trianglelefteq G$ so that $|X| = p^3$. Suppose for a moment that X is of type (p^2, p). Let $X < Y \trianglelefteq G$ with $|Y : X| = p$. Then $Y \cong V \cong C_p \times M_{p^3}$ and so $\Omega_1(Y) = X_1 \cong E_{p^3}$. Since X_1 is characteristic in $Y \triangleleft G$, we have $X_1 \trianglelefteq G$. Then we replace X with X_1 so that we may assume from the start that $X \cong E_{p^3}$. Since X is self-centralizing in G, we have that G/X is isomorphic to a subgroup of $S(p^3)$ and $|G/X| \geq p^2$. Hence $\exp(G/X) = p$ and for each $g \in G - X$, $g^p \in X$ and $X\langle g\rangle \cong C_p \times M_{p^3}$ so that $X = \Omega_1(X\langle g\rangle)$ and therefore $o(g) = p^2$. We have proved that all elements in the set $G - X$ are of order p^2 implying that

$$X = \Omega_1(G) \cong E_{p^3}.$$

Suppose now $V \cong M_p(2,2)$ which is the metacyclic minimal nonabelian group of order p^4 and exponent p^2. Since $|G| \geq p^5$ and $\exp(G) \leq p^2$, it follows that G is nonmetacyclic. Let L be a minimal nonmetacyclic subgroup in G. Since $p > 2$, it follows from Theorem 69.1 that L is either of order p^3 and exponent p or L is a 3-group of order 3^4 and maximal class. In both cases L is not isomorphic to a subgroup of $V \cong \mathcal{H}_p$, a contradiction.

Finally, let V be the nonmetacyclic minimal nonabelian group $M_p(2,1,1)$ of order p^4 and exponent p^2 so that $\Omega_1(V) \cong E_{p^3}$ and all other maximal subgroups of V are abelian of type (p^2, p). Let X be a maximal normal abelian subgroup in G containing a normal subgroup $U \cong E_{p^2}$ so that $|X| = p^3$. Suppose for a moment that X is of type (p^2, p). Let $X < Y \trianglelefteq G$ with $|Y : X| = p$. Then $Y \cong M_p(2,1,1)$ with $X_1 = \Omega_1(Y) \cong E_{p^3}$ so that $X_1 \trianglelefteq G$. Then we replace X with X_1 so that we may assume from the start that $X \cong E_{p^3}$. If $\exp(G/X) = p^2$, then there is $g \in G - X$ with $o(g) = p^2$

and $\langle g\rangle \cap X = \{1\}$ so that $o(g^p) = p$ and $g^p \in G - X$. But then we must have

$$X\langle g^p\rangle \cong M_p(2, 1, 1),$$

contrary to

$$r\Omega_1(X\langle g^p\rangle) = X \cong E_{p^3}.$$

It follows that $\exp(G/X) = p$ and all elements $x \in G - X$ are of order p^2 and $x^p \in X$. Hence $X = \Omega_1(G) \cong E_{p^3}$ is self-centralizing in G and so again $|G| \le p^6$. Our theorem is proved. □

Problem 1. Classify the p-groups all of whose abelian (nonabelian, metacyclic, nonmetacyclic) subgroups of order p^4 are isomorphic.

Problem 2. Study the p-groups all of whose subgroups of order p^5 are isomorphic.

§ 304 Non-Dedekindian p-groups in which each nonnormal subgroup has a cyclic complement in its normalizer

We solve here Problem 3448 by proving the following result.

Theorem 304.1. *Let G be a non-Dedekindian p-group in which each nonnormal subgroup has a cyclic complement in its normalizer. Then for each $x \in G$, the subgroup $\langle x^p \rangle$ is normal in G and G has no elementary abelian subgroup of order p^3. Moreover, if $\mathrm{Z}(G)$ is cyclic, then G has an abelian subgroup of index p and if $\mathrm{Z}(G)$ is noncyclic, then $\Omega_1(\mathrm{Z}(G)) = \Omega_1(G) \cong \mathrm{E}_{p^2}$.*

Proof. Let G be a non-Dedekindian p-group in which each nonnormal subgroup has a cyclic complement in its normalizer.

Let $x \in G$ be of order $> p$ and let $\langle x^p \rangle$ be not normal in G. Then $\mathrm{N}_G(\langle x^p \rangle) = \langle x^p \rangle S$, where S is cyclic and $S \cap \langle x^p \rangle = \{1\}$. But

$$\langle x \rangle \le \mathrm{N}_G(\langle x^p \rangle), \quad S \cap \langle x \rangle \cong \mathrm{C}_p \quad \text{and} \quad S \cap \langle x \rangle \ne \Omega_1(\langle x \rangle) \le \langle x^p \rangle$$

(the second congruence follows by the product formula), contrary to the fact that $\langle x \rangle$ is cyclic. We have proved that for each $x \in G$, the subgroup $\langle x^p \rangle$ is normal in G. It follows that, whenever Z is a cyclic nonnormal subgroup in G, then Z is maximal cyclic in G.

If G has no normal subgroup isomorphic to E_{p^2}, then Lemma 1.4 implies that $G \cong \mathrm{D}_{2^m}$, $m \ge 4$, and so the group G has a cyclic subgroup of index 2 and $\mathrm{Z}(G)$ is cyclic (of order 2).

Assume for a moment that G has an elementary abelian subgroup E of order p^3. In that case, any subgroup of E of order p has no cyclic complement in its normalizer and so it is G-invariant. It follows that $E \le \mathrm{Z}(G)$. Then any non-G-invariant cyclic subgroup L has no cyclic complement in its normalizer since $E < \mathrm{N}_G(L)$. This contradicts our basic assumption. We have proved that G does not possess any elementary abelian subgroup of order p^3.

Suppose that $\mathrm{Z}(G)$ is cyclic. If G has no normal subgroup isomorphic to E_{p^2}, then we know that G is a 2-group of maximal class and G has a cyclic subgroup of index 2. Now assume that the group G has a normal subgroup $U \cong \mathrm{E}_{p^2}$. Then, setting $H = \mathrm{C}_G(U)$, we get $|G : H| = p$. Let V be a nonnormal subgroup of order p contained in U so that $\mathrm{N}_G(V) = H$. By our assumption, $H = V \times X$, where X is cyclic. Then H is an abelian subgroup of index p in G.

Assume that $\mathrm{Z}(G)$ is noncyclic. Then, noting that G has no elementary abelian subgroup of order p^3, we have (note that $\mathrm{Z}(G)$ has no subgroup isomorphic to E_{p^3})

$$\Omega_1(\mathrm{Z}(G)) = \Omega_1(G) \cong \mathrm{E}_{p^2}.$$

Our theorem is proved. □

https://doi.org/10.1515/9783110533149-048

Exercise 1. Classify the non-Dedekindian p-groups in which every nonnormal subgroup (cyclic subgroup) has a cyclic complement.

Exercise 2. Classify the 3-groups of maximal class satisfying the condition of Theorem 304.1.

Exercise 3. Classify the non-Dedekindian p-groups in which the normalizer of each nonnormal cyclic subgroup is metacyclic.

§ 305 Nonabelian p-groups G all of whose minimal nonabelian subgroups M satisfy $Z(M) \leq Z(G)$

We solve here Problem 2 from §203 by proving the following result.

Theorem 305.1. *Let G be a nonabelian p-group all of whose minimal nonabelian subgroups M satisfy $Z(M) \leq Z(G)$. Then either G has an abelian subgroup of index p or $\mho_1(G) \leq Z(G)$ and, if $p = 2$, then G is of class 2 and G' is elementary abelian.*

Proof. Let G be a nonabelian p-group all of whose minimal nonabelian subgroups M satisfy $Z(M) \leq Z(G)$.

Suppose that G has no abelian subgroup of index p. Let A be a maximal normal abelian subgroup in G. Then $Z(G) < A$ and $|G : A| > p$. Suppose for a moment that $\exp(G/A) > p$. Then there is a subgroup B such that $A < B \leq G$ and $B/A \cong C_{p^2}$. Let $b \in B - A$ be such that $\langle b \rangle$ covers B/A and so $b^p \notin A$. By Lemma 57.1, there is $a \in A$ such that $M = \langle a, b \rangle$ is minimal nonabelian. By the hypothesis, $Z(M) \leq Z(G)$ and so

$$b^p \in \Phi(M) = Z(M) \leq Z(G).$$

But then $b^p \notin A$ centralizes A, a contradiction (since $C_G(A) = A$). We have proved that $\exp(G/A) = p$.

By Lemma 57.1, for each $g \in G - A$, there is $h \in A$ such that the subgroup $\langle g, h \rangle$ is minimal nonabelian. Then $g^p \in A$ and $g^p \in Z(\langle g, h \rangle) \leq Z(G)$. Assume for a moment that $\exp(A/Z(G)) > p$ and consider the factor-group $G/Z(G)$. For each $g \in G - A$, we have $g^p \in Z(G)$, by the second sentence of this paragraph, i.e., all elements of the set $G/Z(G) - A/Z(G)$ have order p. It follows that

$$H_p(G/Z(G)) = A/Z(G) > \{1\}.$$

Let $A < H \leq G$, where $|H : A| = p^2$ so that $H/A \cong E_{p^2}$ (recall that G/A has order $> p$ and exponent p). Clearly, $H_p(H) = A$. It follows from $H' \leq A$ that H is a metabelian a metabelian. We apply a result of Hogan–Kappe [HogK] on the group $H/Z(G)$ stating that in a metabelian p-group a nontrivial Hughes subgroup has index at most $p < p^2 = |H : A|$. This gives here a contradiction. Hence we have proved that $\exp(A/Z(G)) = p$, and so we have $\mho_1(G) \leq Z(G)$.

Finally, we have proved that either G has an abelian subgroup of index p or $\mho_1(G) \leq Z(G)$. If in the second case $p = 2$, then

$$\Phi(G) = \mho_1(G) \leq Z(G)$$

and so G is of class 2. Then for all $x, y \in G$, we have

$$1 = [x^2, y] = [x, y]^2,$$

and so the abelian subgroup G' is elementary abelian. The theorem is proved. □

https://doi.org/10.1515/9783110533149-049

Exercise 1. Suppose that the centers of all minimal nonabelian subgroups of a p-group G coincide. Then G satisfies the condition of Theorem 301.1.

Solution. Let $S, T \leq G$ be minimal nonabelian. Then $Z(S) = Z(T)$ so $C = C_G(Z(S))$ contains all minimal nonabelian subgroups of G. By Proposition 10.28, $C = G$ so that $Z(S) \leq Z(G)$. □

Let G be a p-group of maximal class with abelian subgroup A of index p. Let $S \leq G$ be minimal nonabelian. Then $|S| = p^3$. As $|S : (S \cap A)| = p$, we have $Z(S) \leq S \cap A < A$ so that $C_G(Z(S)) \geq AS = G$ hence $Z(S) = Z(G)$, and we conclude that G satisfies the condition of Theorem 305.1.

§ 306 Nonabelian 2-groups all of whose maximal subgroups, except one, are quasi-Hamiltonian or abelian

A p-group G is called *quasi-Hamiltonian* if $G = S \times E$, where S is nonabelian of order p^3 and E is elementary abelian. Not every quasi-Hamiltonian 2-group is Hamiltonian (=nonabelian Dedekindian).

We solve here Problem 3773 for $p = 2$ by proving the following result.

Theorem 306.1. *Let G be a nonabelian 2-group all of whose maximal subgroups, except one denoted with M, are quasi-Hamiltonian or abelian. Then G is of order 2^5,*

$$\mathrm{d}(G) = 2, \quad G' \cong \mathrm{C}_4, \quad \mathrm{cl}(G) = 3, \quad \exp(G) = 8, \quad \mathrm{Z}(G) \cong \mathrm{E}_4$$

and the maximal subgroups of G are A, H, M, where A is abelian of type $(8, 2)$, *$H = Q \times Z$, where $Q \cong \mathrm{Q}_8$ or D_8, $Z \cong \mathrm{C}_2$ and M is the metacyclic minimal nonabelian group of order* 16 *and exponent* 4 *(i.e., isomorphic to $\mathrm{M}_2(2, 2)$). Moreover, if $Q \cong \mathrm{D}_8$, then G is isomorphic to the group defined in Theorem 258.1* (b) *and if $Q \cong \mathrm{Q}_8$, then G is isomorphic to the group defined in Theorem 258.2* (b).

Proof. Let G be a nonabelian 2-group all of whose maximal subgroups, except one denoted with M, are quasi-Hamiltonian or abelian.

(i) First assume that $|M'| > 2$. In this case we may use Theorem 145.8 about p-groups with exactly one maximal subgroup whose derived subgroup is of order $> p$. By that theorem $\mathrm{d}(G) = 2$ and $G' \cong \mathrm{C}_4 \times \mathrm{C}_2$. Also, $[G', G] = \Omega_1(G') \le \mathrm{Z}(G)$, $\Phi(G) = \mathrm{C}_G(G')$ is abelian and $\mho_2(G) \le \mathrm{Z}(G)$. If $\Gamma_1 = \{H_1, H_2, M\}$, then $H_1' = \langle z_1 \rangle$ and $H_2' = \langle z_2 \rangle$ are both of order 2,

$$\langle z_1, z_2 \rangle = \Omega_1(G') = M' \cong \mathrm{E}_4, \quad \mathrm{d}(M) = 3 \quad \text{and} \quad \mho_1(G') = \langle z_1 z_2 \rangle$$

and we have the following possibilities:

(i1) One has $\mathrm{d}(H_1) = \mathrm{d}(H_2) = 2$ in which case H_1 and H_2 are minimal nonabelian (by Lemma 65.2) and so H_1 and H_2 are isomorphic to D_8 or Q_8. But then $|G| = 2^4$ and so $|M| = 2^3$ and this contradicts the fact that $M' \cong \mathrm{E}_4$.

(i2) One has $\mathrm{d}(H_1) = \mathrm{d}(H_2) = 3$ in which case H_1 and H_2 are isomorphic to $\mathrm{C}_2 \times \mathrm{Q}_8$ or $\mathrm{C}_2 \times \mathrm{D}_8$ and $|G| = 2^5$. But then $|M| = 2^4$ and a result of Taussky (Lemma 1.6) implies that M is of maximal class with $M' \cong \mathrm{C}_4$. This contradicts again the fact that $M' \cong \mathrm{E}_4$.

(ii) We have proved that $|M'| \le 2$ and so $|M'| = 2$ since M is nonabelian.

(ii1) First assume that G' is noncyclic. Then we may use Theorem 139.A implying that either $\mathrm{d}(G) = 2$, $\mathrm{cl}(G) = 3$ and $G' \cong \mathrm{E}_4$ or $\mathrm{d}(G) = 3$, $\mathrm{cl}(G) = 2$, $G' \cong \mathrm{E}_4$ or E_8 and $\Phi(G) = \mathrm{Z}(G)$.

https://doi.org/10.1515/9783110533149-050

(ii1a) Assume that $\mathrm{d}(G) = 2$, $\mathrm{cl}(G) = 3$ and $G' \cong \mathrm{E}_4$. Let $\{H_1, H_2, M\}$ be the set of maximal subgroups in G. Since $|G'| > 2$, at most one of $\{H_1, H_2\}$ is abelian and so we may assume that H_2 is nonabelian. Then H_2 is quasi-Hamiltonian, i.e., $H_2 = Q \times V$, where $Q \cong \mathrm{D}_8$ or Q_8 and $\exp(V) \le 2$. But H_2' and M' are central in G and since $\mathrm{cl}(G) = 3$ (and so $\mathrm{E}_4 \cong G' \nleq \mathrm{Z}(G)$), we must have $H_2' = M' \ge H_1'$, where H_1 is either abelian or quasi-Hamiltonian. Note that $G' \cong \mathrm{E}_4$ lies in each maximal subgroup of G since $G' \le \Phi(G)$. By Schreier's inequality (Theorem A.25.1), we have for each maximal subgroup X of G, $\mathrm{d}(X) \le 1 + 2(\mathrm{d}(G) - 1) = 3$ and so $|V| \le 2$. If $|V| = 1$, then $|G| = 2^4$ and since $|G'| = 4$, a result of Taussky (Lemma 1.6) implies that G is of maximal class. This is a contradiction since $G' \cong \mathrm{E}_4$. Hence $V \cong \mathrm{C}_2$ and $G = 2^5$.

Assume that $Q \cong \mathrm{Q}_8$. Then $G' \cong \mathrm{E}_4$ and $G' \le H_2$ imply that

$$G' = \Omega_1(H_2) = Q' \times V = \mathrm{Z}(H_2).$$

But $G' \nleq \mathrm{Z}(G)$ and so $\mathrm{C}_G(G') = H_2$. On the other hand, $G' \le H_1$ and so if H_1 is abelian or if $H_1 \cong \mathrm{D}_8 \times \mathrm{C}_2$, then $\mathrm{C}_{H_1}(G')$ contains an abelian subgroup of order 2^4 or an elementary abelian subgroup of order 8, contrary to the fact that $\mathrm{C}_G(G') = H_2$. It follows that $H_1 \cong \mathrm{Q}_8 \times \mathrm{C}_2$ and then $G' = \mathrm{Z}(H_1)$. But then $\mathrm{C}_G(G') \ge \langle H_1, H_2 \rangle = G$, contrary to $\mathrm{cl}(G) = 3$. We have proved that $H_2 = D \times V$, where $D \cong \mathrm{D}_8$ and $V \cong \mathrm{C}_2$.

Suppose that $H_1 \cong \mathrm{Q}_8 \times \mathrm{C}_2$. Then $G' = \mathrm{Z}(H_1)$ and $\mathrm{C}_G(G') = H_1$. But $G' \le H_2$ and so $\mathrm{C}_{H_2}(G')$ contains an elementary abelian subgroup of order 8, a contradiction. We have proved that H_1 is either abelian or $H_1 \cong \mathrm{D}_8 \times \mathrm{C}_2$.

Now we prove that M is minimal nonabelian of order 2^4. Indeed, assume that M is not minimal nonabelian. Let $M_1 \cong \mathrm{D}_8$ or Q_8 be a minimal nonabelian subgroup in M. Since $M' = M_1' \cong \mathrm{C}_2$, we have $M = M_1 * \mathrm{Z}(M)$, where $M_1 \cap \mathrm{Z}(M) = M_1'$. But (by our hypothesis) M is not quasi-Hamiltonian and so $\mathrm{Z}(M) \cong \mathrm{C}_4$. It is known that such a group M has in any case a unique subgroup $R \cong \mathrm{Q}_8$. It follows that $R \trianglelefteq G$ and so G' (of order 4) is contained in R and so G' is cyclic, a contradiction. Hence M is minimal nonabelian.

We have thus proved that all minimal nonabelian subgroups of G, except one, are isomorphic to D_8. Then by Theorem 258.1, G' is cyclic of order ≤ 4, a final contradiction.

(ii1b) Suppose that $\mathrm{d}(G) = 3$, $\mathrm{cl}(G) = 2$, $G' \cong \mathrm{E}_4$ or E_8 and $\Phi(G) = \mho_1(G) = \mathrm{Z}(G)$. Since $|G'| > 2$, G has at most one abelian maximal subgroup. Then G has at least five quasi-Hamiltonian maximal subgroups $H_1, H_2, \dots, H_5$ and H_6 is either abelian or quasi-Hamiltonian and $M = H_7$ is neither abelian nor quasi-Hamiltonian with $M' \cong \mathrm{C}_2$. We have $|G : H_1| = 2$ and $H_1 = Q_1\mathrm{Z}(G)$, where $Q_1 \cong \mathrm{Q}_8$ or D_8 and $\mathrm{Z}(G) = \Phi(G) = \mathrm{Z}(H_1)$ is elementary abelian so that $\exp(G) = 4$. We set $Q_1 \cap \mathrm{Z}(G) = Q_1' = \langle z_1 \rangle \cong \mathrm{C}_2$. There is $a_1 \in Q_1$ with $a_1^2 = z_1$ and $\mho_1(H_1) = H_1' = \langle z_1 \rangle \trianglelefteq G$.

Note that there are exactly three maximal subgroups X_1, X_2, X_3 of G containing $\langle a_1 \rangle \mathrm{Z}(G) = \langle a_1 \rangle \Phi(G)$. Since at least two of them, say X_1, X_2, are distinct from M, we have $X_1' \le \langle z_1 \rangle$ and $X_2' \le \langle z_1 \rangle$. Hence $G/\langle z_1 \rangle$ has at least two abelian maximal subgroups $X_1/\langle z_1 \rangle$ and $X_2/\langle z_1 \rangle$. This implies that $(G/\langle z_1 \rangle)'$ is of order 2 and so $G' \cong \mathrm{E}_4$.

Next we show that $\mho_1(G) \le G'$ and so $\Phi(G) = G' = Z(G) \cong E_4$ implying that $|G| = 2^5$. Indeed, let $g \in G - \Phi(G)$ be any element of order 4. Considering three maximal subgroups of G containing $\langle g\rangle\Phi(G)$, we know that at least one of them X is quasi-Hamiltonian and so $X' = \langle g^2\rangle \le G'$.

Suppose that G has a quasi-Hamiltonian maximal subgroup $H_1 = Q_1 Z(G)$, where $Q_1 \cong Q_8$ and $H_1' = \langle z_1\rangle$. Let $H_i \in \langle H_2, \dots, H_5\rangle$. Note that all elements contained in $(H_1 \cap H_i) - Z(G)$ are of order 4 whose square is equal to z_1 and so $H_i' = \langle z_1\rangle$ for $i = 2, 3, \dots, 5$. But then the nonabelian group $G/\langle z_1\rangle$ has at least five abelian maximal subgroups, a contradiction. We have proved that all quasi-Hamiltonian maximal subgroups of G are of the form $D_8 \times C_2$.

Finally, we show that the maximal subgroup $M = H_7$ is minimal nonabelian. Indeed, we have $E_4 \cong Z(G) = \Phi(G) \le M$ and $M/Z(G) \cong E_4$ with $M' \cong C_2$. Hence we have $M = M_1 Z(G)$, where M_1 is minimal nonabelian and $|M_1 \cap Z(G)| \ge 2$. If $|M_1| = 8$, then M would be quasi-Hamiltonian, a contradiction. Hence $M = M_1$ and so M is minimal nonabelian of order 2^4.

We have thus proved that all minimal nonabelian subgroups in G, except one, are isomorphic to D_8. By Theorem 258.1, the group G' is cyclic of order ≤ 4, a final contradiction.

(ii2) We assume that G' is cyclic and so, by Theorem 137.7, we have either $G' \cong C_2$ or $d(G) = 2$ and $G' \cong C_4$.

(ii2a) First assume $G' \cong C_2$. If in addition $d(G) = 2$, then Lemma 65.2 implies that G is minimal nonabelian, contrary to our hypothesis. Hence $d(G) \ge 3$ so that G has at least three maximal subgroups which are quasi-Hamiltonian and let H be one of them. Then $H = QZ(H)$, where $Q \cong Q_8$ or D_8 and $\exp(Z(H)) = 2$ with $Z(Q) = Q' = G' = \langle z\rangle \cong C_2$. We have $Q \trianglelefteq G$ and $C = C_G(Q)$ covers G/Q so that $Q \cap C = \langle z\rangle$ and $C \cap H = Z(H)$. Also we have $|C : Z(H)| = 2$ and $C \trianglelefteq G$. Let a be an element of order 4 in Q so that $a^2 = z$ and let $b \in Q - \langle a\rangle$ so that $Q = \langle a, b\rangle$.

Assume for a moment that C is nonabelian. Then $G/C \cong E_4$ implies that there are exactly three nonabelian maximal subgroups $\langle a\rangle C$, $\langle b\rangle C$ and $\langle ab\rangle C$ of G containing C. Since $a \in Z(\langle a\rangle C)$ and $o(a) = 4$, we see that $\langle a\rangle C$ is not quasi-Hamiltonian and so $\langle a\rangle C = M$. Then $\langle b\rangle C$ and $\langle ab\rangle C$ are quasi-Hamiltonian and so C is also quasi-Hamiltonian implying that $|C| \ge 2^3$ and $|G| \ge 2^5$. Since $Z(H)$ is elementary abelian, there is $c \in C - Z(H)$ of order 4 and $Q\langle c\rangle < G$. Let K be a maximal subgroup of G containing $Q\langle c\rangle$ so that K must be quasi-Hamiltonian (since K is nonabelian and $K \ne M = \langle a\rangle C$). But $C_K(Q)$ contains c of order 4 and so $C_K(Q)$ is not elementary abelian, a contradiction.

We have proved that C is abelian. If C is elementary abelian, then G is quasi-Hamiltonian. But then all nonabelian maximal subgroups are quasi-Hamiltonian, contrary to our hypothesis. Hence C is not elementary abelian and so $\Omega_1(C) = Z(H)$ so that all elements in $C - Z(H)$ are of order 4 and let c be one of them. Let W be a complement of $\langle z, c^2\rangle$ in $Z(H)$.

Assume that $c^2 \neq z$. Then we consider distinct maximal subgroups $H_1 = \langle bc, a\rangle$ and $H_2 = \langle b, ac\rangle$ of $Q \times \langle c\rangle$ which are both minimal nonabelian of order 2^4. But then $H_1 \times W$ and $H_2 \times W$ are distinct nonabelian maximal subgroups of G and both are not quasi-Hamiltonian, a contradiction.

Finally, suppose that $c^2 = z$. If $|W| = 1$, then each maximal subgroups of G is either abelian or isomorphic to D_8 or Q_8 , a contradiction. If $|W| \geq 4$, then there are at least three nonabelian maximal subgroups of G containing $Q * \langle c\rangle$ and they all are not quasi-Hamiltonian, a contradiction. Hence we must have $W = \langle t\rangle \cong C_2$. But then $Q * \langle c\rangle$ and $Q * \langle ct\rangle$ are two distinct nonabelian maximal subgroups of G and both are not quasi-Hamiltonian, a contradiction.

(ii2b) Assume $d(G) = 2$ and $G' \cong C_4$. Since $|G'| > 2$, G has at most one abelian maximal subgroup and therefore G has at least one maximal subgroup H which is quasi-Hamiltonian. By Schreier's inequality (Theorem A.25.1), we get $d(H) \leq 3$. This implies that $H = Q \times Z$, where $Q \cong D_8$ or Q_8 and $|Z| \leq 2$. However, if $|Z| = 1$, then $|G| = 2^4$ and since $G' \cong C_4$, a result of Taussky (Lemma 1.6) implies that G is of maximal class. But then each maximal subgroup of G is either abelian or is isomorphic to D_8 or Q_8, a contradiction. Hence we have

$$H = Q \times \langle t\rangle, \quad \text{where } Q \cong D_8 \text{ or } Q_8,\ Z(Q) = Q' = \langle z\rangle \cong C_2,\ o(t) = 2.$$

It follows that $|G| = 2^5$ and since $G' \cong C_4$ and $G' \leq H$, we may assume that $G' = \langle a\rangle < Q$, where $Q = \langle a, b\rangle$ for some $b \in Q - G'$. We have $\Phi(G) < H$, $\Phi(G) > G'$ and $|\Phi(G)| = 2^3$. By a result of Burnside, $\Phi(G)$ cannot be nonabelian of order 8 and so

$$\Phi(G) = \langle a, t\rangle \cong C_4 \times C_2.$$

Since $\Phi(G) = \mho_1(G) < H$, there is $g \in G - H$ such that $g^2 \in \Phi(G) - \langle z, t\rangle$ and so $o(g) = 8$ implying that $\exp(G) = 8$. Also, $G' \nleq Z(G)$ and so G is of class 3. Note that $\Omega_1(\Phi(G)) = \langle z, t\rangle = Z(H)$.

Assume for a moment that G has another quasi-Hamiltonian maximal subgroup H_1 so that $H \cap H_1 = \Phi(G)$. In that case,

$$\Omega_1((\Phi(G)) = \langle z, t\rangle = Z(H_1) \quad \text{implying} \quad t \in Z(G).$$

But then $g \in G - H$ centralizes $g^2 \in \Phi(G) - \langle z, t\rangle$ and so $\Phi(G)\langle g\rangle$ is an abelian maximal subgroup of type $(8, 2)$. In that case all maximal subgroups of G would be either quasi-Hamiltonian or abelian, contrary to our hypothesis.

We have proved that G has only one quasi-Hamiltonian maximal subgroup H and so G must possess an abelian maximal subgroup A with $A \cap H = \Phi(G)$ implying that $\langle z, t\rangle \leq Z(G)$. By Lemma 1.1, $|Z(G)| = 4$ and so $Z(G) = \langle z, t\rangle \cong E_4$. Let $g \in G - H$ be such that

$$g^2 \in \Phi(G) - \langle z, t\rangle \quad \text{and so} \quad o(g) = 8.$$

Set $g^2 = v$. Then $A = \langle g\rangle \times \langle t\rangle$ is abelian of type $(8, 2)$ and $\langle v\rangle = \langle a\rangle$ or $\langle v\rangle = \langle at\rangle$.

Since for each $x \in G - A$, one has

$$x^2 \in Z(G) = \langle z, t \rangle \cong E_4,$$

it follows that all elements in $G - A$ are of order ≤ 4. We note that $\mho_1(A) = \langle v \rangle$ and so if for each $x \in G - A$, $x^2 \in \langle z \rangle$, then $\mho_1(G) = \langle v \rangle$, a contradiction. It follows that there is $w \in G - A$ such that $w^2 = t$, where we have replaced t with tz, if necessary.

We have $C_4 \cong G' \leq A$. However, if $G' = \langle a \rangle = \langle v \rangle$, then $\langle g \rangle \trianglelefteq G$ and since $G = \langle g, w \rangle$, G would be metacyclic. This is not possible since G has a nonmetacyclic maximal subgroup H. It follows that $G' = \langle vt \rangle$ and so we may set $vt = [w, g]$.

Since $\langle vt \rangle \trianglelefteq G$ and $\Phi(G) = \langle v, t \rangle \trianglelefteq G$, we get $\langle v \rangle \trianglelefteq G$ (noting that $\langle v, t \rangle$ has exactly two cyclic subgroups of order 4). Since $v \notin Z(G)$, we get $v^w = v^{-1}$ and so

$$M = \langle v, w \mid v^4 = w^4 = 1,\ v^w = v^{-1} \rangle$$

and this is a metacyclic minimal nonabelian group of order 16 and exponent 4.

We have proved that all minimal nonabelian subgroups of G, except one, are isomorphic to Q (where $Q \cong D_8$ or Q_8) and so G is isomorphic to a group defined in Theorem 258.1 (b) or to a group defined in Theorem 258.2 (b). Our theorem is proved. □

Problem. Classify the p-groups G such that any maximal subgroup M of G is either abelian or $M = S \times A$, where S is nonabelian of order p^3 and A is abelian.

§ 307 Nonabelian p-groups, $p > 2$, all of whose maximal subgroups, except one, are quasi-Hamiltonian or abelian

We recall that a p-group G is called quasi-Hamiltonian if $G = S \times E$, where S is nonabelian of order p^3 (then $S \in \{S(p^3), M_{p^3}\}$) and E is elementary abelian.

We solve here Problem 3773 for $p > 2$ by proving the following result.

Theorem 307.1. *Let G be a nonabelian p-group, $p > 2$, all of whose maximal subgroups, except one denoted with M, are quasi-Hamiltonian or abelian. Then G is of order p^5,*

$$d(G) = 2, \quad G' = \Phi(G) \cong E_{p^3}, \quad cl(G) = 3, \quad \exp(G) = p^2, \quad Z(G) \cong E_{p^2}$$

and p maximal subgroups of G are quasi-Hamiltonian of the form $S(p^3) \times C_p$ or $M(p^3) \times C_p$ and M is nonmetacyclic minimal nonabelian group of order p^4, i.e., $M \cong M_p(2, 1, 1)$.

Proof. Let G be a nonabelian p-group, $p > 2$, all of whose maximal subgroups, except one denoted with M, are quasi-Hamiltonian or abelian.

If $|M'| > p$, then Theorem 145.8 gives a contradiction (in that theorem were considered the p-groups in which exactly one maximal subgroup has the derived subgroup of order $> p$). Hence we have $|M'| = p$.

(i) First assume that the derived subgroup G' is noncyclic. In that case we may use Theorem 139.A which gives the following three possibilities:
(i1) $d(G) = 2$, $cl(G) = 3$, $G' \cong E_{p^2}$.
(i2) $d(G) = 2$, $cl(G) = 3$, $G' \cong E_{p^3}$, $\mho_1(G) \le Z(G)$.
(i3) $d(G) = 3$, $cl(G) = 2$, $G' \cong E_{p^2}$ or $G' \cong E_{p^3}$, and $\Phi(G) = Z(G)$.

(i1) One assumes that

$$d(G) = 2, \quad cl(G) = 3, \quad G' \cong E_{p^2}.$$

It follows from Proposition 145.5 that G has exactly one abelian maximal subgroup H_1. Set $\Gamma_1 = \{H_1, H_2, \dots, H_p, H_{p+1} = M\}$, where $H_2, \dots, H_p$ are quasi-Hamiltonian. Since $p > 2$, the group G has at least two quasi-Hamiltonian maximal subgroups. Due to Lemma 1.1,

$$|G| = p|G'||Z(G)| \quad \text{and so} \quad |Z(G)| = \frac{1}{p^3}|G|.$$

One has $H_2 = QZ(H_2)$, where $Q \cong S(p^3)$ or $M(p^3)$ and $Z(H_2)$ is elementary abelian with $Q \cap Z(H_2) = Q' \cong C_p$. Since $d(G) = 2$, we have $Z(G) \le \Phi(G)$ and so $Z(G) \le Z(H_2)$. It follows that

$$|G : Z(G)| = |G : H||H : Z(G)| = p \cdot p^2 = p^3,$$

and so the subgroup $Z(G) = Z(H_2)$ is elementary abelian. Since $|G/\Phi(G)| = p^2$ and $G' \nleq Z(G)$, we get $\Phi(G) = G'Z(G)$ is elementary abelian and $G' \cap Z(G) = Q' \cong C_p$. It follows that $\Phi(G)$ is elementary abelian implying that $\exp(G) \le p^2$.

https://doi.org/10.1515/9783110533149-051

Suppose that $Z(H_2) = Q'$ so that $|H_2| = p^3$ and $|G| = p^4$. But then $|M| = p^3$ and so M, being nonabelian, is quasi-Hamiltonian which contradicts our hypothesis. Hence $|Z(H_2)| \geq p^2$ implying that $|G| \geq p^5$. We have $\mho_1(H_2) \leq Q' = \langle z \rangle$ and since $E_{p^2} \cong G' \nleq Z(H_2)$, one may assume that $G' < Q$. Then $C_{H_2}(G') = G'Z(H_2)$ and so $C_G(G')$ covers G/H_2 implying that $C_G(G') = H_1$ is the unique abelian maximal subgroup in G. Since $G' \nleq Z(G)$, we get

$$H_2' = H_3' = \cdots = H_p' = M' = \langle z \rangle = Q' \quad \text{and} \quad \mho_1(H_2) \leq \langle z \rangle, \ldots, \mho_1(H_p) \leq \langle z \rangle.$$

We have proved that all maximal subgroups of the group $G/\langle z \rangle$ are abelian and $(G/\langle z \rangle)' = G'/\langle z \rangle$ is of order p and so $G/\langle z \rangle$ is minimal nonabelian (Lemma 65.1 (a)). We see that if we take $h_2 \in H_2 - \Phi(G)$ and $h_3 \in H_3 - \Phi(G)$ (noting that $H_2 \cap H_3 = \Phi(G)$), then $G/\langle z \rangle = \langle h_2, h_3 \rangle / \langle z \rangle$. But $h_2^p \in \langle z \rangle$ and $h_3^p \in \langle z \rangle$ implying that $G/\langle z \rangle \cong S(p^3)$ and therefore $|G| = p^4$, a final contradiction.

(i2) Assume that

$$d(G) = 2, \quad cl(G) = 3, \quad G' \cong E_{p^3}, \quad \mho_1(G) \leq Z(G).$$

Since $|G'| > p$, it follows that G has at most one abelian maximal subgroup and so G has at least $p - 1 \geq 2$ quasi-Hamiltonian maximal subgroups $H_1, H_2, \ldots, H_{p-1}$. Set $\{1\} \neq [G, G'] = Z$ so that G/Z is a two-generator p-group of class 2 and so G/Z has a cyclic commutator subgroup G'/Z. Hence $|G'/Z| = p$ and $Z \leq Z(G)$ with $Z \cong E_{p^2}$. Since $G' \nleq Z(G)$ and for any maximal subgroup X of G we have $|X'| \leq p$ and so X' is central in G, it follows that $X' \leq Z$. Hence all maximal subgroups of G/Z are abelian and $(G/Z)' \cong C_p$ and so G/Z is minimal nonabelian. But for quasi-Hamiltonian maximal subgroups H_1 and H_2 we have $\mho_1(H_1) \leq H_1' \leq Z$ and $\mho_1(H_2) \leq H_2' \leq Z$ and so the minimal nonabelian group G/Z is generated with two elements of order p. By Lemma 65.1, $G/Z \cong S(p^3)$ and so $|G| = p^5$ with $Z(G) = [G, G'] \cong E_{p^2}$ and $\Phi(G) = G' \cong E_{p^3}$.

If G has an abelian maximal subgroup, then by Lemma 1.1, $|G| = p^5 = p|G'||Z(G)|$ implying that $|Z(G)| = p$, a contradiction. Hence G has p quasi-Hamiltonian maximal subgroups $H_1, H_2, \ldots, H_p$ and $H_{p+1} = M$. Also, all H_i', $i = 1, 2, \ldots, p+1$, are pairwise distinct subgroups of order p in $Z(G)$.

It is now easy to determine the structure of $M = H_{p+1}$. To this end, let M_1 be a minimal nonabelian subgroup of M so that M_1 covers M/Z, where $Z = Z(G)$ and $M_1 \cap Z = Z(M_1) = \Phi(M_1)$. If $|M_1 \cap Z| = p$, then M_1 is nonabelian of order p^3 and so M would be quasi-Hamiltonian, a contradiction. Hence $|M_1 \cap Z| = p^2$ and so $M = M_1$ is minimal nonabelian of order p^4. Since $M > G'$ and $G' \cong E_{p^3}$, it follows that M is nonmetacyclic. In particular, $\exp(M) = p^2$ and so also $\exp(G) = p^2$. We have obtained a p-group described in our theorem.

(i3) Suppose that

$$d(G) = 3, \quad cl(G) = 2, \quad G' \cong E_{p^2} \text{ or } G' \cong E_{p^3}, \quad \Phi(G) = Z(G).$$

Then G has at least $p^2 + p - 1$ quasi-Hamiltonian maximal subgroups and let H be one of them so that $H = QZ(H)$, where Q is nonabelian of order p^3 and $Q \cap Z(H) = Q' \cong C_p$, where $Z(H)$ is elementary abelian. Since $d(G) = 3$ and $\Phi(G) = Z(G)$, we have

$$Z(H) = Z(G) = \Phi(G) \quad \text{and} \quad \exp(G) \le p^2.$$

But $M/Z(G) \cong E_{p^2}$ and if M_1 is a minimal nonabelian subgroup of M, then M_1 covers $M/Z(G)$ and $M_1 \cap Z(G) = \Phi(M_1)$. Since M is not quasi-Hamiltonian, $|M_1| \ge p^4$ implying that $\exp(M_1) = p^2$ and so $\exp(G) = p^2$.

Suppose that G has two distinct quasi-Hamiltonian maximal subgroups H_1 and H_2 of exponent p. Then take two elements $h_1, h_2 \in H_1 - Z(G)$ so that $\langle h_1, h_2 \rangle$ covers $H_1/Z(G) \cong E_{p^2}$ and take $h_3 \in H_2 - H_1$. Then $\langle h_1, h_2, h_3 \rangle = G$ and then $cl(G) = 2$ implies that $\exp(G) = p$, a contradiction. Indeed, note that our group G is p-abelian since for all $x, y \in G$,

$$(xy)^p = x^p y^p [y, x]^{\binom{p}{2}} = x^p y^p.$$

Hence we have at least $p^2 + p - 2$ quasi-Hamiltonian maximal subgroups of exponent p^2 and let $K_1 \ne K_2$ be two of them.

Let $k_1, k_2 \in K_1 - Z(G)$ so that $o(k_1) = p$, $o(k_2) = p^2$ and $\langle k_1, k_2 \rangle$ covers $K_1/Z(G)$ and let $k_3 \in K_2 - K_1$. Then $\langle k_1, k_2, k_3 \rangle = G$ and then the fact that G is p-abelian implies that $\mho_1(G) \le \langle k_2^p, k_3^p \rangle$ which is elementary abelian of order $\le p^2$. Since $\mho_1(G)$ has at most $p^2 - 1$ nontrivial elements and $p^2 + p - 2 > p^2 - 1$, we may choose above the quasi-Hamiltonian maximal subgroups K_1 and K_2 so that

$$\mho_1(K_1) = \mho_1(K_2) = \langle k_2^p \rangle \cong C_p.$$

This implies that $\mho_1(G) = \langle k_2^p \rangle \cong C_p$. Note that $k_2^p \in Z(G)$ and we consider the nonabelian group $G/\langle k_2^p \rangle$, where for each of $p^2 + p - 2$ maximal subgroups X of G which are quasi-Hamiltonian of exponent p^2 we have $X' = \mho_1(X) = \langle k_2^p \rangle$ and so each $X/\langle k_2^p \rangle$ is abelian. But a nonabelian p-group has at most $p + 1$ abelian maximal subgroups so that this fact gives in this case a final contradiction.

(ii) Assume now that G' is cyclic. By Theorem 137.7, we have either $G' \cong C_p$ or $d(G) = 2$ and $G' \cong C_{p^2}$.

(ii1) Suppose that $G' \cong C_p$. If in addition $d(G) = 2$, then by Lemma 65.2, G is minimal nonabelian, contrary to our hypothesis. Hence $d(G) \ge 3$ and so G has at least $p^2 - 1 \ge 8$ maximal subgroups which are quasi-Hamiltonian and let H be one of them. Then $H = QZ(H)$, where $Q = \langle a.b \rangle$ is nonabelian of order p^3 and $Z(H)$ is elementary abelian with $Q \cap Z(H) = Q' = \langle z \rangle \cong C_p$ and $Q' = G'$. We have $Q \trianglelefteq G$ and $C = C_G(Q)$ covers G/Q so that

$$Q \cap C = Q' = \langle z \rangle, \quad C \cap H = Z(H), \quad |C : Z(H)| = p, \quad C \trianglelefteq G.$$

Assume for a moment that C is nonabelian. Since $C' = Q' = G' = \langle z \rangle$ and C has an abelian maximal subgroup $Z(H)$, it follows from Lemma 1.1 that $Z(C) < Z(H)$ and $|C : Z(C)| = p^2$. But C is contained in $p + 1$ maximal subgroups of G which are all nonabelian and so at least p of them are quasi-Hamiltonian. Hence C is quasi-Hamiltonian and so $C = Q_1 Z(C)$, where Q_1 is nonabelian of order p^3 and $Q_1 \cap Z(C) = \langle z \rangle$. If $Q \cong M_{p^3}$, then Q has exactly p maximal subgroups L_i isomorphic to C_{p^2}. But then p nonabelian maximal subgroups $L_i C$ of G have a noncyclic center (because they contain L_i which centralize C), a contradiction. Similarly, if $Q_1 \cong M_{p^3}$, then Q_1 has exactly p maximal subgroups K_i isomorphic to C_{p^2}. Let V be a complement of $\langle z \rangle$ in $Z(C)$ so that $(QK_i) \times V$ are p nonabelian maximal subgroups of G with a noncyclic center (because they contain K_i), a contradiction. Hence also $Q_1 \cong S(p^3)$ and since $G = (Q * Q_!) \times V$, we have $\exp(G) = p$. Suppose that $|V| \geq p^2$. Then G contains at least $p + 1$ maximal subgroups containing $Q * Q_1$ and so all of them are not quasi-Hamiltonian, a contradiction. Suppose that $V = \langle v \rangle \cong C_p$ and set $Q_1 = \langle a_1, b_1 \rangle$. Then $Q_1^* = \langle a_1, b_1 v \rangle \cong S(p^3)$ and so $Q * Q_1$ and $Q * Q_1^*$ are two distinct nonabelian maximal subgroups of G and both are not quasi-Hamiltonian, a contradiction. Hence we must have $G = Q * Q_1$ with $Q \cap Q_1 = \langle z \rangle = G'$. But then each maximal subgroup of G is isomorphic to $S(p^3) \times C_p$, contrary to our hypothesis,

We have proved that C must be abelian. If C is elementary abelian, then G is quasi-Hamiltonian and then each maximal subgroup of G is either abelian or quasi-Hamiltonian, contrary to our hypothesis. Hence $\Omega_1(C) = Z(H)$ and all elements in $C - Z(H)$ are of order p^2 and let c be one of them. Let W be a complement of $\langle z, c^p \rangle$ in $Z(H)$, where W is elementary abelian.

Assume that $c^p \notin \langle z \rangle$ so that $G = Q \times \langle c \rangle \times W$, where $Q = \langle a, b \rangle$. We consider the distinct maximal subgroups $\langle a, bc \rangle \times W$ and $\langle ac, b \rangle \times W$ of G. But $\langle a, bc \rangle$ and $\langle ac, b \rangle$ are both minimal nonabelian of order p^4 and so both $\langle a, bc \rangle \times W$ and $\langle ac, b \rangle \times W$ are not quasi-Hamiltonian, a contradiction.

Finally, suppose that $c^p = z$. If $|W| = 1$, then each maximal subgroups of G is either abelian or is nonabelian of order p^3, a contradiction. If $|W| \geq p^2$, then there are at least $p + 1$ nonabelian maximal subgroups of G containing $Q * \langle c \rangle$ (central product) and they all are not quasi-Hamiltonian, a contradiction. Hence we must have $W = \langle t \rangle \cong C_p$. But then $Q * \langle c \rangle$ and $Q * \langle ct \rangle$ are two distinct nonabelian maximal subgroups of G and both are not quasi-Hamiltonian, a final contradiction.

(ii2) Suppose that $d(G) = 2$ and $G' \cong C_{p^2}$. Since $|G'| > p$, it follows that G has at most one abelian maximal subgroup and so G has at least $p - 1 \geq 2$ quasi-Hamiltonian maximal subgroups and let $H_1 \neq H_2$ be two of them so that $H_1 \cap H_2 = \Phi(G)$. We have

$$C_{p^2} \cong G' \leq \Phi(G)$$

and set

$$\Omega_1(G') = \langle z \rangle \cong C_p.$$

By Lemma 65.2, $G/\langle z \rangle = P$ is minimal nonabelian. Take two elements $h_1 \in H_1 - \Phi(G)$ and $h_2 \in H_2 - \Phi(G)$. Since $\mho_1(H_1) = \mho_1(H_2) = \langle z \rangle$, we have $h_1^p \in \langle z \rangle$ and $h_2^p \in \langle z \rangle$. Hence P is generated by two elements of order p and so (by Lemma 65.1) $P \cong S(p^3)$ implying that $|G| = p^4$. But then each maximal subgroup of G is either abelian or nonabelian of order p^3, a contradiction. Our theorem is proved. □

Problem 1. Classify the p-groups all of whose maximal subgroups, except one, have the form $S \times A$, where S is minimal nonabelian and $A' = \{1\}$.

Problem 2. Classify the p-groups G all of whose (i) maximal subgroups, (ii) maximal subgroups, except one, have the form $M \times A$, where M is of maximal class and $A' = \{1\}$.

Problem 3. Classify the p-groups G all of whose (i) maximal subgroups, (ii) maximal subgroups, except one, have the form $M \times A$, where M is metacyclic and $A' = \{1\}$.

§ 308 Nonabelian p-groups with an elementary abelian intersection of any two distinct maximal abelian subgroups

We solve Problem 3748 by proving the following result.

Theorem 308.1. *A nonabelian p-group G has the property that the intersection of any two distinct maximal abelian subgroups is elementary abelian if and only if the center of each nonabelian subgroup in G is elementary abelian.*

Proof. Let G be a nonabelian p-group with an elementary abelian intersection of any two distinct maximal abelian subgroups. Then each nonabelian subgroup M has an elementary abelian center $Z(M)$. Indeed, suppose that $Z = Z(M)$ is not elementary abelian. Let A_1 be a maximal abelian subgroup of M so that $Z < A_1 < M$ and let $m \in M - A_1$; then the subgroup $\langle m, Z\rangle$ is abelian. Let A_2 be a maximal abelian subgroup in M containing $\langle m, Z\rangle$ (such an A_2 exists since $\langle x, Z(M)\rangle$, being abelian, is a proper subgroup of M). Let B_1 be a maximal abelian subgroup in G containing A_1 and let B_2 be a maximal abelian subgroup in G containing A_2. Then $B_1 \cap M = A_1$ and $B_2 \cap M = A_2$, where $A_1 \neq A_2$ implying that $B_1 \neq B_2$. But $B_1 \cap B_2 \geq Z$ and Z is not elementary abelian, which is a contradiction.

Conversely, we will prove that if the center of each nonabelian subgroup in G is elementary abelian, then $S_1 \cap S_2 = D$ is elementary abelian for any two distinct maximal abelian subgroups S_1 and S_2 in G. Indeed, the subgroup $H = \langle S_1, S_2\rangle > S_1$ is nonabelian and $Z(H) = D$ (in fact, $C_G(D) \geq \langle S_1, S_2\rangle = H$ so that $D \leq Z(H)$ and the reverse containment is obvious). □

Exercise 1. Is it true that a nonabelian p-group G has the property that the intersection of any two distinct maximal abelian subgroups is of exponent $\leq p^e$ if and only if the center of each nonabelian subgroup in G has exponent $\leq p^e$?

Problem 1. Study the irregular p-groups G such that $\exp(R_1 \cap R_2) = p$ for any two distinct maximal regular subgroups R_1, R_2 of G.

https://doi.org/10.1515/9783110533149-052

§ 309 Minimal non-p-central p-groups

Recall that a p-group G is p-central if either $p > 2$ and $\Omega_1(G) \leq Z(G)$ or $p = 2$ and $\Omega_2(G) \leq Z(G)$.

In this section we determine the structure of minimal non-p-central p-groups (Problem 3754), where we use the following two characterizations of p-central p-groups.

Theorem 152.1. *A p-group G with $p > 2$ is p-central if and only if the following condition holds:*
(∗) *Whenever $x, y \in G$ and $x^p = y^p$, then $[x, y] = 1$.*

Theorem 152.2. *A 2-group G is 2-central if and only if the following condition holds:*
(∗∗) *Whenever $x, y \in G$ and $x^4 = y^4$, then $[x, y] = 1$.*

First we consider the case $p > 2$.

Theorem 309.1. *A p-group G with $p > 2$ is minimal non-p-central if and only if G is minimal nonabelian and there is $g \in G - Z(G)$ of order p.*

Proof. Let G be a minimal non-p-central p-group with $p > 2$. Then by Theorem 152.1, there are $a, b \in G$ such that $a^p = b^p$ but $[a, b] \neq 1$. Since the subgroup $\langle a, b\rangle$ is non-p-central, we get $G = \langle a, b\rangle$ so that

$$C_G(a^p) = C_G(b^p) \geq \langle a, b\rangle = G$$

hence $a^p = b^p \in Z(G)$.

If x is any element of order p in $\Phi(G)$, then x is central in each maximal subgroup of G and so $x \in Z(G)$. Hence $Z = \Omega_1(\Phi(G)) \leq \Omega_1(M) \leq Z(M)$ for all $M \in \Gamma_1$ and this implies that $Z \leq Z(G)$. Since the group G is not p-central, one has $\Omega_1(G) \nleq Z(G)$; therefore there is $g \in G - \Phi(G)$ of order p such that $g \notin Z(G)$.

Set $H = \langle g\rangle\Phi(G)$ so that $g \in \Omega_1(H) \leq Z(H)$ (because H is p-central) and, since $H \lhd G$, we get

$$W = Z \times \langle g\rangle = \Omega_1(Z(H)) \quad \text{with } W \trianglelefteq G.$$

Let $h \in G - C_G(g)$ so that $g \notin Z(\langle g, h\rangle)$ hence the subgroup $\langle g, h\rangle$ is not p-central hence $\langle g, h\rangle = G$. Note that W/Z is a normal subgroup of order p in G/Z so that $1 \neq [g, h] \in Z$. Since $\exp(Z) = p$, it follows that $G' = \langle [g, h]\rangle \cong C_p$ implying that, by Lemma 65.2 (a), the group G is minimal nonabelian.

We have proved that a minimal non-p-central p-group G with $p > 2$ is minimal nonabelian and there is $g \in G - \Phi(G)$ of order p. Since the converse of this statement is clear, the theorem is proved. □

We turn now to the case $p = 2$.

Theorem 309.2. *A 2-group G is minimal non-2-central if and only if G is a two-generator 2-group of class 2, $G' \neq \{1\}$ is cyclic of order ≤ 4, $\Omega_2(\Phi(G)) \leq Z(G)$ and $\langle g, h\rangle = G$, where $o(g) \leq 4$ and $C_G(g) = \Phi(G)\langle g\rangle$.*

https://doi.org/10.1515/9783110533149-053

Proof. Let G be a minimal non-2-central 2-group. Then by Theorem 152.2, there are $a, b \in G$ such that $a^4 = b^4$ but $[a, b] \neq 1$. Again by Theorem 152.2, $G = \langle a, b\rangle$ and so $\mathrm{d}(G) = 2$.

If x is any element of order ≤ 4 in $\Phi(G)$, then x is central in each maximal subgroup of G and so $x \in Z(G)$. Hence $Z = \Omega_2(\Phi(G)) \leq Z(G)$. Since G is not 2-central, there is $g \in G - \Phi(G)$ of order ≤ 4 (indeed, $\Omega_2(G) \nleq Z(G)$), with $g^2 \in Z$ (recall that $\exp(G/\Phi(G)) = 2$) and $g \notin Z(G)$.

Set $H = \langle g\rangle\Phi(G)$ so that $g \in Z(H)$ (because $H \lhd G$ is 2-central) and

$$W = Z * \langle g\rangle = \Omega_2(Z(H)) \quad \text{with } W \trianglelefteq G.$$

Let $h \in G - H$ so that $\langle g, h\rangle = G$ (see Exercise 1 below), and we conclude that $H \in \Gamma_1$. In that case, $C_G(g) = H$. Since $g^2 \in Z(\leq Z(G))$, it follows that W/Z is a normal subgroup of order 2 in G/Z so that $1 \neq [g, h] \in Z$. Hence the cyclic subgroup $G' = \langle [g, h]\rangle$ of order ≤ 4 is contained in $Z(G)$ implying that G is of class 2.

We have proved that a minimal non-2-central 2-group G is two-generated of class 2, $G' \neq \{1\}$ is cyclic of order ≤ 4, $\Omega_2(\Phi(G)) \leq Z(G)$ and G has an element $g \in G - \Phi(G)$ of order ≤ 4, where $C_G(g) = \Phi(G)\langle g\rangle$. Since the converse of this statement is clear, our theorem is proved. □

Exercise 1. Present another proof of the statement that a minimal non-p-central p-group G is two-generator.

Solution. Let $g \in G - Z(G)$ be of order p if $p > 2$ and of order ≤ 4 if $p = 2$ and let $h \in G - C_G(g)$. Set $K = \langle g, h\rangle$. Since $g \notin Z(K)$, we conclude that the subgroup K is not p-central implying $K = G$. It follows that $\mathrm{d}(G) = 2$. □

Exercise 2. Let G be a p-group, $g \in G - Z(G)$. If any proper subgroup H of G containing x satisfies $x \in Z(H)$, then $\mathrm{d}(G) = 2$.

Hint. Indeed, if $y \in G - C_G(x)$, then $G = \langle x, y\rangle$. □

Problem 1. Classify the 2-groups all of whose subgroups of index 4 are 2-central. (If $p > 2$ and all subgroups of index $\leq p^2$ in a p-group G are p-central, then G is an $\mathcal{A}_n$-group, $n \leq 2$, by Theorem 309.1.)

Problem 2. Classify the non-Dedekindian p-groups all of whose nonnormal subgroups are p-central.[1]

1 Note that the non-Dedekindian p-groups with all abelian nonnormal subgroups are not classified.

§ 310 Nonabelian p-groups in which each element in any minimal nonabelian subgroup is half-central

Recall that an element x in a p-group G is half-central in G if $|G : C_G(x)| \leq p$. We solve here Problem 4018 by proving the following result.

Theorem 310.1. *Let G be a nonabelian p-group such that whenever $x \in S \leq G$, where S is minimal nonabelian, then x is half-central in G. Then each element of G is half-central in G and so $|G'| = p$.*

Proof. Let G be a nonabelian p-group such that whenever $x \in S \leq G$, where S is minimal nonabelian, then x is half-central in G. If $y \in K < G$ and y is half-central in G, then y is half-central in K. This is clear if $y \in Z(K)$. Otherwise, $C_G(y) \in \Gamma_1$. As $C_K(y) = K \cap C_G(y)$ has index p is K, by the product formula, y is half-central in K.

Let A be a maximal normal abelian subgroup in G. By Lemma 57.1, each $g \in G - A$ is contained in a minimal nonabelian subgroup and so g is half-central in G.

Let $a \in A - Z(G)$. Then there is $g \in G - A$ such that $[a, g] \neq 1$. Set $H = A\langle g\rangle$. Since g is half-central in G and $C_G(A) = A$, we have (by the result of the first paragraph, g is half-central in H) $|H : C_H(g)| = p$ and $C_A(g) = Z(H)$ with $|A : Z(H)| = p$ and hence $a^p \in Z(H)$. Since $H' \leq A \cap C_H(g) = Z(H)$, the subgroup H is of class 2. Therefore we get

$$[a, g]^p = [a^p, g] = 1.$$

It follows that $\mho_1(H) \leq Z(H)$. As $H' \leq Z(H)$, the quotient group $H/Z(H)$ is abelian. As $\Omega_1(H/Z(H)) = H/Z(H)$, we get $|H/Z(H)| = p^2$. Now it follows from Exercise A.132.2 that the element a is contained in minimal nonabelian subgroup $S \leq G$. By our hypothesis, the element a is half-central in G.

We have proved that all elements in G are half-central in G. Then by a result of Knoche (Proposition 121.9), we get $|G'| = p$ and we are done. □

Exercise 1. Let G be a p-group with derived subgroup of order p. Is it true that all elements of G are half-central in G?

Problem 1. Study the p-groups G containing two distinct maximal subgroups F, H such that all elements of the set $F \cup H$ are half-central in G.

Problem 2. Study the p-groups G of exponent $> p$ all of whose elements of orders $> p$ are half-central.

Problem 3. Study the p-groups G containing a maximal subgroup H such that all elements of the set $G - H$ are half-central in G.

https://doi.org/10.1515/9783110533149-054

§ 311 Nonabelian p-groups G of exponent p in which $C_G(x) = \langle x \rangle^G$ for all noncentral $x \in G$

We prove here the following result.

Theorem 311.1. *Let G be a nonabelian p-group of exponent p in which $C_G(x) = \langle x \rangle^G$ for all noncentral $x \in G$. Then G is a special p-group of order $p|Z(G)|^2$ and for each $y \in G - Z(G)$, we have $C_G(y) = Z(G)\langle y \rangle$. (A smallest example of such groups is any nonabelian group of order p^3 generated by elements of order p.)*

Proof. Let G be a nonabelian p-group of exponent p in which $C_G(x) = \langle x \rangle^G$ for all noncentral $x \in G$. Then we have $p > 2$.

Let $x \in G - Z(G)$ and set $H = C_G(x)$. Since $H = \langle x \rangle^G$, we have $H \trianglelefteq G$. All G-conjugates of $\langle x \rangle$ are central in H and they generate H and so H is abelian. But $H = C_G(x)$ and so H is also a maximal normal abelian subgroup in G. Let $Z(G) < K \leq H$ with $K \trianglelefteq G$ and $|K : Z(G)| = p$. Let $k \in K - Z(G)$ so that $\langle k \rangle^G \leq K$. On the other hand, $\langle k \rangle^G = C_G(k) \geq H$ implying that $K = H$ and so $|H : Z(G)| = p$.

We have proved that for each $x \in G - Z(G)$, the centralizer $C_G(x) = Z(G)\langle x \rangle$ is a maximal normal abelian subgroup in G. This implies at once that $\mathrm{cl}(G) = 2$. Indeed, let $a, b \in G$ with $[a, b] \neq 1$. Then $A = Z(G)\langle a \rangle$ and $B = Z(G)\langle b \rangle$ are two distinct maximal normal abelian subgroups in G. This gives $[a, b] \in A \cap B = Z(G)$ and so $G' \leq Z(G)$, and we are done.

Now it is easy to show that G is a special p-group of order $p|Z(G)|^2$. Indeed, let $a \in G - Z(G)$ so that $A = Z(G)\langle a \rangle$ is a maximal normal abelian subgroup in G. We consider the homomorphism $G \to G/A$ with $g \in G$:

$$g \mapsto [g, a] \quad (a \in A)$$

with kernel A so that $G/A \cong [G, A]$. But

$$\langle a \rangle^G = Z(G)\langle a \rangle \implies [G, A] = Z(G)$$

and so, since $|A| = p|Z(G)|$, one has

$$G' = Z(G) \text{ and } |G/A| = |Z(G)| \implies |G| = |A||Z(G)| = p|Z(G)|^2$$

completing the proof. Indeed, the group G is special since $\Phi(G) = G' = Z(G)$ (recall that $\exp(G) = p$). □

Problem 1. Study the nonabelian p-groups G of exponent p such that $|G : A^G| = p$ for each noncyclic nonnormal $A < G$.

Problem 2. Study the non-Dedekindian p-groups G such that $|G : A^G| = p$ for all nonnormal $A < G$ of order $> p$.

https://doi.org/10.1515/9783110533149-055

§ 312 Nonabelian 2-groups all of whose minimal nonabelian subgroups, except one, are isomorphic to $\mathrm{M}_2(2,2) = \langle a, b \mid a^4 = b^4 = 1,\ a^b = a^{-1}\rangle$

We prove here the following result.

Theorem 312.1. *Let G be a nonabelian 2-group all of whose minimal nonabelian subgroups, except one denoted with M, are isomorphic to*

$$\mathrm{M}_2(2,2) = \langle a, b \mid a^4 = b^4 = 1,\ a^b = a^{-1}\rangle.$$

Then $M \trianglelefteq G$ and if M_0 is a maximal subgroup of M with $M_0 \trianglelefteq G$, then $C_G(M_0) = M_0$. All elements in $G - M$ are of order 4, $\Omega_1(M) \cong E_4$ or E_8, $\Omega_1(M) = Z(G)$ and one of the following holds:

(a) *If $\exp(M) > 4$, then $|G : M| = 2$.*

(b) *If $\exp(M) = 4$, then $1 \neq |G : M| \le 4$, G is of class 2 and M is the nonmetacyclic minimal nonabelian group of exponent 4 and order 2^5.*

Proof. Let G be a nonabelian 2-group all of whose minimal nonabelian subgroups, except one denoted with M, are isomorphic to

$$\mathrm{M}_2(2,2) = \langle a, b \mid a^4 = b^4 = 1,\ a^b = a^{-1}\rangle.$$

We have $M \trianglelefteq G$ and let $M_0 < M$ with $|M : M_0| = 2$ and $M_0 \trianglelefteq G$. Let A be a maximal normal abelian subgroup of G containing M_0.

(i) First assume that $A = M_0$ so that $C_G(M_0) = M_0$. Let $x \in G - M$. By Lemma 57.1, there is $a \in M_0$ such that $\langle a, x\rangle$ is minimal nonabelian (noting that here M_0 is a maximal normal abelian subgroup in G, by assumption). Since $\langle a, x\rangle \neq M$, we have $\langle a, x\rangle \cong \mathrm{M}_2(2,2)$ and so $o(x) = 4$. We have proved that all elements in the set $G - M$ are of order 4.

If $|M_0| = 4$, then the fact that $C_G(M_0) = M_0$ implies that $|G : M_0| = 2$. But then $M = G$, contrary to our assumption. Hence we have $|M| \ge 2^4$ and so setting $E = \Omega_1(M)$, we get $E \cong E_4$ or E_8 (Lemma 65.1).

Let $x \in G - M$ so that $o(x) = 4$ and $x^2 \in E = \Omega_1(M)$. We act with $\langle x\rangle$ on E. If $\langle x\rangle$ does not centralize E, then there is $e \in E$ such that $\langle x, e\rangle$ is minimal nonabelian (Lemma 57.1). Since $\langle x, e\rangle \neq M$, we must have $\langle x, e\rangle \cong \mathrm{M}_2(2,2)$, a contradiction because each involution in $\mathrm{M}_2(2,2)$ is contained in $\Phi(\mathrm{M}_2(2,2))$. Hence all elements in the set $G - M$ centralize E implying that $E \le Z(G)$. Since $C_G(M_0) = M_0$, we have $Z(G) \le M_0$ and so $Z(G) \le Z(M)$, where $|M_0 : Z(M)| = 2$.

Let again $x \in G - M$, where $o(x) = 4$ and $x^2 \in E < M_0$ and consider the subgroup $K = M_0\langle x\rangle$. Since $C_G(M_0) = M_0$, the subgroup K is nonabelian. On the other hand,

https://doi.org/10.1515/9783110533149-056

$M \nleq K$ (clearly, $|K| = 2|M_0| = |M|$ and $x \notin M$), and so each minimal nonabelian subgroup in K is isomorphic to $M_2(2, 2)$. Hence K is isomorphic to one of the groups from conclusion of Theorem 57.3. In particular, we see that $Z(K)$ is elementary abelian. But $Z(G) \leq M_0$ and so $Z(G) \leq Z(K)$ and therefore $Z(G)$ is elementary abelian. Hence $Z(G) \leq E = \Omega_1(M)$ and so we have proved that $Z(G) = E = \Omega_1(M)$.

(i1) Assume that $\exp(M) > 4$. Then M/E is abelian of exponent ≥ 4, where $E = \Omega_1(M)$. For each $x \in G - M$, $x^2 \in E$ and so $H_2(G/E) = M/E$. By a result of Burnside, $|G : M| = 2$ and so we have obtained the groups stated in part (a) of our theorem.

(i2) Suppose that $\exp(M) = 4$. Then $Z(M)$ is elementary abelian and so $E = Z(M) = Z(G)$, where $E \cong E_4$ or E_8. But M is not isomorphic to $M_2(2, 2)$ and so M is nonmetacyclic with $E = \Omega_1(M) \cong E_8$ and $E = Z(M)$ implying that $|M| = 2^5$. Since $\mho_1(G) \leq E = Z(G)$, it follows that G is of class 2. Since G/M_0 acts faithfully on M_0 stabilizing the chain $M_0 > E > \{1\}$, we get that G/M_0 is elementary abelian of order $\leq 2^3 = |E|$. We get $1 \neq |G : M| \leq 4$ and so we have obtained the groups stated in part (b) of our theorem.

(ii) Assume that $A > M_0$, where A is a maximal normal abelian subgroup of G containing M_0. Set $G_0 = AM$ so that $G_0/M \cong A/M_0$ is abelian and therefore $G_0' \leq A \cap M = M_0$. Our aim is to obtain a contradiction. It is surprising that this will be a very difficult task.

For each $x \in G - (A \cup M)$, there is (by Lemma 57.1) $a \in A$ such that $\langle x, a\rangle$ is minimal nonabelian. Since $\langle x, a\rangle \neq M$, we have $\langle x, a\rangle \cong M_2(2, 2)$ and so $o(a) = o(x) = 4$. Hence all elements in $G - (A \cup M)$ are of order 4 and A is not elementary abelian.

Let $E = \Omega_1(A)$ so that $A \trianglelefteq G$ and let $x \in G_0 - (A \cup M)$, where $x^2 \in E$. If x does not centralize E, then there is $e \in E$ such that $\langle x, e\rangle$ is minimal nonabelian (noting that in this case E is a maximal normal abelian subgroup in $E\langle x\rangle$ and then we use again Lemma 57.1). Since $\langle x, e\rangle \neq M$, we must have $\langle x, e\rangle \cong M_2(2, 2)$, a contradiction. Hence x centralizes E and so $E \leq Z(G_0)$. Let $y \in G - G_0$ (in case that $G_0 < G$) so that $o(y) = 4$. Then y^2 is an involution and so either $y^2 \in A$ and so $y^2 \in E$ or $y^2 \in M - M_0$ and y^2 centralizes E (since $E \leq Z(G_0)$. Hence in any case y^2 centralizes E and $\langle y^2\rangle E$ is elementary abelian. If y does not centralize E, then there is $e' \in \langle y^2\rangle E$ such that $\langle y, e'\rangle$ is minimal nonabelian. But $\langle y, e'\rangle \neq M$ and so $\langle y, e'\rangle \cong M_2(2, 2)$, a contradiction. We have proved that $E = \Omega_1(A) \leq Z(G)$.

Suppose that $E = \Omega_1(A) \nleq M_0$ and let $f \in E - M_0$. Set $M = \langle a, b\rangle$. But then (noting that f is an involution in $Z(G)$) $\langle fa, b\rangle \cong M$ and $\langle fa, b\rangle \neq M$, a contradiction. We have proved that $E \leq M_0$ and so

$$E = \Omega_1(A) = \Omega_1(M_0) \leq Z(M) = \Phi(M).$$

We consider $G_0/Z(M)$ and note that all elements in $G_0 - A$ are involutions in $G_0/Z(M)$ and so each $x \in G_0 - A$ inverts each element in $A/Z(M)$. But we know that $G_0' \leq M_0$ and so A/M_0 is elementary abelian.

It is now easy to see that $Z(M)$ is elementary abelian and so $E = \Omega_1(M_0) = Z(M)$. Indeed, let $x \in G_0 - (A \cup M)$, where $x^2 \in E = \Omega_1(A) \leq Z(M)$. Noting that $M_0 \trianglelefteq G$, we see that $K = M_0\langle x\rangle$ is nonabelian with $Z(K) = Z(M)$ (because if x would centralize M_0, then $M_0 \le Z(G_0)$ and so $M_0 \le Z(M)$, a contradiction). Since $M \nleq K$, all minimal nonabelian subgroups of K are isomorphic to $M_2(2,2)$. By inspecting the groups from Theorem 57.3, we see that $Z(K)$ is elementary abelian. Then $Z(M) = Z(K)$ is elementary abelian. But $Z(M) = \Phi(M)$ and so M is of exponent 4 and all elements in $M_0 - Z(M)$ are of order 4.

Suppose that $|Z(M)| = 2$ in which case $M \cong D_8$ or Q_8. Since $Z(M) = \Omega_1(A)$, A is cyclic, $M_0 \cong C_4$ and since A/M_0 is elementary abelian, we get that $|A/M_0| = 2$. Then $A \cong C_8$ and $|G_0| = 2^4$. But G_0 contains a subgroup isomorphic to $M_2(2,2)$ and so $G_0 \cong M_2(2,2)$ would be minimal nonabelian, a contradiction.

We have proved that we have $|Z(M)| \ge 4$. But M is not isomorphic to $M_2(2,2)$ and so M is nonmetacyclic minimal nonabelian of exponent 4 and order 2^4 or 2^5. In particular, D_8 is not a subgroup in G and so G does not possess non-commuting involutions.

Suppose that there is an involution $i \in M - M_0$. Suppose also that i commutes with all elements x in $G_0 - (A \cup M)$ (of order 4), where $x^2 \in Z(M)$. But then i commutes with all elements in M_0x and so i centralizes M_0, a contradiction. We have proved that there is $x_0 \in G_0 - (A \cup M)$ which does not commute with i, where $1 \ne x_0^2 \in Z(M)$ so that $o(x_0) = 4$ and x_0^2 commutes with i. A nonabelian group $\langle x_0, i\rangle$ (generated by an involution and an element of order 4) is described in Theorem A.105.1. Since D_8 is not a subgroup in G, it follows from this theorem that $\langle x_0, i\rangle$ is nonmetacyclic minimal nonabelian of order 2^4. But $x_0 \notin M$ and so $\langle x_0, i\rangle \ne M$, contrary to our hypothesis. We have proved that there is no involution in $M - M_0$ and so M is nonmetacyclic minimal nonabelian of exponent 4 and order 2^5. Also we have proved that $Z(G) = Z(M) = \Omega_1(G) \cong E_8$.

Suppose that for an element $x \in A - M$, we have $x^2 \in Z(M)$. Set $L = M\langle x\rangle$ and note that in this case $\mho_1(L) \le Z(M)$ implying that $L' \le Z(M)$ and so $\mathrm{cl}(L) = 2$ with elementary abelian L'. Then any two non-commuting elements a and b in L generate a minimal nonabelian subgroup $S = \langle a, b\rangle$ (by Lemma 65.2), where $\Phi(S) = \langle a^2, b^2, [a,b]\rangle$. If $|\Phi(S)| = 2$ or $|\Phi(S)| = 8$ and $S \ne M$, then we get a contradiction. Take $m \in M - M_0$ and $n \in M_0 - Z(M)$ so that

$$M = \langle m, n \mid m^4 = n^4 = 1,\ [m,n] = s,\ s^2 = [s,m] = [s,n] = 1\rangle,$$

where $Z(M) = \langle m^2, n^2, s\rangle \cong E_8$. We have to examine all possibilities for

$$x^2 \in \{Z(M) - \langle n^2\rangle\} = \{m^2, s, m^2s, n^2s, m^2n^2, m^2n^2s\}$$

(where we have excluded the possibility $x^2 = n^2$ because in that case xn would be an involution in $A - M$, a contradiction) and all possibilities for

$$[x,m] \in \{m^2, s, n^2, m^2s, n^2s, m^2n^2, m^2n^2s\}.$$

Hence there are exactly $6 \cdot 7 = 42$ possibilities to be examined for the structure of $L = M\langle x\rangle$.

If $x^2 = m^2$ and $[x, m] = m^2$, then

$$|\Phi(\langle m, x\rangle)| = |\langle m^2\rangle| = 2,$$

a contradiction.

If $x^2 = s$ and $[x, m] = m^2$, then we consider $S = \langle xn, m\rangle$ and get

$$|\Phi(S)| = |\langle sn^2, m^2, m^2 s\rangle| = 8,$$

where $S \neq M$, a contradiction.

If $x^2 = m^2 s$ and $[x, m] = m^2$, then we consider $S = \langle xn, m\rangle$ and get

$$|\Phi(S)| = |\langle m^2 sn^2, m^2, m^2 s\rangle| = 8,$$

where $S \neq M$, a contradiction.

If $x^2 = n^2 s$ and $[x, m] = m^2$, then we consider $S = \langle xm, mn\rangle$ and get

$$|\Phi(S)| = |\langle sn^2, m^2 n^2 s, m^2 s\rangle| = 8,$$

where $S \neq M$, a contradiction.

If $x^2 = m^2 n^2$ and $[x, m] = m^2$, then we consider $S = \langle x, mn\rangle$ and get

$$|\Phi(S)| = |\langle m^2 n^2, m^2 n^2 s, m^2\rangle| = 8,$$

where $S \neq M$, a contradiction.

If $x^2 = m^2 n^2 s$ and $[x, m] = m^2$, then we consider $S = \langle xnm, n\rangle$ and get

$$|\Phi(S)| = |\langle m^2, n^2, s\rangle| = 8,$$

where $S \neq M$, a contradiction.

If $x^2 = m^2$ and $[x, m] = s$, then we consider $S = \langle x, mn\rangle$ and get

$$|\Phi(S)| = |\langle m^2, m^2 n^2 s, s\rangle| = 8,$$

where $S \neq M$, a contradiction.

If $x^2 = s$ and $[x, m] = s$, then we consider $S = \langle xnm, n\rangle$ and get

$$|\Phi(S)| = |\langle m^2 n^2 s, n^2, s\rangle| = 8,$$

where $S \neq M$, a contradiction.

If $x^2 = m^2 s$ and $[x, m] = s$, then we consider $S = \langle x, mn\rangle$ and get

$$|\Phi(S)| = |\langle m^2 s, m^2 n^2 s, s\rangle| = 8,$$

where $S \neq M$, a contradiction.

If $x^2 = n^2 s$ and $[x, m] = s$, then we consider $S = \langle x, m\rangle$ and get

$$|\Phi(S)| = |\langle sn^2, m^2, s\rangle| = 8,$$

where $S \neq M$, a contradiction.

If $x^2 = m^2n^2$ and $[x, m] = s$, then we consider $S = \langle x, m\rangle$ and get

$$|\Phi(S)| = |\langle m^2n^2, m^2, s\rangle| = 8,$$

where $S \neq M$, a contradiction.

If $x^2 = m^2n^2s$ and $[x, m] = s$, then we consider $S = \langle x, m\rangle$ and get

$$|\Phi(S)| = |\langle m^2n^2s, m^2, s\rangle| = 8,$$

where $S \neq M$, a contradiction.

If $x^2 = m^2$ and $[x, m] = n^2$, then we consider $S = \langle xn, m\rangle$ and get

$$|\Phi(S)| = |\langle m^2n^2, m^2, n^2s\rangle| = 8,$$

where $S \neq M$, a contradiction.

If $x^2 = s$ and $[x, m] = n^2$, then we consider $S = \langle x, m\rangle$ and get

$$|\Phi(S)| = |\langle s, m^2, n^2\rangle| = 8,$$

where $S \neq M$, a contradiction.

If $x^2 = m^2s$ and $[x, m] = n^2$, then we consider $S = \langle x, m\rangle$ and get

$$|\Phi(S)| = |\langle m^2s, m^2, n^2\rangle| = 8,$$

where $S \neq M$, a contradiction.

If $x^2 = n^2s$ and $[x, m] = n^2$, then we consider $S = \langle x, m\rangle$ and get

$$|\Phi(S)| = |\langle sn^2, m^2, n^2\rangle| = 8,$$

where $S \neq M$, a contradiction.

If $x^2 = m^2n^2$ and $[x, m] = n^2$, then we consider $S = \langle x, mn\rangle$ and get

$$|\Phi(S)| = |\langle m^2n^2, m^2n^2s, n^2\rangle| = 8,$$

where $S \neq M$, a contradiction.

If $x^2 = m^2n^2s$ and $[x, m] = n^2$, then we consider $S = \langle x, m\rangle$ and get

$$|\Phi(S)| = |\langle m^2sn^2, m^2, n^2\rangle| = 8,$$

where $S \neq M$, a contradiction.

If $x^2 = m^2$ and $[x, m] = m^2s$, then we consider $S = \langle xnm, n\rangle$ and get

$$|\Phi(S)| = |\langle m^2n^2, n^2, s\rangle| = 8,$$

where $S \neq M$, a contradiction.

If $x^2 = s$ and $[x, m] = m^2s$, then we consider $S = \langle x, m\rangle$ and get

$$|\Phi(S)| = |\langle s, m^2n^2s, m^2s\rangle| = 8,$$

where $S \neq M$, a contradiction.

If $x^2 = m^2s$ and $[x, m] = m^2s$, then we consider $S = \langle xnm, n\rangle$ and get

$$|\Phi(S)| = |\langle m^2n^2s, n^2, s\rangle| = 8,$$

where $S \neq M$, a contradiction.

If $x^2 = m^2n^2$ and $[x, m] = m^2s$, then we consider $S = \langle x, m\rangle$ and get

$$|\Phi(S)| = |\langle m^2n^2, m^2, m^2s\rangle| = 8,$$

where $S \neq M$, a contradiction.

If $x^2 = m^2n^2s$ and $[x, m] = m^2s$, then we consider $S = \langle x, m\rangle$ and get

$$|\Phi(S)| = |\langle m^2n^2s, m^2, m^2s\rangle| = 8,$$

where $S \neq M$, a contradiction.

If $x^2 = m^2$ and $[x, m] = n^2s$, then we consider $S = \langle xn, mn\rangle$ and get

$$|\Phi(S)| = |\langle m^2n^2, m^2n^2s, n^2\rangle| = 8,$$

where $S \neq M$, a contradiction.

If $x^2 = s$ and $[x, m] = n^2s$, then we consider $S = \langle x, m\rangle$ and get

$$|\Phi(S)| = |\langle s, m^2, n^2s\rangle| = 8,$$

where $S \neq M$, a contradiction.

If $x^2 = m^2s$ and $[x, m] = n^2s$, then we consider $S = \langle x, m\rangle$ and get

$$|\Phi(S)| = |\langle m^2s, m^2, n^2s\rangle| = 8,$$

where $S \neq M$, a contradiction.

If $x^2 = n^2s$ and $[x, m] = n^2s$, then we consider $S = \langle xn, m\rangle$ and get

$$|\Phi(S)| = |\langle s, m^2, n^2\rangle| = 8,$$

where $S \neq M$, a contradiction.

If $x^2 = m^2n^2$ and $[x, m] = n^2s$, then we consider $S = \langle x, m\rangle$ and get

$$|\Phi(S)| = |\langle m^2n^2, m^2, n^2s\rangle| = 8,$$

where $S \neq M$, a contradiction.

If $x^2 = m^2n^2s$ and $[x, m] = n^2s$, then we consider $S = \langle xn, m\rangle$ and get

$$|\Phi(S)| = |\langle sm^2, m^2, n^2\rangle| = 8,$$

where $S \neq M$, a contradiction.

If $x^2 = m^2$ and $[x, m] = m^2n^2$, then we consider $S = \langle xn, m\rangle$ and get

$$|\Phi(S)| = |\langle m^2n^2, m^2, m^2n^2s\rangle| = 8,$$

where $S \neq M$, a contradiction.

If $x^2 = s$ and $[x, m] = m^2n^2$, then we consider $S = \langle x, m\rangle$ and get

$$|\Phi(S)| = |\langle s, m^2, m^2n^2\rangle| = 8,$$

where $S \neq M$, a contradiction.

If $x^2 = m^2s$ and $[x, m] = m^2n^2$, then we consider $S = \langle x, m\rangle$ and get

$$|\Phi(S)| = |\langle m^2s, m^2, m^2n^2\rangle| = 8,$$

where $S \neq M$, a contradiction.

If $x^2 = n^2s$ and $[x, m] = m^2n^2$, then we consider $S = \langle x, m\rangle$ and get

$$|\Phi(S)| = |\langle sn^2, m^2, m^2n^2\rangle| = 8,$$

where $S \neq M$, a contradiction.

If $x^2 = m^2n^2$ and $[x, m] = m^2n^2$, then we consider $S = \langle xnm, n\rangle$ and get

$$|\Phi(S)| = |\langle m^2sn^2, n^2, s\rangle| = 8,$$

where $S \neq M$, a contradiction.

If $x^2 = m^2n^2s$ and $[x, m] = m^2n^2$, then we consider $S = \langle x, m\rangle$ and get

$$|\Phi(S)| = |\langle m^2sn^2, m^2, m^2n^2\rangle| = 8,$$

where $S \neq M$, a contradiction.

If $x^2 = m^2$ and $[x, m] = m^2n^2s$, then we consider $S = \langle xnm, n\rangle$ and get

$$|\Phi(S)| = |\langle m^2, n^2, s\rangle| = 8,$$

where $S \neq M$, a contradiction.

If $x^2 = s$ and $[x, m] = m^2n^2s$, then we consider $S = \langle x, m\rangle$ and get

$$|\Phi(S)| = |\langle s, m^2, m^2n^2s\rangle| = 8,$$

where $S \neq M$, a contradiction.

If $x^2 = m^2s$ and $[x, m] = m^2n^2s$, then we consider $S = \langle x, m\rangle$ and get

$$|\Phi(S)| = |\langle sm^2, m^2, m^2n^2s\rangle| = 8,$$

where $S \neq M$, a contradiction.

If $x^2 = n^2s$ and $[x, m] = m^2n^2s$, then we consider $S = \langle xn, m\rangle$ and get

$$|\Phi(S)| = |\langle s, m^2, m^2n^2\rangle| = 8,$$

where $S \neq M$, a contradiction.

If $x^2 = m^2n^2$ and $[x, m] = m^2n^2s$, then we consider $S = \langle x, m\rangle$ and get

$$|\Phi(S)| = |\langle m^2n^2, m^2, m^2n^2s\rangle| = 8,$$

where $S \neq M$, a contradiction.

If $x^2 = m^2n^2s$ and $[x, m] = m^2n^2s$, then we consider $S = \langle xn, m\rangle$ and get

$$|\Phi(S)| = |\langle sm^2, m^2, m^2n^2\rangle| = 8,$$

where $S \neq M$, a final contradiction for the 42 possibilities for the structure of the group $L = \langle a\rangle M$ of order 2^6.

We have proved that for each $a \in A - M_0$, $a^2 \in M_0 - Z(M)$ and so $o(a) = 8$, $A = \langle a\rangle Z(M)$ (noting that A/M_0 is elementary abelian) and therefore we have obtained that $|A : M_0| = |G_0 : M| = 2$ and $\exp(A) = 8$.

We get $H_2(G/Z(M)) = A/Z(M)$ and so by a result of Burnside, $|G : A| = 2$ implying that $G = G_0$ and so $|G| = 2^6$.

By Lemma 1.1, $2^6 = |G| = 2|Z(G)||G'|$, and since $Z(G) = Z(M) \cong E_8$ we get $|G'| = 4$. Note that all elements in $G - A$ are of order 4 and they generate G. If G is of class 2, then $\exp(G) = 4$, a contradiction. Hence $cl(G) = 3$ and (noting that $G' \leq M_0$ and $|M_0 : Z(M)| = 2$), we see that G' covers $M_0/Z(M)$ and so $G' \cong C_4$. But $M' < G'$ and so M' is a square in M, contrary to the fact that the minimal nonabelian subgroup M is nonmetacyclic. This is a final contradiction and our theorem is proved. □

§ 313 Non-Dedekindian 2-groups all of whose maximal Dedekindian subgroups have index 2

We prove here the following result which solves Problem 4002.

Theorem 313.1. *Let G be a non-Dedekindian 2-group all of whose maximal Dedekindian subgroups have index 2. Then either G has an abelian subgroup of index 2 or G is of the form* $G = (Q * D) \times V$, *where* $Q \cong \mathrm{Q}_8$, $D \cong \mathrm{D}_8$ *and* $\exp(V) \le 2$.

Proof. Let G be a non-Dedekindian 2-group all of whose maximal Dedekindian subgroups have index 2. Assume that G has no abelian subgroup of index 2 and let A be a maximal normal abelian subgroup in G so that $|G : A| \ge 4$.

Let H be a maximal Dedekindian subgroup in G containing A so that by our hypothesis $|G : H| = 2$. Since H is nonabelian (noting that $H > A$), it follows that H is Hamiltonian (nonabelian Dedekindian) and so $|H : A| = 2$ implying that $|G : A| = 4$ (indeed, $H \triangleleft$ has exactly three abelian subgroups of index 2 so at least one of them is G-invariant). We have $H = QA$, where $Q \cong \mathrm{Q}_8$ and $Q \cap A \cong \mathrm{C}_4$ so that $\mathrm{Z}(H) = \Omega_1(H)$ and $H/\Omega_1(H) \cong \mathrm{E}_4$ with $\mho_1(H) = \mathrm{Z}(Q) = \langle z \rangle \cong \mathrm{C}_2$.

Let $g \in G - H$ and let K be a maximal Dedekindian subgroup in G containing $\langle g \rangle$ so that $|G : K| = 2$ and K is Hamiltonian since K is nonabelian, by assumption. Since $|H : (H \cap K)| = 2$ and $|Q \cap K| > 2$, it follows that $H \cap K$ contains elements of order 4 and therefore $\mho_1(H \cap K) = \langle z \rangle = \mho_1(H)$. But then $\mho_1(K) = \langle z \rangle$ so that $g^2 \in \langle z \rangle$ for all $g \in G - H$. We have proved that $\mho_1(G) = \langle z \rangle \cong \mathrm{C}_2$ and so the group $G/\langle z \rangle$ is elementary abelian implying that $G' = \langle z \rangle$ and $Q \trianglelefteq G$.

Set $C = \mathrm{C}_G(Q)$ so that (noting that $G' = Q' = \langle z \rangle$) $G = Q * C$ (see also Lemma 4.3) and $|C : \Omega_1(H)| = 2$. If C is abelian, then G would have an abelian subgroup of index 2, contrary to our assumption. Hence C is nonabelian and so $C' = Q' = \langle z \rangle$ and there is $c \in C - \Omega_1(H)$ such that $c^2 = z$ (noting that $\mho_1(G) = \langle z \rangle$). Also, applying Lemma 1.1 on the group C, we get $|C| = 2|C'||\mathrm{Z}(C)|$ and so $|C : \mathrm{Z}(C)| = 4$. But $\mathrm{Z}(C) \le \Omega_1(H)$ and so $|\Omega_1(H) : \mathrm{Z}(C)| = 2$ and $\mathrm{Z}(C)$ is elementary abelian. Also, $\mathrm{Z}(C) \ge \langle z \rangle$ and $C = \mathrm{Z}(C)D$, where D is minimal nonabelian covering $C/\mathrm{Z}(C)$. But we have $\mho_1(D) = D' = \langle z \rangle$ and so $\Phi(D) = \langle D', \mho_1(D) \rangle = \langle z \rangle$ implying $|D| = 8$ with $D \cap \mathrm{Z}(C) = \langle z \rangle$.

Let V be a complement of D in $\mathrm{Z}(C)$ so that $C = D \times V$, where V is elementary abelian. We get $G = (Q * D) \times V$ and note that D also covers $\Omega_1(H)/\mathrm{Z}(C)$ and so D has a noncentral involution implying that $D \cong \mathrm{D}_8$. Our theorem is proved. □

Problem 1. Classify the 2-groups all of whose maximal subgroups, except one, are Dedekindian.

Problem 2. Classify the 2-groups all of whose subgroups of index 4 are Dedekindian.

Note that, for $p = 2$, the $\mathcal{A}_2$-groups satisfy the condition of Problem 2.

https://doi.org/10.1515/9783110533149-057

§ 314 Theorem of Glauberman–Mazza on p-groups with a nonnormal maximal elementary abelian subgroup of order p^2

We prove here a deep result of Glauberman and Mazza [GM] about p-groups, $p > 2$, with a nonnormal maximal elementary abelian subgroup of order p^2. On related results, see also §§ 127, 134.

Theorem 314.A. *Let G be a p-group with an odd prime p. If G has a nonnormal maximal elementary abelian subgroup of order p^2, then G is of rank at most p (i.e., G does not possess an elementary abelian subgroup of order p^{p+1}).*

The p-groups of Theorem 314.A appear in some different areas of finite group theory, in classifying of simple groups and in representation theory (for details, see the paper [GM]). As the group $\Sigma_{p^2} \in \mathrm{Syl}_p(\mathrm{S}_{p^2})$ shows, there exists a p-group from Theorem 314.A that has an elementary abelian subgroup of order p^p.

We first state some preliminary results.

The following basic result was used many times in our book.

Lemma 314.1. *Suppose that E is a nonnormal maximal elementary abelian subgroup of rank 2 of a p-group G. Let H be a subgroup of order p^n and exponent p in G that is normalized by E. Then:*

(a) *For each positive integer $k < n$, H contains a subgroup H_k of order p^k that is normalized by E.*

(b) $n \le p$.

(c) *If E is not contained in H, then the subgroup H_k in part (a) is unique for each k.*

Proof. Obviously, $E \cap Z(G) = \Omega_1(Z(G))$ has order p, i.e., G is a monolith.

Set $T = EH$. One may assume that $T > E$. As any element of the set $H - E$ does not centralize E, we get $C_T(E) = E$ (indeed, by the modular law, $C_T(E) = E(H \cap E))$), so that the subgroup T is of maximal class (Proposition 1.8). By the product formula, $|T : H| = p$. Now (a) and (c) follow from Exercise 9.1 (b) and Lemma 9.1. Assertion (b) follows from Theorems 9.5 and 9.6. Assertion (c) follows from Exercise 9.1 (b). (It suffices to assume in the lemma that $EH = HE$.) □

Proposition 314.2. *Let G be a p-group with $p > 2$. If G has a nonnormal maximal elementary abelian subgroup E of rank 2, then G has a unique normal elementary abelian subgroup of rank 2.*

Proof. Assume, by way of contradiction, that G possesses two distinct normal elementary abelian subgroups U, V of order p^2. As in Lemma 314.1, $E \cap Z(G) = \Omega_1(Z(G)) \cong C_p$, i.e., G is a monolith. Set $Z_0 = \Omega_1(Z(G))$; then $U \cap V \cap E = Z_0$. Assume that $E < UV$. Then $UV \cong S(p^3)$ (in that case, UV is nonabelian) contains exactly $p + 1$ subgroups

https://doi.org/10.1515/9783110533149-058

isomorphic to E_{p^2}, and at least two of them are G-invariant. It follows that all $p+1$ maximal subgroups of UV are G-invariant, a contradiction. Set $H = EUV$; then $|H| = p^4$ since $E \nleq UV$. In that case $C_H(E) = E(UV \cap E) = E$ so that H is of maximal class. However, U, V are distinct normal subgroups of order p^2 and index $p^2 > p$ in H, contrary to Exercise 9.1. □

Lemma 314.3. *Suppose that N is a normal subgroup of a p-group G and k is an integer, $k \geq 1$.*

(a) *If $N \cap Z_k(G) = N \cap Z_{k+1}(G)$, then $N \leq Z_k(G)$.*

(b) *If $|N| = p^k$, then $N \leq Z_k(G)$.*

Proof. (b) We know that $|N \cap Z(G)| \geq p$ so one may assume that $k > 1$. By induction,

$$|(N \cap Z_{k-1}(G/((N \cap Z(G))| \geq p^{k-1} \implies |N \cap Z_k(G)| \geq p^k,$$

completing the proof of (b).

Clearly, (a) follows from the proof of (b). □

Lemma 314.4 (Alperin–Gorenstein). *If a prime $p \geq 5$ and $A \cong E_{p^n}$ is a subgroup of a p-group G, then there is $B \cong E_{p^n}$ in G such that $B \lhd B^G$ (here B^G is the normal closure of B in G).*

Proof of Theorem 314.A. Let G be a p-group, $p > 2$. Assume, by way of contradiction, that G has a nonnormal maximal elementary abelian subgroup E of order p^2 but G is of rank $> p$ so that G possesses an elementary abelian subgroup A of order p^{p+1}.

Suppose that $p = 3$. Then $|A| = 3^4$ and then, by Corollary 10.7 (or Theorem 10.4), G has a normal elementary abelian subgroup A_0 of order 3^4. Then $H = EA_0 (> A_0)$ since $E \nleq A_0$. As $C_H(E) = EC_{A_0}(E) = E$ (indeed, $C_{A_0}(E) = \{1\}$), the subgroup H is of maximal class (Proposition 1.8). But this contradicts to Exercise 9.1 (b) (by this exercise, our group must have exactly one normal subgroup of order p^2). Thus, our theorem is proved for $p = 3$.

In the sequel we may suppose that $p \geq 5$ and we set $|G| = p^n$. By Lemma 314.4, we may choose the elementary abelian subgroup A of order p^{p+1} to be normal in its normal closure $N = A^G$. Since $A \trianglelefteq N$, there is $B < A$ and $B \lhd N$ of order p^{p-1}.

Let $M = \Omega_1(Z_{p-1}(N))$; then $M \lhd G$ and $B \leq M$, by Lemma 314.3. Since $Z_{p-1}(N)$ has class at most $p-1$, it is a regular p-group (Theorem 7.1 (b)). Therefore, $\exp(M) = p$ because it is a regular p-group generated by elements of order p (Theorem 7.2 (b)). Since $M \trianglelefteq G$ (and so obviously E normalizes M), Lemma 314.1 yields that $|M| \leq p^p$. Hence $|M : B| \leq p$.

Let $Y = \Omega_1(Z_2(N))$ and $W = \Omega_1(Z(N))$. In that case, since $Z_2(N) \leq Z_{p-1}(N)$, we get

$$W \leq Y \leq M \quad \text{and} \quad W, Y \trianglelefteq G, \tag{1}$$

since Y and W are characteristic in $N \lhd G$.

Assume first that $Y \leq A$. Then $C_G(Y) \trianglelefteq G$ and $A \leq C_G(Y)$. Therefore, $N = A^G \leq C_G(Y)$ and $Y \leq Z(N)$. More generally, observe similarly that any normal abelian subgroup of G

that is contained in any G-conjugate of A is necessarily contained in $Z(N)$. Then, since $\exp(A) = p = \exp(Y)$, we get

$$\begin{aligned} A \cap Z_2(N) &= A \cap \Omega_1(Z_2(N)) = A \cap Y = Y \le \Omega_1(Z(N)) \\ &= W = A \cap \Omega_1(Z(N)) = A \cap Z(N) \le A \cap Z_2(N), \end{aligned} \tag{2}$$

and therefore (2) implies $A \le Z(N)$, by Lemma 314.3. Since $\exp(A) = p$, it follows that $A \le \Omega_1(Z(N))(= W)$. But then,

$$A \le \Omega_1(Z_{p-1}(N)) = M \quad \text{and} \quad p^{p+1} = |A| \le |M| \le p^p,$$

which a contradiction. Thus, $Y \nleq A$. Therefore, $B < BY \le M$ (see (1)).

Since $|M : B| \le p$, we have $M = BY$. Moreover, $Y/W \le Z(N/W)$. Thus, $M/W = BY/W$ is centralized by AW/W. As $M \trianglelefteq G$, it follows that M/W is centralized by $A^G W/W$, i.e., by N/W. Therefore, $M \le Z_2(N)$ (recall that $W \le Z(N)$). But now (recall that $p \ge 5$)

$$A \cap Z_3(N) \le A \cap Z_{p-1}(N) = A \cap M \le A \cap Z_2(N) \le A \cap Z_3(N).$$

So $A \cap Z_3(N) = A \cap Z_2(N)$. By Lemma 314.3, we have $A \le Z_2(N)$. Hence $A \le M$, and we obtain a contradiction as in the previous paragraph since $|A| > |M|$. Theorem 314.A is proved. □

Remark. The situation for $p = 2$ is very different. Suppose that a 2-group G has a nonnormal maximal elementary abelian subgroup E of order 4. Then G has no normal elementary abelian subgroup A of order 8. Assume, however, that A is normal in G. Indeed, $C_{EA}(E) = E$ so that EA is of maximal class, a contradiction since 2-groups of maximal class has no subgroup isomorphic to E_8. Then by the 4-generator theorem (Theorem 50.3), every subgroup of G is generated by four elements and so G has rank at most 4 and there are examples of such 2-groups of rank 4.

Problem 1. Suppose that a p-group G, $p > 2$, has a maximal nonnormal elementary subgroup of order p^2. Is it true that G has no subgroup of order p^{p+1} and exponent p?

Problem 2. Suppose that a p-group G, $p > 2$, has a maximal nonnormal elementary subgroup of order p^3. Is it true that the order of maximal elementary abelian (maximal subgroup of exponent p) of G is bounded?

Problem 3. Study the p-groups containing a maximal abelian subgroup which is of type (i) (p^2, p), (ii) (p^2, p^2).

Problem 4. Suppose that a p-group G possesses a subgroup R of order p^p and exponent p. Study the structure of G if $R < \Omega_1(G)$ is a maximal subgroup of G of exponent p.

Problem 5. Study the irregular p-groups containing a maximal absolutely regular subgroup of order p^p.

Problem 6. Is it true that if a 2-group G has some nonnormal maximal elementary abelian subgroup E of rank 2, then G has at most two normal elementary abelian subgroups of rank 2?

§ 315 *p*-groups with some non-*p*-central maximal subgroups

For a prime p, set $\epsilon(p) = 1$ if $p > 2$ and $\epsilon(2) = 2$. We recall that a p-group G is p-central if $\Omega_{\epsilon(p)}(G) \le Z(G)$.

The minimal non-p-central p-groups are classified, up to isomorphism, in Appendix 132. It is interesting to know the minimal number of non-p-central maximal subgroups in a p-group which is not minimal non-p-central. In this section we show that p-groups in Problem A.132.3 do not exist by proving that the number in the previous sentence is > 1. Namely, we prove the following result.

Theorem 315.1. *Let G be a p-group possessing at least one non-p-central maximal subgroup. Then G has at least two distinct non-p-central maximal subgroups.*

Proof. Let G be a p-group all of whose maximal subgroups, except one, are p-central. Let A be a maximal normal abelian subgroup of exponent $p^{\epsilon(p)}$.

First assume that G/A is noncyclic. Then there is in G a normal subgroup G_0 such that $A \le G_0 < G$ and $G/G_0 \cong E_{p^2}$. Considering $p + 1$ maximal subgroups of G containing G_0, we see that there are at least two distinct maximal subgroups G_1 and G_2 of G which contain G_0 and which are p-central. It follows from $A \le \Omega_{\epsilon(p)}(G_i) \le Z(G_i)$, $i = 1, 2$, that $C_G(A)$ contains $G_1 G_2 = G$ and so $A \le Z(G)$. By Corollary 10.2, $\Omega_{\epsilon(p)}(G) = A$ and so $\Omega_{\epsilon(p)}(G) \le Z(G)$ and therefore G would be p-central. In that case, all maximal subgroups of G are p-central. This is a contradiction since, by our hypothesis, G has a maximal subgroup which is not p-central.

We have proved that the quotient group G/A is cyclic. Let H be any p-central maximal subgroup in G. Since $\exp(H \cap A) \le p^{\epsilon(p)}$, we get $H \cap A \le Z(H)$. If $A \le H$, then H is abelian since H/A, being subgroup of G/A, is cyclic. If $A \nleq H$, then $G = HA$ so $H/(H \cap A) \cong G/A$ is cyclic so H is abelian again. Thus, any p-central maximal subgroup of G is abelian. It follows that G has exactly one nonabelian maximal subgroup, a contradiction since a nonabelian p-group has either 0, or 1, or $p + 1$ abelian maximal subgroups which implies that $|\Gamma_1| \in \{1, 2, p + 2\}$, a contradiction since $|\Gamma_1| \equiv 1 \pmod{p}$ and $|\Gamma_1| > 1$ since G is noncyclic. The theorem is proved. □

Problem 1. Study the p-groups containing at most p non-p-central maximal subgroups.

Problem 2. Classify the non-p-central p-groups G all of whose maximal p-central subgroups have index p in G.

Problem 3. Classify the nonmetacyclic p-groups all of whose maximal metacyclic subgroups are p-central.

https://doi.org/10.1515/9783110533149-059

§ 316 Nonabelian p-groups, $p > 2$, of exponent $> p^3$ all of whose minimal nonabelian subgroups, except one, have order p^3

Here we solve Problem 3894 for $p > 2$ by proving the following result. This problem was solved for $p = 2$ in § 264.

Theorem 316.1. *Let G be a nonabelian p-group, $p > 2$, of exponent $> p^3$ which is not minimal nonabelian but all its minimal nonabelian subgroups, except one denoted with M, have order p^3. Then one of the following holds:*

(a) *G has an abelian subgroup A of index p and all elements in $G - (A \cup M)$ are of order $\leq p^2$.*

(b) *We have $|G : M| = p$ and all elements in $G - M$ are of order $\leq p^2$.*

Proof. Let G be a nonabelian p-group, $p > 2$, of exponent $> p^3$ which is not minimal nonabelian but all its minimal nonabelian subgroups, except one denoted with M, have order p^3.

Let A be a maximal normal abelian subgroup in G. Suppose that $\exp(G/A) > p$. Then there is a subgroup C in G such that $A < C \leq G$ with $C/A \cong C_{p^2}$ and let D/A be the subgroup of order p in C/A. If all elements $x \in C - D$ lie in M, then $C \leq M$ and since M is minimal nonabelian, D would be abelian, contrary to $C_G(A) = A$. Hence there is $g \in C - D$ such that $g \notin M$. By Lemma 57.1, there is $a \in A$ such that $\langle a, g\rangle$ is minimal nonabelian. But $\langle a, g\rangle \neq M$ and so $|\langle a, g\rangle| = p^3$. Noting that $\langle g\rangle$ covers $C/A \cong C_{p^2}$, we get $o(g) = p^2$, $\langle g\rangle \cap A = \{1\}$ and $\langle a, g\rangle = \langle a\rangle \times \langle g\rangle$ with $o(a) = p$, a contradiction. We have proved that $\exp(G/A) = p$.

Suppose that $|G : A| = p$. Then $G = AM$ and all elements in $G - (A \cup M)$ are of order $\leq p^2$. Indeed, if $x \in G - (A \cup M)$, then by Lemma 57.1, there is $a \in A$ such that $\langle a, x\rangle$ is minimal nonabelian. But $\langle a, x\rangle \neq M$ and so $|\langle a, x\rangle| = p^3$ implying that $o(x) \leq p^2$. This is possibility (a) in our theorem.

From now on assume that G has no abelian subgroup of index p and A will always denote a maximal normal abelian subgroup.

Now assume that setting $M_0 = M \cap A$, we have $|M/M_0| > p$. Note that $M \trianglelefteq G$ so that $M' \cong C_p$ and so $M' \leq Z(G)$ implying that M' centralizes A and so $M' \leq A$ and $M' \leq M_0$. Hence M/M_0 is elementary abelian so that $M/M_0 \cong E_{p^2}$ and then $M_0 = \Phi(M) = Z(M)$ and $|M_0| \geq p^2$ (since by our hypothesis $|M| \geq p^4$). Set $C = AM$ and let C_1/A be a subgroup of order p in C/A. We have $M < C$ and M covers C/A. Indeed, if $M = C$, then $C_1(< C)$ would be abelian, a contradiction. Note that $M_0 = Z(M) \leq Z(C)$ and so for each $m \in M - M_0$ (using Lemma 57.1) there is $a \in A - M_0$ such that $\langle m, a\rangle$ is minimal nonabelian. But $\langle m, a\rangle \neq M$ and therefore $|\langle m, a\rangle| = p^3$ and so $o(m) \leq p^2$. Hence if $m_1, m_2 \in M - M_0$ are such that $M = \langle m_1, m_2\rangle$, then

$$M_0 = \Phi(M) = \langle m_1^p, m_2^p, M'\rangle$$

https://doi.org/10.1515/9783110533149-060

implying that M_0 is elementary abelian and so $\exp(M) = p^2$. We have proved that each minimal nonabelian subgroup in G is of exponent $\leq p^2$. By our hypothesis, $\exp(G) > p^3$ and so $\exp(A) > p^3$ (noting that by Lemma 57.1, all elements in $G - A$ are of order $\leq p^2$). We consider now the group $C/\Omega_1(A)$ which is metabelian since $C' \leq A$ and for each $c \in C - A$, $c^p \in \Omega_1(A)$. Hence

$$\mathrm{H}_p(C/\Omega_1(A)) = A/\Omega_1(A) \neq \{1\}$$

and this contradicts a result of Hogan–Kappe [HogK] stating that in a metabelian p-group the nontrivial Hughes subgroup is of index at most p.

We have proved that setting $M_0 = M \cap A$ we have $|M : M_0| = p$. Set $H = MA$ so that $\mathrm{Z}(M) \leq \mathrm{Z}(H)$. If $\exp(M) \leq p^2$, then (by Lemma 57.1) all elements in $G - A$ are of order $\leq p^2$ and so (by our hypothesis) $\exp(A) > p^3$. Then again we consider a subgroup C_0 such that $A < C_0 \leq G$, where $C_0/A \cong \mathrm{E}_{p^2}$ and C_0 is metabelian. Here we have

$$\mathrm{H}_p(C_0/\Omega_1(A)) = A/\Omega_1(A) \neq \{1\}$$

and this again contradicts a result of Hogan–Kappe [HogK]. Hence $\exp(M) \geq p^3$ and so there are elements of order $\geq p^3$ in $M - M_0$.

Assume that $M < H = AM$ so that the set $S = H - (A \cup M)$ is nonempty. Observe that by Lemma 57.1, this set consists of elements of order $\leq p^2$. For any $h \in S$ and $m_0 \in M_0 - \mathrm{Z}(M)$, we have $[h, m_0] \neq 1$. Indeed, if $[h, m_0] = 1$, then $m_0 \in \mathrm{Z}(H)$, a contradiction. Set $X = M_0\langle h\rangle$ so that $X' \leq \mathrm{Z}(M) \leq \mathrm{Z}(H)$ and therefore $\mathrm{cl}(X) = 2$. We get $[h, m_0] \in \mathrm{Z}(M)$ and $[h, m_0]^p = [h^p, m_0] = 1$ and so

$$\langle h, m_0\rangle' = \langle [h, m_0]\rangle \cong \mathrm{C}_p.$$

Therefore by Lemma 65.2, $\langle h, m_0\rangle$ is minimal nonabelian. Since $\langle h, m_0\rangle \neq M$, we get $\mathrm{o}(m_0) \leq p^2$. It follows that all elements in $M_0 - \mathrm{Z}(M)$ are of order $\leq p^2$ and so $\exp(M_0) \leq p^2$ and so $\exp(M) = p^3$.

Let $m \in M - M_0$ with $\mathrm{o}(m) = p^3$ and set $v = m^p$ and $v^p = z$. Since

$$\mathrm{C}_{p^2} \cong \langle v\rangle \leq \mathrm{Z}(M) \leq \mathrm{Z}(H),$$

it follows that there are elements of order p^2 in the set S and let h (fixed) be one of them. Let k be any element of order p^2 in $M_0 - \mathrm{Z}(M)$. By the previous paragraph, $\langle h, k\rangle$ is minimal nonabelian distinct from M. Hence $\langle h, k\rangle \cong \mathrm{M}_{p^3}$ and so $h^p = k^p \in \mathrm{Z}(M)$ for a fixed element h of order p^2 in S and an arbitrary element k of order p^2 in $M_0 - \mathrm{Z}(M)$. It follows that

$$\mho_1(M_0) = \langle h^p\rangle = \langle z\rangle.$$

Then $\langle k, v\rangle$ contains an element of order p in $M_0 - \mathrm{Z}(M)$. It follows that in any case there is an element t of order p in $M_0 - \mathrm{Z}(M)$. We get $M = \langle m, t\rangle$ and so we have exactly two possibilities for the structure of M (depending on whether M is metacyclic or nonmetacyclic):

$$M \cong \mathrm{M}_{p^4} = \langle m, t \mid m^{p^3} = t^p = 1,\ [m, t] = m^{p^2} = z\rangle$$

or

$$M \cong \mathrm{M}_p(3,1,1) = \langle m, t \mid m^{p^3} = t^p = 1,\ [m,t] = z',\ (z')^p = [z',m] = [z',t] = 1 \rangle.$$

By the above structure of M in both cases M has exactly p conjugate classes of cyclic subgroups of order p^3 with the representatives mt^i, $i = 0, 1, \dots, p-1$. Set $Y = \langle m \rangle$. If $\mathrm{C}_G(Y) \nleq M$, then there is $c \in G - M$ such that c centralizes Y. Then $ct \notin M$ and $\langle ct, Y \rangle \neq M$ is minimal nonabelian of order $\geq p^4$, a contradiction. It follows that $\mathrm{C}_G(Y) \leq M$.

Assume that $|A : M_0| = p$ and then the fact that $\exp(M_0) = p^2$ implies that we have $\exp(A) \leq p^3$. Note that (using Lemma 57.1) for each $g \in G - H$, there is a minimal nonabelian subgroup L with $|L| = p^3$ implying that $\mathrm{o}(g) \leq p^2$. It follows that $\exp(G) = p^3$, contrary to our hypothesis.

We have proved that $|A : M_0| = |H : M| > p$ and note that $M \trianglelefteq G$ so that we can act with G/M of order $\geq p^3$ on the p conjugate classes of cyclic subgroups of order p^3 in M. It follows that

$$|\mathrm{N}_G(Y) : \mathrm{N}_M(Y)| = p^2$$

and (noting that $\mathrm{C}_G(Y) \leq M$) there is $b \in \mathrm{N}_G(Y) - \mathrm{N}_M(Y)$ such that $\langle Y, b \rangle$ is minimal nonabelian of order $\geq p^4$ and $\langle Y, b \rangle \neq M$, a contradiction.

Finally, we have proved that $A < M$ and so $|M : A| = p$. Since G is not minimal nonabelian, we have $M < G$. All elements in $G - M$ are of order $\leq p^2$ (where we use again Lemma 57.1) and so by our hypothesis $\exp(M) > p^3$ implying that $\exp(A) > p^2$. Assume that $|G/M| > p$ so that there is a subgroup G_0 such that $M < G_0 \leq G$ and $|G_0/M| = p^2$. Since $\exp(G/A) = p$, it follows that $G_0/M \cong \mathrm{E}_{p^2}$ and so $G_0' \leq M$. For each $x \in G_0 - M$ we have $x^p \in \Omega_1(M)$ and $\exp(\Omega_1(M)) = p$ (noting that $p > 2$ and $M' \cong \mathrm{C}_p$ with $M' \leq \Omega_1(M)$). We apply a result of Hogan–Kappe [HogK] on the group $G_0/\Omega_1(M)$ noting that $G_0/\Omega_1(M)$ is metabelian and

$$\mathrm{H}_p(G_0/\Omega_1(M)) = M/\Omega_1(M) \neq \{1\}$$

and so we get a contradiction. We have proved that $|G : M| = p$ and since we already know that all elements in $G - M$ are of order $\leq p$ our theorem is proved. □

Problem 1. Let $\mathcal{M}_1$ be the set of all minimal nonabelian subgroups of a p-group G that have order p^3 and let $\mathcal{M}_2 = \mathcal{A}_1(G) - \mathcal{M}_1$, where $\mathcal{A}_1(G)$ is the set of all minimal nonabelian subgroups of G. Study the structure of (a nonabelian) G if $\langle S \mid S \in \mathcal{M}_i \rangle < G$ for $i = 1, 2$.

Problem 2. Study the nonabelian p-groups G, $p > 2$, all of whose minimal nonabelian subgroups are isomorphic to $\mathrm{M}_p(2,2)$. (For $p = 2$, this was solved in Theorem 57.3.)

Problem 3. Study the nonabelian p-groups G all of whose minimal nonabelian subgroups, except one, are isomorphic to $\mathrm{M}_p(2,2)$.

§ 317 Nonabelian p-groups, $p > 2$, all of whose minimal nonabelian subgroups are isomorphic to $M_p(2,2)$

Here we solve Problem 316.2 for $p > 2$. In case $p = 2$ this problem was solved with Theorem 57.3.

Theorem 317.1. *Let G be a nonabelian p-group, $p > 2$, all of whose minimal nonabelian subgroups are isomorphic to*

$$M_p(2,2) = \langle a, b \mid a^{p^2} = b^{p^2} = 1,\ [a,b] = a^p \rangle.$$

Then G is p-central of class 2 with $\exp(G) = p^2$ *and* $\Omega_1(G) = Z(G)$.

Proof. Let G be a nonabelian p-group, $p > 2$, all of whose minimal nonabelian subgroups are isomorphic to $M_p(2,2)$ (given as in the statement).

Let A be a maximal normal abelian subgroup in G. Let $g \in G - A$. By Lemma 57.1, there is $a \in A$ so that $\langle g, a\rangle$ is minimal nonabelian and therefore $\langle g, a\rangle \cong M_p(2,2)$. It follows that $o(g) = p^2$ and $g^p \in A$ (by Lemma 57.1, there is in $G - A$ no element of order p). Hence all elements in $G - A$ are of order p^2 and so $\Omega_1(A) = \Omega_1(G)$.

Suppose that there is an element $x \in G - A$ such that x does not centralize $\Omega_1(A)$. Then $x^p \in \Omega_1(A)$ and so $\Omega_1(A)$ is a maximal normal abelian subgroup in $\Omega_1(A)\langle x\rangle$, where

$$|(\Omega_1(A)\langle x\rangle) : \Omega_1(A)| = p.$$

By Lemma 57.1, there is $y \in \Omega_1(A)$ such that the subgroup $\langle x, y\rangle$ is minimal nonabelian. Since $\langle x, y\rangle \ncong M_p(2,2)$ (indeed, $\Phi(M_p(2,2)) = \Omega_1(M_p(2,2))$), we have a contradiction. Hence all elements in the set $G - A$ centralize $\Omega_1(A)$ and so

$$\Omega_1(A) = \Omega_1(G) \le Z(G)$$

implying that G is p-central (see §152). In particular, we have $\Omega_1(A) < A$.

By Corollary 157.2, $\Omega_2(G)$ is of exponent p^2 and since $\Omega_2(G) = G$, we get that G is of exponent p^2. By Theorem 157.6,

$$\exp(G') = \exp(G/Z(G)) = p$$

and so

$$G' \le \Omega_1(A) \le Z(G)$$

and therefore G is of class 2.

Any two non-commuting elements $x, y \in G$ generate a minimal nonabelian subgroup. Indeed, $1 \neq [x,y] \in \Omega_1(A) \le Z(G)$ and so

$$\langle x, y\rangle' = \langle [x,y]\rangle \cong C_p$$

and, by Lemma 65.2, $\langle x, y\rangle$ is minimal nonabelian.

https://doi.org/10.1515/9783110533149-061

Let $W \cong \mathrm{M}_p(2,2)$ with $p > 2$ be a minimal nonabelian subgroup in G. Then we have $W' = \langle w \rangle \cong \mathrm{C}_p$, $\mathrm{Z}(W) = \Omega_1(W) \cong \mathrm{E}_{p^2}$ and each element in $\mathrm{Z}(W)$ is a p-th power of an element in W. Assume that $\Omega_1(A) < \mathrm{Z}(G) < A$ and let $\langle v \rangle \leq \mathrm{Z}(G)$ with $\langle v \rangle \cong \mathrm{C}_{p^2}$ and we set $z = v^p$.

First suppose that $W \cap \langle v \rangle = \{1\}$ so that $\langle W, v \rangle = W \times \langle v \rangle$ is of order p^6 with

$$\langle W, v \rangle' = W' = \langle w \rangle \quad \text{and} \quad \Omega_1(\langle W, v \rangle) = \mathrm{Z}(W) \times \langle z \rangle \cong \mathrm{E}_{p^3}.$$

Let $\langle w \rangle$, $\langle a_1 \rangle$, $\langle b_1 \rangle$ be three pairwise distinct subgroups of order p in $\mathrm{Z}(W)$ and let $a, b \in W$ be such that $a^p = a_1$ and $b^p = b_1$. Let $Y = \langle a, bv \rangle$ so that $[a, bv] \neq 1$ and therefore (by the above) Y is minimal nonabelian with $Y' = \langle w \rangle$. But then Y contains w, $a^p = a_1$ and $b^p v^p = b_1 z$ and

$$\langle w, a_1, b_1 z \rangle \cong \mathrm{E}_{p^3},$$

a contradiction.

Now assume that $W \cap \langle v \rangle \neq \{1\}$ and so $W \cap \langle v \rangle = \langle z \rangle \cong \mathrm{C}_p$ and $\langle W, v \rangle = W * \langle v \rangle$ is of order p^5. Let $d_1 \in \mathrm{Z}(W) - \langle z \rangle$ and let $d, c \in W$ be such that $c^p = z$ and $d^p = d_1$. We have $\langle c, d \rangle = W$ and so

$$\langle [c, d] \rangle = \langle w \rangle \cong \mathrm{C}_p.$$

We may choose $c_0 \in \langle c \rangle - \langle c_1 \rangle$ and $v_0 \in \langle v \rangle - \langle z \rangle$ so that $h = c_0 v_0$ is of order p. We have

$$[h, d] = [c_0 v_0, d] \neq 1$$

and therefore $\langle h, d \rangle$ is minimal nonabelian. But the minimal nonabelian subgroup $\langle h, d \rangle$ contains a noncentral element h of order p and this is a contradiction since we must have

$$\langle h, d \rangle \cong \mathrm{M}_p(2,2).$$

We have proved that $\Omega_1(A) = \mathrm{Z}(G)$ and this completes the proof of our theorem. □

Problem 1. Classify the nonabelian p-groups all of whose minimal nonabelian subgroups are isomorphic and have noncyclic centers.

Problem 2. Classify the nonmetacyclic 2-groups all of whose minimal nonmetacyclic subgroups have order 2^5.

Problem 3. Classify the nonabelian p-groups all of whose minimal nonabelian subgroups are isomorphic to M_{p^n} or $\mathrm{M}_p(2,2)$.

Problem 4. Classify the nonabelian p-groups all of whose minimal nonabelian subgroups have order p^4.

§ 318 Nonabelian p-groups, $p > 2$, of exponent $> p^2$ all of whose minimal nonabelian subgroups, except one, are isomorphic to $M_p(2, 2)$

We are continuing to study influence of minimal nonabelian subgroups on the structure of finite p-groups.

Here we solve Problem 316.3 for $p > 2$ by proving the following result. For $p = 2$ this problem was already solved in § 312.

Theorem 318.1 (Janko). *Let G be a nonabelian p-group, $p > 2$, of exponent $> p^2$ which is not minimal nonabelian but all its minimal nonabelian subgroups, except one denoted with M, are isomorphic to*

$$M_p(2, 2) = \langle a, b \mid a^{p^2} = b^{p^2} = 1,\ [a, b] = a^p \rangle.$$

Then G has an abelian subgroup A of index p and all elements in $G - (A \cup M)$ are of order p^2.

Proof. Let G be a nonabelian p-group, $p > 2$, of exponent $> p^2$ which is not minimal nonabelian but all its minimal nonabelian subgroups, except one denoted with M, are isomorphic to $M_p(2, 2)$. Note that if a, b is a minimal system of generators of $M_p(2, 2)$, then $o(a) = o(b) = 4$ and $\langle a \rangle \cap \langle b \rangle = \{1\}$.

Let A be a maximal normal abelian subgroup in G and set $M_0 = M \cap A$. If $|G : A| = p$, then all elements in the set $G - (A \cup M)$ are of order p^2. Indeed, by Lemma 57.1, for each $x \in G - (A \cup M)$ there is $a \in A$ so that the subgroup $\langle x, a \rangle$ is minimal nonabelian and so $\langle x, a \rangle \cong M_p(2, 2)$ implying that $o(x) = p^2$. In what follows we assume that G has no abelian subgroup of index p.

(i) Assume that $\exp(G/A) > p$. Then there is a subgroup B in G such that $A < B \leq G$ and $B/A \cong C_{p^2}$. Let C/A be the subgroup of order p in B/A so that $|C : A| = p$. If all elements of the set $B - C$ are contained in M, then $B = \langle B - C \rangle \leq M$ and so $C(< B)$ is abelian, contrary to $C_G(A) = A$. Hence there is an element $g \in B - C$ such that $g \notin M$. By Lemma 57.1, there is $a \in A$ such that $\langle g, a \rangle$ is minimal nonabelian and so (by our hypothesis) $\langle g, a \rangle \cong M_p(2, 2)$. In particular, $o(g) = o(a) = p^2$ and so $h = g^p$ is of order p, $\langle g \rangle \cong C_{p^2}$ and $\langle g \rangle \cap A = \{1\}$. Note that h centralizes $\langle a \rangle$. Again by Lemma 57.1, there is $b \in A - \langle a \rangle$ such that $\langle b, h \rangle$ is minimal nonabelian. Since $\langle b, h \rangle \ncong M_p(2, 2)$, we must have $\langle b, h \rangle = M$. Setting $M_0 = M \cap A$, we have $|M : M_0| = p$. For each $x \in B - C$, $x \notin M$ and so (by Lemma 57.1) each such element lies in a minimal nonabelian subgroup which is isomorphic to $M_p(2, 2)$. In particular, each $x \in B - C$ is of order p^2 with $x^p \in M - M_0$. Note that M is characteristic in G.

Set $Z = \Omega_1(A)$ and suppose that there is $x \in B - C$ which does not centralize Z, where $x^p \in M - M_0$ (see the previous paragraph) and $Z \cap \langle x \rangle = \{1\}$. Set $X = Z\langle x \rangle$. If x^p centralizes Z, then $Z \times \langle x^p \rangle$ is a maximal normal abelian subgroup of X. By Lemma 57.1,

https://doi.org/10.1515/9783110533149-062

there is $x_0 \in Z \times \langle x^p \rangle$ of order p such that $\langle x, x_0 \rangle$ is minimal nonabelian. But $x \notin M$ and so $\langle x, x_0 \rangle \cong \mathrm{M}_p(2,2)$, a contradiction. If x^p does not centralize Z, then Z is a maximal normal abelian subgroup of X. By Lemma 57.1, there is $x_1 \in Z$ of order p such that $\langle x, x_1 \rangle$ is minimal nonabelian. But $x \notin M$ and so $\langle x, x_1 \rangle \cong \mathrm{M}_p(2,2)$, a contradiction. We have proved that in any case each element in $B - C$ centralizes Z implying that $Z \leq \mathrm{Z}(B)$. Note that all elements of order p in $B - A$ lie in M and let i be one of them. Then for each $z \in Z = \Omega_1(A)$, zi is an element of order p in $B - A$ and so $zi \in M$ implying $z \in M$. Hence we get $Z = \Omega_1(A) = \Omega_1(M_0)$. It follows that $\Omega_1(B) = \Omega_1(M)$ and $|\Omega_1(M) : \Omega_1(M_0)| = p$. If each element in $B - C$ centralizes each element in $\Omega_1(M) - \Omega_1(M_0)$, then B centralizes $\Omega_1(M)$, contrary to $C_G(A) = A$. Hence there are $k \in B - C$ and $j \in \Omega_1(M) - \Omega_1(M_0)$ such that $[k, j] \neq 1$. But $[k, j] \in \Omega_1(M_0) \leq \mathrm{Z}(B)$ and so $\langle k, j \rangle$ is minimal nonabelian. Since $k \notin M$, it follows that $\langle k, j \rangle \cong \mathrm{M}_p(2,2)$, a contradiction. We have thus proved that $\exp(G/A) = p$.

(ii) Assume that $|M/M_0| > p$. Note that $M \trianglelefteq G$ so that $M' \cong C_p$ and $M' \leq \mathrm{Z}(G)$ implies that M' centralizes A and therefore $M' \leq A$ and so $M' \leq M_0 = M \cap A$. Hence the quotient group M/M_0 is elementary abelian so that (noting that $\mathrm{d}(M) = 2$) $M/M_0 \cong \mathrm{E}_{p^2}$ and then $M_0 = \Phi(M) = \mathrm{Z}(M)$.

Set $C = AM$ and let C_1/A be any subgroup of order p in C/A. We have $M < C$ and M covers C/A. Indeed, if $M = C$, then C_1 would be a proper subgroup of M and so C_1 would be abelian, contrary to $C_G(A) = A$. Note that $M_0 = \mathrm{Z}(M) \leq \mathrm{Z}(C)$ and $C' \leq A \cap M = M_0$ so that C is of class 2.

For each $m \in M - M_0$, there is $a \in A - M_0$ so that (by Lemma 57.1) $\langle m, a \rangle$ is minimal nonabelian. Since $\langle m, a \rangle \neq M$, we have $\langle m, a \rangle \cong \mathrm{M}_p(2,2)$ and so $o(m) = p^2$. Hence all elements in $M - \Phi(M)$ are of order p^2 which implies (see Lemma 65.1) that $M_0 = \Phi(M)$ is elementary abelian and so $\exp(M) = p^2$.

For each $g \in G - (A \cup M)$, there is (by Lemma 57.1) $a \in A$ such that $\langle g, a \rangle$ is minimal nonabelian and so $\langle g, a \rangle \cong \mathrm{M}_p(2,2)$ and therefore $o(g) = p^2$. Hence all elements in $G - A$ are of order p^2 and so $\Omega_1(G) = \Omega_1(A)$. For each $c \in C_1 - A$, there is $a \in A$ (by Lemma 57.1) such that $\langle c, a \rangle \cong \mathrm{M}_p(2,2)$.

Suppose that an element $l \in G - A$ does not centralize $\Omega_1(A)$. Then considering $\Omega_1(A)\langle l \rangle$, where $l^p \in \Omega_1(A)$, we see that there is $a_0 \in \Omega_1(A)$ such that $\langle l, a_0 \rangle$ is minimal nonabelian. This is a contradiction because $\langle l, a_0 \rangle$ has a generator a_0 of order p, whereas neither $\mathrm{M}_p(2,2)$ nor M have any such generator. We have proved that $\Omega_1(A) = \Omega_1(G) \leq \mathrm{Z}(G)$ and so G is p-central. By Corollary 157.2, $\exp(\Omega_2(G)) = p^2$. But $\Omega_2(G) = G$ and so G is of exponent p^2, contrary to our basic assumption.

(iii) We have proved that $|M/M_0| = p$. Here $M_0 = M \cap A$ and we set $H = AM$ so that $|H : A| = p$. Since $|G : A| > p$, there is a subgroup K in G such that $H < K \leq G$ and $|K : H| = p$ so that $K/A \cong \mathrm{E}_{p^2}$.

Let L/A be any subgroup of order p in K/A such that $H/A \neq L/A$ implying that $H \cap L = A$. Then each $x \in L - A$ is of order p^2. Indeed, $x \notin M$ and so by Lemma 57.1 there is $y \in A$ such that $\langle x, y \rangle$ is minimal nonabelian and then $\langle x, y \rangle \cong \mathrm{M}_p(2,2)$

and therefore $o(x) = p^2$. Hence $\Omega_1(A) = \Omega_1(L)$. Assume that there is $l \in L - A$ such that l does not centralize $\Omega_1(A)$. Then we consider the subgroup $\Omega_1(A)\langle l\rangle$, where $l^p \in \Omega_1(A)$. It follows that $\Omega_1(A)$ is a maximal normal abelian subgroup in $\Omega_1(A)\langle l\rangle$ and so (by Lemma 57.1) there is $a \in \Omega_1(A)$ such that $\langle l, a\rangle$ is minimal nonabelian and so $\langle l, a\rangle \cong \mathrm{M}_p(2, 2)$, a contradiction. Hence each element in $L - A$ centralizes $\Omega_1(A) = \Omega_1(L)$ implying that $\Omega_1(L) \leq Z(L)$ and so L is p-central. By Corollary 157.2, $\Omega_2(L)$ is of exponent p^2. But $\Omega_2(L) = L$ and so $\exp(L) = p^2$; in particular, we have $\exp(A) = p^2$ (noting that $\Omega_1(A) \leq Z(L)$ and so $\Omega_1(A) \neq A$).

For each $g \in G - (A \cup M)$ (using Lemma 57.1), there is $a \in A$ such that $\langle g, a\rangle$ is minimal nonabelian and so $\langle g, a\rangle \cong \mathrm{M}_p(2, 2)$ and therefore $o(g) = p^2$. Hence all elements in $G - (A \cup M)$ are of order p^2. By our hypothesis, $\exp(G) > p^2$ and so $\exp(M) > p^2$. But $\exp(M_0) \leq p^2$ and so $\exp(M) = p^3$ and all elements in $M - M_0$ are of order p^3. Indeed, if there is $m \in M - M_0$ with $o(m) \leq p^2$, then take an element $n \in M_0 - \Phi(M)$ so that $o(n) \leq p^2$ and $M = \langle m, n\rangle$. But then $\exp(M) \leq p^2$ (see Lemma 65.1), a contradiction. We have proved that $\Omega_1(A) = \Omega_1(G)$.

Let $g \in G - H$ and we act with $\langle g\rangle$ on $\Omega_1(A)$ noting that $g^p \in \Omega_1(A) = \Omega_1(G)$. If g does not centralize $\Omega_1(A)$, then (using Lemma 57.1) there is $v \in \Omega_1(A)$ such that $\langle g, v\rangle$ is minimal nonabelian and then $\langle g, v\rangle \cong \mathrm{M}_p(2, 2)$, a contradiction (since $o(v) = p$). Hence each $g \in G - H$ centralizes $\Omega_1(A)$ implying that $\Omega_1(A) \leq Z(G)$. It follows that G is p-central and so by Corollary 57.2 we have $\exp \Omega_2(G)) = p^2$. But $\langle G - H\rangle = G$ and so $\Omega_2(G) = G$ and G is of exponent p^2, contrary to our basic hypothesis. Our theorem is proved. □

Problem 1. Classify the p-groups all of whose minimal nonabelian subgroups are nonmetacyclic of order p^4. Moreover, consider the p-groups all of whose minimal nonabelian subgroups are nonmetacyclic and isomorphic.

Problem 2. Classify the nonmetacyclic p-groups all of whose minimal nonabelian subgroups are isomorphic metacyclic with noncyclic center.

Problem 3. Classify the nonmetacyclic p-groups with an abelian subgroup of index p all of whose minimal nonabelian subgroups are (i) metacyclic, (ii) nonmetacyclic.

Problem 4. Study the non-Dedekindian p-groups all of whose minimal nonabelian subgroups are quasinormal.

§ 319 A new characterization of p-central p-groups

In the book there are many results showing the essential influence of minimal nonabelian subgroups on the structure of a finite p-group. In this section we prove that a nonabelian p-group is p-central if and only if all its minimal nonabelian subgroups are p-central.

For a prime p, set $\epsilon(p) = 1$ if $p > 2$ and $\epsilon(2) = 2$. Recall that a p-group G is termed p-central if $\Omega_{\epsilon(p)}(G) \le Z(G)$. Here we shall present a new characterization of p-central p-groups.

Theorem 319.1. *A nonabelian p-group G is p-central if and only if each minimal nonabelian subgroup of G is p-central.*

Proof. Assume that each minimal nonabelian subgroup of a nonabelian p-group G is p-central. Let A be a maximal normal abelian subgroup in G. By Lemma 57.1, there is no any element of order $\le p^{\epsilon(p)}$ in $G - A$ and so we get

$$Z = \Omega_{\epsilon(p)}(A) = \Omega_{\epsilon(p)}(G).$$

If G is not p-central, then $Z \nleq Z(G)$ and so in that case there is $g \in G - A$ such that g does not centralize Z but g^p centralizes Z. Set $X = Z\langle g\rangle$ so that $Y = Z\langle g^p\rangle$ is a maximal normal abelian subgroup of X. By Lemma 57.1, there is $y \in Y$ such that $M = \langle g, y\rangle$ is minimal nonabelian. We have $y = zy'$, where $z \in Z$ and $y' \in \langle g^p\rangle$. It follows

$$M = \langle g, y\rangle = \langle g, zy'\rangle = \langle g, z\rangle,$$

where $o(z) \le p^{\epsilon(p)}$ and $z \notin Z(M)$, contrary to our assumption that M is p-central. Since the converse is obvious, our theorem is proved. □

Corollary 319.2 (Berkovich). *A nonabelian p-group is minimal non-p-central if and only if G is non-p-central minimal nonabelian.*

Proof. Let G be a minimal non-p-central p-group. If G is not minimal nonabelian, then each minimal nonabelian subgroup of G is p-central and so by Theorem 319.1, G would be p-central, a contradiction. □

Problem 1. Study the nonabelian p-groups all of whose minimal nonabelian subgroups, except one, are p-central.

Problem 2. Classify the prime power $\mathcal{A}_n$-groups, $n > 1$, all of whose $\mathcal{A}_2$-subgroups are p-central.

Problem 3. Study the p-groups in which the intersection of any two nonincident subgroups is p-central.

https://doi.org/10.1515/9783110533149-063

§ 320 Nonabelian p-groups with exactly one non-p-central minimal nonabelian subgroup

For a prime p, set $\epsilon(p) = 1$ if $p > 2$ and $\epsilon(2) = 2$. We know that a p-group G is p-central if $\Omega_{\epsilon(p)}(G) \leq Z(G)$. We solve here Problem 319.1 by proving the following result.

Theorem 320.1. *Let G be a nonabelian p-group which possesses exactly one minimal nonabelian subgroup M which is not p-central. Then $G = M$ is minimal nonabelian.*

Proof. Let G be a nonabelian p-group which has exactly one minimal nonabelian subgroup M which is not p-central. Assume that G is not minimal nonabelian so that all minimal nonabelian subgroups in G which are distinct from M are p-central.

We have $M \trianglelefteq G$ and let M_0 be a G-invariant maximal subgroup of M. Let A be a maximal normal abelian subgroup of G containing M_0. Set

$$H = AM$$

so that

$$H \trianglelefteq G, \quad |H : A| = p \quad \text{and} \quad A \cap M = M_0.$$

(i) Assume that G has no abelian subgroup of index p. Then $H < G$ and let G_0 be a maximal subgroup of G such that $H \leq G_0$. Set $Z = \Omega_{\epsilon(p)}(A) \trianglelefteq G$ and we act with elements in $G - G_0$ on Z. Suppose that there is $x \in G - G_0$ which does not centralize Z. Set $X = Z\langle x\rangle$ and $Y = Z\langle x^p\rangle$ so that X is nonabelian, X/Z is cyclic and $|X : Y| = p$. Since X is generated by its minimal nonabelian subgroups, there is a minimal nonabelian subgroup N in X with $N \nleq Y$. Then N covers X/Y and since X/Z is cyclic, N also covers X/Z and set $Z_0 = N \cap Z$. If $Z_0 \leq Z(N)$, then N would be abelian, a contradiction. Hence $Z_0 \nleq Z(N)$ and so N is not p-central. On the other hand, N covers G/G_0 and so $N \neq M$, contrary to our basic assumption. It follows that each element in $G - G_0$ centralizes Z and so we get $Z \leq Z(G)$.

Suppose that $Z \nleq M_0$ and let $z \in Z - M_0$. Since M is not p-central, we may set $M = \langle m_1, m_2\rangle$ with $o(m_1) \leq p^{\epsilon(p)}$. Then we consider $K = \langle zm_1, m_2\rangle$, where

$$K' = M' = \langle [m_1, m_2]\rangle \cong C_p$$

and so, by Lemma 65.2, K is minimal nonabelian with $o(zm_1) \leq p^{\epsilon(p)}$. Hence K is not p-central and $K \neq M$, a contradiction. We have proved that

$$Z = \Omega_{\epsilon(p)}(A) = \Omega_{\epsilon(p)}(M_0) \leq Z(G).$$

Set $V = \Omega_{\epsilon(p)}(M)$ so that (computing in the group M of class 2 with $M' \cong C_p$), we have $\exp(V) \leq p^{\epsilon(p)}$. If $V = Z$, then (noting that $Z \leq Z(G)$) M would be p-central, a contradiction. Hence $V \neq Z$ and so $|V : Z| = p$ and V is abelian. Since $V \nleq Z(G)$, there is $g \in G - G_0$ such that g does not centralize an element $v \in V - Z$. Since

$$[V, \langle g\rangle] \leq Z \leq Z(G),$$

https://doi.org/10.1515/9783110533149-064

we have that $V\langle g\rangle$ is of class 2. Then $[v, g]^p = [v^p, g] = 1$ and so $\langle v, g\rangle' = \langle [v, g]\rangle \cong C_p$ implying that (by Lemma 65.2) $\langle v, g\rangle$ is minimal nonabelian. But $\mathrm{o}(v) \le p^{\epsilon(p)}$ and so $\langle v, g\rangle$ is not p-central. On the other hand, $g \in G - G_0$ and so $\langle v, g\rangle \neq M$, a contradiction.

(ii) We have proved that G has an abelian subgroup A of index p. Set $M_0 = M \cap A$ so that $G = AM$. Since (by our assumption) G is not minimal nonabelian, we have $M < G$ and $M_0 < A$ so that the set $S - (A \cup M)$ is nonempty.

Set $Z = \Omega_{\epsilon(p)}(A) \trianglelefteq G$ and assume that there is $s \in S$ which does not centralize Z. Set $X = Z\langle s\rangle$ and $Y = Z\langle s^p\rangle$, where $|X : Y| = p$ and $Y = X \cap A$. It follows that Y is a maximal normal abelian subgroup of X and so by Lemma 57.1, there is $y \in Y$ such that $\langle s, y\rangle$ is minimal nonabelian. We have $y = zs'$, where $z \in Z$ and $s' \in \langle s^p\rangle$ and so

$$\langle s, y\rangle = \langle s, zs'\rangle = \langle s, z\rangle.$$

But $s \notin M$ and so $\langle s, y\rangle \neq M$ with $\mathrm{o}(z) \le p^{\epsilon(p)}$ and $z \notin \mathrm{Z}(\langle s, y\rangle)$ implying that $\langle s, y\rangle$ is not p-central, contrary to the uniqueness of M. Hence each $s \in S$ centralizes Z and therefore $Z \le \mathrm{Z}(G)$.

Assume that $Z \nleq 0M$ and let $z_0 \in Z - M$. Since M is not p-central, we may set $M = \langle m_1, m_2\rangle$ with $\mathrm{o}(m_1) \le p^{\epsilon(p)}$. But $\langle m_1 z_0, m_2\rangle$ is minimal nonabelian, where $\mathrm{o}(m_1 z_0) \le p^{\epsilon(p)}$. Hence $\langle m_1 z_0, m_2\rangle \neq M$ is non-p-central, a contradiction. Thus we have $Z \le M$ and so

$$Z = \Omega_{\epsilon(p)}(A) = \Omega_{\epsilon(p)}(M_0) \le \mathrm{Z}(G).$$

Set $V = \Omega_{\epsilon(p)}(M)$ so that $\exp(V) \le p^{\epsilon(p)}$ and $|V : Z| \le p$. But if $V = Z$, then $Z \le \mathrm{Z}(M)$ implies that M is p-central, a contradiction. Hence $|V : Z| = p$ and (since $Z \le \mathrm{Z}(G)$) the subgroup V is abelian.

Let J be a maximal subgroup of G containing M and let $i \in V - Z$. If all elements in $G - J$ centralize i, then $V \le \mathrm{Z}(G)$, a contradiction. Hence there is $k \in G - J$ such that $[k, i] \neq 1$. Note that $V \trianglelefteq G$ and $\langle k\rangle V$ is of class 2 since $[V, k] \le Z \le \mathrm{Z}(G)$. We have

$$[k, i]^p = [k, i^p] = 1 \quad \text{and so} \quad \langle k, i\rangle' = \langle [k, i]\rangle \cong C_p.$$

By Lemma 65.2, $\langle k, i\rangle$ is minimal nonabelian. But we have $\langle k, i\rangle \neq M$, $\mathrm{o}(i) \le p^{\epsilon(p)}$ and $i \notin \mathrm{Z}(\langle k, i\rangle)$ and so $\langle k, i\rangle$ is not p-central, contrary to the uniqueness of M. This is a final contradiction and so our group G must be minimal nonabelian and our theorem is proved. □

Problem 1. Study the nonabelian p-groups in which any minimal nonabelian subgroup has order p^4.

Problem 2. Study the nonabelian p-groups in which all minimal nonabelian subgroups are nonmetacyclic.

Problem 3. Study the nonabelian p-groups G containing a maximal subgroup H such that all minimal nonabelian subgroups of G not contained in H are metacyclic (nonmetacyclic).

§ 321 Nonabelian p-groups G in which each element in $G - \Phi(G)$ is half-central

Recall that an element x in a p-group G is half-central in G if $|G : C_G(x)| \leq p$. We consider here Problem 4024 and prove the following result.

Theorem 321.1. *Let G be a nonabelian p-group such that all elements in the set $G - \Phi(G)$ (or $G - \mho_1(G)$ or $G - G'$) are half-central. Then each element in G is half-central and so $|G'| = p$.*

Proof. Let G be a nonabelian p-group such that all elements in the set $G - \Phi(G)$ are half-central. Let $x \in G - \Phi(G)$. If x does not centralize $\Phi(G)$, then $C_G(x)$ covers $G/\Phi(G)$ and so

$$G = C_G(x)\Phi(G) = C_G(x),$$

a contradiction. Hence each $x \in G - \Phi(G)$ centralizes $\Phi(G)$ and this implies that we have $\Phi(G) \leq Z(G)$.

We have proved that all elements in G are half-central in G. Then by a result of Knoche (Proposition 121.9), we get $|G'| = p$ and we are done.

If all elements in $G - \mho_1(G)$ are half-central, then all elements in $G - \Phi(G)$ are half-central and also if all elements in $G - G'$ are half-central, then all elements in $G - \Phi(G)$ are half-central in G. Then by the above, all elements are half-central in G and so our theorem is proved. □

https://doi.org/10.1515/9783110533149-065

§ 322 Nonabelian p-groups G such that $C_G(H) = Z(G)$ for any nonabelian $H \le G$

The centralizer of any nonabelian subgroup of a p-group G contains $Z(G)$. In this section we consider the nonabelian p-groups G satisfying the following extremal property: $C_G(H) = Z(H)$ for any nonabelian $H \le G$. The 2-groups of maximal class satisfy this property.

We solve here Problem 3800 and prove the following result.

Theorem 322.1. *Let G be a nonabelian p-group such that $C_G(H) = Z(G)$ for any nonabelian $H \le G$. Then either G has an abelian subgroup of index p or $\mho_1(G) \le Z(G)$.*

Proof. Let G be a nonabelian p-group such that $C_G(H) = Z(G)$ for any nonabelian $H \le G$. Then the center of any nonabelian subgroup of G is contained in $Z(G)$. Let A be a maximal normal abelian subgroup in G so that $Z(G) < A < G$. Let $g \in G - A$; then, by Lemma 57.1, there is $a \in A$ such that the subgroup $M = \langle a, g\rangle$ is minimal nonabelian. Since

$$g^p \in Z(M) \le Z(G),$$

it follows that $g^p \in Z(G) < A$ and so $\exp(G/A) = p$. In particular, $\mho_1(G) \le A$.

Assume that $|G : A| > p$. Suppose in addition that $\exp(A/Z(G)) > p$ and let

$$A < H \le G \quad \text{with } |H : A| = p^2.$$

Then the group $H/Z(G)$ is metabelian and, by the result of the previous paragraph, we have

$$H_p(H/Z(G)) = A/Z(G).$$

But then we get a contradiction with a result of Hogan–Kappe [HogK]: the H_p-subgroup of a metabelian p-group of exponent $> p$ has index p. Hence, in the case under consideration, we have $\exp(A/Z(G)) = p$ and so $\mho_1(G) \le Z(G)$. The theorem is proved. □

Problem 1. Study the nonabelian p-groups G such that $C_G(S) = Z(G)$ for any minimal nonabelian $S \le G$.

Problem 2. Study the nonabelian p-groups G such that $N_G(H) = HZ(G)$ for any nonabelian $H \le G$.

https://doi.org/10.1515/9783110533149-066

§ 323 Nonabelian p-groups that are not generated by its noncyclic abelian subgroups

We solve here Problem 3828 and prove the following result.

Theorem 323.1. *Let G be a nonabelian p-group which has a noncyclic abelian subgroup and all its noncyclic abelian subgroups generate a proper subgroup H of G. Then $|G : H| = p$ and for each $g \in G - H$, the subgroup $\langle g \rangle$ is maximal abelian in G. The converse conclusion is also true.*

Proof. Let G be a nonabelian p-group which has a noncyclic abelian subgroup and all its noncyclic abelian subgroups generate a proper subgroup H of G. (If all abelian subgroups of a nonabelian p-group G are cyclic, then $p = 2$ and G is a generalized quaternion group.)

If G has no normal subgroup isomorphic to E_{p^2}, then by Lemma 1.4 we have $p = 2$ and G is of maximal class. In this case we get $G \cong SD_{2^n}$, $n \geq 4$, $H \cong D_{2^{n-1}}$ so that $|G : H| = 2$.

Now assume that G has a normal subgroup $U \cong E_{p^2}$ and we have $H \trianglelefteq G$ and $U \leq H$. Note that $|G : C_G(U)| \leq p$. If an element $g \in G - H$ centralizes U, then the noncyclic abelian subgroup $U\langle g \rangle \leq H$, a contradiction. It follows that $C_G(U) = H$ and so $|G : H| = p$.

In any case, if $g \in G - H$, then $\langle g \rangle$ is a self-centralizing cyclic subgroup in G so that $\langle g \rangle$ is a maximal abelian subgroup in G.

Conversely, suppose that a p-group has a maximal subgroup H such that all maximal abelian subgroups of G not contained in H are self-centralizing cyclic. Then all noncyclic abelian subgroup of G are contained in H. We have to prove that if all noncyclic abelian subgroups of G generate the subgroup T, then $T = H$. Assume that this is false; then $T < H$. One may assume that G is not of maximal class. Then there is in G a normal subgroup $R \cong E_{p^2}$ (Lemma 1.4); then $R \leq T$, by definition of T. One has $|G : C_G(R)| \leq p$. Therefore, there is $x \in C_G(R) = T$. However, the noncyclic abelian subgroup $\langle R, x \rangle$ is not contained in T, which is a contradiction. Thus, $T = H$, completing the proof. □

Problem. Study the non-Dedekindian p-groups not generated by nonnormal abelian subgroups.

https://doi.org/10.1515/9783110533149-067

§ 324 A separation of metacyclic and nonmetacyclic minimal nonabelian subgroups in nonabelian p-groups

By Theorem 10.28, minimal nonabelian subgroups generate a nonabelian p-group. In this section we consider the p-group that is not generated by metacyclic and nonmetacyclic minimal nonabelian subgroups. It appears that this is a fairly strong restriction as Theorem 324.1 shows.

We solve here Problem 3882 and prove the following result.

Theorem 324.1 (Janko). *Let G be a nonabelian p-group, let M be the subgroup generated by all metacyclic minimal nonabelian subgroups of G and let N be the subgroup generated by all nonmetacyclic minimal nonabelian subgroups of G, where we assume that $M < G$ and $N < G$. Then $G = MN$ and the three groups $M \cap N$, $M/(M \cap N)$ and $N/(M \cap N)$ are abelian.*

Proof. Let G be a nonabelian p-group, let M be the subgroup generated by all metacyclic minimal nonabelian subgroups of G and let N be the subgroup generated by all nonmetacyclic minimal nonabelian subgroups of G, where we assume that $M < G$ and $N < G$. Since (by Proposition 10.28) G is generated by all its minimal nonabelian subgroups, we get $G = MN$, $M \neq \{1\}$, $N \neq \{1\}$ and $M \trianglelefteq G$, $N \trianglelefteq G$.

Let A be a maximal normal abelian subgroup in G. Since each minimal nonabelian subgroup of G is contained in M or N and, by Lemma 57.1, each $g \in G - A$ is contained in a minimal nonabelian subgroup of G, it follows that $S = G - (M \cup N) \le A$. For any $s \in S$, $(M \cap N)s \le S$ and so $(M \cap N)s \le A$ and $M \cap N \le A$. In particular, the G-invariant subgroup $M \cap N$ is abelian.

Suppose that $AM < G$. Then for each $g \in G - (AM)$, we have (by Lemma 57.1) that g is contained in a minimal nonabelian subgroup X and so $X \le N$. Hence all $g \in G - (AM)$ are contained in N and therefore, since $\langle G - AM \rangle = G$, it follows that $G = N$, a contradiction. We have proved that $AM = G$ and similarly we get also $AN = G$. This shows that the quotient groups $G/M \cong A/(A \cap M)$ and $G/N \cong A/(A \cap N)$ are abelian and then the quotient groups $M/(M \cap N)$ and $N/(M \cap N)$ are abelian (indeed, the group $G/(M \cap N)$ is isomorphic to a subgroup of the direct product of two abelian groups G/M and G/N and $M/(M \cap N)$ and $N/(M \cap N)$, being subgroups of $G/(M \cap N)$ are abelian). The proof is complete. □

Problem 1. Study the irregular p-groups that are not generated by nonmetacyclic regular subgroups.

Problem 2. Study the irregular p-groups containing exactly one nonmetacyclic maximal regular subgroup.

https://doi.org/10.1515/9783110533149-068

§ 325 *p*-groups which are not generated by their nonnormal subgroups, 2

We improve here the following result:

Theorem 231.1. *Let G be a non-Dedekindian p-group and let G_0 be the subgroup generated by all nonnormal subgroups of G, where we assume that $G_0 < G$. Then G is of class 2, G/G_0 is cyclic and for each $g \in G - G_0$, one has $\{1\} \neq \langle g\rangle \cap G_0 \trianglelefteq G$ and $G/(\langle g\rangle \cap G_0)$ is abelian so that G' is cyclic.*

The p-groups of Theorem 231.1 are not necessarily metacyclic but they are in some way close to them as the following result shows.

Theorem 325.1. *Let G be a non-Dedekindian p-group which is not generated by its nonnormal subgroups. Then G is generated by its normal metacyclic minimal nonabelian subgroups.*

Proof. Let G be a non-Dedekindian p-group and let G_0 be the subgroup generated by all nonnormal subgroups of G, where we assume $G_0 < G$. Then we may use Theorem 231.1 together with the notation introduced there.

Let G_1 be the subgroup of G such that $G_0 \leq G_1 < G$ with $|G : G_1| = p$. Note that G/G_0 is cyclic and so for each $x \in G - G_1$, $\langle x\rangle$ covers G/G_0. If for each $x \in G - G_1$, we have $x \in Z(G)$, then (noting that $\langle G - G_1\rangle = G$) G would be abelian, a contradiction. Hence there is $g \in G - G_1$ such that $g \notin Z(G)$. Set $A = \langle g\rangle$ so that A covers G/G_0, $A \trianglelefteq G$ and setting $B = C_G(A)$, we have $B \trianglelefteq G$ and $B < G$.

By Theorem 231.1 we have $\{1\} \neq G' \leq A \cap G_0$ and $G' \leq Z(G)$. Then Theorem 285.1 implies that for each $h \in G - B$ there is an element $a \in A$ such that $M = \langle h, a\rangle$ is minimal nonabelian. Since G is of class 2, we have $G' \leq Z(G)$ and so $a \in A - G'$. Note that $\langle z\rangle = M' = \Omega_1(G')$. It follows that $G' \leq M$ and therefore $M \trianglelefteq G$. Also, $\langle z\rangle < \langle a\rangle$ and so $M' = \langle z\rangle$ is a p-th power of an element in M and consequently, by Lemma 65.1, M is metacyclic.

Since all elements $h \in G - B$ generate G, it follows that all normal metacyclic minimal nonabelian subgroups of G generate G and we are done. □

https://doi.org/10.1515/9783110533149-069

§ 326 Nonabelian p-groups all of whose maximal abelian subgroups are normal

It is clear that in a p-group of class 2 each maximal abelian subgroup is normal. By an idea of the first author, we consider here a possible converse of this statement and prove the following result.

Theorem 326.1. *Let G be a nonabelian p-group all of whose maximal abelian subgroups are normal. Then G is metabelian and either G is of class 2 or G is a 3-group of class 3, where such 3-groups exist.*

In the proof of this theorem we use the following known result.

Theorem 224.1. *Suppose that the normal closure of each cyclic subgroup in a p-group G is abelian. Then G is metabelian and either G is of class ≤ 2 or G is a 3-group of class 3.*

Proof of Theorem 326.1. Let G be a nonabelian p-group all of whose maximal abelian subgroups are normal. Then obviously the normal closure of each cyclic subgroup in G is abelian. By Theorem 224.1, G is metabelian and either G is of class 2 or G is a 3-group of class 3. It is easy to see that such 3-groups of class 3 exist.

Let G_0 be the 3-subgroup of order 3^7 of the alternating group A_{81} defined in Theorem 27.2, where $d(G_0) = 3$, $G_0' = \Phi(G_0) \cong E_{3^4}$, $Z(G_0) = [G_0, G_0'] \cong C_3$, $cl(G_0) = 3$ and $C_{G_0}(g)$ is normal in G_0 for each $g \in G_0$. Let A be any maximal abelian subgroup in G_0. Then $C_{G_0}(A) = A$ and

$$C_{G_0}(A) = \bigcap_{a \in A} C_{G_0}(a)$$

so that $A \trianglelefteq G_0$. Hence each maximal abelian subgroup in G_0 is normal in G_0 and so G_0 satisfies the assumption of our theorem and we are done. □

Exercise 1. Classify the nonabelian metacyclic p-groups all of whose maximal abelian subgroups are normal.

Problem 1. Study the nonabelian p-groups of exponent $> p$ all of whose maximal abelian subgroups of exponent $> p$ are normal.

Problem 2. Study the non-Dedekindian p-groups all of whose nonnormal maximal abelian subgroups are conjugate. Study in detail the metacyclic p-groups satisfying the above property.

Problem 3. A p-group G is said to be a $(*)$-group if it is nonabelian but all of its maximal abelian subgroups have the same order. Study the nonabelian p-groups all of whose nonabelian sections are $(*)$-groups.

Problem 4. Study the non-Dedekindian p-groups all of whose maximal abelian subgroups are quasinormal.

https://doi.org/10.1515/9783110533149-070

Appendix 110 Non-absolutely regular p-groups all of whose maximal absolutely regular subgroups have index p

Recall that a p-group G is said to be *absolutely regular* if $|G/\mho_1(G)| < p^p$. By Theorem 9.8 (a), absolutely regular p-groups are regular. If $p > 2$, then metacyclic p-groups are absolutely regular. Absolutely regular 2-groups are cyclic. If G is minimal non-absolutely regular, then either G is of order p^p and exponent P of G is of maximal class and order p^{p+1} with $|\Omega_1(G)| = p^{p-1}$.

In this appendix we prove the following result.

Theorem A.110.1. *Suppose that a p-group G of order $> p^p$ is not absolutely regular. If all maximal absolutely regular subgroups have index p in G, then one and only one of the following holds:*

(a) *G is a minimal non-absolutely regular p-group of order p^{p+1} (that group is of maximal class and minimal irregular, $|\Omega_1(G)| = p^{p-1}$). In particular, if $p = 2$, then $G \cong Q_8$.*

(b) *One has $|G| > p^{p+1}$, the group G has a subgroup R of order p^p and exponent p. Then $p > 2$, all maximal subgroups of R are G-invariant, $R = \Omega_1(G)$ (in particular, G is not of maximal class). Next, $R/\Phi(R) \le Z(G/\Phi(R))$, $\Phi(R) = R \cap \Phi(G)$, $|G/\mho_1(G)| = p^p$. Every maximal subgroup of G that does not contains R, is absolutely regular.*

Proof. Since G is not absolutely regular, we have $|\Omega_1(G)| \ge p^{p-1}$ (Theorem 12.1 and Theorem 9.8 (a)). If $p = 2$, then $G \cong Q_8$, by Theorem 1.2 (indeed, absolutely regular 2-groups are cyclic).[1] Next we assume that $p > 2$.

(i) Assume that $R < G$ is of order p^p and exponent p. If $K < R$ is maximal, then there is in the set Γ_1 an absolutely regular member $H > K$. This is true for any choice of K. It follows that all maximal subgroups of R are G-invariant, $R \not\le \Phi(G)$ and $R \lhd G$. By Exercise 9.1 (b), the group G is not of maximal class. Therefore, if $|G| = p^{p+1}$, then G is regular and $R = \Omega_1(G)$. Obviously, then $d(R) = d(G)$. Now assume that G is minimal irregular; then $|G| > p^{p+1}$ and $d(G) = 2$. As R has at least $p + 1$ maximal subgroups and all of then are contained in absolutely regular members of the set Γ_1, it follows that G is minimal non-absolutely regular, a contradiction since $R < G$ is not absolutely regular. Now let G contains a proper minimal irregular subgroup H. By the above, H is as in (a).

(ii) If G has no subgroup of order p^p and exponent p, then G is of maximal class (Theorem 12.1 (a)). Assume that $|G| > p^{p+1}$. Then G_1, the fundamental subgroup of G,

1 The group M_{2^n}. $n > 3$, does not satisfy the hypothesis since it contains a maximal cyclic subgroup of index 2^2. Indeed, G has exactly two cyclic subgroups of index 2^2 and one of them is contained in $\Phi(G)$ and the second one, say C, is such that G/C is cyclic. Besides, that group G contains a maximal cyclic subgroup of order 2.

https://doi.org/10.1515/9783110533149-071

is the unique absolutely regular member of the set Γ_1. Set $\overline{G} = G/\Omega_1(G_1)$ and take the subgroup $\overline{K} < \overline{G}$ of order p such that $\overline{K} \not\leq \overline{G}_1$. Then its inverse image K is G is absolutely regular. Since $K \not\leq G_1$, K is not contained in an absolutely regular member of the set Γ_1, a contradiction. Thus, $|G| = p^{p+1}$ hence G is minimal absolutely regular so that $|G| = p^{p+1}$ and $|\Omega_1(G)| = p^{p-1}$.

(iii) We claim that the set Γ_1 has no members of maximal class. Assume that $M \in \Gamma_1$ is of maximal class. Since G is not of maximal class, the quotient group $G/\mho_1(G)$ is of order p^{p+1} and exponent p (Theorem 12.12 (b)). In that case, there is no in the set Γ_1 absolutely member, a contradiction.

It follows from (i) and (iii) that all members of the set Γ_1, that are not contain R are absolutely regular. □

Theorem A.110.2. *Suppose that a 2-group G is not metacyclic. If each maximal metacyclic subgroup has index 2 in G, then G is a minimal nonmetacyclic group.*

Proof. Let $M \leq G$ be minimal nonmetacyclic; then $\mathrm{d}(M) = 3$, by Theorem 66.1. Take in M a maximal subgroup K and let $K \leq K_1 \in \Gamma_1$, where K_1 is metacyclic. If $L < M$ is another maximal subgroup of M and $L \leq L_1 \in \Gamma_1$, where L_1 is metacyclic, then $K_1 \neq L_1$ (otherwise, the nonmetacyclic $M = KL \leq K_1$, a contradiction). Note that $\mathrm{d}(G) \leq \mathrm{d}(K_1) + 1 = 3$. By the above, $\mathrm{d}(G) \geq \mathrm{d}(M) = 3$ so that $\mathrm{d}(G) = 3$. It follows that all members of the set Γ_1 are metacyclic so that G is minimal nonmetacyclic. □

Theorem A.110.3. *Suppose that G is nonmetacyclic p-group of exponent $> p$, $p > 2$, and suppose that all maximal metacyclic subgroups have index p in G. Then one of the following holds:*
(a) *G is minimal nonmetacyclic (see §§ 66, 69).*
(b) *$G = \Omega_1(G) * C$, there $\Omega_1(G)$ is of order p^3 and exponent p, C is cyclic of order $> p$.*

Proof. Let $H \leq G$ be minimal nonmetacyclic. If $H \cong \mathrm{E}_{p^3}$, then $G = H$ (see the proof of Theorem A.110.2). Assume that G has no subgroup isomorphic to E_{p^3}. Then G is one of groups from conclusion of Theorem 13.7. Let $H \cong \mathrm{S}(p^3)$. Then all maximal subgroups of H are G-invariant so that $H' = G'$ has order p. By Lemma 4.3, $G = \Omega_1(G) * C$, where C is cyclic of order $> p$. This G satisfies the condition. It remains to consider the case when H is a 3-group of maximal class and order $3^n > 3^3$. In that case, G has no subgroups isomorphic to $\mathrm{S}(3^3)$. It easily follows that $G = H$ has order 3^4 and $|\Omega_1(G)| = 3^3$. □

Problem 1. Classify the non-absolutely regular p-groups all of whose maximal absolutely regular subgroups are normal (nonnormal).

Problem 2. Classify the nonmetacyclic p-groups all of whose maximal metacyclic subgroups are normal (nonnormal).

Problem 3. Classify the irregular p-groups all of whose maximal regular subgroups have index p (are normal, nonnormal).

Appendix 111 Nonabelian p-groups of exponent $> p$ all of whose maximal abelian subgroups of exponent $> p$ are isolated

We recall that a subgroup H is isolated in a p-group G if for each cyclic $Z \leq G$ we have either $Z \leq H$ or $Z \cap H = \{1\}$. We start with the following known result (see the commentary of A. Mann to Problem 115). The presented proof is due to the first author.

Proposition A.111.1 (Mann, personal communication). *In a nonabelian p-group G of exponent $> p$, $p > 2$, all its minimal nonabelian subgroups are isomorphic to $\mathrm{S}(p^3)$ if and only if G has an abelian subgroup $A = \mathrm{H}_p(G)$ of index p.*

Proof. Let G be a nonabelian p-group, $p > 2$, of exponent $> p$, all of whose minimal nonabelian subgroups are isomorphic to $\mathrm{S}(p^3)$. Let A be a maximal abelian subgroup of exponent $> p$ in G. Let $x \in G - A$ be such that x normalizes A. Then Lemma 57.1 implies that there is in $\langle x, A\rangle$ a minimal nonabelian subgroup containing x and so $\mathrm{o}(x) = p$. Let $A < B \leq G$ with $|B : A| = p$. Then, by the above, all elements in the set $B - A$ are of order p so that $A = \mathrm{H}_p(B)$. Assume, by way of contradiction, that $B < G$ and let $B < C \leq G$ with $|C : B| = p$. Since A is characteristic in B, we have $A \trianglelefteq C$ and so $A = \mathrm{H}_p(C)$ because all elements in the set $C - A$ are of order p. Note that $C' \leq A$ and so C is metabelian. By [HogK], the index of a nontrivial Hughes subgroup in a metabelian p-group is at most p. This is a contradiction and therefore we have $B = G$ which implies that $|G : A| = p$.

Conversely, let A be an abelian maximal subgroup of exponent $> p$ in a nonabelian p-group G, $p > 2$, so that $A = \mathrm{H}_p(G)$. If M is any minimal nonabelian subgroup in G, then $M = \langle M - A\rangle$, where all elements in the set $M - A$ are of order p (indeed, $M - A \subseteq G - A$) hence $\Omega_1(M) = M$. By Lemma 65.1, $M \cong \mathrm{S}(p^3)$ and we are done. □

Now we shall prove the following result in which the title p-groups will be classified. This solves Problem 3368 stated by the first author.

Theorem A.111.2 (Janko). *Let G be a nonabelian p-group of exponent $> p$ all of whose maximal abelian subgroups of exponent $> p$ are isolated in G. Then G has an abelian maximal subgroup A such that $A = \mathrm{H}_p(G)$.*

Proof. Let G be a nonabelian p-group of exponent $> p$ all of whose maximal abelian subgroups of exponent $> p$ are isolated. We note that this assumption is hereditary for all nonabelian subgroups of G of exponent $> p$ (Exercise A.101.7). Let M be any minimal nonabelian subgroup in G. If $\exp(M) = p$, then $p > 2$ and $M \cong \mathrm{S}(p^3)$ (Lemma 65.1). If all minimal nonabelian subgroups of G are isomorphic to $\mathrm{S}(p^3)$, the result follows from Proposition A.111.1. Suppose that $\exp(M) > p$ and let X be a maximal subgroup in M

https://doi.org/10.1515/9783110533149-072

of exponent $> p$. Let $X \le B$, where $B < G$ is maximal abelian. As $X = M \cap B$ is isolated in M (Exercise A.101.7), all elements in the set $M - X$ are of order p. By Lemma 65.1, $p = 2$ and $M \cong D_8$.

If $p = 2$, then all minimal nonabelian subgroups in G are isomorphic to D_8 and so Theorem 10.33 implies that G is quasidihedral (generalized dihedral); then $H_2(G)$ is abelian of index 2 in G. □

Appendix 112 Metacyclic p-groups with an abelian maximal subgroup

There is a lot of information on metacyclic p-groups in our book and in published papers. This section is devoted to the investigation of metacyclic p-groups with large abelian subgroups.

The nonabelian p-groups containing an abelian subgroup of index p are described in [NRSB]. This does not mean that it is easy to obtain nontrivial information on any proper subclass of such p-groups. Below we consider a special class of such p-groups and classify the nonabelian metacyclic p-groups with abelian subgroup of index p and also the metacyclic p-groups with abelian Frattini subgroup. Some related open problems are stated.

Theorem A.112.1. *Suppose that a nonabelian metacyclic p-group G possesses an abelian subgroup A of index p. Then G is either minimal nonabelian or $p = 2$ and $G/Z(G)$ is of maximal class (in that second case $G' \cong A/Z(G)$ is cyclic). The converse is also true.*

Proof. Let $L \lhd G$ be cyclic such that G/L is cyclic. By hypothesis, $G = BL$, where the cyclic subgroup B of G covers G/C.

(i) Assume that $L \nleq A$; then $AL = G$ and $A \cap L \leq Z(G)$, We have

$$G/(A \cap L) = A/(A \cap L) \times L/(A \cap L)$$

has the cyclic subgroup $A/(A \cap L)$ of index p. If $T/(A \cap L) \neq A/(A \cap L)$ is cyclic of index p in $G/(A \cap L)$, then A and T are distinct abelian subgroups of index p in G. In that case, $A \cap T = Z(G)$ has index p^2 is G, and we conclude that the group G is minimal nonabelian since $d(G) = 2$. If $A/(A \cap L)$ is the unique cyclic subgroup of index p, then $p = 2$ and $G/(A \cap L)$ is a 2-group of maximal class. It follows that $G/Z(G)$ is also of maximal class in the case under consideration.

(ii) Now let $L < A$. As G is nonabelian, it follows that $C_G(L) = A$. In that case, $A \cap B \leq Z(G)$. Write $\overline{G} = G/(A \cap B)$. Then $\overline{G}$ contains a cyclic subgroup $\overline{A} \cong A/(A \cap L)$ of index p. As in (i), we shall prove that G is either minimal nonabelian or $G/Z(G)$ is a 2-group of maximal class. □

In particular, if G is a nonabelian metacyclic p-group, $p > 2$, contains an abelian subgroup of index p, then it follows from Theorem A.112.1 that G is minimal nonabelian.

Exercise 1. Suppose that a nonabelian two-generator p-group G possesses a normal abelian subgroup C and an abelian subgroup A of index p. If G/C is cyclic and $C \nleq A$, then G is minimal nonabelian.

Hint. Since $G = AC$, we get $A \cap C \leq Z(G)$. One has $|C : (A \cap C)| = p$, by the product formula, and

$$G/(A \cap C) = A/(A \cap C) \times C/(A \cap C).$$

https://doi.org/10.1515/9783110533149-073

Since $d(G/(A \cap C)) = 2$ and G/C is cyclic, it follows that $A/(A \cap C) \cong AC/C = G/C$ is also cyclic. Now the solution is finished as in part (i) of the proof of Theorem A.112.1. □

Corollary A.112.2. *Let G be a nonabelian metacyclic p-group. If the Frattini subgroup $\Phi(G)$ of G is abelian, then one of the following holds:*
(a) *G is minimal nonabelian.*
(b) *G is an $\mathcal{A}_2$-group.*
(c) *$p = 2$ and there is in G a normal abelian subgroup R such that G/R is of maximal class.*

Proof. If G has an abelian subgroup of index p, then one of the two assertions (a) and (c) takes place, by Theorem A.112.1. In what follows we assume that G has no abelian subgroup of index p. If any maximal subgroup of G is minimal nonabelian, then G is an $\mathcal{A}_2$-group (in that case, $G' \cong C_{p^2}$, by Proposition 72.1). Next we assume that G is not an $\mathcal{A}_2$-group. Then, by Theorem A.112.1, $p = 2$ and there is in G a maximal subgroup H such that $H/Z(H)$ is of maximal class. The subgroup $R = Z(H)$ is characteristic in $H \lhd G$ so that $R \lhd G$. As $d(G/R) = 2$, it follows from Theorem 12.12 (a) that G/R is of maximal class. □

Exercise 2. Let G be a nonabelian metacyclic p-group. If $|G/G'| = p^3$, then the Frattini subgroup of G is abelian.

Solution. Obviously the quotient group G/G' is abelian of type (p^2, p) hence we have $|\Phi(G) : G'| = p$ so that $\Phi(G)$ has the cyclic subgroup G' of index p (recall that G is metacyclic). By Lemma 1.4, $Z(\Phi(G))$ is noncyclic. Then, by Theorem 1.2, $\Phi(G)$ is abelian. In that case, G is one of the groups listed in Corollary A.112.2. In particular, if $p > 2$, then G is an $\mathcal{A}_n$-group, $n \in \{1, 2\}$. □

Problem 1. Study the metacyclic p-groups containing a normal abelian subgroup of index p^2.

Problem 2. Classify the metacyclic p-groups G such that $|G/G'| = p^3$.

Problem 3. Classify the metacyclic p-groups G such that $|G/G'| = p^4$.

Problem 4. Study the two-generator p-groups G such that $G/Z(G)$ is minimal nonabelian of order $\leq p^4$.

Appendix 113 Nonabelian p-groups with a cyclic intersection of any two distinct maximal abelian subgroups

In the following theorem Problem 2370 is solved for all p (for $p = 2$, see Theorem 91.1).

Theorem A.113.1. *A nonabelian p-group G has a cyclic intersection of any two distinct maximal abelian subgroups if and only if G has an abelian subgroup of index p and* $Z(G)$ *is cyclic.*

Proof. Let G be a nonabelian p-group with a cyclic intersection of any two distinct maximal abelian subgroups. If G has no abelian normal subgroup of type (p, p), then Lemma 1.4 implies that G is a 2-group of maximal class and then G has an abelian subgroup of index 2 and $Z(G)$ is cyclic (Theorem 1.2).

Now assume that G has a normal abelian subgroup U of type (p, p). Let A be a maximal normal abelian subgroup in G containing U. If $|G : A| > p$, then there is $g \in G - A$ so that g centralizes U (indeed, $C_G(U) > A$ since $|G : C_G(U) \leq p$). Let B be a maximal abelian subgroup of G containing (the abelian subgroup) $U\langle g\rangle$. But then $A \cap B > U$ is noncyclic, a contradiction. Hence $|G : A| = p$ and in this case $Z(G) < A$ obviously must be cyclic (indeed, if $C < G$ is another maximal abelian, then $Z(G) = A \cap C$).

The theorem is proved because the converse, obviously, is clear. □

Exercise 1. Let A be a maximal normal abelian subgroup of a p-group G. Suppose that, whenever $B \neq A$ is maximal abelian subgroup of G, then $B \cap A$ is cyclic. Then one of the following holds:

(a) A is cyclic.

(b) $|G : A| = p$ and $Z(G)$ is cyclic.

See the proof of Theorem A.113.1 (that theorem follows from the exercise).

Theorem A.113.2. *Let G be a nonmetacyclic nonabelian p-group,* $p > 2$*. If the intersection of any two distinct maximal abelian subgroups of G is metacyclic, then G has a normal abelian subgroup of index* p^3.

Proof. We proceed by induction on $|G|$. Assume that G has no normal subgroup isomorphic to E_{p^3}. Then, by Theorem 13.7, there is in the nonmetacyclic group G a normal abelian subgroup of index p^2. Indeed, if G is a 3-group of maximal class, then its Frattini subgroup $\Phi(G)$ of index 3^2 is abelian. Now let $G = SC$, where $S = \Omega_1(G) \cong S(p^3)$ and C is cyclic. Take in S a G-invariant subgroup R of order p^2. Then $C_G(R)$ is abelian of index p in G.

In what follows we assume that there is in G a normal subgroup $E \cong E_{p^3}$. Note that a Sylow p-subgroup of $\mathrm{Aut}(E)$ is isomorphic to $S(p^3)$. Let $E \leq A < G$, where A is a maximal normal abelian subgroup of G. Assume that $|G : A| > p^3$. Then there is in

https://doi.org/10.1515/9783110533149-074

the set $C_G(E) - A$ an element x. Set $B = \langle x, E\rangle$; then B is abelian. Let $B \leq T$, where T is a maximal abelian subgroup of G. Clearly, $T \neq A$. Since $E \leq A \cap T$, the intersection of distinct maximal abelian subgroups A and T is nonmetacyclic, a contradiction. Thus, $|G : A| \leq p^3$. □

Exercise 2. Suppose that a nonabelian 2-group G has a normal subgroup isomorphic to E_8. If it true that if the intersection of any two distinct maximal abelian subgroups of G is metacyclic, then G has a normal abelian subgroup of index 2^3?

Exercise 3. Suppose that a nonabelian p-group G has a normal subgroup isomorphic to $E_{p^{d+1}}$. If it true that if the intersection of any two distinct maximal abelian subgroups of G is of rank $\leq d$, then G has a normal abelian subgroup of index $p^{d(d+1)/2}$?

Problem 1. Study the nonabelian p-groups G containing a maximal abelian subgroup A such that the intersection $A \cap S$ is cyclic for any minimal nonabelian $S < G$.

Problem 2. Study the nonabelian p-groups G containing a maximal abelian subgroup A of exponent $> p$ such that the intersection $A \cap S$ has exponent p for any minimal nonabelian $S < G$.

Problem 3. Study the nonabelian p-groups all of whose distinct maximal abelian subgroups have intersection of exponent p.

Problem 4. Study the non-absolutely regular p-groups G, $p > 2$, all of whose distinct maximal absolutely regular subgroups have cyclic intersection.

Appendix 114 An analog of Thompson's dihedral lemma

The following analog of Thompson's dihedral lemma holds (see [GLS, Lemma 34.2]).

Theorem A.114.1. *Let p, q be primes such that $q \equiv 1 \pmod p$. Suppose that the group $P \cong \mathrm{E}_{p^n}$ acts faithfully on an abelian q-group Q. Let the group $G = P \cdot Q$ be the natural semidirect product of P and Q. Then there is in G a subgroup $H = D_1 \times \cdots \times D_n$, where all D_i are nonabelian of order pq.*

Proof. By Frobenius' normal p-complement Theorem A.52.2, P acts faithfully on $\Omega_1(Q)$. Let Q_0 be a minimal subgroup of Q, on which P acts faithfully. Then $Q_0 \leq \Omega_1(Q)$ so that Q_0 is elementary abelian. Without loss of generality one may assume that $G = P \cdot Q_0$ is the natural semidirect product of P and Q_0. Write $Q_0 = Q$. By Exercise 293.335, the group $G = PQ$ is supersolvable. One may assume that $n > 1$ (otherwise, if $H \leq G$ is minimal nonnilpotent, it is nonabelian of order pq, by Theorem A.22.1, and we are done); then $|Q| > q$. In view of supersolvability of G, there is in Q a G-invariant subgroup Q_1 of index q. By the choice of Q, the subgroup P does not act faithfully on Q_1 so there is in P a subgroup L_1 of order p centralizing Q_1. One has $P = L_1 \times P_1$ and write $U = P_1 Q_1$. By Maschke's theorem, $Q = Q_1 \times M_1$, where $M_1 \lhd G$ is of order q (recall that Q is elementary abelian). Set $D_1 = L_1 M_1$. As L_1 does not centralize $Q = Q_1 \times M_1$ and centralizes Q_1, the subgroup D_1 is nonabelian of order pq. As $\mathrm{Aut}(M_1)$ is cyclic, M_1 centralizes a maximal subgroup of P; we denote that subgroup by P_1 again (this is legitimate since L_1 does not centralize M_1). In that case, D_1 centralizes the subgroup $U = P_1 Q_1$ and $D_1 \cap U = \{1\}$; then $D_1 U = D_1 \times U$. We claim that the subgroup $P_1 (\cong \mathrm{E}_{p^{n-1}})$ acts faithfully on Q_1. Assume, however, that $K_1 \leq P_1$ of order p centralizes Q_1. Then $K_1 \times L_1 \cong \mathrm{E}_{p^2}$ centralizes Q_1. In that case, $\mathrm{C}_{K_1 \times L_1}(M_1)$ contains a subgroup K_2 of order p (indeed, consider the semidirect product $(K_1 \times L_1) \cdot M_1$ and take into account that $\mathrm{Aut}(M_1)$ is cyclic of order $q - 1$). Note that $K_2 \neq L_1$. In that case, $K_2 < P$ centralizes $Q = Q_1 \times M_1$ so the action of P on Q is not faithful, contrary to the hypothesis. Then, by induction, $P_1 Q_1$ contains a subgroup $D_2 \times \cdots \times D_n$, where the subgroups $D_2, \dots, D_n$ are nonabelian of order pq. It follows that $D_1 \times D_2 \cdots \times D_n$ is the required subgroup, and we are done.[1] □

Exercise 1. Let p, q be primes such that $q \equiv 1 \pmod p$. Suppose that an elementary abelian p-group $P \cong \mathrm{E}_{p^n}$, $n > 1$, acts faithfully on a q-group Q. Then $\mathrm{d}(Q) \geq n$. Is it true that $\Omega_1(Q)$ contains a subgroup isomorphic to E_{q^n}?

---

1 In particular, if $p = 2$, then all subgroups D_i are dihedral of order $2p$, and we obtain a partial case of Thompson's dihedral theorem [Tho3, Lemma 5.34] for solvable G.

https://doi.org/10.1515/9783110533149-075

Hint. Apply Theorem A.114.1 to $Q/\Phi(Q)$ (on which P acts faithfully, by Burnside) and the subgroup $\Omega_1(Q)$. □

In Theorem A.114.1 and Exercise 1 one may assume that P is abelian of arbitrary exponent. Then $\Omega_1(P)$ acts faithfully on Q, and our claim follows. Thus, the following analog of [Tho3, Lemma 5.34] holds:

Theorem A.114.2. *Let p, q be primes such that $q \equiv 1 \pmod p$. Suppose that an abelian group P of rank $n > 1$ acts faithfully on an abelian q-group Q. Let G be a natural semidirect product of P and Q. Then there is in G a subgroup $H = D_1 \times \cdots \times D_n$, where all D_i are nonabelian of order pq.*

As we have noted, in the case under consideration, $\mathrm{d}(Q) \geq n$.

Appendix 115 Some results from Thompson' papers and the Odd Order paper

The following remark is a basic result. Let $H < G$, $C = \mathrm{C}_G(H)$ and $N = \mathrm{N}_G(H)$; then C is normal in N. Indeed, let $h \in H$, $c \in C$ and $x \in N$. It suffices to prove that $c^x \in C$. One has $h^{x^{-1}} \in H$. Therefore,

$$h^{c^x} = ((h^{x^{-1}})^c)^x = (h^{x^{-1}})^x = h \implies c^x \in C,$$

and we are done.

Proposition A.115.1. *Suppose that A is a maximal abelian normal subgroup of a Sylow p-subgroup P of a group G. Then any p-element of G which centralizes A lies in A, i.e., $\mathrm{C}_G(A) = A \times L$ for some p'-subgroup L.*

Proof. Put $N = \mathrm{N}_G(A)$ and $C = \mathrm{C}_G(A)$; then C is normal in N. Let $P_1 \in \mathrm{Syl}_p(C)$. We have to show that $P_1 = A$; then $C = A \times L$ for some p'-subgroup L, by the Burnside normal p-complement theorem. Let $P_1 \le P_2 \in \mathrm{Syl}_p(N)$; then $P_1 = P_2 \cap C$ is normal in P_2. Since $P \in \mathrm{Syl}_p(N)$ too, one has $P_2 = P^t$ for some $t \in N$. Then $P = P_2^{t^{-1}}$ normalizes the group $P_1^t \in \mathrm{Syl}_p(C^t) = \mathrm{Syl}_p(C)$ since $t \in N$. As A is normal in C, we get $A \le P_1^t (\le P)$. Since $P_1^t \in \mathrm{Syl}_p(C)$ centralizes A, these two subgroups coincide (recall that A is maximal abelian normal subgroup in P) hence $A \in \mathrm{Syl}_p(C)$, and we are done. □

Corollary A.115.2 (PhD thesis of Thompson). *Let $A < B \le G$ and $[G, B] \le A$. Then B normalizes every subgroup of G containing A.*

Proof. If $A \le H \le G$, then $[H, B] \le [G, B] \le A \le H$ implies $B \le \mathrm{N}_G(H)$. □

Proposition A.115.3 ([Tho3, Lemma 5.24]). *Suppose that G is a p-solvable group without subgroups isomorphic to $\mathrm{E}(p^3)$, $p > 2$. Then each chief p-factor of G has order $< p^3$.*

Sketch of proof. We proceed by induction on $|G|$. One may assume that $\mathrm{O}_{p'}(G) = \{1\}$. Then $P \in \mathrm{Syl}_p(G)$ is a p-group of Theorem 13.7. Let R be a minimal normal subgroup of G; then R is a p-group. If P is metacyclic, then the result is obvious. If P is a 3-group of maximal class, the result follows from Theorems 9.5 and 9.6. Let, at last, G be as in Theorem 13.7 (c). Taking into account that $\mathrm{C}_G(\mathrm{O}_p(G)) \le \mathrm{O}_p(G)$, one can take $|R| = p^2$ (not necessary minimal normal subgroup of G); then P/R is metacyclic and the result follows. □

Proposition A.115.4 ([FT, Lemma 8.4 (ii)]). *If a p-group P, $p > 2$, has no normal subgroup isomorphic to E_{p^3} and ϕ is an automorphism of P of a prime order $q \ne p$, then q divides $p^2 - 1$.*

Proof. Let $G = \langle \phi \rangle \cdot P$ be the natural semidirect product and let $S \le G$ be minimal nonnilpotent such that $\phi \in S$. By Theorems A.22.1 and 13.7, $S \cap P \in \{\mathrm{C}_p, \mathrm{E}_{p^2}, \mathrm{S}(p^3)\}$, and therefore, $q \mid p^2 - 1$ (if $|S \cap P| = p$, then $q \mid p - 1 \mid p^2 - 1$). □

https://doi.org/10.1515/9783110533149-076

Proposition A.115.5 ([FT, Lemma 8.5]). *If G is a group of odd order, $p \in \pi(G)$ is minimal and if, in addition, G has no subgroup isomorphic to E_{p^3}, then G is p-nilpotent.*

Proof. Assume that there is in G a minimal nonnilpotent subgroup S with the property that $S' = P \in \mathrm{Syl}_p(S)$. By Theorem 65.1, $\exp(P) = p$ and, as P has no subgroup isomorphic to E_{p^3}, we get $P \in \{\mathrm{C}_p, \mathrm{E}_{p^2}, \mathrm{S}(p^3)\}$. If $\pi(S) = \{p, q\}$, then $q \mid p^2 - 1$ (Proposition A.115.4), which is impossible since $q \geq p + 2$. Thus, S does not exist so G is p-nilpotent, by Frobenius' normal p-complement theorem (Theorem A.52.2). □

Exercise 1 ([FT, Lemma 8.7]). Let A be a p'-group of automorphisms of the p-group P, $[P, A] = P$ and $[\Phi(P), A] = \{1\}$. Then $\Phi(P) \leq \mathrm{Z}(P)$.

Solution. We have $[\Phi(P), A, P] = \{1\}$ and $[P, \Phi(P), A] \leq [\Phi(P), A] = \{1\}$. Therefore, by the Three Subgroups Lemma,

$$\{1\} = [A, P, \Phi(P)] = [P, \Phi(P)] \implies \Phi(P) \leq \mathrm{Z}(P).$$ □

Exercise 2 ([FT, Lemma 8.9]). Let G be a p-group containing a normal subgroup E isomorphic to E_{p^3}. Then any G-invariant elementary abelian subgroup, say A, of order $< p^3$ contained in G-invariant subgroup isomorphic to E_{p^3}.

Solution. Set $H = AE$; then H is normal in G. One may assume that $A \nleq E$. If $|A| = p$, then $H = A \times E$; then $A \times E_1$, where $E_1 < E$ is G-invariant of order p^2, is the required subgroup. Now let $|A| = p^2$. If B/A is a G-invariant subgroup of order p in $\mathrm{C}_H(A)/A$, then $B = A \times (B \cap E)$ is a required subgroup too. (Here we use the fact that p^2 does not divide $|\mathrm{Aut}(\mathrm{E}_{p^2})|$ and the modular law.) □

Appendix 116 On normal subgroups of a p-group

We begin with the following result.

Theorem A.116.1. *Let G be a non-Dedekindian p-group of order $p^n > p^3$. If all subgroups of G of order $\leq p^{n-3}$ are normal, then $G \in \{Q_{2^4}, M_p(2,2)\}$.*

Proof. First assume that G is minimal nonabelian. Note that all nonnormal subgroups have index p^2 in G. If H is not normal in G, then it is cyclic since H is not generated by its maximal subgroups.

(i) Let $G = \langle a, b \mid o(a) = p^r,\ o(b) = p^s,\ a^b = a^{1+p^{r-1}} \rangle \cong M_p(r,s)$ be metacyclic minimal nonabelian, $r + s = n > 3$, $r > 1$. Then the nonnormal cyclic subgroups $B = \langle b \rangle$ and $B_1 = \langle a^{p^{r-2}} \rangle$ have index p^2 in G so $r = 2 = s$, and we have $G = M_p(2,2)$.

Let $G \cong M_p(r,s,1)$. Then G has nonnormal cyclic subgroups of orders p^r and p^s of indices p^{s+1} and p^{r+1}, respectively, and this implies $r, s = 1$ so $n = r + s + 1 = 3$, contrary to the hypothesis. It follows that if G is minimal nonabelian, then $G \cong M_p(2,2)$.

(ii) Next we assume that G is not minimal nonabelian. Let $H < G$ be minimal nonabelian. In that case, all subgroups of index $\leq p^2$ in H are G-invariant so H-invariant; then H is Dedekindian hence $H \cong Q_8$ (Theorem 1.20). In that case, by Corollary A.17.3, $G = Q \times E$, where $Q \cong Q_{2^n}$ and $\exp(E) \leq 2$. One has $n = 4$ and $E = \{1\}$ since G is non-Dedekindian. Thus, if G is not minimal nonabelian, then $Q \cong Q_{2^4}$. □

Exercise 1 ([FT, Lemma 8.3]). Let G be a p-group, $p > 2$, and let $A \triangleleft G$ be abelian with maximal possible number $d(A)$. Then $\Omega_1(C_G(\Omega_1(A)) = \Omega_1(A)$.

Solution. Let $x \in \Omega_1(C_G(\Omega_1(A))$ be of order p. Assume that $x \notin \Omega_1(A)$. In that case, $\langle x, \Omega_1(A)\rangle = \langle x \rangle \times \Omega_1(A)$ is abelian of rank $d(A) + 1$. As $\Omega_1(A) \triangleleft G$, it follows from Theorem 10.1 (by that theorem, the number of elementary abelian subgroups of G of order $p^{d(A)+1}$ containing $\Omega_1(A)$ is $\equiv 1 \pmod p$) that there is in G a normal elementary abelian subgroup $B > \Omega_1(A)$ of rank $1 + d(A)$, contrary to the choice of A. Thus, x does not exist, and we are done. □

Let S of order $> p^3$ be a minimal nonabelian p-group. Assume that $|S : X^S| = p$ for any nonnormal cyclic $X < S$. Then $X^S = XS'$ has order $p|X|$, and we conclude that $|S : X| = p^2$. It follows from Lemma 65.1 that $|S| = p^3$ (see also part (i) in the proof of Theorem A.116.1).

Exercise 2. Let G be a nonabelian p-group. If for any $x \in G - Z(G)$ one has $|G : \langle x \rangle^G| = p$, then $|G| = p^3$.

Solution. Assume that $|G| > p^3$. Obviously, $G/Z(G) \cong E_{p^2}$. By Lemma 1.1, $|G'| = p$. Let $T \triangleleft G$ be of index p^2. If $x \in T$ and $X = \langle x \rangle$, then $|G : X^G| \geq |G : T| = p^2 > p$ so that $X \leq Z(G)$, and we conclude that $T = Z(G)$ since $|G : Z(G)| > p$. It follows that $Z(G)$ is the unique normal subgroup of index p^2 in G. In particular, the abelian group G/G' has

https://doi.org/10.1515/9783110533149-077

only one subgroup of index p^2. It follows that $|G/G'| = p^2$ so that

$$|G| = |G : G'||G'| = p^3. \qquad \square$$

Exercise 3. Let a p-group G be an $\mathcal{A}_n$-group, $n > 1$. Suppose that $S^G \in \Gamma_1$ for any $\mathcal{A}_1$-subgroup $S < G$. If $A \lhd G$ is maximal abelian, then $|G : A| \leq p^2$.

Solution. Assume that $|G : A| > p^2$. Let $B/A \lhd G/A$ be of order p; then B is nonabelian. Let $S \leq B$ be minimal nonabelian. One has $S^G \leq B$ so that

$$p = |G : S^G| \geq |G : B| \geq p^2,$$

a contradiction. Thus, $|G : A| \leq p^2$. □

Exercise 4. Let a group G be nonabelian of exponent p. If $Z(S) \leq \Phi(S)$ for each nonabelian $S \leq G$, then G is of maximal class with abelian subgroup of index p, and the converse is also true.

Solution. One has $p > 2$. Let $S \leq G$ be minimal nonabelian; then $S \cong S(p^3)$ (Lemma 65.1). Assume that $C_G(S) \nleq S$. Take $x \in C_G(S) - S$. Then $H = \langle x, S\rangle = S \times \langle x\rangle$ is such that $Z(H) \nleq \Phi(H)$, contrary to the hypothesis. Thus, $C_G(S) < S$ so that G is of maximal class (Proposition 10.17). Then, by Theorem 9.5, $|G| \leq p^p$. Applying Proposition 10.17 to the fundamental subgroup G_1 of G, we prove that G_1 is abelian. □

Exercise 5. Let G be a p-group, $p > 2$. If $\Omega_1(G)$ is nonabelian, then there is in G a subgroup isomorphic to $S(p^3)$.

Solution. Let $E \lhd G$ be elementary abelian of maximal order; then $E < \Omega_1(G)$. Take $x \in \Omega_1(G) - E$ of order p and put $H = \langle x, E\rangle$. Then H is nonabelian, by Theorem 10.1. By Lemma 57.1, there is $a \in E$ such that $S = \langle a, x\rangle$ is minimal nonabelian. As $\Omega_1(S) = S$, we get $S \cong S(p^3)$ (Lemma 65.1). □

It follows from the solution of Exercise 5 that if $\Omega_1(G)$ is nonabelian, then it is generated by subgroups isomorphic to $S(p^3)$.

Exercise 6. Let G be a nonmetacyclic two-generator p-group, $p > 2$. Then there is in G a characteristic subgroup N such that $G/N \cong S(p^3)$.

Solution. Let $L < G'$ be G-invariant of index p. Then L is determined uniquely so characteristic in G (Lemma 36.5 (b)). By Lemma 65.2 (a), $\overline{G} = G/L$ is a nonmetacyclic minimal nonabelian group. Then $\overline{G}/\mho_1(\overline{G}) \cong S(p^3)$. As the inverse image, say N, of $\mho_1(\overline{G})$ is characteristic in G, we are done. □

Exercise 7. Let G be a p-group.

(a) If $k \leq p$ and G has a subgroup of order p^k and exponent p, then G has a normal subgroup of order p^{k-1} and exponent p.
(b) If G has a subgroup of order p^{p+1} and exponent p, then G has a normal subgroup of order p^p and exponent p.

Solution. (a) Assume that the assertion is false; then G has no normal subgroup of order p^{p-1} and exponent p. It follows from Theorem 12.1 (a) that G is absolutely regular. In that case, $\exp(\Omega_1(G)) = p$ and $|\Omega_1(G)| \geq p^k$, contrary to the assumption.

(b) By Theorem 9.5 and 9.6, the group G is neither absolutely regular nor of maximal class. Then G has a normal subgroup of order p^p and exponent p (Theorem 12.1 (a)), as asserted. □

Exercise 8 ([Man18]). All minimal nonabelian subgroups of a nonabelian p-group G of exponent p are normal if and only if either $|G'| = p$ or G is of order p^4 of maximal class.

Solution. One may assume that $|G| > p^4$ (indeed, if $|G| = p^4$ and $|G'| > p$, then G is of maximal class). Let $S < G$ be minimal nonabelian; then $S \triangleleft G$ is of order p^3 (Lemma 65.1.). In that case, all subgroups of G containing S are G-invariant (Theorem 10.28) hence $S \geq G' = \Phi(G)$. By Lemma 1.4, $G' < S$ so that $|G'| < |S| = p^3$. Assume that $|G'| = p^2$. In that case, G' is contained in all nonabelian subgroups of G and $G' \nleq Z(G)$. Then $C = C_G(G') \in \Gamma_1$. As C has no minimal nonabelian subgroup (see the previous sentence), it is abelian. By Lemma 1.1, one has $|G : Z(G)| = p|G'| = p^3$ so that $|Z(G)| > p$. Take $x \in Z(G) - S$ and set $H = S \times X$, where $X = \langle x \rangle$. One has $\alpha_1(H) = p^2$, where $\alpha_1(H)$ is the number of minimal nonabelian subgroups in H. As G' is contained in exactly $p + 1 < p^2$ maximal subgroups of H, there is in H a minimal nonabelian subgroup S_1 such that $G' \nleq S_1$, which is a contradiction since all nonabelian subgroups of order p^3 contain G', by the above. Thus, $C_G(S) < S$ so that G is of maximal class (Proposition 10.17). However, $|G'| = \frac{1}{p^2}|G| > p^2$, a contradiction. Thus, G is of maximal class and order p^4. □

Remark 1. Suppose that $p > 2$ and all minimal nonabelian subgroups of a nonabelian p-group G have order p^3 (here one may assume, in view of Exercise 8, that $\exp(G) > p$). If all minimal nonabelian subgroups of G are normal, then either $|G'| = p$ or G is of maximal class and order p^4. Indeed. assume that $|G| > p^4$ and let $S < G$ be minimal nonabelian; then $S \triangleleft G$. Moreover, all subgroups of G containing S are G-invariant, by Theorem 10.28, so that G/S is abelian. By Lemma 1.4, $G' < S$ for any choice of S ("<" since $\alpha_1(G) > 1$). Assume, in addition, that $|G'| = p^2$. As $|\mathrm{Aut}(S)|_p = p^3$, one has $C_G(S) \nleq S$. Take $x \in C_G(S) - S$ such that $x^p \in S$, and set $H = \langle x, S \rangle$; then $d(H) = 3$ and $|H| = p^4$. As H has exactly $p + 1$ abelian subgroups of index p, it follows that

$$\alpha_1(H) = (1 + p + p^2) - (1 + p) = p^2.$$

Therefore, there is in H a nonabelian subgroup S_1 of index p such that $G' \nleq S_1$ (indeed, as H/G' is of order p^2, it has at most $p + 1 < p^2$ maximal subgroups), contrary to what has been said above. Now we get the result, as in the last two sentences in the solution of Exercise 8.

Exercise 9 ([Gor1, Exercise 5.3]). A p-group P of class $m \geq 3$ possesses a characteristic abelian subgroup not contained in $Z(G)$.

Solution. We have
$$[\mathrm{K}_2(P), \mathrm{K}_{m-1}(P)] \le \mathrm{K}_{m+1}(P) = \{1\}.$$
It follows from $m \ge 3$ that $\mathrm{K}_{m-1}(P) \le \mathrm{K}_2(P)$, and we conclude that $\mathrm{K}_{m-1}(P) \le \mathrm{Z}(\mathrm{K}_2(P))$ hence $\mathrm{K}_{m-1}(P)$ is a characteristic abelian subgroup of G. As $\mathrm{K}_{m-1}(P) \nleq \mathrm{Z}(P)$, we are done. □

Exercise 10. Let a p-group G, $p > 2$, be nonabelian. If each maximal abelian subgroup of G has a cyclic complement, then $G/\mathrm{Z}(G)$ is metacyclic.

Solution. Let $A < G$ be maximal abelian. Then $G = AB$, where B is cyclic and $A \cap B = \{1\}$. In that case, $\mathrm{Z}(G) < A$ and $H = B\mathrm{Z}(G) = B \times \mathrm{Z}(G)$ is a maximal abelian subgroup of G. Then $G = HC$, where $C < G$ is cyclic and $H \cap C = \{1\}$. It follows that $F = C\mathrm{Z}(G) = C \times \mathrm{Z}(G)$ is a maximal abelian subgroup of G. One has $G = HF$. Set $\overline{G} = G/\mathrm{Z}(G)$. Then $\overline{G} = \overline{HF}$ is the product of two cyclic subgroups $\overline{H} \cong B$ and $\overline{F} \cong C$. Since $\overline{G}/\mho_1(\overline{G}) \cong \mathrm{E}_{p^2}$, the group $\overline{G}$ is metacyclic (Theorem 9.11). □

Below a partial case of Theorem 13.7 follows.

Exercise 11. If a regular p-group G has no normal subgroup isomorphic to E_{p^3}, then one of the following holds:
(a) G is metacyclic.
(b) One has $\Omega_1(G) \cong \mathrm{S}(p^3)$ and $G/\Omega_1(G)$ is cyclic, $G = L\Omega_1(G)$, where L is cyclic, and $\mho_1(L) = \mathrm{Z}(G)$.

Solution. One may assume that G is nonabelian and $|G| > p^3$; then $p > 2$ (indeed, regular 2-groups are abelian). If $|\Omega_1(G)| \le p^2$, then G is metacyclic since
$$|G/\mho_1(G)| = |\Omega_1(G)| \le p^2$$
(Theorem 7.2 (d) and Theorem 9.11 or Corollary 36.8). Now let $|\Omega_1(G)| > p^2$ and, since G has no normal subgroup isomorphic to E_{p^3}, then it has no subgroup isomorphic to E_{p^3} (Theorem 10.4). Then $|\Omega_1(G)| = p^3$ and $\Omega_1(G) \cong \mathrm{S}(p^3)$. Let $R < \Omega_1(G)$ be G-invariant of index p and set $H = \mathrm{C}_G(R)$; then $H \in \Gamma_1$. As $\Omega_1(H) = R$, it follows that H is metacyclic (Theorem 7.2 (d) and Corollary 36.8). Set $C = \mathrm{C}_G(\Omega_1(G))$; then $C \lhd G$ and, obviously, C is cyclic and $|G/C| \le |\mathrm{Aut}(\Omega_1(G))|_p = p^3$. If $|G/C| = p^2$, then $C * \Omega_1(G) = G$, by the product formula, and we are done. Now let $G/C \cong \mathrm{S}(p^3)$ (note that a Sylow p-subgroup of $\mathrm{Aut}(\mathrm{S}(p^3))$ is isomorphic to $\mathrm{S}(p^3)$). Take $x \in G - C\Omega_1(G)$; then $x^p \in C$ since $\exp(G/C) = p$. Let $F = \langle x, C\rangle$ and $T = \Omega_1(F)$. Assume that $|T| > p$. Then $T \cong \mathrm{E}_{p^2}$ and $F = CT$, by the product formula. On the other hand,
$$G = F\Omega_1(G) = CT\Omega_1(G) = C\Omega_1(G) < G,$$
a contradiction. Thus, $|T| = p$ hence F is cyclic (Proposition 1.3). Since $G = F\Omega_1(G)$, by the product formula, the quotient group $G/\Omega_1(G)$ is cyclic. By Lemma 4.3,
$$C\Omega_1(G) = C * \Omega_1(G).$$
It follows from $\mathrm{C}_G(C) \ge T\Omega_1(G) = G$ that $C = \mathrm{Z}(G)$. □

Exercise 12. Suppose that a p-group G is such that $\Omega_1(H)$ is abelian for any $H \in \Gamma_1$. If $\Omega_1(G)$ is nonabelian, then $G \in \{\mathrm{S}(p^3), \mathrm{D}_8\}$.

Solution. Let $H \in \Gamma_1$ be such that $|\Omega_1(H)|$ is maximal possible. The subgroup $\Omega_1(H) \triangleleft G$. Then there is $x \in G - H$ of order p (indeed, $\Omega_1(H)$ is abelian and $\Omega_1(G)$ is nonabelian). Set $F = \langle x, \Omega_1(H)\rangle$. As $\Omega_1(F) = F > \Omega_1(H)$, it follows that $F = G$, and we conclude that $\Omega_1(H) = H \in \Gamma_1$ is elementary abelian. By Lemma 57.1, there is $a \in H$ such that $S = \langle a, x\rangle$ is minimal nonabelian, and therefore $S = G$ since $\Omega_1(S) = S$ is nonabelian. Now the result follows from Lemma 65.1. □

Exercise 13 (P. Hall). If all abelian characteristic subgroups of a nonabelian p-group G, $p > 2$, are cyclic, then $G = C * E$ (central product), where C is cyclic and $E = \Omega_1(G)$ is extraspecial. In particular, $C = \mathrm{Z}(G)$.

Solution. As $\mathrm{Z}(\Phi(G))$ is a characteristic abelian subgroup of G, it is cyclic so that $\Phi(G)$ is cyclic (Lemma 1.4). By the paragraph following Exercise 293.20, $\Phi(G) \le \mathrm{Z}(G)$ and $|G'| = p$. Let $\Phi(G) \le C < M$. where $C \triangleleft G$ is maximal cyclic in G and $|M : C| = p$. Then $M = C\Omega_1(M)$ (Lemma 1.4 and Theorem 1.2). For any $x \in G - \Phi(G)$ of order p, one has $|\langle C, x\rangle : C| = p$ so that $x \in \Omega_1(M)$ for $M = \langle C, x\rangle$. Therefore, since such subgroups M generate G, we conclude that $G = CE$, where $E = \Omega_1(G)$. As E is characteristic in G, so is $\mathrm{Z}(E)$, and we conclude that $\mathrm{Z}(E)$ is cyclic. As $\exp(E) = p$ (Theorem 7.1 (b)), it follows that $|\mathrm{Z}(E)| = p$ and $E/\mathrm{Z}(E)$ is elementary abelian which implies that E is extraspecial and $\mathrm{Z}(E) \le \mathrm{Z}(G)$ and $E/\mathrm{Z}(E) \cong G/C$. By Lemma 4.3, $G = E\mathrm{Z}(G)$ since $|G'| = p$ and E is a central product of minimal nonabelian subgroups of order p^3 (clearly, those subgroups are G-invariant). □

By more stronger Hall's theorem, if all characteristic abelian subgroups of a nonabelian 2-group G are cyclic, then $G = E * M$, where E is either $\{1\}$ or extraspecial and M is either cyclic or a 2-group of maximal class. Such groups are said to be groups of *symplectic type*. They play important role of the Odd Order paper [FT].

Exercise 14. Let G be an extraspecial p-group of exponent p^2, $p > 2$. There are in G characteristic subgroups (i) $\Omega_1(G)$ of exponent p and index p in G, (ii) $\mathrm{Z}(\Omega_1(G)) \cong \mathrm{E}_{p^2}$, (iii) G' of order p. In particular, if a p'-group of automorphisms acts irreducibly on G/G', where G is an extraspecial p-group, then either $\exp(G) = p$ or $p = 2$.

Hint. Let $|G| = p^{2m+1}$. One has, by Theorem 7.2 (b),

$$|\Omega_1(G)| = |G/\mho_1(G)| = p^{2m} \implies \exp(\Omega_1(G)) = p, \ |G : \Omega_1(G)| = p.$$

The subgroup $\Omega_1(G)$ of order p^{2m} is not extraspecial. By Lemma 4.3,

$$\Omega_1(G) = A_1 * \cdots * A_s\mathrm{Z}(\Omega_1(G)) \ (s \le m-1) \quad \text{and} \quad A_1 \cong \cdots \cong A_s \cong \mathrm{S}(p^3).$$

Since $|A_1 * \cdots * A_s| \le p^{2(m-1)+1}$, we have $|\mathrm{Z}(\Omega_1(G)| \ge p^2$. As $\mathrm{Z}(\Omega_1(G))$ is characteristic of the characteristic in G subgroup $\Omega_1(G)$, it follows that a noncyclic subgroup $\mathrm{Z}(\Omega_1(G))$ is characteristic in G. In fact, $|\mathrm{Z}(\Omega_1(G))| = p^2$. Indeed, $G = S_1 * \cdots * S_{m-1} * M$, where

$S_i \cong \mathrm{S}(p^3)$ for $i < m$ and $M \cong \mathrm{M}_{p^3}$. Write $H = (S_1 * \cdots * S_{m-1}) * \Omega_1(M)$. As $|H| = |\Omega_1(G)|$ and $\Omega_1(H) = H$, we get $H = \Omega_1(G)$. It is clear that $\mathrm{Z}(H) = \Omega_1(M) \cong \mathrm{E}_{p^2}$ is characteristic in G. □

Exercise 15 ([Gor1, Theorem 5.3]). Any p-group G possesses a characteristic subgroup $D > \{1\}$ of class ≤ 2 and exponent $\leq p^e$, where $e = 1$ if $p > 2$ and $e = 2$ if $p = 2$, such that any nontrivial p'-automorphism of P induces a nontrivial automorphism of D.

Solution. Let H be a critical subgroup of G (see § 269). Set $D = \Omega_e(H)$. Then we have $\mathrm{cl}(D) \leq \mathrm{cl}(H) \leq 2$. Let ϕ be a nontrivial p'-automorphism of G; then $\phi_H \neq \mathrm{id}_H$, by Frobenius' normal p-complement theorem. As the natural semidirect product $\langle \phi \rangle \cdot H$ is nonnilpotent, it contains a minimal nonnilpotent subgroup S. By Theorem A.22.1, $S' \leq \Omega_e(H) = D$. Let $\pi(S) = \{q, p\}$. One may assume that a q-Sylow subgroup of S is generated by some power ψ of ϕ. It follows that $\psi_{S'} \neq \mathrm{id}_{S'}$, and this implies that $\phi_D \neq \mathrm{id}_D$. □

Note that the subgroup D from Exercise 15 need not necessarily be critical in G (generally speaking, $\mathrm{C}_G(D) \nleq D$). Also, if D is nonabelian, it may be non-special.

Exercise 16 (Burnside). Let D be a nonabelian p-group such that $|D/D'| = p^2$. Then D cannot be a G-invariant subgroup of the Frattini subgroup of any p-group G.

Hint. Take in D' a G-invariant subgroup L of index p and apply Lemma 1.4 to the quotient group $D/L < G/L$. □

Appendix 117 Theorem of Mann

The following theorem was reported by Mann.

Theorem A.117.1. *Let G be a p-group such that the subgroup $A = Z_k(G)$ of G satisfies $|A : A'| = p^2 < |A|$ for some $k > 1$. Then $A = G$.*

This theorem follows easily from the following special case.

Lemma A.117.2. *Let G be a p-group such that $A = Z_2(G)$ is nonabelian of order p^3. Then $G = A$.*

Proof. Assume, by way of contradiction, that $A < G$. Write $C = C_G(A)$; then $C \lhd G$ and $C \cap A = Z(A)$ has order p. Assume now that $C < A$. In that case, G is of maximal class, by Proposition 10.17. Then $|Z_2(G)| = p^2 < p^3 = |A|$, a contradiction. Thus, $|C| > p$. Let $B \leq C$ be G-invariant of order p^2; then $B \nleq A$. This is a contradiction since all G-invariant subgroups of order p^2 are contained in $Z_2(G) = A$. □

Second proof of Lemma A.117.2 for $p = 2$. One has $A \in \{D_8, Q_8\}$. As all normal cyclic subgroups of order 4 are contained in A and A contains either one or three such subgroups, the number $c_2(G)$ is odd, and therefore Theorem 1.17 (b) implies that G is of maximal class. In that case, if $A < G$, then $|Z_2(G)| = 4 < 8 = |A|$, a contradiction. □

Proof of Theorem A.117.1. We proceed by induction on $|G|$. One may assume that $k > 2$. By hypothesis, A is nonabelian and $Z(G) < A$. Then $A/Z(G) = Z_{k-1}(G/Z(G))$. If $A/Z(G)$ is nonabelian then, by induction, $A/Z(G) = G/Z(G)$ so that $A = G$, and we are done. Now assume that $A/Z(G)$ is abelian. As, by hypothesis, $A/Z(G) \cong E_{p^2}$, the subgroup A is minimal nonabelian; then $|A'| = p$ (Lemma 65.1). In that case, $|A| = |A : A'||A'| = p^3$ hence, by Lemma A.117.2, we have $A = G$. □

https://doi.org/10.1515/9783110533149-078

Appendix 118 On p-groups with given isolated subgroups

Recall that a p-group X is said to be absolutely regular provided $|X/\mho_1(X)| < p^p$. By Theorem 9.8 (a), such p-groups are regular. A subgroup $H \leq G$ is said to be *isolated* if $C \cap H > \{1\}$ implies $C \leq H$ for any cyclic $C \leq G$ (in that case, all maximal cyclic subgroups of H are maximal cyclic subgroups of G). If $H < G$ is isolated and $H < M \leq G$, where $|M : H| = p$, then all elements of the set $M - H$ have order p.

Theorem A.118.1. *If all maximal absolutely regular subgroups of a non-absolutely regular nonabelian p-group G are isolated, then one of the following holds:*
(a) $\exp(G) = p$.
(b) $p = 2$ *and G is a dihedral group.*
(c) $p > 2$ *and G is an irregular group of maximal class such that for the unique absolutely regular* $H \in \Gamma_1$ *all elements of the set* $G - H$ *have order* p.

Proof. All groups of exponent p satisfy the hypothesis. Therefore we assume in what follows that $\exp(G) = p^e > p$. We proceed by induction on $|G|$.

Let $p = 2$; then absolutely regular 2-groups are cyclic. Let $C < G$ be maximal cyclic of order > 2 and $C < M \leq G$, where $|M : C| = 2$. Then all elements of the set $M - C$ have order 2 so that $M \cong D_{2|C|}$ is dihedral (Theorem 1.2). In that case, by Exercise 10.10, G is of maximal class since M is an arbitrary subgroup of G containing C as a subgroup of index 2, and it is easily seen that $G \cong D_{2^n}$ is dihedral (note that SD_{2^n} has no nontrivial isolated subgroups of order > 2 and Q_{2^n} has no nontrivial isolated subgroups at all).

Next we assume that $p > 2$. As $\exp(G) > p$ and G is not absolutely regular, we get $|G| > p^p$. Let $A < G$ be a maximal absolutely regular subgroup of exponent $> p$ and let $A < H \leq G$, where $|H : A| = p$. Since A is isolated in H, all elements of the set $H - A$ have order p which implies that $\Omega_1(H) = H$ so that the subgroup H is irregular (Lemma 7.2 (b)). If $|G| = p^{p+1}$, then G is of maximal class and $H = H_p(G)$ is the Hughes subgroup of G. Next we assume that $|G| > p^{p+1}$. By Theorem 12.1 (b), H is of maximal class. Since all subgroups of G containing A as a subgroup of index p are of maximal class, it follows that G is also of maximal class (Exercise 10.10). Let G_1 be the fundamental subgroup of G. As $|G : G_1| = p$ and the absolutely regular subgroup G_1 is isolated in G, all elements of the set $G - G_1$ have order p (in that case, $G_1 = H_p(G)$. This G satisfies the condition. □

Now we offer alternate proof of the following theorem of Janko.

Theorem A.118.2 (Theorem 252.1). *If all maximal abelian subgroups of a nonabelian p-group G are isolated, then* $\exp(G) = p$.

Proof. Let G be a counterexample of minimal order. Then there is in G a maximal abelian subgroup A of exponent $p^e > p$. Assume that $|G : A| > p$. Let $A < H < G$,

https://doi.org/10.1515/9783110533149-079

where $|H : A| = p$; then A is isolated in H. Let $A_1 < H$ be maximal abelian, $A_1 \neq A$ and $A_1 \leq B < G$, where B is maximal abelian in G. Then $A_1 = H \cap B$ is isolated in H (Exercise A.101.7). It follows that all maximal abelian subgroups of a nonabelian group H are isolated and so, by induction, $\exp(H) = p < \exp(A)$, a contradiction. Thus, $|G : A| = p$. It follows that $A = \mathrm{H}_p(G)$ and any $\neq A$ maximal abelian subgroup C of G has exponent p. Then, since $A \in \Gamma_1$, $A \cap C = \mathrm{Z}(G)$ and this intersection is isolated in A (Exercise A.101.7) and has exponent p. Then

$$\langle A - (A \cap C)\rangle = A \implies \Omega_1(A) = A \implies \exp(A) = p,$$

a contradiction. Therefore, A does not exist so that $\exp(G) = p$. □

If the group $G = \mathrm{M}_p(m, n, 1)$ is minimal nonabelian, then the quotient group G/G' is abelian of type (p^m, p^n).

Lemma A.118.3. *Let $S \cong \mathrm{M}_p(m, 1, 1)$ be a minimal nonabelian subgroup of a p-group G. Suppose that all minimal nonabelian subgroups are isolated in G. If $|G : S| = p$, then $m = 1$ and the group G possesses an abelian subgroup of index p.*

Proof. As $|G : S| = p$, we get $S = \mathrm{H}_p(G)$. By Burnside, $p > 2$. If $U \neq S$ is minimal nonabelian subgroup of G, then the set $U - S \subset G - S$ consists of elements of order p so that

$$\langle U - S\rangle = U \implies U = \Omega_1(U) \implies U \cong \mathrm{S}(p^3)$$

by Lemma 65.1. Thus, all minimal nonabelian subgroups of G that $\neq S$, are isomorphic with $\mathrm{S}(p^3)$.

Let $L < S$ be cyclic of order p^m. Below we keep this notation.

Set $R = \Omega_1(\mathrm{Z}(S))$; then $R \cong \mathrm{E}_{p^2}$ is not isolated in S since $R \cap L > \{1\}$ so in G. Therefore, if $y \in G - S$, then $T = \langle y, R\rangle$ is not isolated in G so $T \cong \mathrm{E}_{p^3}$ since T of order p^3 is not minimal nonabelian. Then

$$G = \langle G - S\rangle \leq \mathrm{C}_G(R) \implies R \leq \mathrm{Z}(G).$$

Assume, by way of contradiction, that $m > 1$. As all elements of the set $G - S$ have order p so that $\Omega_1(G) = G$, G is irregular. We claim that then $p = 3$. Indeed, otherwise $G/\mho_1(G)$ is of order $\geq p^p \geq p^5$ and exponent p. Since $|S/\mho_1(S)| = p^3$, such S cannot be maximal subgroup of G, a contradiction. Thus, $p < 5$ so that $p = 3$. Since all minimal nonabelian subgroups are isolated in G, we get $L \cap U = \{1\}$ for any minimal normal subgroup $U \neq S$ of G. The intersection $S \cap U \cong \mathrm{E}_{3^2}$.

Assume that $m > 2$. Then $L_1 = \mho_1(L) = \mho_1(S)$ is characteristic in $S \lhd G$ so normal in G. Let $x \in G - S$ and $H = \langle x, L_1\rangle$; then $H \nleq S$ and H contains a cyclic subgroup $L_2 \neq L_1$ of order 3^{m-1}. Then $L_2 \nleq S$ (otherwise, $L_1 L_2 = H \leq S$), a contradiction since all cyclic subgroups of G not contained in S have order 3. Thus, $m = 2$, $S \cong \mathrm{M}_3(2, 1, 1)$ and hence

$$|G| = 3|S| = 3^5, \quad |\mho_1(S)| = |L_1| = 3, \quad S/L_1 \cong \mathrm{S}(3^3).$$

Therefore, since all elements of the set $G - S$ have order 3, we get $\exp(G/L_1) = 3$

and the group G/L_1 of order 3^4 is nonabelian since $S(3^3) \cong S/L_1 < G/L_1$). One has $L_1 < R = Z(S) \triangleleft G$ and $R \cong E_{3^2}$, by the above.

Let $V/L_1 \cong E_{3^3}$ be a subgroup of G/L_1 (clearly, V/L_1 exists); then $cl(V) \leq 2$ so V is regular (Theorem 7.1 (b)). As $S = H_3(G)$ (Hughes' subgroup of G) and $\Omega_1(V) = V$ since $V \not\leq S$, it follows that $\exp(V) = 3$. Assume that V is abelian; then $V \cong E_{3^4}$. As $|Z(G)| = 3^2$ (indeed, G of order 3^5 is irregular and $E_{3^2} \cong R \leq Z(G)$, we get $|G'| = \frac{1}{3}|G : Z(G)| = 3^2$ (Lemma 1.1). As $cl(G) > 2$ since G is irregular and $\Omega_1(G) = G$, we get $G' \neq Z(G)$. Set $\overline{G} = G/S'$; then $\overline{G}$ is regular since its derived subgroup has order 3. As $S/S' = H_3(G/S')$ (that group is abelian of type $(3^2, 2)$), it follows that $\Omega_1(\overline{G}) = \overline{G}$ so that $\exp(\overline{G}) = 3$, a contradiction since $\overline{G}$ contains the abelian subgroup S/S' of type $(3^2, 3)$. Thus, the subgroup V is nonabelian. Take in V a minimal nonabelian subgroup W (isomorphic to $S(3^3)$). Since $V/L_1 \cong E_{3^3}$, it follows that $L_1 < W$ so that $L \cap W = L_1 > \{1\}$ hence W is not isolated in G, a contradiction. Thus, $m = 1$.

As all $\mathcal{A}_1$-subgroups of G are isomorphic to $S(p^3)$, it follows from Mann's commentary to Problem 115 that $H_p(G)$ is abelian of index p in G. □

If S is a minimal nonabelian p-group of order $> p^3$, then any maximal subgroup of S is non-isolated. Indeed, if maximal $B < S$ is isolated, then $\Omega_1(S) \geq \langle S - B \rangle = S$ so that $|S| = p^3$ (Lemma 65.1), a contradiction.

Exercise 1. Suppose that G is a nonabelian p-group with an abelian subgroup A of index p. If G is not minimal nonabelian and all $\mathcal{A}_1$-subgroups of G are isolated, then all $\mathcal{A}_1$-subgroups of G are isomorphic to $S(p^3)$.

Solution. All elements of the set $G - A$ have order p. Let $S < G$ be minimal nonabelian. Then $S - A \subset G - A$ so $\Omega_1(S) = \langle S - A \rangle$, and Lemma 65.1 implies that $|S| = p^3$ hence $S \cong S(p^3)$ (indeed, by Burnside, $p > 2$). □

Now we are ready to offer an alternate proof of Theorem 251.1.

Theorem A.118.4. *Suppose that a p-group G of exponent $> p$ is nether abelian nor minimal nonabelian. If all minimal nonabelian subgroups of G are isolated, then G has an abelian subgroup of index p, $p > 2$ and all minimal nonabelian subgroups of G are isomorphic to* $S(p^3)$.

Proof. Assume that G is a counterexample of minimal order; then G has no abelian subgroup of index p (Exercise 2). By Burnside, $p > 2$. By Proposition A.111.1 (see also Mann's commentary to Problem 115), there is in G an $\mathcal{A}_1$-subgroup S of exponent $> p$. Let $S < H \leq G$, where $|H : S| = p$. As $\exp(H) \geq \exp(S) > p$ and all $\mathcal{A}_1$-subgroups are isolated in H, there is in H an abelian subgroup $A = H_p(H)$ of index p, by induction, if $H < G$. Then, by Exercise 2, all $\mathcal{A}_1$-subgroups of H must have exponent p. As $\exp(S) > p$, we get a contradiction. Thus, $H = G$ so that $S \in \Gamma_1$ and $S = H_p(G)$. If $S_1 < G$ is minimal nonabelian and $S_1 \neq S$, then $S_1 - S \leq G - S$ so that $\Omega_1(S_1) = S_1$, and we conclude that $S_1 \cong S(p^3)$. As $S \cap S_1$ is isolated in S (Exercise A.101.7)), it follows that $S \cong M_p(m, 1, 1)$, $m > 1$, contrary to Lemma A.118.3. Thus, all $\mathcal{A}_1$-subgroups of G

are isomorphic to $S(p^3)$. Then, by Mann's Proposition A.111.1, there is in G an abelian subgroup of index p. □

Exercise 2. Suppose that $S < G$ is a minimal nonabelian subgroup of exponent $> p$ in a p-group G. Check the following assertion: If S is isolated in G, then there is a minimal nonabelian subgroup $U < G$ with $|U \cap S| > p$ such that U is not isolated in G.

Exercise 3. If all minimal nonabelian subgroups of exponent $> p > 3$ of a nonabelian p-group G are isolated, then G has an abelian subgroup of index p.

Solution. Assume that G has no abelian subgroup of index p. In that case, by Mann's commentary to Problem 115 (see also Proposition A.111.1), there is in G a (proper) minimal nonabelian subgroup S of exponent $> p$. Let $S < T \leq G$, where $|T : S| = p$. As $\Omega_1(T) \geq \langle T - S\rangle = T$ and $\exp(T) > p$, it follows from Theorem 7.2 (b), that the subgroup T is irregular. As $p > 3$, we get $|S/\mho_1(S)| \leq p^3 < p^{p-1}$. It follows that $|G/\mho_1(G)| < p^p$ so that G is regular (Theorem 9.8 (a), a contradiction. Thus, S does not exist hence G possesses an abelian subgroup of index p. □

The following result is due to the second author.

Theorem A.118.5. *If all maximal metacyclic subgroups of a nonmetacyclic p-group G are isolated, then* $\exp(G) = p$.

Proof. We proceed by induction on $|G|$. Assume that $\exp(G) > p$.

(i) Let $p > 2$ and let $M < G$ be maximal metacyclic of exponent $> p$. Take $M < H \leq G$, where $|H : M| = p$. Let $N \neq M$ be maximal metacyclic in H and $N \leq T < G$, where T is a maximal metacyclic subgroup of G. By Exercise A.101.7, $N = H \cap T$ is isolated in H. Thus, all maximal metacyclic subgroups are isolated in the nonmetacyclic group H of exponent $> p$, so, by induction, $H = G$. In that case, $M = \mathrm{H}_p(G)$ since M of exponent $> p$ is isolated of index p in G. Then all maximal metacyclic subgroups of G, that are $\neq M$ have exponent p so they are isomorphic to E_{p^2}. Let $K \neq M$ be maximal metacyclic in G. By Exercise A.101.7 again, $M \cap K$ is isolated in M. Then M has a cyclic subgroup C of index p. The subgroup $C_1 = \mho_1(C)$ is characteristic in $M \lhd G$; then $C_1 \lhd G$. Take $x \in G - M$. Then $U = \langle C_1, x\rangle$ is metacyclic of order $p|C_1| = |C| > p$. Let $U \leq V < G$, where V is maximal metacyclic in G. By the above, $V \cong \mathrm{E}_{p^2}$ hence $U = V \cong \mathrm{E}_{p^2}$ and $|C_1| = p$, $|C| = p^2$. As $C \cap U = C_1 > \{1\}$ and $C \nleq U$, the maximal metacyclic subgroup U is not isolated in G, a contradiction. Thus, if $p > 2$, then $\exp(G) = p$.

(ii) Now let $p = 2$. If $M < H \leq G$ are as above, then, provided $H < G$, all maximal metacyclic subgroups of H are isolated so, by induction, $\exp(H) = 2(< \exp(M))$, a contradiction. Thus, $H = G$. In that case, M is abelian of index 2 in G (Burnside) hence $M = \mathrm{H}_2(G)$. Then all $\neq M$ maximal metacyclic subgroups of H are isomorphic to E_4. As in (i), M has a cyclic subgroup Z of index 2. Let $x \in H - M$ and $M_1 = \mho_1(Z)\langle x\rangle$; then $M_1 \cong \mathrm{E}_4$. As in (i), M_1 is a maximal metacyclic subgroup of G. Since $M_1 \cap Z > \{1\}$ and $Z \nleq M_1$, we get a contradiction. □

Exercise 4. Suppose that a p-group G contains a minimal nonabelian subgroup S of exponent $> p$. Then there is in G a non-isolated maximal abelian subgroup.

Solution. Suppose that $B < S$ is non-isolated maximal subgroup (as $\exp(S) > p$, all maximal subgroups of S are non-isolated). If $B \leq A$, where A is a maximal abelian in G, then $B = S \cap A$ is non-isolated in S so that, by Exercise A.101.7, A is non-isolated in G. □

Exercise 5. If each maximal subgroup of a p-group G is isolated, so is $\Phi(G)$.

Solution. One may assume that $\exp(G) > p$; then G is noncyclic. Assume that $C < G$ is cyclic such that $C \cap \Phi(G) > \{1\}$ but $C \nleq \Phi(G)$. Then we have $|C\Phi(G) : \Phi(G)| = p$. Let $G/\Phi(G) = (C\Phi(G)/\Phi(G)) \times (H/\Phi(G))$; then $H \in \Gamma_1$ and $C \nleq H$. In that case, we obtain $C \cap H \geq C \cap \Phi(G) > \{1\}$ so that H is not isolated, contrary to the hypothesis. Thus, $\Phi(G)$ is isolated in G.[1] □

Proposition A.118.6. *An irregular p-group G contains a non-isolated maximal regular subgroup.*

Proof. We assert that there is no irregular p-group all of whose maximal regular subgroups are isolated. Let $M \leq G$ be minimal irregular and let $H < M$ be maximal; then H is regular and $d(M) = 2$. Let $H \leq R < G$, where R is maximal regular in G; then $H = M \cap R$ is isolated in M (Exercise A.101.7). Thus, all maximal subgroups of M are isolated. Taking in mind our aim, one may assume that $G = M$; then $d(G) = 2$. By Exercise 5, $\Phi(G)$ is isolated in $H \in \Gamma_1$. Then $\Omega_1(H) \geq \langle H - \Phi(G)\rangle = H$ so $\exp(H) = p$ since H is regular (Theorem 7.2 (b)). Thus, all members of the set Γ_1 have exponent p which implies that $\exp(G) = p$ and G is regular (Theorem 7.1 (b)), contrary to the assumption. Thus, G has no minimal irregular subgroup so it is regular, contrary to the hypothesis. □

1 The result also follows from Exercise A.101.7.

Appendix 119 Two-generator normal subgroups of a p-group G that contained in $\Phi(G)$ are metacyclic

Below we offer a new proof of Theorem 44.12 for $p > 2$.

Theorem A.119.1. *Let N be a two-generator normal subgroup of a p-group G, $p > 2$. If $N \leq \Phi(G)$, then N is metacyclic. In particular, $\Phi(G)$ is metacyclic if and only if* $\mathrm{d}(\Phi(G)) \leq 2$.

Proof. Assume that the two-generator G-invariant subgroup $N \leq \Phi(G)$ is nonmetacyclic; then N is nonabelian. Let $L < N'$ be G-invariant of index p. Set $\overline{G} = G/L$; then $|\overline{N}'| = p$. The subgroup $\overline{N} = N/L$ is minimal nonabelian (Lemma 65.2 (a)) and nonmetacyclic (Theorem 36.1). In that case,

$$\begin{aligned}\overline{N} &= \langle a, b \mid \mathrm{o}(a) = p^m,\ \mathrm{o}(b) = p^n,\ [a, b] = c,\ c^p = 1,\ [a, c] = [b, c] = 1\rangle \\ &= \mathrm{M}_p(m, n, 1).\end{aligned}$$

The subgroup $\mho_1(\overline{N}) = \langle a^p, b^p\rangle$ is characteristic in $\overline{N} \lhd \overline{G}$ so it is $\overline{G}$-invariant. Besides, we have $\overline{N}/\mho_1(\overline{N}) \cong \mathrm{S}(p^3)$. As $\overline{N}/\mho_1(\overline{N})$ is $\overline{G}$-invariant subgroup with center of order p in $\Phi(\overline{G})/\mho_1(\overline{N})$, this is a contradiction with Lemma 1.4. □

Exercise 1. Prove a result, similar to Theorem A.119.1, for 2-groups.

In particular, the two-generator Frattini subgroup of a p-group is metacyclic. Also, the two-generator derived subgroup of a p-group is metacyclic.

https://doi.org/10.1515/9783110533149-080

Appendix 120 Alternate proofs of some counting theorems

Counting theorems play an important role in our book.

Below we offer more detailed proofs of both parts of Theorem 1.17. In Remarks 1 and 2 we present surprisingly short proofs those theorems for $p > 2$. We also prove Theorem 10.4 and a partial case of Theorem 10.5. Some additional material is contained in exercises.

Theorem A.120.1 (Theorem 1.17 (a)). *If a p-group G is neither cyclic nor a 2-group of maximal class, then* $c_1(G) \equiv 1 + p \pmod{p^2}$.

Proof. Setting $|G| = p^n$, we proceed by induction on n. One may assume that G is nonabelian. The theorem holds for $n \leq 3$. Next we assume that $n > 3$ and G has no cyclic subgroup of index p. By Lemma 1.4, G has a normal subgroup $R \cong E_{p^2}$. One may assume that $\Omega_1(G) > R$ (otherwise, $c_1(G) = 1 + p$). If G/R is cyclic, then $|\Omega_1(G)| \leq p^3$ and $\Omega_1(G)$ is elementary abelian; in that case, $c_1(G) \in \{1 + p, 1 + p + p^2\}$. Next we assume that $|\Omega_1(G)| > p^3$; then G/R is noncyclic. In that case, there is in G/R a normal subgroup T/R such that $G/T \cong E_{p^2}$. Let $M_1/T, \dots, M_{p+1}/T$ be all $p + 1$ subgroups of order p in G/T. For any $C < T$ of order p all M_i contain C and for $C \nleq T$ only one $M_i = CT$ contains C. Therefore,

$$c_1(G) = \sum_{i=1}^{p+1} c_1(M_i) - pc_1(T). \tag{1}$$

Let $n = 4$; then $T = R$. If $p > 2$, the result follows by (1). Now let $p = 2$. If $R \leq Z(G)$, the result follows as in the previous sentence. Now let $R \nleq Z(G)$. Then among $p + 1$ members of the set Γ_1 containing R there is exactly one abelian member. Again formula (1) yields the result. Now let $n > 4$. Then all (noncyclic) M_i are not 2-groups of maximal class (Theorem 9.6 (c)) hence $c_1(M_i) \equiv 1 + p \pmod{p^2}$ for all i, by induction. By Sylow, $pc_1(T) \equiv p \pmod{p^2}$. Therefore, by (1), one has

$$c_1(G) \equiv (p + 1)(1 + p) - p \equiv (1 + p)^2 - p \equiv 1 + p \pmod{p^2},$$

completing the proof. □

Remark 1. Let a p-group G be noncyclic, $p > 2$. Then there is in G a normal subgroup $R \cong E_{p^2}$. Let $x \in G - R$ be of order p; then $H_1 = \langle x \rangle R$ is of order p^3 and exponent p. Therefore, $c_1(H_1) - c_1(R) = p^2$. If $y \in G - H_1$ is of order p, then $H_2 = \langle y \rangle R$ is of order p^3 and exponent p. We also have $H_1 \cap H_2 = R$. Let us continue so until we exhaust all elements of order p in the set $G - R$. As a result we obtain $c_1(G) \equiv c_1(R) \pmod{p^2}$.

Theorem A.120.2 (Theorem 1.17 (b)). *If a p-group G is neither cyclic nor a 2-group of maximal class and $k > 1$, then* $c_k(G) \equiv 0 \pmod{p}$.

https://doi.org/10.1515/9783110533149-081

Proof. Let $|G| = p^n$. We proceed by induction on n. The theorem holds for $n = k+1$ for nonabelian (Theorem 1.2) and for abelian G. In what follows we assume that $n > k+1$ and G is nonabelian. By Lemma 1.4, G has a normal subgroup $R \cong \mathrm{E}_{p^2}$. Let G/R be cyclic; then $\Omega_1(G)$ is elementary abelian of order $p^s \le p^3$. In that case,

$$c_k(G) = \frac{|\Omega_k(G) - \Omega_{k-1}(G)|}{(p-1)p^{k-1}},$$

and it is easy to check that the above number of a multiple of p. Next assume that G/R is noncyclic.

Let $n = 4$; then $k = 2$. If $R < H \in \Gamma_1$, then $c_2(H) \equiv 0 \pmod p$ provided $p > 2$. Then

$$c_2(G) = \sum_{R<H\in\Gamma_1} c_2(H) \equiv 0 \pmod p.$$

Let $p = 2$. If G is minimal nonabelian, the result follows from Theorem 65.1 since $\exp(\Omega_1(G)) = 2$. Now let $H \in \Gamma_1$ be nonabelian. Then, by Proposition 10.17, $G = HZ(G)$, and it is easy to check that $c_2(G) \equiv 0 \pmod 2$.

Now let $n > 4$. Then G/R contains a normal subgroup T/R such that $G/T \cong \mathrm{E}_{p^2}$. Let $H_1/T, \dots, H_{p+1}/T$ be all maximal subgroups of G/T. One has, by induction, $c_k(H_i) \equiv 0 \pmod p$ for all i since H_i is neither cyclic nor a 2-group of maximal class (see Theorem 9.6 (c)). In that case,

$$c_k(G) = \sum_{i=1}^{s} c_k(H_i) - pc_1(T) \equiv 0 \pmod p,$$

and the proof is complete. □

In particular, if a 2-group G is as in Theorem A.120.2, then the number of noncyclic subgroups of order $p^n > p$ in G is $\equiv 1 \pmod p$, by Sylow's theorem. Therefore, if G is neither cyclic nor a 2-group of maximal class, it possesses a proper normal noncyclic subgroup of order p^k for any $k > 1$.

Remark 2. Let a p-group G be noncyclic, $p > 2$. Then there is in G a normal subgroup $R \cong \mathrm{E}_{p^2}$. Let $C_1 < G$ be cyclic of order $p^k > p$ and set $H_1 = C_1R$. Then $c_k(H_1) \equiv 0 \pmod p$ (check!). Let $C_2 < G$ be cyclic of order p^k such that $C_2 \nleq H_1$ and set $H_2 = C_2R$. Then $\exp(H_1 \cap H_2) < p^k$ and $c_k(H_2) \equiv 0 \pmod p$. Let us continue until we exhaust all cyclic subgroups of order p^k in G. As a result we obtain $c_k(G) \equiv 0 \pmod p$, as desired. A similar argument allows us to prove that $c_k(G) \equiv 0 \pmod 2$ if a 2-group G is neither cyclic nor of maximal class and $k > 2$.

Exercise 1. Let G be a noncyclic p-group, $n > 1$. It is known that, if $c_n(G) = 1$, then $p = 2$ and G is of maximal class (Theorem 1.17 (b)). Present an independent proof.

Solution. Let $C < G$ be maximal cyclic of order $\ge p^n$ and $C < B \le G$, where $|B : C| = p$. Then $c_n(B) = 1$ so that B is a 2-group of maximal class (Theorem 1.2). By Exercise 10.10, G is of maximal class. If $n = 2$, then G is dihedral (Theorem 1.2). □

Exercise 2. Let G be a noncyclic p-group, $n > 1$ and $c_n(G) \not\equiv 0 \pmod p$. Present a proof, independent of Theorem 1.17 (b), that $p = 2$.

Hint. Use induction on $|G|$, Hall's enumeration principle and Theorem 1.2. □

Exercise 3. Let G be a nonmetacyclic p-group of order $> p^n$, $p > 3$, $n > 3$, and let $\mu_n(G)$ be the number of metacyclic subgroups of order p^n in G. Then $\mu_n(G) \equiv 0 \pmod p$.

Hint. Let $M < G$ be metacyclic of order p^n. If $|G : M| = p$, then $G = M\Omega_1(G)$ since G is regular (Theorem 12.1 (a)); clearly, $|\Omega_1(G)| = p^3$. In that case, all maximal subgroups of G not containing $\Omega_1(G)$ are metacyclic so that $\mu_n(G) \equiv 0 \pmod p$. If $|G : M| > p$, apply the obtained result, induction and Hall's enumeration principle. □

The result of Exercise 3 is not true for $p < 5$. If $p = 2$ and $n = 4$, the counterexample yields the minimal nonmetacyclic group of order 2^5. If $p = 3$, the counterexample yields any 3-group of maximal class and order 3^{n+1}, $n > 3$.

Exercise 4. State and prove an analog of Exercise 3 for absolutely regular subgroups instead of metacyclic ones.

Hint. Use Theorems 7.2, 12.1 and Hall's enumeration principle. □

Exercise 5. Classify the 2-groups G such that $\Omega_1(G)$ is nonabelian of order 8.

Solution. Obviously, we have $\Omega_1(G) \cong D_8$ so that $c_1(G) = 5 \not\equiv 3 \pmod 4$. It follows from Theorem A.120.1 that G is a 2-group of maximal class, and we conclude that $G \in \{D_8, SD_{16}\}$. □

If $\Omega_1(G) \cong D_{2^n}$, then $G \in \{D_{2^n}, SD_{2^{n+1}}\}$.

Exercise 6 (Theorem 13.5). If a p-group G is neither absolutely regular nor of maximal class, then $e_p(G) \equiv 1 \pmod p$, where $e_n(G)$ is the number of subgroups of order p^n and exponent p in G.

Hint. One may assume that G is irregular (otherwise, the result follows from Theorem 7.2 (b) and Sylow's theorem); then $|G| \geq p^{p+2}$ since G is not of maximal class. One has

$$e_p(G) \equiv \sum_{H\in\Gamma_1} e_p(H) \pmod p.$$

If $H \in \Gamma_1$ is absolutely regular, then, by Theorem 12.1 (a), $G = H\Omega_1(G)$, where $\Omega_1(G)$ is of order p^p and exponent p so $e_p(G) = 1$. One may assume, therefore, that the set Γ_1 has no absolutely regular members. If $H \in \Gamma_1$ is not of maximal class, then, by induction, $e_p(H) \equiv 1 \pmod p$, and the number of such H is $|\Gamma_1| \equiv 1 \pmod p$. Let $H \in \Gamma_1$ be of maximal class. Then, by Theorem 12.12 (b,c), the number of such H is a multiple of p^2. If $|H| > p^{p+1}$, then $e_p(H) \equiv 0$ (Theorem 9.6 (c)), and it follows from the above and the displayed formula that $e_p(G) \equiv 1 \pmod p$. We suggest the reader to consider the case when $|G| = p^{p+2}$ using Theorem 12.12 (b)–(c). □

Exercise 7. Let R be a normal subgroup of order p^2 of a group G of exponent p and order $p^n > p^3$. Then the number of nonabelian subgroups containing R as a subgroup of index p, is a multiple of p. Consider the case when R is nonnormal.

Solution. Write $C = C_G(R)$. If $C = G$, then the required number is equal to 0. Now let $|G : C| = p$. If $x \in G - C$, then $H = \langle x, R\rangle \cong S(p^3)$. One has $|H - R| = p^2(p-1)$ and $|G - C| = p^{n-1}(p-1)$. Therefore, the required number is equal to $\frac{p^{n-1}(p-1)}{p^2(p-1)} = p^{n-3}$. □

Let $\epsilon_k(G)$ denote the number of subgroups isomorphic to E_{p^k} is a p-group G.

Exercise 8 (Theorem 10.4). If a p-group G contains a subgroup isomorphic to E_{p^3}, $p > 2$, then $\epsilon_3(G) \equiv 1 \pmod p$.

Solution. We proceed by induction on $|G|$. One may assume that $|G| > p^3$. By Hall's enumeration principle,

$$\epsilon_3(G) \equiv \sum_{H \in \Gamma_1} \epsilon_3(H) \pmod p. \tag{2}$$

If each $H \in \Gamma_1$ possesses a subgroup isomorphic to E_{p^3}, then, by induction, $\epsilon_3(H) \equiv 1 \pmod p$, and the result follows from (2) since $|\Gamma_1| \equiv 1 \pmod p$. Next we assume that there is $F \in \Gamma_1$ that has no subgroup isomorphic to E_{p^3}. In that case, F has no subgroup of order p^4 and exponent p and G has no subgroup isomorphic to E_{p^4}. It follows that G has no subgroup of order p^5 and exponent p. By Exercise 293.212, there is in G a normal subgroup $E \cong E_{p^3}$. One may assume that there is in G a normal $E_1 \neq E$ such that $E_1 \cong E_{p^3}$ (otherwise, we are done). Set $R = EE_1$. Then $|R| = p^4$ (by the above) and $\exp(R) = p$ (Theorem 7.2 (b)) and R is nonabelian (also by the above) and $|R'| = p$ (Lemma 1.1). By Exercise 1.6, $\epsilon_3(R) \equiv 1 \pmod p$ hence one may assume that $R < G$. Therefore, there is in G a normal $E_2 \cong E_{p^3}$ and $E_2 \nleq R$. Write $S = RE_2$. As S has no subgroup isomorphic to E_{p^4}, we get $R' < E_2$ (recall that $|R'| = p$) so that $S/R' = RE_2/R'$ is of exponent p. By Theorem 13.6, F is either 3-group of maximal class or $\Omega_1(F) \cong S(p^3)$. Therefore, consideration of the intersection $F \cap S$ leads us to contradiction. □

Exercise 9. If a p-group G contains a subgroup isomorphic to E_{p^4}, $p > 2$, then G contains a normal subgroup isomorphic to E_{p^4}.

Hint. Let G be a counterexample of minimal order. Let $G > E \cong E_{p^4}$ and $E < M \in \Gamma_1$. By induction, one may assume that $E \lhd M$. Next, $\epsilon_4(M) \equiv 0 \pmod p$ (otherwise, we are done). Therefore, there is in M a normal subgroup $E_1 \cong E_{p^4}$, $E_1 \neq E$. Write $R = EE_1$; then $\exp(R) = p$ since $\mathrm{cl}(R) \leq 2$ (Fitting's lemma). By Exercise 8, there is a G-invariant subgroup $S \cong E_{p^3}$. We have $SC_R(S) > S$ (let us consider $SC_E(S)$ and $SC_{E_1}(S)$) so $S < S_1$, where $S_1 \cong E_{p^4}$. Now the result follows from Theorem 10.1.[1] □

1 We have proved a partial case of Theorem 10.5 containing the following essentially more stronger result: $\epsilon_4(G) \equiv 1 \pmod p$. Konvisser [Kon1] has also proved that if a p-group G, $p > 2$, contains a subgroup isomorphic to E_{p^5}, then $\epsilon_5 \equiv 1 \pmod p$.

Exercise 10. Classify the p-groups G of order $> p^{p+1}$ such that $|\Omega_1(G)| = p^{p+1}$ and $\exp(\Omega_1(G)) > p$.

Hint. By Theorem 7.2 (b), $\Omega_1(G)$ is irregular so it is of maximal class (Theorem 7.1 (b)). It follows that $\mathrm{e}_p(G) = \mathrm{e}_p(\Omega_1(G)) \in \{2, \dots, p\}$ so G is of maximal class (Theorem 13.5). If G has a normal subgroup of order p^p and exponent p, then $|G| = p^{p+1}$ (Theorem 9.6 (c)). Now assume that $R < \Omega_1(G)$ is non-G-invariant of order p^p and exponent p. Then $|G : \mathrm{N}_G(R)| = |\mathrm{N}_G(R) : R| = p$ so that $\mathrm{N}_G(R) = \Omega_1(G)$. As $|\mathrm{N}_G(R)| = p^{p+1}$, we get $|G| = p^{p+2}$. □

Exercise 11. Let G be a p-group of order $> p^3$, $p > 2$. Is it true that the number of abelian subgroups of order p^3 in G is $\equiv 1 \pmod p$?

Appendix 121 On p-groups of maximal class

The p-groups of maximal class play an important role not only in our book but also in the whole p-group theory. In this section we, together with some new results, reprove a number of old theorems.

We begin with the following result.

Theorem A.121.1. *Let G be a p-group of order $p^{n+1} \geq p^{p+3}$. If G is not of maximal class, then the number of subgroups $D \triangleleft G$ such that G/D is of maximal class and order p^n is a multiple of p.*

Proof. One may assume that G/D is of maximal class for some $D \triangleleft G$ of order p; then one has $|Z(G)| = p^2$ (otherwise, G is of maximal class: to see this, it suffices to consider the upper central series of G). Assume that $Z(G)$ is cyclic. By Lemma 1.4, there is in G a normal subgroup $K \cong E_{p^2}$. Then $Z(G) \cap K = D$ since G is a monolith so $KZ(G)/D \cong E_{p^2}$ is a central subgroup hence G/D is not a p-group of maximal class, a contradiction.

Thus, $Z(G) \cong E_{p^2}$. As $n \geq p + 2$, the quotient group G/D, being of maximal class, is irregular (Theorem 9.5). By Theorem 12.1 (a), there is $R \triangleleft G$ of order p^p and exponent p. One has $R > D$ (otherwise, the p-group G/D of maximal class and order $> p^{p+1}$ has a normal subgroup $RD/D \cong R$ of order p^p and exponent p, contrary to Theorem 9.6 (c)). Consideration of $Z(G)R/D$ shows that $Z(G) \leq R$ (otherwise, as in the previous sentence, the normal subgroup $Z(G)R/D$ is G/D-invariant of order $\geq p^p$ and exponent p so G/D of order $\geq p^{p+2}$ is not of maximal class, by Theorem 9.6 (c)) which is a contradiction. Therefore, by Exercise 9.1 (b), R is the unique G-invariant subgroup of order p^p and exponent p in G. Let $S/D \triangleleft G/D$ be of order p^p; then $R/D < S/D$ since $|S| = p^{p+1} > p^p = |R|$ (see Exercise 9.1 (b)), and S/D is absolutely regular, by Theorem 9.6 (c). Since $R < S$, the subgroup S is not absolutely regular, hence $S/\mho_1(S) \triangleleft G/\mho_1(S)$ is of order p^p and exponent p (see Theorem 9.8 (a); then $\mho_1(S) \triangleleft G$ has order p, and therefore the group $G/\mho_1(S)$, being of order $> p^{p+1}$, is not of maximal class, by Theorem 9.6 (c). Note that $\mho_1(S) < Z(G)$ is of order p. Since $\mho_1(S) \neq D$ (indeed, S/D is absolutely regular of order p^p) is normal in G, we get $Z(G) \cong E_{p^2}$ again.

Let $D_1 \notin \{D, \mho_1(S)\}$ be a subgroup of order p in $Z(G)$. It suffices to show that G/D_1 is of maximal class for each such D_1 which is the same as to show that G/D_1 has no normal subgroup of order p^p and exponent p (see Theorems 9.6 (c), (e) and 12.1 (a)). Since $\exp(S) = p^2$, it follows that $S/D_1 \triangleleft G/D_1$ is absolutely regular of order p^p (otherwise, $\exp(S) = p$ since S, in the case under consideration, is isomorphic to a subgroup of $(S/D) \times (S/D_1)$, the direct product of two groups of exponent p). Therefore, since S/D_1 is the unique normal subgroup of order p^p in G/D_1, it follows that G/D_1 is of maximal class (Theorem 12.1 (a)). Since $\mho_1(S)$ is the unique central subgroup of order p such that $G/\mho_1(S)$ is not of maximal class, we are done (indeed, $Z(G)$ has exactly $p + 1$ subgroups of order p.) □

https://doi.org/10.1515/9783110533149-082

By using Theorems A.121.1 and 12.5 (assertion dual to Hall's enumeration principle), it is easy to prove the following:

Theorem A.121.2. *Let G be a p-group of order $> p^n \geq p^{p+2}$. If G is not of maximal class, then the number of normal subgroups D such that G/D is of maximal class and order p^n is a multiple of p.*

Recall that $\alpha_1(G)$ is the number of minimal nonabelian subgroups in a p-group G.

Exercise 1. Let G be the nonabelian p-group G of order p^n such that, whenever $H \leq G$ is nonabelian of order $p^k \leq p^n$, then $\alpha_1(H) = p^{k-3}$. Then G is of maximal class with an abelian subgroup of index p.

Solution. Let $S \leq G$ be minimal nonabelian and let $|S| = p^k$. Then $1 = \alpha_1(S) = p^{k-3}$ so that $k = 3$, $|S| = p^3$. Thus, all minimal nonabelian subgroups of G have order p^3. Next we assume that $S < G$. Let $S < H \leq G$, where $|H : S| = p$. Then all proper nonabelian subgroups of H are minimal nonabelian and their number is $\alpha_1(H) = p^{4-3} = p$, by hypothesis. Assume that $\mathrm{d}(H) > 2$; then $\mathrm{d}(H) = 3$. In that case, there are in H exactly $(1+p+p^2)-\alpha_1(H) = 1+p^2 > 1+p$ distinct abelian subgroups of index p so H is abelian (Exercise 1.6), a contradiction. Thus, $\mathrm{d}(H) = 2$. By Theorem 12.12 (a), H is of maximal class. Thus, any subgroup of G, containing S as a subgroup of index p, is of maximal class. By Exercise 10.10, G is also of maximal class. As any nonabelian subgroup of G is of maximal class, the fundamental subgroup of G (see § 9) must be abelian. □

Exercise 2. Prove that a p-group G is not covered by $p + 1$ proper subgroups of maximal class.

Solution. We use results of § 116. Assume that the conclusion is false. Let $G = \bigcup_{i=1}^{p+1} M_i$, where all M_i are of maximal class; then $M_i \in \Gamma_1$ and $D = \bigcap_{i=1}^{p+1} M_i$ has index p^2 in G (Remark 116.1 and Lemma 116.3 (b)). It follows that all subgroups of G containing D as a subgroup of index p are of maximal class. By Exercise 10.10, G is of maximal class so $\mathrm{d}(G) = 2$. But not all members of the set Γ_1 are of maximal class (for example, the fundamental subgroup G_1), and this is a contradiction. □

Exercise 3. Let H be a nonnormal subgroup of a p-group G. If $|G : H^G| = p$, then $H^G = HG'$. (Note that $HG' < G$ always since $G' \leq \Phi(G)$.)

Exercise 4. Let C be a nonnormal subgroup of order 2 of a 2-group G. If $|G : C^G| = 2$, then $G \in \{\mathrm{D}_{2^n}, \mathrm{SD}_{2^n}\}$ (see Theorem 1.2).

Solution. By Exercise 3, we have $C^G = CG'$ and $C \cap G' = \{1\}$ so that $|C^G : G'| = 2$, and we conclude that $|G : G'| = |G : CG'||CG' : G'| = 4$. By Taussky's Proposition 1.6, G is a 2-group of maximal class, and the result follows from Theorem 1.2. □

Exercise 5. Let G be a nonabelian p-group. If $\mathrm{C}_G(A) < A$ for any minimal nonabelian $A \leq G$ of minimal order, then either $\Omega_1(G)$ is elementary abelian or G is of maximal class.

Solution. Let $E < G$, where E is elementary abelian of maximal possible order. Assume that $E < \Omega_1(N_G(E))$. Let $x \in \Omega_1(N_G(E)) - E$ be of order p; then $H = \langle E, x\rangle$ is nonabelian, by a choice of E. By Lemma 57.1, there is $a \in E$ such that $S = \langle a, x\rangle$ is minimal nonabelian. One has $\Omega_1(S) = S$ so that $S \in \{D_8, S(p^3)\}$ (Lemma 65.1). By hypothesis, $C_G(S) < S$ so that G is of maximal class (Proposition 10.17). Now let $E = \Omega_1(N_G(E))$. Then E is characteristic in $N_G(E)$). In that case, $N_G(E) = G$ so that $\Omega_1(G) = E$ is elementary abelian. □

Exercise 6. If a p-group G contains exactly p nonabelian subgroups of order p^3, then it is of maximal class and order p^4.

Solution. By Theorem 76.6 (a), a group G of order p^4 satisfies $\alpha_1(G) = p$ if and only if it is of maximal class. Let $S < G$ be nonabelian of order p^3 and let $S < H \le G$, $|H : S| = p$. By hypothesis, $\alpha_1(H) \le p$. Therefore, $\alpha_1(H) = p$ (Theorem 76.6) hence H is of maximal class, by the above. By Exercise 10.10, G is of maximal class. It is easy to prove that $|G| = p^4$ (take into account that one of $\alpha_1(H)$ minimal nonabelian subgroups of H is not G-invariant so its normalizer in G coincides with H; use Proposition 10.17). □

Exercise 7. If G is an irregular p-group such that $R = \Omega_1(G) < G$ is of maximal class and order $\ge p^p$, then one of the following holds:

(i) If $p = 2$, then $G \cong SD_{2^n}$.

Next we assume that $p > 2$. Then $|\Omega_1(G)| \ge p^p$.

(ii) Let $|\Omega_1(G)| = p^p$; then all members of the set Γ_1 not containing $\Omega_1(G)$, are absolutely regular.

(iii) If $|\Omega_1(G)| > p^p$, then G is of maximal class.

Solution. (i) If $p = 2$, then $c_1(G) = c_1(\Omega_1(G)) \equiv 1 \pmod 4$ so that G is of maximal class, and we conclude that $G \cong SD_{2^n}$ is semi-dihedral (Theorem 1.17 (a), Theorem 1.2).

Set $R = \Omega_1(G)$. By Lemma 1.4, $R \nleq \Phi(G)$ (otherwise, $|Z(R)| > p$).

(ii) Now let $p > 2$.

(ii1) Let $|R| = p^p$; then $\exp(R) = p$ (Theorem 7.1 (b)). Assume that $|G| = p^{p+1}$. Then all members of the set $\Gamma_1 - \{R\}$ are absolutely regular. Now let $|G| > p^{p+1}$. Then, since $R \nleq \Phi(G)$ (Lemma 1.4), there is $M \in \Gamma_1$ such that $R \nleq M$. By Theorem 12.1 (b), M is either absolutely regular or of maximal class. Let S be a G-invariant subgroup of order p^2 and set $C = C_G(S)$. If $R \nleq C$, we get

$$|\Omega_1(C)| = p^{p-1}$$

and C is absolutely regular (Theorem 13.5). Let us prove that if $T \in \Gamma_1 - \{C\}$ and $R \nleq T$, then T is absolutely regular. Assume that this is false. Then T is of maximal class, by Theorem 12.1 (b). By Theorem 12.12 (b), one has $G/\mho_1(G)$ has order p^{p+1} and exponent p. As such a group G has no absolutely regular subgroup of index p, we get a contradiction.

(ii2) Now let $|R| > p^p$. Then the number of subgroups of order p^p and exponent p in R (and so in G) is $\not\equiv 1 \pmod p$ so that G is of maximal class (Theorem 13.5). □

Exercise 8. Let $p > 2$ and let M be a metacyclic subgroup of index p in a p-group G, $|G| > p^3$. If G does not split over M, then G is either metacyclic or a 3-group of maximal class.

Solution. Obviously, G has no normal subgroup of order p^3 and exponent p. Then G is either metacyclic or 3-group of maximal class, by Theorem 13.7. (As we know, in the second case, M is either abelian or minimal nonabelian.) □

Exercise 9 (Exercise 10.10). Let R be a proper subgroup of a p-group G such that, whenever $R < S \leq G$ and $|S : R| = p$, then S is of maximal class. Then G is also of maximal class.

Solution. Obviously, $C_G(R) \leq R$. Therefore, in view of Proposition 1.8, one may assume that $|R| > p^2$. We proceed by induction on $|G|$. If $R < S \leq G$, where $|S : R| = p$, then $Z(G) = Z(S) \cong C_p$ since S is of maximal class, and we conclude that $|Z(G)| = p$. Write $\overline{G} = G/Z(G)$. Then any subgroup of $\overline{G}$, containing $\overline{R}$ as a subgroup of index p, is of maximal class. In that case, $\overline{G}$ is of maximal class, by induction. □

Exercise 10 (Theorem 12.12 (a)). Let a p-group G be not of maximal class and let $M \in \Gamma_1$ be of maximal class. Then $d(G) = 3$.

Solution. Assume that $d(G) = 2$. Let $F < M$ be of index p. Then, by Remark 10.5, $F \triangleleft G$ (indeed, by Remark 10.5, $N_G(F) \neq M$ implies $N_G(F) = G$). Therefore, $G/\Phi(M)$ is abelian of type (p^2, p). Let $L < \Phi(M)$ be G-invariant of index p. Then G/L is minimal nonabelian (Lemma 65.2 (a)). It follows that M/L is abelian of order p^3 so M is not of maximal class, contrary to the hypothesis. Thus, $d(G) = d(M) + 1 = 3$. If $L = Z(G)$, then G is of maximal class, as in the previous paragraph. If $L = Z(G)$, then G is of maximal class, as in the previous paragraph. □

Exercise 11. Let G be a 2-group of maximal class. Study the 2-groups H such that $cd(H) = cd(G)$ and $k(H) = k(G)$. It is true that H is also of maximal class?

Solution. Let $|G| = 2^n$, $|H| = 2^m$, $|H : H'| = 2^s$. One has

$$cd(H) = cd(G) = \{1, 2\}, \quad |G : G'| = 4, \quad k(H) = k(G) = 2^{n-2} + 3$$

and

$$k(H) = |\mathrm{Irr}(H)| = 2^s + \frac{2^m - 2^s}{2^2} = 2^{m-2} + 3 \cdot 2^{s-2}.$$

It follows that

$$2^{n-2} + 3 = 2^{m-2} + 3 \cdot 2^{s-2} \implies 2^{n-2} - 2^{m-2} = 3(2^{s-2} - 1). \tag{1}$$

It follows from (1) that $s = 2$ so that H is a 2-group of maximal class (Taussky's Proposition 1.6). In that case, $m = n$. Thus, H is a 2-group of maximal class and order $|G|$. □

Exercise 12. Describe the p-groups G containing a maximal subgroup H such that, whenever $A < G$ is abelian of order $> p^2$, then $A \leq H$.

Solution. As maximal abelian subgroups cover G, there is a maximal abelian $A < G$ such that $A \nleq H$. It follows from $C_G(A) = A$, $|A| = p^2$ and Proposition 1.8 that G is of maximal class. □

Exercise 13. Suppose that a p-group G, $p > 2$, contains a proper minimal nonabelian subgroup S of order p^3. If all minimal nonabelian subgroups of G are G-invariant, then either $|G'| = p$ or G is of maximal class and order p^4.

Solution. One may assume that $S < G$. By Theorem 10.28, all subgroups of the group G containing S are G-invariant hence the quotient group G/S is abelian, i.e., $G' \leq S$. By Lemma 1.4, however, $G' < S$. Assume that $C_G(S) < S$. Then G is of maximal class (Proposition 10.17). By Theorem 9.6 (f), $|G| = p^4$. Now let $C_G(S) \nleq S$. Let $S < H \leq SC_G(S)$ be such that $|H : S| = p$. Then $|H'| = |S'| = p$. As G' is contained in all minimal nonabelian subgroups of H of order p^3 and the intersection of these subgroups coincides with S', it follows that $G' = S'$ so that $|G'| = p$. □

Exercise 14. Let A be a noncyclic abelian subgroup of index p in a nonabelian p-group G and suppose that $\Omega_2^\#(G) \nleq A$. If all maximal abelian subgroups of G that $\neq A$, are cyclic, then G is of maximal class.

Solution. Let $U < G$ be cyclic of order p^2 and let $U \nleq A$ (such a U exists, by hypothesis); then $U \nleq Z(G)$ since $Z(G) < A$. Set $B = UZ(G)$. As $B \nleq A$ is abelian, it is cyclic hence $Z(G) < U$, and we conclude that $|Z(G)| = p$. One has $G = UA$. If $U \leq V$, where V is maximal abelian subgroup of G, then $A \cap V = Z(G)$. Since $|Z(G)| = p$, we get $|V| = p^2 = |U|$. In that case, $C_G(U) = U$ and, by Proposition 1.8, G is of maximal class. □

Exercise 15. Suppose that a nonabelian p-group G, $p > 2$, contains a maximal elementary abelian subgroup A of type (p, p) and a normal subgroup $R \cong E_{p^p}$. Then either $A < R$ (in that case, R is of maximal class) or $AR \cong \Sigma_{p^2} \in \mathrm{Syl}_p(S_{p^2})$.

Solution. Assume that $A \nleq R$. Set $H = AR$. Then $C_H(A) = A$ so that, by Proposition 1.8, subgroup H of order p^{p+1} is of maximal class. By Exercise 9.13, $H \cong \Sigma_{p^2}$. □

Exercise 16. Let a group G be nonabelian of exponent p. If $Z(H) \leq \Phi(H)$ for each nonabelian $H \leq G$, then G is of maximal class (so $|G| \leq p^p$, by Theorem 9.5) with abelian subgroup of index p.

Solution. One has $p > 2$. Let $S \leq G$ be minimal nonabelian; then, by Lemma 65.1, $S \cong S(p^3)$. Assume that $C_G(S) \nleq S$. Take $x \in C_G(S) - S$. Then $H = \langle x, S\rangle = S \times \langle x\rangle$ is such that $Z(H) \nleq \Phi(H)$, contrary to the hypothesis. Thus, $C_G(S) < S$ so that G is of maximal class (Proposition 10.17). Then, by Theorem 9.5, $|G| \leq p^p$. Applying the above argument to the fundamental subgroup G_1 of G, we prove that it is abelian.[1] □

1 Moreover, the following assertion holds: Suppose that p-group G contains a nonabelian subgroup S of order p^3 and $Z(H) \leq \Phi(H)$ for any $H \leq G$ such that $S \leq H$. Then G is of maximal class and $|G| \leq p^{p+1}$.

Exercise 17. Given $n > 1$, classify the 2-groups G such that $\Omega_n^{\#}(G) = \langle x \mid o(x) = 2^n \rangle$ is of maximal class.

Solution. Let $M = \Omega_n^{\#}(G)$; then $c_n(G) = c_n(M) \equiv 1 \pmod 4$ unless $M \cong Q_8$. By Theorem 1.17 (b), G is of maximal class. By Theorem 1.2, $n = 2$ and $G \in \{Q_{2^m}, SD_{2^m}\}$. □

Exercise 18. Show that a 2-group G of maximal class is not a product of two nonnormal subgroups. Consider a similar problem for p-groups of maximal class, $p > 2$.

Solution. We proceed by induction on $|G|$. Obviously, $|G| > 2^3$. Suppose that $G = AB$, where $A, B < G$ are nonnormal; then $|G : A| > 2$, $|G : B| > 2$, $|A| > 2$, $|B| > 2$. Let $C < G$ be cyclic of index 2. Then $A \cap C \geq \Omega_1(C)$ and $B \cap C \geq \Omega_1(C)$. Set $\overline{G} = G/\Omega_1(C)$. Then $\overline{G} = \overline{AB}$ is of maximal class. Therefore, one of subgroups $\overline{A}$, $\overline{B}$, say $\overline{A}$, is normal in G, by induction. If $\overline{A} \lhd G$, then $A \lhd G$. □

Exercise 19. If a nonabelian 2-group $G = \Omega_1(G)$ has no sections isomorphic to E_8, then it is of maximal class.

Solution. Let $|G| = 2^n$. We proceed by induction on n. The theorem is true for $n = 3$ so we assume in what follows that $n > 3$. Let $L < G$ be normal of order 2. We claim that $\Omega_1(G/L) = G/L$. Assume that $\Omega_1(G/L) = S/L$. Then $S \geq \Omega_1(G) = G$ so that $S = G$, and our claim follows. Since $|G/L| \geq 8$, it follows that G/L is nonabelian, by hypothesis. As G/L satisfies the hypothesis, it is of maximal class, by induction. One can choose L by $\equiv 1 \pmod 2$ ways so, by Theorem A.121.1, G is of maximal class if $n > 4$. It is easy to check that, if $n = 4$, G is also of maximal class. □

Exercise 20. Classify the noncyclic p-groups of order $> p^{p+1}$ containing only one absolutely regular subgroup of index p.

Solution. By Sylow, G is not absolutely regular. Let $H \in \Gamma_1$ be absolutely regular. Suppose that G is regular. Then $G = H\Omega_1(G)$, where $|\Omega_1(G)| = p|\Omega_1(H)|$ and all members of the set Γ_1 not containing $\Omega_1(G)$ are absolutely regular and their number is a multiple of p. Thus, G is irregular. Assume that G is not of maximal class. Then again, by Theorem 12.1 (a), $G = H\Omega_1(G)$, and as above, we get a contradiction using Theorem 12.12 (b) (or Theorem 13.6). Thus, G is irregular of maximal class. (Note that some irregular groups of order p^{p+1} do not satisfy the hypothesis.)[2] □

Exercise 21. Let G be a p-group. If $Z_p(G)$ is absolutely regular, then G is either absolutely regular or of maximal class.

Hint. Note that all G-invariant subgroups of order p^p are contained in $Z_p(G)$. Use Theorem 12.1 (a) (see also Theorem 13.6). □

---

2 It is possible to prove, using results of § 12 and Hall's enumeration principle, that if the p-group G is neither absolutely regular nor of maximal class and $n > p$, then the number of absolutely regular subgroups of order p^n in G is a multiple of p. We suggest the reader to consider the case $n = p$ in detail.

Exercise 22. Let G be a 2-group with cyclic $\Phi(G)$, let G be not of maximal class and $\Phi(G) \nleq Z(G)$. Then there is in G a subgroup M of maximal class containing $\Phi(G)$ as a subgroup of index 4. If E is generated by all G-invariant subgroups of type $(2, 2)$, containing $\Omega_1(\Phi(G))$, then $G = EM$, $[E, \Phi(G)] = \{1\}$, $[E, M] \leq \Omega_1(\Phi(G))$. If, in addition, every characteristic abelian subgroup of G is cyclic, then E is extraspecial.

Hint. As $\Phi(G) = \mho_1(G)$, there is in G a cyclic subgroup X containing $\Phi(G)$ as a subgroup of index 2. Let $X < M \leq G$, where $|M : X| = 2$. Assume that all such M are not of maximal class. Then, by Theorem 1.2, $\Phi(G) \leq Z(M)$. It follows that $M < G$ and all such M generate G. In that case, $\Phi(G) \leq Z(G)$, contrary to the hypothesis. Thus, one such M is of maximal class and $|M : \Phi(G)| = 4$. Let $M < R \leq G$, where $|R : M| = 2$; then $R/\Phi(G) \cong E_8$ so $R \lhd G$ is not of maximal class. Then R has a G-invariant subgroup $K \cong E_4$ (Theorem 1.17 (b)), and let E be generated by all such K. Clearly, one has $\Omega_1(\Phi(G)) < E \leq C_G(\Phi(G))$ since all G-invariant abelian subgroups of type $(2, 2)$ centralize $\Phi(G)$, hence $[E, \Phi(G)] = \{1\}$. □

Exercise 23. Let G be an irregular p-group. Suppose that, whenever $R < G$ is regular, then it is either absolutely regular or of exponent p. Then either G is of maximal class or has a subgroup of order p^{p+1} and exponent p.

Solution. Assume that G is not of maximal class. Then it has a normal subgroup R of order p^p and exponent p (Theorem 12.1 (a)). Let $R < M \leq G$, where $|M : R| = p$. If M is regular, it is of exponent p, by hypothesis. Now assume that G has no subgroup of order p^{p+1} and exponent p. Then M is irregular so of maximal class (Theorem 7.1 (b)). Then, by Exercise 10.10, G is of maximal class, contrary to the assumption. □

Exercise 24. Classify the nonabelian p-groups G such that, whenever a nonidentity $N \lhd G$, then N has exactly one G-invariant subgroup of index p.

Solution. We claim that G is of maximal class. We proceed by induction on $|G|$. By hypothesis, G is a monolith. Let $L \lhd G$ be of order p; then G/L is noncyclic. Assume that G/L is abelian. If, in addition, G is minimal nonabelian, then it is easy to show that $|G| = p^3$; then G is of maximal class. If, however, $S < G$, then any maximal subgroup of S is G-invariant, a contradiction. Next assume that G/L is nonabelian. Then, by induction, G/L is of maximal class. If $L = Z(G)$, then G is of maximal class. Now assume that $Z(G) > L$; then $Z(G) \cong C_{p^2}$. By Lemma 1.4, there is in G a normal abelian subgroup $R \cong E_{p^2}$ (Lemma 1.4). One has $R \cap Z(G) = L$ and $RZ(G)/L \cong E_{p^2}$ is a central subgroup of the p-group of maximal class G/L, a final contradiction. □

Exercise 25. Let G be a 3-group and $H \in \Gamma_1$ of maximal class and order $> 3^4$. If G is not of maximal class, there exist an element $x \in G - H$ of order 3 centralizing $\Omega_1(\Phi(H))$.

Solution. Note that the fundamental subgroup of H is metacyclic (Theorems 9.6 (c) and 9.11) so is $\Phi(H)$. By Theorem 13.7, there is in G a normal subgroup $E \cong E_{p^3}$. By Theorem 9.6 (e), $E \nleq H$. As $H \cap E = \Omega_1(\Phi(H))$, we are done with $x \in E - \Omega_1(\Phi(H))$. □

Exercise 26. If an irregular p-group $G = \Omega_1(G)$, $p > 2$, has no sections of order p^{p+1} and exponent p, then it is of maximal class. (See Exercise 19 for $p = 2$.)

Solution. Setting $|G| = p^n$, we proceed by induction on n. The theorem is true for $n = p + 1$, by Theorem 7.2 (b) (indeed, being irregular, G is of maximal class and exponent p^2, by Theorem 9.5); therefore we assume in what follows that $n > p + 1$. Let $L \lhd G$ be of order p. Set $M/L = \Omega_1(G/L)$; then $M \geq \Omega_1(G) = G$ so that $M = G$, i.e., $\Omega_1(G/L) = G/L$. As, in addition, $|G/L| = p^{n-1} > p^p$, it follows that the group G/L is irregular (indeed, if G/L is regular, then $\exp(G/L) = p$, by Theorem 7.2 (b); then $|G/L| = p^p$ implies $|G| = p^{p+1}$ and so G is of maximal class, by the above). By hypothesis, the quotient group G/L has no section of order p^{p+1} and exponent p. Therefore, by induction, G/L is a group of maximal class. Assume, however, that G is not of maximal class. Then one has $|\mathrm{Z}(G)| = p^2$ (to see this, it suffices to consider the upper central series of G). By Lemma 1.4, there is in G a normal subgroup $K \cong \mathrm{E}_{p^2}$. Assume that $\mathrm{Z}(G)$ is cyclic; then G is a monolith hence $K \cap \mathrm{Z}(G) = L$. In that case, $K\mathrm{Z}(G)/L \cong \mathrm{E}_{p^2}$ is contained in $\mathrm{Z}(G/L)$ hence G/L is not of maximal class, a contradiction. Thus, $\mathrm{Z}(G) \cong \mathrm{E}_{p^2}$. By Theorem 12.1 (a), there is in G a normal subgroup R of order p^p and exponent p. As $\mathrm{cl}(R\mathrm{Z}(G)) = \mathrm{cl}(R) < p$ and $\Omega_1(R\mathrm{Z}(G)) = R\mathrm{Z}(G)$, the subgroup $R\mathrm{Z}(G)$ is regular so of exponent p, by Theorem 7.2 (b), we get $\mathrm{Z}(G) \leq R$ (otherwise, $R\mathrm{Z}(G)$ is of exponent p and order $\geq p^{p+1}$, contrary to the hypothesis). Take $x \in G - R$ of order p (x exists since $G = \Omega_1(G) > R$). Then $T = \langle x, R\rangle = \Omega_1(T)$, being of order p^{p+1} and of class $< p$ (recall that $|\mathrm{Z}(T)| > p$), is regular (Theorem 7.1 (b)), and we conclude that $\exp(T) = p$ (Theorem 7.2 (b)), contrary to the hypothesis. Thus, R does not exist so G, which is not of maximal class, is absolutely regular (Theorem 12.1 (a)), a contradiction since absolutely regular p-groups are regular (Theorem 9.8 (a)). Thus, $|\mathrm{Z}(G)| = p$ hence G is of maximal class. □

Appendix 122 Criteria of regularity

We begin with the following result.

Exercise 1. If all maximal abelian subgroups of a nonabelian p-group G are normal, then all maximal abelian subgroups of any proper nonabelian subgroup of G are also normal. If $p > 2$, then G is regular.

Solution. Let $H < G$ be nonabelian and let $L < H$ be maximal abelian. If $L \le A$, where A is a maximal abelian subgroup of G, then $L = H \cap A \lhd H$, proving the first assertion. Let $p > 2$ and $x, y \in G$ and $x \in U$, $y \in V$, where U, V are maximal abelian in G. By Fitting's lemma, $\mathrm{cl}(UV) \le 2$ so UV is regular (Theorem 7.1 (b)). Then $\langle x, y\rangle \le UV$ is regular, and this implies the regularity of G (indeed, by definition, a p-group is regular if and only if each its 2-generator subgroup is regular). □

Exercise 2 (Proposition A.83.1). A nonabelian regular group of exponent p^e is generated by minimal nonabelian subgroups of exponent p^e.

Solution. As the regular 2-groups are abelian, we get $p > 2$. Assume that the result holds for all proper nonabelian subgroups of G of exponent p^e. By Theorem 10.28 and Lemma 65.1, $e > 1$. Any member of the set Γ_1 has exponent $\ge p^{e-1}$. Assume that the result does not holds. As $\Omega_{e-1}(G) < G$, there is in the set Γ_1 at most one nonabelian member of exponent $< p^e$ (Theorem 7.2 (b)). Let the distinct nonabelian $F, H \in \Gamma_1$ have exponent p^e (note that the set Γ_1 contains at least $p > 2$ nonabelian members). By induction, F and H are generated by their minimal nonabelian subgroups of exponent p^e so $G = FH$ is generated by its minimal nonabelian subgroups of exponent p^e. □

Exercise 3 (Theorem 9.4). Let $N \lhd G$ and let G/N be cyclic, $|N| = p^w$ and $\exp(N) = p$. Then:

(a) If G is irregular, then $w \ge p$.

(b) If $w = p$ and $\Omega_1(G) = N$, then all proper subgroups of G are regular.

Solution. (a) If $w < p$, then

$$|G'| < |N| \le p^{p-1},$$

and so G is regular, by Theorem 9.8 (c). Thus, $w \ge p$.

(b) Let $|G : N| > p$ (otherwise, we are done, by Theorem 7.1 (b)). Take $M \in \Gamma_1$ and suppose that $N \not\le M$. Then

$$MN = G, \quad |\Omega_1(M)| = |N \cap M| < |N| = p^p,$$

and so the subgroup M is regular, by (a). Now let, in addition, $N < M \in \Gamma_1$. Let D be a G-invariant subgroup of index p^2 in N. Set $\overline{G} = G/D$. Then $\overline{M}$ is abelian as a subgroup of the abelian group $C_{\overline{G}}(\overline{N})$, and so $M' \le D$. Since $|D| = p^{p-2}$, we get $\mathrm{cl}(M) \le p - 1$ so it is regular (Theorem 7.1 (b)). (It follows that $\mathrm{d}(G) = 2$ if G is irregular.) □

https://doi.org/10.1515/9783110533149-083

Exercise 4. If each two-generator subgroup H of a p-group G, $p > 2$, satisfies

$$|H' : \mho_1(H')| < p^{p-1},$$

then G is regular.[1]

Solution. Assume that G is a counterexample of minimal order. If a two-generator $H < G$, then $|H' : \mho_1(H')| < p^{p-1}$, so H is regular (Theorem 9.8 (c)). By induction, all maximal subgroups of G are regular so that G is minimal irregular. It follows that $\mathrm{d}(G) = 2$. In that case, we have $|G' : \mho_1(G')| \geq p^{p-1}$, by Theorem 9.8 (a), contrary to the assumption. □

1 In particular, if all two-generator subgroups of a p-group G, $p > 2$, have cyclic derived subgroups, then G is regular. By [Alp1, Theorem 2], G in that case is metabelian.

Appendix 123 Nonabelian p-groups in which any two nonincident subgroups have an abelian intersection

We solve here Problem 3715 by proving the following theorem, where we note that all A_1-groups and A_2-groups satisfy the assumption of that theorem.

Theorem A.123.1. *Let G be a nonabelian p-group in which any two nonincident subgroups have an abelian intersection. Then $\Phi(G)$ is abelian, $d(G) \leq 3$ and each maximal normal abelian subgroup A of G is of index $\leq p^2$. Moreover, if $d(G) = 3$, then $\Phi(G) \leq Z(G)$ and so G is of class 2 and G/A is elementary abelian of order $\leq p^2$.*

Proof. Let G be a nonabelian p-group in which the intersection of any two nonincident subgroups is abelian.

Let $A \lhd G$ be maximal abelian. Let $A < B \trianglelefteq G$ with $|B : A| = p$. Since B is nonabelian, it follows that G/B has at most one maximal subgroup and so G/B is cyclic. Since G/A is an extension of a central subgroup B/A of order p by a cyclic group G/B, we conclude that G/A is abelian and so the group G is metabelian. In fact, G/A is either cyclic or abelian of type (p, p^m), where $m \geq 1$. If in the second case $m > 1$, then G/A has a subgroup C/A of order p such that $C/A \neq B/A$ and the abelian group G/C is noncyclic. This gives a contradiction since C is nonabelian and the intersection of two distinct subgroups of order p. Hence G/A is either cyclic of order p^n, $n \geq 1$, or $G/A \cong E_{p^2}$.

Assume for a moment that G/A is cyclic of order p^n with $n \geq 3$. Let $g \in G - A$ be such that the cyclic subgroup $\langle g \rangle$ covers G/A. Then $A_0 = A\langle g^p \rangle$ is a maximal subgroup of G containing A. Let G_0 be a maximal subgroup of G containing $\langle g \rangle$ so that $A_1 = A \cap G_0$ is a maximal subgroup of A. By our hypothesis, $A_0 \cap G_0$ is abelian and so g^p centralizes A_1. By Lemma 57.1, there is $a \in A - A_1$ such that $M = \langle a, g^p \rangle$ is minimal nonabelian. But then $g^{p^2} \in Z(M)$ and so g^{p^2} centralizes $\langle a, A_1 \rangle = A$ (note that $|A : A_1| = p$). This is a contradiction since $g^{p^2} \notin A$. We have proved that, whenever A is a maximal normal abelian subgroup in G, then either $G/A \cong E_{p^2}$ or $\{1\} \neq G/A$ is cyclic of order $\leq p^2$.

The Frattini subgroup $\Phi(G)$ is abelian since it lies in the intersection of two distinct maximal subgroups of G. Suppose that $d(G) > 2$ and let $x \in G - \Phi(G)$. Then $\Phi(G)\langle x \rangle$ is the intersection of two distinct maximal subgroups of G since $G/(\Phi(G)\langle x \rangle)$ is elementary abelian of order $> p$ and so $\Phi(G)\langle x \rangle$ is abelian. It follows that $C_G(\Phi(G)) = G$ hence $\Phi(G) \leq Z(G)$ and so $cl(G) = 2$. In this case, $G/A \cong C_{p^2}$ is not possible, where A is a maximal normal abelian subgroup in G. Indeed, if $G/A \cong C_{p^2}$, then consider an element $h \in G - A$ so that the cyclic subgroup $\langle h \rangle$ covers G/A. But then $h^p \notin A$ and $h^p \in \Phi(G) \leq Z(G)$, a contradiction.

Finally, we show here that $d(G) \geq 4$ is not possible. Indeed, if in this case we consider $k \in G - \Phi(G)$ and $y \in G - (\Phi(G)\langle k \rangle)$, then $\Phi(G)\langle k, y \rangle$ is contained in the

https://doi.org/10.1515/9783110533149-084

intersection of two distinct maximal subgroups of G and so $\Phi(G)\langle k, y\rangle$ is abelian. Hence $C_G(\Phi(G)\langle k\rangle) = G$ and so $k \in Z(G)$. Since $G = \langle G - Z(G)\rangle$, the group G is abelian, a contradiction. We have proved that $d(G) \leq 3$ and our theorem is proved. □

Problem 1. Study the p-groups G in which the intersection of any two distinct members of the set Γ_1 is abelian.

Problem 2. Study the non-Dedekindian p-groups in which the intersection of any two nonincident subgroups (subgroups of equal order) is Dedekindian. In particular, consider the case where the intersection of any two distinct maximal subgroups of a p-group G is Dedekindian.

Problem 3. Classify the non-Dedekindian p-groups G such that $N_G(K)/K$ is cyclic for any nonnormal maximal abelian $K < G$.

Problem 4. Classify the nonabelian p-groups G such that $N_G(K)/K$ is cyclic for any minimal nonabelian $K < G$.

Appendix 124 Characterizations of the p-groups of maximal class and the primary ECF-groups

In the book a number of characterizations of p-groups of maximal class is presented.

In this appendix we offer a theorem characterizing the p-groups of maximal class. The proof of that characterization is very easy and use, mainly, definitions. As a generalization, we offer a similar characterization of ECF-groups (see Definition 1 below). ECF-groups are natural generalizations of p-groups of maximal class. However, our information on ECF-groups is not so detailed (see Problems 2–4 below).

Theorem A.124.1. *A nonabelian p-group G is of maximal class if and only if it satisfies the following conditions:*
(a) $|G : G'| = p^2$.
(b) *The center of each nonabelian epimorphic image of G is cyclic.*

Proof. We proceed by induction on $|G|$.

Let $|G| = p^m$, where $m > 2$. One may assume that $m > 3$. Then $|G'| > p$ (otherwise, $|G| = |G : G'||G'| = p^3$ so that $m = 3$, a contradiction). If $m = 4$, then, not being minimal nonabelian, the group G has center of order p. Since $G/\mathrm{Z}(G)$ is nonabelian of order p^3 (indeed, $|G : G'| = p^2$) so of maximal class, we conclude that G is also of maximal class since $|\mathrm{Z}(G)| = p$. Next we assume that $m > 4$.

By hypothesis, $\mathrm{Z}(G)$ is cyclic. If $L = \mathrm{Z}(G)$, then G is of maximal class, as in the previous paragraph. Let $L < \mathrm{Z}(G)$ be of order p. By induction, the quotient group G/L is of maximal class. Then $\mathrm{Z}(G) \cong C_{p^2}$. By Lemma 1.4, there is in G a normal abelian subgroup R of type (p, p). As G is a monolith, we get $R \cap \mathrm{Z}(G) = L$. In that case, the quotient group $R\mathrm{Z}(G)/L$ is central of type (p, p) in nonabelian G/L, contrary to the hypothesis. Thus, $\mathrm{Z}(G) = L$ and G is of maximal class, completing the proof. □

We are working further to generalize Theorem A.130.1.

Definition 1. A nonabelian p-group G is said to be an ECF-group if it satisfies the following conditions:
(a) $\Phi(G) = G'$.
(b) The center of each nonabelian epimorphic image $\overline{G}$ of G is cyclic and contained in $\overline{G}'$.

Obviously, the p-groups of maximal class are ECF-groups. A two-generator primary ECF-group is of maximal class. If G is an ECF-group and L is a G-invariant subgroup of index p in G', then the quotient group G/L is extraspecial. To prove this, one may assume that $L = \{1\}$. Then, by Lemma 4.3, $G = A_1 * \cdots * A_n$, where all A_i are minimal nonabelian, and, by (b), one has $|A_i| = p^3$ for all i. In that case, G is extraspecial, as asserted.

https://doi.org/10.1515/9783110533149-085

Theorem A.124.2. *The following conditions for a nonabelian p-group G are equivalent:*
(a) *G is an* ECF*-group.*
(b) *The center of each nonabelian epimorphic image* $\overline{G}$ *of G is cyclic and contained in* $\overline{G}'$.

Proof. Obviously (a) implies (b).

It remains to prove that (b) implies (a). By (b), the center $Z(G)$ is cyclic. As above, using Lemma 1.4, we prove that $|Z(G)| = p$. If $Z(G) = G'$, then G is extraspecial so ECF-group. If $Z(G) < G'$, then $G/Z(G)$ is an ECF-group, by induction. Then G is also an ECF-group since $Z(G)$ is of order p. □

Exercise 1. Let G be an ECF-group with $|G'| > p$. Then all maximal subgroups, except one, have centers of order p.

Solution. One may assume that G is not a 2-group of maximal class. Then there is in G a normal subgroup isomorphic to E_{p^2}. As $|\Phi(G)| = |G'| > p$, we get $R \leq \Phi(G)$ so that R is contained in all maximal subgroups of G. We also have $H = C_G(R) \in \Gamma_1$ (recall that $Z(G)$ is cyclic). Moreover, R is the unique normal subgroup of order p^2 in G (see Definition 1). Take $F \in \Gamma_1 - \{H\}$ and let R_1 be a G-invariant subgroup of order p^2 in F. Then $R = R_1$. Assume that $Z(F)(\lhd G)$ has order $> p$. Then $Z(F)$ contains a G-invariant subgroup R_2 of order p^2. As before, $R_2 = R$. Since $R \leq Z(G)$, we get $C_G(R) \geq FH = G$, a contradiction. Thus, $|Z(F)| = p$ for all $F \in \Gamma_1 - \{H\}$, and the solution is complete. □

In the following problems we assume that if p-group G is an ECF-group, then $d(G) > 2$ (otherwise, it is of maximal class). The nonabelian epimorphic images of ECF-groups are ECF-groups.

Problem 1. Describe the nonabelian p-groups with cyclic centers of all its nonabelian epimorphic images.

Problem 2. Does there exist a positive integer n such that any ECF-group G with $|G'| \geq p^n$ is irregular?

Problem 3. Is it true that an ECF-group contains a regular maximal subgroup?

Problem 4. Let G be an irregular ECF-group. Is it true that the set Γ_1 contains two distinct regular members?

Problem 5. Is it true that a nonmetabelian ECF-group contains an ECF-subgroup of index p?

Problem 6. Does there exist an ECF-group whose automorphism group is a p-group?

Problem 7. Is it true that the order of elementary abelian Frattini subgroup of a primary ECF-group is bounded?

Problem 8. Let H be a maximal subgroup of an ECF-group whose center has order $> p$. Is it true that H is regular?

Appendix 125 Nonabelian p-groups all of whose proper nonabelian subgroups have exponent p

Below we prove the following result.

Theorem A.125.1. *Suppose that a p-group G is neither abelian nor minimal nonabelian and* $\exp(G) > p$. *All proper nonabelian subgroups of G have exponent p if and only if $A = \mathrm{H}_p(G)$ is abelian of index p in G and* $\mathrm{d}(G) = 2$.

Proof. Clearly, $p > 2$ (if $p = 2$, then all members of the set Γ_1 are abelian so G is minimal nonabelian, contrary to the hypothesis). Note that G satisfies the hypothesis if and only if all its nonabelian maximal subgroups have exponent p. All minimal nonabelian subgroups of G have exponent p so there is in G an abelian subgroup A of index p, by Mann's commentary to Problem 115 (see also Proposition A.111.1). As $\exp(G) > p$, we get $\exp(A) > p$ (since, using Lemma 57.1, we prove that any element of the set $G - A$ has order p). Let $H \in \Gamma_1$ be nonabelian (such an H exists since G is not minimal nonabelian); then $\exp(H) = p$ which implies $\exp(G) = p^2$. In that case, $A \cap H$ is a maximal subgroup of A and $\exp(A \cap H) = p$. As $A \cap H \le \Omega_1(A)$ is maximal in A, it follows that $\Omega_1(A) = A \cap H$. If $F \in \Gamma_1 - \{H\}$ is nonabelian (such F exists, by Exercise 1.6), then, by the above, $A \cap F = \Omega_1(A) = A \cap H$. Assume that $B \in \Gamma_1 - \{A\}$ is abelian; then $A \cap B = \mathrm{Z}(G)$ has index p^2 in G. In that case, G is regular since $p > 2$ (Theorem 7.1 (b)). As $HF = G$, we get $\Omega_1(G) = G$ so that $\exp(G) = p$ (Theorem 7.2 (b)), contrary to the hypothesis. Thus, A is the unique abelian member of the set Γ_1, and we conclude that $A = \mathrm{H}_p(G)$, the Hughes subgroup of G (indeed, if $x \in G - A$, then, as we have noted, $\mathrm{o}(x) = p$). As we have established, $A \cap H = \Omega_1(A) = A \cap F$ has index p^2 in G for any two distinct nonabelian members H, F of the set Γ_1 and A is the unique abelian member of the set Γ_1. It follows that $H \cap F = \Phi(G)$, the Frattini subgroup of G hence $|G : \Phi(G)| = p^2$ so that $\mathrm{d}(G) = 2$.

Conversely, if $\mathrm{d}(G) = 2$, the abelian subgroup $A = \mathrm{H}_p(G)$ has index p in G, A is the unique abelian member of the set Γ_1 and $\Omega_1(A) = \Phi(G)$, then all nonabelian members of the set Γ_1 have exponent p. Indeed, if $M \in \Gamma_1 - \{A\}$, then M is nonabelian since $|\Gamma_1| = p + 1 > 1$. Assume that $\exp(M) > p$. In that case, there is in M an element x of order p^2; then $x \in \mathrm{H}_p(G) = A$. We have

$$M = \langle x, \Phi(G)\rangle = \langle x, \Omega_1(A)\rangle = A,$$

which is a contradiction. Thus, all members of the set $\Gamma_1 - \{A\}$ are nonabelian of exponent p. □

Another proof. Let $A < G$ be maximal abelian of exponent $> p$. Assume that $|G : A| > p$. If $A < B < G$ is such that $|B : A| = p$, then B is a proper nonabelian subgroup of G of exponent $> p$, a contradiction. It follows that $|G : A| = p$. If $x \in G - A$, then there is an

https://doi.org/10.1515/9783110533149-086

element $a \in A$ such that $S = \langle a, x\rangle$ is minimal nonabelian. By hypothesis, $p > 2$ and $S \cong \mathrm{S}(p^3)$. It follows that all elements of the set $G - A$ have order p so that $A = \mathrm{H}_p(G)$. All remaining assertions are proved as in Theorem A.125.1. □

If $\mathrm{H}_p(G) < G$ is the abelian subgroup of a p-group G, then $|G : \mathrm{H}_p(G)| = p$, by [HogK].

Recall that a subgroup H of a p-group G is isolated in G, if $C \cap H > \{1\} \Rightarrow C \le H$ for all cyclic $C \le G$.

As an easy consequence of Theorem A.125.1, we produce a short proof the following theorem due to the second author.

Theorem A.125.2 (Theorem 251.1). *If all maximal abelian subgroups of a nonabelian p-group G are isolated, then* $\exp(G) = p$.

Proof. Assume that G is minimal nonabelian. Then all its maximal subgroups are isolated. If $B \in \Gamma_1$, then $\Omega_1(G) = \langle G - B\rangle$ so $|G| = p^3$ (Lemma 65.1). If $p > 2$, then $G \cong \mathrm{S}(p^3)$ has exponent p (in that case, G is regular). Now let $p = 2$. Then $G \cong \mathrm{D}_8$ does not satisfy the hypothesis: any noncyclic subgroup of index 2 is not isolated in G.

Next we assume that G is not minimal nonabelian.

Let $H \in \Gamma_1$ be nonabelian. If $U < H$ is maximal abelian and $U \le A < G$, where A is a maximal abelian subgroup of G, then the subgroup $U = A \cap H$ is isolated in H (Exercise A.101.7). Thus, all maximal abelian subgroups of H are isolated. By induction on $|G|$, we get $\exp(H) = p$. Therefore, all proper nonabelian subgroups of G have exponent p. Then, by Theorem A.125.1, if $\exp(G) > p$, then the abelian subgroup $A = \mathrm{H}_p(G)$ has index p in G, $|A : \Omega_1(A)| = p$ (note that $A \cap H = \Omega_1(A)$ has index p in A) so $A = X\Omega_1(A)$, where $X < A$ is cyclic of order p^2. It follows that $\Omega_1(A) = \Phi(G)$. We have $\mho_1(A) = \mho_1(X) \le Z(G)$ so $\mho_1(X)$ is contained in all maximal abelian subgroups of G. Let $B \ne A$ be a maximal abelian subgroup of G. Then $B = \langle B - A\rangle \le \Omega_1(B)$ so that $\exp(B) = p$. As $X \cap B = \mho_1(X) > \{1\}$ and $X \nleq B$, the subgroup B is not isolated in G, a contradiction. Thus, $\exp(G) = p$. □

There are in the book some different proofs of Theorem A.125.2 (see, for example, §§ 251 and 254).

Problem 1. Describe the irregular p-groups all of whose nonabelian regular subgroups have exponent p.

Problem 2. Classify the non-Dedekindian p-groups all of whose nonnormal subgroups have exponent p.

Problem 3. Classify the non-Dedekindian p-groups all of whose $\mathcal{A}_2$-subgroups have exponent p.

Problem 4. Let G be a p-group of exponent $> p^k$, $k > 1$, which is neither abelian nor minimal nonabelian. Suppose that all nonabelian subgroups of G have exponent $\le p^k$. Study the structure of G.

Appendix 126 On p-groups with abelian automorphism groups

Below we report without proofs some results on the structure of nonabelian p-groups whose automorphism groups are abelian. Here we use a comprehensive survey [KY] devoted to this subject. The first p-group with an abelian automorphism group was presented by G. A. Miller. Now there are a number of papers devoted to this topic. Following [KY], we call such nonabelian groups *Miller p-groups*. In what follows, G denotes a Miller p-group. Obviously, a Miller p-group has class 2 since $\mathrm{Inn}(G) \cong G/\mathrm{Z}(G)$, being a subgroup of the abelian group $\mathrm{Aut}(G)$, is abelian. Below we state a number of results on the structure of Miller p-groups stated and commented in [KY].

Theorem A.126.1 ([Hop2]). *Every automorphism of a Miller p-group centralizes G'.*

It follows from Theorem A.126.1 that the derived subgroup of G is contained in the center of the holomorph of G.

Theorem A.126.2 ([Hop2]). *A Miller p-group, $p > 2$, is not a direct product of two nonidentity groups.*

As Theorem A.126.3 shows, a Miller 2-group can have nontrivial abelian direct factors.

Theorem A.126.3 ([Ear]). *A p-group $G = N \times A$, where $A > \{1\}$ is abelian and N has no nontrivial direct factor, is a Miller p-group if and only if $p = 2$ and*
(a) *A is cyclic of order > 2,*
(b) *N is a special Miller group.*

Theorem A.126.4 ([Hop2]). *If G is a Miller p-group, then $\mathrm{Aut}(G)$ is a p-group.*

Theorem A.126.5 ([Ear]). *If G is a Miller p-group, then:*
(a) $\exp(G) > p$.
(b) *If $p > 2$, then $\mathrm{Z}(G) \cap \Phi(G)$ is noncyclic.*

Theorem A.126.6 ([KY]). *If G is a Miller p-group, $p > 2$, then G' possesses at least two cyclic factors of maximal order in the direct product decomposition.*

The analog of Theorem A.126.6 holds for $p = 2$ except $|G'| = 2$.

Theorem A.126.7. *Let G be a Miller p-group. Then:*
(a) $\mathrm{d}(G) \geq 3$ ([Ear]).
(b) *If $p > 2$, then* $\mathrm{d}(G) \geq 4$ (Morigi).

Thus, if G is a nonabelian two-generator p-group, then $\mathrm{Aut}(G)$ is nonabelian.

Theorem A.126.8. *Let G be a Miller p-group. Then:*
(a) $|G| \geq p^6$ ([Ear]).
(b) *If $p > 2$, then $|G| \geq p^7$ and this estimate is attained* ([Mori4]).

https://doi.org/10.1515/9783110533149-087

A special Miller p-group, $p > 2$, of order p^7 with elementary abelian automorphism group of order p^{12} is presented in [Mor4].

Theorem A.126.9 ([H, BY]). *If G is a Miller p-group, p* > 2, *then* $|\mathrm{Aut}(G)| \geq p^{12}$.

Theorem A.126.10 ([JRY]). *If G is a Miller p-group, p* > 2, *with elementary abelian automorphism group* $\mathrm{Aut}(G)$, *then* $\Phi(G)$ *is elementary abelian and one of the following holds:*
(a) $Z(G) = \Phi(G)$.
(b) $G' = \Phi(G)$.
Moreover, there is G in which exactly one of conditions (a) *and* (b) *holds.*

Recall that a group G is said to be *Camina group*, if the coset xG' is a G-class for all $x \in G - G'$.

Theorem A.126.11 ([KY]). *If a p-group G is simultaneously a Miller group and a Camina group, then all automorphisms of G are class-preserving. In particular, all normal subgroups of G are characteristic.*

Problem 1 ([KY]). Can a finite 2-group G with $|G'| = 2$ be a Miller p-group?

Problem 2 ([KY]). Does there exist a Miller p-group in which $Z(G)$ and $\Phi(G)$ are non-incident?

Problem 3. Study the p-groups G such that a Sylow p-subgroup of $\mathrm{Aut}(G)$ is abelian.

Problem 4. Study the p-groups G such that $\mathrm{Aut}(G)$ has an abelian subgroup of index p.

Problem 5. Study the holomorphs of Miller p-groups.

Appendix 127 Alternate proof of Proposition 1.23

Donald Passman [Pas1] has proved the following result playing important role in the book.

Proposition A.127.1 (Proposition 1.23). *If a p-group G is non-Dedekindian, then there exists a G-invariant subgroup K of index p in G' such that G/K is not Dedekindian.*

For $p > 2$ or $|G'| = 2$ this is trivial.

Proof. Assume that G is a counterexample of minimal order. Without loss of generality one may assume that $|G'| = p^2$. Take $K \leq G' \cap Z(G)$ of order p. As G/K is nonabelian Dedekindian, one has $p = 2$. Let $Q/K \cong Q_8$ be a (normal) subgroup of G/K (see Theorem 1.20); then $|Q| = 2^4$. As G/Q is elementary abelian, we get $G' < Q$. Assume that $Q' = G'$. Then $|Q : Q'| = 4$ so, by Proposition 1.6, Q is of maximal class. In that case, $Q/K \cong D_8$, contrary to the assumption. Thus, we get $|Q'| = 2$ which implies that $Q' \neq K$ (indeed, Q/K is nonabelian). In that case, $G' = K \times Q' \leq Z(G) = \Omega_1(Q)$. If $L < G'$ is of order 2, then G/L is (nonabelian) Dedekindian (nonabelian since $|G'/L| = 2$). Then, if $H < G$ is nonnormal of minimal order, we must have $H \cap G' = \{1\}$. Therefore, since all maximal subgroups of H are G-invariant and so do not generate H, it is cyclic. Then $H \times K$ and $H \times Q'$ are distinct G-invariant hence $H = (H \times K) \cap (H \times Q') \lhd G$, contrary to a choice of H. Thus, H does not exist so G is Dedekindian, a contradiction since $|G'| > 2$ (see Theorem 1.20). Thus, for some G-invariant subgroup K of index 2 in G' the quotient group G/K must be non-Dedekindian. □

https://doi.org/10.1515/9783110533149-088

Appendix 128 Alternate proof of the theorem of Passman on p-groups all of whose subgroups of order $\leq p^s$ ($s \geq 1$ is fixed) are normal

Theorem 1.20 describes the so-called Dedekindian p-groups, i.e., p-groups all of whose subgroups are normal. It is interesting to describe the p-groups in which only some subgroups are normal. Such p-groups can be considered as generalizations of Dedekindian p-groups. In this section we consider one of such sets of p-groups.

Donald Passman [Pas, Proposition 1.6] has proved the following nice result.

Theorem A.128.1. *Let $s \geq 1$ be a fixed positive integer. If all subgroups of order $\leq p^s$ of a non-Dedekindian p-group G are G-invariant, then $\Omega_s(G) \leq \mathrm{Z}(G)$.*

Here we present an alternate proof of this theorem, based on properties of minimal nonabelian subgroups which are important in the offered proof. Note that the theorem is not true if G is nonabelian Dedekindian 2-group and $s = 2$.

Proof. We proceed by induction on $|G|$ assuming that G is a counterexample of minimal order. One may assume that $s > 1$. A p-group, satisfying the hypothesis, is called a D_s-group. The property D_s is inherited by subgroups. Therefore, if $H < G$, then, by induction, either $\Omega_s(H) \leq \mathrm{Z}(H)$ or H is nonabelian Dedekindian and $s = 2$. All subgroups of exponent $\leq p^s$ are G-invariant since they are generated by cyclic subgroups of orders $\leq p^s$. Any subgroup of G of exponent $\leq p^s$ is Dedekindian. By assumption, $\exp(G) > p^s$.

(i) Let $p = 2$ and assume that there is in G a subgroup $Q \cong \mathrm{Q}_8$. As $\exp(Q) = 2^2 < 2^s$, it follows that $Q \triangleleft G$. Set $C = \mathrm{C}_G(Q)$. Then G/C is a subgroup of $\mathrm{D}_8 \in \mathrm{Syl}_2(\mathrm{Aut}(\mathrm{Q}_8))$. If C possesses a cyclic subgroup L of order 4, then the subgroup $Q * L$ of order 2^4 or 2^5 (by the product formula) is non-Dedekindian since it has a section isomorphic to D_8, a contradiction (indeed, $\exp(Q * L) = 2^2 < 2^s$). Thus, $\exp(C) = 2$ and $C \leq \mathrm{Z}(G)$ since all elements of C lie in $\mathrm{Z}(G)$, by hypothesis. Then $|G/C| > 2$ (otherwise, $\exp(G) = 4$) and G/C is noncyclic since G is nonabelian. If $G/C \cong \mathrm{D}_8$ and $B/C < G/C$ is nonnormal of order 2, then B is non-G-invariant of exponent ≤ 4, a contradiction. Thus, $|G/C| \cong \mathrm{E}_4$, $G = C * Q$. If $C = \mathrm{Z}(Q) \times E$, then $G = Q \times E$ is Dedekindian, contrary to the hypothesis.

Thus, G has no subgroup isomorphic to Q_8.

(ii) Now we omit the restriction $p = 2$. In that case, if $H < G$, then $\Omega_s(H) \leq \mathrm{Z}(H)$, by induction (it is important that G has no subgroup isomorphic to Q_8). Let A be a normal abelian subgroup of exponent $\leq p^s$ of maximal possible order. Let $A < B \leq G$, where $B = \langle x, A\rangle$ for some $x \in G - A$. Assume that the quotient group G/A is noncyclic; then $B < G$ and $A \leq \Omega_s(B)$ is contained in $\mathrm{Z}(B)$, by induction, so B is abelian. As x is an

https://doi.org/10.1515/9783110533149-089

arbitrary element of the set $G - A$, it follows that $A \leq Z(G)$ since $\langle G - A\rangle = G$. Therefore, G/A is cyclic. If $y \in G - A$ is of order $\leq p^s$ such that $y^p \in A$, then the subgroup $D = \langle y, A\rangle$ is abelian. By Theorem 10.1, $A < D \lhd G$, where $|D| = p|A| > |A|$ and D is abelian of exponent $\leq p^s$, contrary to the choice of A. Thus, $\Omega_s(G) = A$ is abelian. By Lemma 57.1, for any $t \in G - A$ there is an $a \in A$ such that $T = \langle a, t\rangle$ is minimal nonabelian. Then $\exp(T) > p^s$ since $t \notin A = \Omega_s(G)$, and since $\Omega_s(T) = A \cap T$ is such that $T/(A \cap T)$ is cyclic, one has $A \cap T = \Omega_s(T) \nleq Z(T)$ (otherwise, T will be abelian), hence the subgroup T is not a D_s-group. In that case, by induction, $T = G$, i.e., G is minimal nonabelian. Let $x \in G - A$ be such that $G = \langle x\rangle A$; then $o(x) = p^n \geq p^{s+1}$, and set $X = \langle x\rangle$. By hypothesis, X normalizes all subgroups of A but there is a cyclic $L \leq A$ such that X does not centralize $L = \langle t\rangle$ (otherwise, $G = XA$ will be abelian); then XL is a nonabelian subgroup. As $L \nleq Z(XL)$ and $|L| \leq \exp(A) = p^s$, the subgroup XL is not a D_s-group. Therefore, by induction, $G = XL$. As G is a metacyclic minimal nonabelian p-group, one may assume that $X \cap L = \{1\}$ (Lemma 65.1). Let $|L| = p^k (\leq \exp(A) = p^s)$ and set $y = x^{p^{n-k}}$; then $o(y) = p^k \leq p^s$ so that $Y = \langle y\rangle \lhd G$. In that case, $G = XY = X \times Y$ is abelian, a final contradiction. □

Below we present a variant of the proof of Theorem A.128.1.

Theorem A.128.2 (Theorem 128.1 = Theorem298.1). *If all subgroups of order $\leq p^i$ are normal in a non-Dedekindian p-group G, then $\Omega_i(G) \leq Z(G)$.*

Proof. If $\exp(G) \leq p^i$, then our group G is abelian, a contradiction. Next we assume that $\exp(G) > p^i$.

(a) We have $\exp(\Omega_i(G)) \leq p^i$, which will be proved by induction on i. This is true for $i = 1$. Suppose that $\exp(\Omega_{i-1}(G)) = p^{i-1}$. Let $x \in G$ be of order p^i. Set $X = \langle x\rangle$; then $X \lhd G$ so that $\overline{X} \lhd \overline{G}$, where $\overline{G} = G/\Omega_{i-1}(G)$. Let $\overline{Y} \leq \overline{G}$ be of order p. Take $y \in Y - \Omega_{i-1}(G)$; then $o(y) = p^i$ and therefore $\overline{Y} \leq Z(\overline{G})$. It follows that $\Omega_1(\overline{G}) \leq Z(\overline{G})$ so that $\Omega_i(G)$, the inverse image of $\Omega_1(\overline{G}))$ in G, is a normal subgroup of exponent p^i in G.

Next we assume that $\exp(G) > p^i$.

(b) If Q_8 is not involved in G, then $\Omega_i(G) \leq Z(G)$. Indeed, the subgroup $\Omega_i(G)$ is Dedekindian since all its subgroups of order $\leq p^i$ so all its subgroups are normal (here we use (a)). As $\Omega_i(G)$ has no subgroup isomorphic to Q_8, it is abelian (Theorem 1.20).

It remains to prove that $\Omega_i(G) \leq Z(G)$. We prove this by induction on $|G|$. Let $\Omega_i(G) \leq H \in \Gamma_1$. Then, by induction, $\Omega_i(G) \leq Z(H)$ since H has no subgroup isomorphic to Q_8. If $F \in \Gamma_1 - \{H\}$ and $\Omega_i(G) \leq F$, then $\Omega_i(G) \leq Z(F)$, by induction. Then one has $C_G(\Omega_i(G)) \geq HF = G$ so that $\Omega_i(G) \leq Z(G)$. Therefore, one may assume that $\Omega_i(G)$ is contained in only one member of the set Γ_1. It follows that the quotient group $G/\Omega_i(G)$ is cyclic. Let $x \in G - \Omega_i(G)$ be such that the subgroup $\langle x\rangle$ covers $G/\Omega_i(G)$. Then there is $a \in \Omega_i(G)$ such that the subgroup $S = \langle a, x\rangle$ is minimal nonabelian. By induction, $S \cap \Omega_i(G) = \Omega_i(S) \leq Z(S)$. Since $S/(S \cap \Omega_i(G)) \cong G/\Omega_i(G)$ is cyclic, it follows that S is abelian, a contradiction. Thus, $\Omega_i(G) \leq Z(G)$ in that case, as required.

(c) Now suppose that Q_8 is involved in G; then $p = 2$. In that case, $\Omega_i(G)$ is nonabelian Dedekindian (recall that $i > 1$) and let $Q \cong Q_8$ be its subgroup; then $Q \triangleleft G$, by hypothesis. Since G is non-Dedekindian, we get $\exp(G) > 2^2$. Therefore there exists an element $z \in G - \Omega_2(G)$ of order 2^3. Set $L = QZ$, where $Z = \langle z \rangle$. As $Z \cap Q \cong C_4$ and all cyclic subgroups of order 4 in $\Omega_2(L)$ contain $Z(Q)$ (see Theorem 1.20), we get $Z(Q) < Z$. Assume that $|Q \cap Z| = 4$; then $|L : Q| = 2$, by the product formula, and so L is of maximal class, by Theorem 1.2. In that case, L has a nonnormal subgroup of order $2^2 = 2^s$, a contradiction. Now let $Q \cap Z = Z(Q)$ be of order 2. Since L is not of maximal class, we get $C_L(Q) \nleq Q$ (Proposition 10.17) so $C_L(Q)$ contains a cyclic subgroup $U \cong C_4$ (one may take $U < Z$). The subgroup $M = QU = Q * U$ has order 2^4. This subgroup M contains a subgroup $D \cong D_8$. This is a contradiction since D has a nonnormal subgroup of order $2 < 2^2 = 2^i$. □

Exercise 1. Let G be a two-generator nonabelian group of order p^n. If all subgroups of order $\leq p^{n-2}$ are G-invariant, then $G \cong Q_8$.

Hint. The group G is Dedekindian. Use Theorem 1.20. □

Exercise 2. Classify the nonabelian groups of order p^n all of whose subgroups of order $\leq p^{n-3}$ are G-invariant.

Hint. If G is non-Dedekindian, it possesses a cyclic subgroup of index p^2. □

Exercise 3. Classify the non-Dedekindian p-groups G such that, whenever $S \leq G$ is minimal nonabelian, then all subgroups of S are G-invariant.

Hint. Use Corollary A.17.3. □

Problem 1. Study the normal structure of the automorphism group of a group from Theorem A.128.1.

Problem 2. Given a positive integer n, study the p-groups all of whose subgroups of order $\leq p^n$ are quasinormal.

Appendix 129 Alternate proofs of Theorems 309.1 and 309.2 on minimal non-p-central p-groups

Recall that a p-group G is p-central if, for $p > 2$ one has $\Omega_1(G) \leq Z(G)$ and $\Omega_2(G) \leq Z(G)$ for $p = 2$. A non-p-central p-group X all of whose proper subgroups are p-central is said to be minimal non-p-central. A minimal nonabelian p-group S is minimal non-p-central if and only if

$$S \in \{M_{p^n}, n > 1; M_p(n, 1, 1), n \geq 1\} \text{ for } p > 2\} \tag{1}$$

and

$$S \in \{D_8; Q_8; M_{2^n}, n > 1; M_2(2, n), n > 1; M_2(n, 2), n > 2; M_2(2, n, 1)\}. \tag{2}$$

For another proof of Theorem A.129.1 see Theorem 309.1. Note that for $p = 2$ the conclusions of Theorems A.129.2 and 309.2 are different: in the first theorem G is minimal nonabelian, in the second one G has a more complicated structure.

Theorem A.129.1. *Suppose that a p-group G, $p > 2$, is minimal non-p-central. Then $G \in \{M_{p^n}, M_p(n, 1, 1), n > 1\}$, i.e., G is minimal nonabelian.*

Proof. We proceed by induction on $|G|$. If $\exp(G) = p$, all proper subgroups of G are abelian, by hypothesis, and so G is minimal nonabelian hence $G \cong S(p^3) = M_1(1, 1, 1)$.

Next we assume that $\exp(G) > p$. Let $E < G$ be a G-invariant abelian subgroup of exponent p of maximal order; then $E < G$. Assume that there is $x \in G - E$ of order p and set $H = \langle x\rangle E$. By Theorem 10.1 and by the choice of E, $H = \Omega_1(H)$ is nonabelian. By Lemma 57.1, there is $a \in E$ such that the subgroup $S = \langle a, x\rangle$ is minimal nonabelian, and $S \cong S(p^3)$ since $\Omega_1(S) = S$. As S is non-p-central, we get $S = G$.

Now assume that $E = \Omega_1(G)$; then $E < G$. As $E \nleq Z(G)$ (otherwise, G is p-central), there is $y \in G - E$ non-centralizing E. Set $F = \langle y\rangle E$; then F is nonabelian. As F/E is cyclic, we get $E = \Omega_1(F) \nleq Z(F)$, hence the subgroup F is noncentral, and we conclude that $G = F$. Let $E \leq A < G$, where $A \in \Gamma_1$. As A is p-central, $E \leq Z(A)$ and, since A/E is cyclic as a subgroup of G/E, it follows that A is abelian. Therefore, by Lemma 57.1, there is $z \in E$ such that $T = \langle y, z\rangle$ is minimal nonabelian. Since T is non-p-central (indeed, the element z of order p is not contained in $Z(T)$), we get, by induction, that $T = G$, i.e., G is minimal nonabelian so it is as in (1). □

The proof of the following theorem essentially repeat the proof of Theorem A.129.1.

Theorem A.129.2. *If a 2-group G is minimal non-2-central, then it is minimal nonabelian.*

Proof. We proceed by induction on $|G|$ assuming that G is a minimal counterexample. Let $E \leq G$ be a normal subgroup of exponent 2^2 of maximal order (such an E exists

https://doi.org/10.1515/9783110533149-090

since $\exp(G) > 2$). If $E = G$ (or, what is the same, $\exp(G) = 2^2$), then all proper subgroups of G are 2-central so abelian, by hypothesis, so that G is minimal nonabelian.

Next we assume that $\exp(G) > 2^2$.

Thus, $E < \Omega_2(G)$. In that case, being 2-central, $E < G$ is abelian. Assume that there is $x \in G - E$ of order ≤ 4 and set $H = \langle x\rangle E$. By Theorem 10.1 and by the choice of E, $H = \Omega_2(H)$ is nonabelian so non-2-central since $o(x) \leq 4$ and $x \notin H$. Therefore, $H = G$, by induction. By Lemma 57.1, there is $a \in E$ such that the subgroup $S = \langle a, x\rangle$ is minimal nonabelian. Then S is not 2-central since $x \notin Z(S)$ and $o(x) \leq 4$. Therefore, we have $S = G$.

Now assume that $E = \Omega_2(G)$; then $E < G$ since $\exp(G) \geq 2^3 > \exp(E)$. As $E \nleq Z(G)$ (otherwise, G is 2-central, by definition), there is $y \in G - E$ that does not centralize E (one has $o(y) \geq 2^3$). Set $F = \langle y\rangle E$. As F/E is cyclic, we have $E = \Omega_2(F) \nleq Z(F)$, and we conclude that the subgroup F is non-2-central so that $G = F$. In that case, the quotient group G/E is cyclic. Let $E \leq A \in \Gamma_1$. As A is 2-central, $E = \Omega_2(A) \leq Z(A)$ and, since A/E as a subgroup of G/E is cyclic, we conclude that A is abelian. Therefore, by Theorem 285.1, there is $z \in E$ such that $T = \langle y, z\rangle$ is minimal nonabelian. Since T is not 2-central (indeed, the element z of order $\leq 2^2$ is not contained in $Z(T)$), we get $T = G$, i.e., G is minimal nonabelian. It follows from Lemma 65.1 (see also the paragraph preceding Theorem A.129.1) that G is as in (2). □

For a shorter proof see Appendix 132.

Exercise 1. Classify the p-groups G all of whose subgroups of index p^2 are p-central.

Hint. All maximal subgroups of G are either Dedekindian or minimal nonabelian, by Theorems A.129.1 and A.129.2. See § 71 in the case where all Dedekindian subgroups of G are abelian (then G is an $\mathcal{A}_2$-group; this is a case provided $p > 2$). □

Exercise 2. Study the irregular p-groups all of whose maximal regular subgroups, except one, are p-central.

Problem 1. Study the p-groups all of whose maximal subgroups, except one, are p-central.

Problem 2. Study the irregular p-groups all of whose maximal regular subgroups, except one, are p-central.

Appendix 130 Non-Dedekindian p-groups all of whose nonnormal maximal cyclic subgroups are conjugate

We obtain in the following theorem a nontrivial additional information on p-groups all of whose nonnormal maximal cyclic subgroups are conjugate. The group M_{p^n} satisfies the above condition.

Theorem A.130.1 (Janko). *Let G be a non-Dedekindian p-group all of whose nonnormal maximal cyclic subgroups are conjugate. Then* $\mathrm{cl}(G) = 2$, *G' is cyclic and* $\mathrm{d}(G) = 2$.

Proof. Let G be a non-Dedekindian p-group all of whose nonnormal maximal cyclic subgroups are conjugate. Let Z be a nonnormal maximal cyclic subgroup in G. Set $F = Z\Phi(G)(< G)$ so that $F \trianglelefteq G$ and $|F : \Phi(G)| \leq p$. Then $Z^G \leq F$ and G/F is elementary abelian. We have $Z^G < G$. Each cyclic subgroup X in G which is not contained in Z^G must be normal in G. Let $M < G$ be nonnormal and assume that $M \nleq Z^G$. Then all cyclic subgroups of M not contained in $M \cap Z^G$, are G-invariant. Since these subgroups generate M, we get $M \triangleleft G$, contrary to the choice of M. Thus, all nonnormal subgroups of G are contained in Z^G and therefore $G_0 = Z^G$ is the subgroup in G generated by all nonnormal subgroups of G. By the above, we have $G_0 = Z^G < G$ and so we may use Theorem 231.1.

By Theorem 231.1, $G/G_0 = G/Z^G$ is cyclic. Since $G/F = G/Z^G\Phi(G)$ is elementary abelian, it follows that $|G : F| = p$ (note that $F < G$), and we conclude that $\Phi(G) < F$. Then $|F : \Phi(G)| = p$ (see the previous paragraph on the definition of F), and we get

$$|G : \Phi(G)| = |G : F||F : \Phi(G)| = p^2 \implies \mathrm{d}(G)| = 2.$$

Also, Theorem 231.1 gives that $\mathrm{cl}(G) = 2$ and G' is cyclic. The proof is complete.[1] □

Exercise 1. Let G be as in Theorem A.130.1. Give an independent proof that $\exp(G) > p$.

Solution. Assume, by way of contradiction, that $\exp(G) = p$. Let $Z < G$ be nonnormal cyclic and $Z^G = G_0$; then $G_0 < G$. Take $x \in G - G_0$ and write $X = \langle x \rangle$; then $X \triangleleft G$. All such X generate G which implies $G = \mathrm{Z}(G)$, a contradiction. □

Exercise 2. Is the following assertion true? If all nonnormal cyclic subgroups of a minimal nonabelian p-group G are conjugate, then $G \cong M_{p^n}$.

Hint. By Lemma 65.1, G is metacyclic. The number of nonnormal cyclic subgroups equals p. □

1 We have $G' < G_0$ and hence $G_0 = Z^G \leq G'Z \leq G_0$ implies $G_0 = G'Z$. Since $G' \leq \mathrm{Z}(G)$, the subgroup $G'Z = G_0$ is abelian.

https://doi.org/10.1515/9783110533149-091

Exercise 3. If G is as in Theorem A.130.1, then any its minimal nonabelian subgroup is metacyclic.

Solution. Assume that this is false. Let $S \leq G$ be nonmetacyclic minimal nonabelian. In that case, there are in S two nonnormal cyclic subgroups U, V such that $U \cap V = \{1\}$ and $\langle U, V\rangle = S$ (Theorem 65.1). Note that $G' \leq Z(G)$ is cyclic (Theorem A.130.1), $G' \cap U = \{1\} = G' \cap V$ since $S' = \Omega_1(G')$, $S' \cap U = \{1\} = S' \cap V$ and hence we have $G' \cap U = \{1\} = G' \cap V$ and $G' \leq Z(G)$ is cyclic (Theorem A.130.1). Therefore $U \times G' \lhd G$ does not contain V, which implies that U and V are not conjugate in G, which is a contradiction. Thus, S is metacyclic. □

Exercise 4. Is it true that if G is as in Theorem A.130.1, so is any its non-Dedekindian epimorphic image?

Exercise 5. If all nonnormal cyclic subgroups are conjugate in a p-group G, then $\Omega_1(G)$ is elementary abelian.

Hint. Let $E \lhd G$ be elementary abelian of maximal order; then all subgroups of G of order p that are not contained in E are not G-invariant (this follows from the maximal choice of E). Let $x \in G - E$ be of order p. By the choice of E the subgroup $T = \langle x, E\rangle$ is nonabelian (in that case, $|E| > p$). Then there is $a \in E$ such that $S = \langle a, x\rangle$ is minimal nonabelian (Lemma 57.1) so, since $\Omega_1(S) = S$, we get $S \in \{D_8, S(p^3)\}$ (Lemma 65.1). In that case, there is in the set $S - (S \cap E)$ a nonnormal cyclic subgroup of order p, a contradiction. Thus, $E = \Omega_1(G)$, and we are done. □

Problem 1. Study the p-groups G with exactly two conjugate classes of nonnormal cyclic (maximal cyclic) subgroups.

Problem 2. Study the p-groups G all of whose nonnormal cyclic subgroups have the same order.

Problem 3. Study the p-groups G such that, whenever $H \in \Gamma_1$, then all non-G-invariant maximal abelian subgroups of H are conjugate in G.

Problem 4. Study the p-groups all of whose nonnormal maximal abelian subgroups are conjugate. (If a p-group G, $p > 2$, is irregular, it has a nonnormal maximal abelian subgroup.)

Problem 5. Classify the p-groups possessing at most $p + 1$ conjugate classes of minimal nonabelian subgroups. (Note that there is no a nonabelian p-group containing less than $p + 1$ conjugate classes of maximal abelian subgroups.)

Problem 6. Study a p-group that is a union of $p + 1$ centralizers of noncentral elements.

Appendix 131 A characterization of some 3-groups of maximal class

In this section we prove the following result.

Theorem A.131.1. *Suppose that a p-group G is nonmetacyclic, $p > 2$. If, whenever $H \le G$ is nonmetacyclic, then any cyclic subgroup of H is contained in a minimal nonmetacyclic subgroup of H. Then one and only one of the following holds:*
(a) $\exp(G) = p$.
(b) *G is minimal nonmetacyclic group of order 3^4.*

Let a p-group G be as in the statement of the theorem. In both cases G is covered by minimal nonmetacyclic subgroups. If G is of order p^3, then $G \in \{E_{p^3}, S(p^3)\}$. If G is of order p^4 and exponent $> p$, then it is of maximal class of order 3^4 and $\Omega_1(G) \cong E_9$. Moreover, if $\exp(G) > p$, then $\exp(G) = 3^2$ (Theorem 69.1). Recall that if X is a minimal nonmetacyclic p-group, $p > 2$, then X is either of order p^3 and exponent p or the 3-group of maximal class mentioned above (Theorem 69.1).

Proof. Clearly, if $\exp(G) = p$, then G is covered by subgroups of order p^3. Therefore, it remains to consider the case when $\exp(G) > p$.

Assume that $\exp(G) = p^e > p$. Let $C < G$ be cyclic such that $|C| = p^e$. Then $C < M \le G$, where M is minimal nonmetacyclic of exponent $> p$. By the classification of minimal nonmetacyclic p-groups, M is a group of maximal class and order 3^4 and therefore $p^e = \exp(M) = 3^2$ (Theorem 69.1) so that $p = 3$ and $e = 2$. Assume, by way of contradiction, that $M < G$. Without loss of generality, one may assume that $|G : M| = 3$; then $|G| = 3^5$. In that case, the set Γ_1 has no member of exponent 3 (if such a subgroup exists, let us consider its intersection with M).

(i) Suppose that G is of maximal class (of order 3^5) and let G_1 be its fundamental subgroup (see § 9). By Theorems 9.6 (a) and 9.7, G_1 is metacyclic of exponent 3^2 and all proper subgroups of G_1 are abelian. All members of the set $\Gamma_1 - \{G_1\}$ must be minimal nonmetacyclic. It follows that G is of maximal class and order 3^5 satisfying $\Omega_1(G) = E_{3^2}$. Next we assume that G is not of maximal class.

(ii) By Theorem 12.12 (a), we have $d(G) = 3$ so that

$$\Phi(G) = \Phi(M) = G' \cong E_{3^2}.$$

In view of Theorem 13.7, there is in G a normal subgroup $E \cong E_{3^3}$; then, by the product formula, $G = ME$ and $E \cap M \cong E_{3^2}$. By Theorem 12.12 (b), we have $\exp(G/\mho_1(G)) = 3$ and $|G/\mho_1(G)| = 3^4$. Since $d(G) = 3$, it follows that $\overline{G} = G/\mho_1(G)$ is nonabelian. Let $E < H \in \Gamma_1$. By Theorem 69.1, H is not a minimal nonmetacyclic subgroup. Considering the intersection $M \cap H$, we conclude that $\exp(H) = 3^2$. Therefore, any cyclic subgroup of H of order 3^2 is not contained in a minimal nonmetacyclic subgroup of H since H has no minimal nonmetacyclic subgroup of exponent 3^2, contrary to the hypothesis. □

https://doi.org/10.1515/9783110533149-092

The following problem generalizes the situation considered in Theorem A.131.1.

Problem 1. Classify the nonmetacyclic p-groups G such that, whenever $C < G$ is cyclic, then C is contained in a minimal nonmetacyclic subgroup of G. (In other words, G is covered by minimal nonmetacyclic subgroups.)

Even for $p > 2$ this problem is fairly difficult.

Problem 2 (Old problem). Classify the nonabelian p-groups G such that, whenever $C < G$ is cyclic, then C is contained in a minimal nonabelian subgroup of G. (This coincides with Problem 860.)

Problem 3. Classify the p-groups G such that, whenever $C < G$ is cyclic, then $C < M \leq G$, where M is a subgroup of maximal class.

Appendix 132 Alternate approach to classification of minimal non-p-central p-groups

For a prime p, set $\epsilon(p) = 1$ if $p > 2$ and $\epsilon(2) = 2$. A p-group G is termed p-central if $\Omega_{\epsilon(p)}(G) \le Z(G)$. The minimal non-$p$-central p-groups are described in Theorems 309.1 (for $p > 2$) and 309.2 (for $p = 2$). In this appendix we describe, up to isomorphism, minimal non-p-central p-groups. Note that if a 2-group G is minimal non-2-central, then $\exp(G) > 2$ since it is nonabelian.

Theorem A.132.1. *A p-group G is minimal non-p-central if and only if G is minimal nonabelian with $\Omega_{\epsilon(p)}(G) \nleq Z(G)$.*

Proof. We consider the cases $p > 2$ and $p = 2$ simultaneously. Let $A < G$ be an abelian subgroup of exponent $p^{\epsilon(p)}$ of maximal order and set $N = N_G(A)$. Assume that there is $x \in N - A$ of order $\le p^{\epsilon(p)}$; then $o(x) = p$ if $p > 2$. Obviously, x does not centralize A. In that case, by Lemma 57.1, there is $a \in A$ such that the subgroup $B = \langle a, x\rangle$ is minimal nonabelian. The subgroup B is not p-central since $\Omega_{\epsilon(p)}(B) = B$ is nonabelian (indeed, a of order $\le p^{\epsilon(p)}$ is not contained in $Z(B)$). It follows that $B = G$, and we are done in this case. Next we assume that an element x with the above properties does not exist. Then it follows that $\Omega_{\epsilon(p)}(N) = A$ so that A is characteristic in N, and we conclude that $N = G$; then $A \trianglelefteq G$. As $\Omega_{\epsilon(p)}(G) = A \nleq Z(G)$ (otherwise, G is p-central), the group G is non-p-central. Therefore, there is $y \in G - A$ which does not centralize A. In that case, there is $a_0 \in A$ such that the subgroup $S = \langle a_0, y\rangle$ is minimal nonabelian (Lemma 57.1 again). Since the element $a_0(\in A)$ of order $\le \exp(A) \le p^{\epsilon(p)}$ is not contained in $Z(S)$, the subgroup S is not p-central, and we conclude that $S = G$. Thus, in any case our group G is a non-p-central minimal nonabelian p-group. □

For another proof see Appendix 129.

It follows from Lemma 65.1 that if a p-group G is minimal non-p-central, then:

(1) If $p = 2$, then

$$G \in \{D_8; Q_8; M_{2^n}, n > 3; M_2(2, n), n > 1; M_2(n, 2), n > 1; M_2(n, 1, 1), n > 1; M_2(n, 2, 1), n > 1\}.$$

(2) If $p > 2$, then

$$G \in \{M_{p^n}, M_p(n, 1, 1)\}.$$

Exercise 1. Classify the p-groups G such that $\Omega_{\epsilon(p)}(G)$ is nonabelian but $\Omega_{\epsilon(p)}(H)$ is abelian for any $H < G$ of exponent $\le p^{\epsilon(p)}$.

Solution. We consider in detail only the case $p > 2$. Let $E \trianglelefteq G$ be a normal elementary abelian of maximal order. By hypothesis, there is $x \in G - E$ of order p (such an x exists since $E < \Omega_{\epsilon(p)}(G)$, by hypothesis). Set $H = \langle x, E\rangle$. By Theorem 10.1, the subgroup H is nonabelian. By Lemma 57.1, there is $a \in E$ such that the subgroup $S = \langle a, x\rangle$ is

https://doi.org/10.1515/9783110533149-093

minimal nonabelian. Since $\Omega_1(S) = S$, it follows that $S \cong \mathrm{S}(p^3)$ (Lemma 65.1). Since $\exp(S) = p$ and S is nonabelian, we get $G = S$. (If $p = 2$, one can take as E a maximal normal abelian subgroup of G of exponent ≤ 4 and, setting $N = \mathrm{N}_G(E)$, argue, as above, with N.) □

Exercise 2. Let G be a p-group such that $|G/\mathrm{Z}(G)| = p^2$. If $S \leq G$ is two-generator nonabelian, then it is minimal nonabelian.

Solution. By the product formula, $S\mathrm{Z}(G) = G$ so that $S/(S \cap \mathrm{Z}(G)) \cong G/\mathrm{Z}(G) \cong \mathrm{E}_{p^2}$. It follows that $S \cap \mathrm{Z}(G) = \Phi(S) = \mathrm{Z}(S)$, and so S is minimal nonabelian since $\mathrm{d}(S) = 2$. □

We have used this in the proof of Theorem A.132.1.

Problem 1. Study the p-groups all of whose subgroups of order $\leq p^n$ are quasinormal.

Problem 2. Classify the p-groups all of whose subgroups of index p^2 are p-central.

Problem 3. Study the p-groups all of whose maximal subgroups, except one, are p-central.

Appendix 133 Nonabelian p-groups all of whose minimal nonabelian subgroups are isomorphic to $M_p(n, n)$ or $M_p(n, n, 1)$ for a fixed natural $n > 1$

The minimal nonabelian subgroups have important influence on the structure of finite p-groups in spite of the fact that these subgroups are comparatively small. A number of results proved in the book justify that assertion. Probably, knowledge of all minimal nonabelian subgroups of a p-group G allows us to obtain some information on $Z(G)$.

To prove Theorem A.133.1, we use the following two theorems.

Theorem 285.1. *Let A be a normal abelian subgroup of a nonabelian p-group G such that $A \nleq Z(G)$. Then for any $x \in G - C_G(A)$ there is $a \in A$ such that the subgroup $\langle a, x\rangle$ is minimal nonabelian.*

Theorem 10.1. *Let $A < B \leq G$, where A and B are abelian subgroups of a p-group G, $|B : A| = p$, $\exp(B) = p^n$ and $p^n > 2$. Let $\mathcal{A}$ be the set of all abelian subgroups T in G such that $A < T$, $|T : A| = p$ and $\exp(T) \leq p^n$. Then $|\mathcal{A}| \equiv 1 \pmod p$. In particular, if $A \triangleleft G$, then the set $\mathcal{A}$ contains a G-invariant member.*

Recall that for $n > 1$,

$$M_p(n, n) = \langle a, b \mid o(a) = o(b) = p^n,\ a^b = a^{1+p^{n-1}}\rangle$$

and

$$M_p(n, n, 1) = \langle a, b, c \mid o(a) = o(b) = p^n,\ c = [a, b],\ o(c) = p,\ [a, c] = [b, c] = 1\rangle.$$

The group $M_p(n.n)$ is metacyclic of order p^{2n}, The group $M_p(n.n, 1)$ is nonmetacyclic of order p^{2n+1}. Both these groups have exponent p^n.

We will use the following assertion:

(∗) Any system of two generators of the minimal nonabelian groups $M_p(n, n)$ and $M_p(n, n, 1)$ consists of elements of order p^n.

Indeed, if S is one of the groups from (∗), then $\Omega_{n-1}(S) = \Phi(S)$.

Our main result is the following:

Theorem A.133.1. *Suppose that all minimal nonabelian subgroups of a nonabelian p-group G are isomorphic to either the metacyclic minimal nonabelian group $M_p(n, n)$ or the nonmetacyclic minimal nonabelian group $M_p(n, n, 1)$ for a fixed natural $n > 1$. Then the following assertions hold:*

(a) *If $B \triangleleft G$ is maximal abelian, then $\exp(G/B) = p$ and $\exp(B) \geq p^n$.*
(b) *If $A \triangleleft G$ is abelian of exponent $< p^n$, then $A \leq Z(G)$. Next, $\exp(Z(G) \geq p^{n-1}$.*
(c) *If either $p > 2$ or $p = 2$ and $n > 2$, then $\Omega_{n-1}(G) \leq Z(G)$; in particular G is a p-central group.*

https://doi.org/10.1515/9783110533149-094

Proof. (a) Let $B \lhd G$ be a maximal abelian subgroup of G and let $x \in G - B$. Then, by Lemma 57.1, there is $a \in B$ such that the subgroup $S = \langle a, x\rangle$ is minimal nonabelian so it is isomorphic to either $M_p(n, n)$ or $M_p(n, n, 1)$. By $(*)$, $o(x) = p^n$. Thus, all the elements of the set $G - B$ have the same order p^n. Assume that $o(yB)$ is an element of order p^2 in the quotient group G/B. By the above, $o(y) = p^n$. Since the element y^p of order $p^{n-1} < p^n$ is also contained in the set $G - B$, it must have order p^n, hence we get a contradiction. Thus, $o(yB) = p$ for all $y \in G - B$ so that $\exp(G/B) = p$. As the above chosen element $a \in B$ has order p^n, by $(*)$, we get $\exp(B) \geq p^n$ and $\exp(G/B) = p$, completing the proof of part (a). (Generally speaking, it is possible that $\exp(B) > p^n$.)

(b) Let $A \lhd G$ be abelian of exponent $< p^n$ and let $A < B \lhd G$, where B is a maximal abelian subgroup of G (by (a), $B > A$ since $\exp(B) \geq p^n > \exp(A)$). Assume that $A \nleq Z(G)$. Since $\langle G - B\rangle = G$, there is an element $z \in G - B$ that does not centralize A. Setting $H = \langle z, A\rangle$, we see that the subgroup H is nonabelian. In that case, by Theorem 285.1, applied to $A < H$, there is $a \in A$ such that the subgroup $T = \langle a, z\rangle$ is minimal nonabelian so that $T = M_p(n, n)$ or $M_p(n, n, 1)$ (in that case, Lemma 57.1 is not applicable since A may not be a maximal abelian subgroup of H). However, $o(a) \leq \exp(A) < p^n$, contrary to assertion $(*)$. Thus, all elements of the set $G - B$ centralize A. Since $\langle G - B\rangle = G$, we get $C_G(A) \geq \langle G - B\rangle = G$ so that $A \leq Z(G)$. It follows from (a) that there is in G a normal abelian subgroup of exponent $\geq p^{n-1}$. Therefore, $\exp(Z(G)) \geq p^{n-1}$. This proves (b).

It remains to prove that $\Omega_{n-1}(G) \leq Z(G)$ if either $p > 2$ or $p = 2$ and $n > 2$. For clarity, we consider these cases together.

(c) Let either $p > 2$ or, if $p = 2$, then $n > 2$. In that case, we prove a stronger assertion that was proved in (a): $\Omega_{n-1}(G) \leq Z(G)$. Assume, however, that $\Omega_{n-1}(G) \nleq Z(G)$. By (a), $\exp(Z(G)) \geq p^{n-1}$ so, if $p = 2$, then $\exp(Z(G)) > 2$, by (b).

Let $v \in G - Z(G)$ be of minimal possible order. If $o(v) \geq p^n$, then $\Omega_{n-1}(G) \leq Z(G)$, contrary to the assumption in the previous paragraph. Thus, we have $o(v) < p^n$. Set $F = \langle v, \Omega_{n-1}(Z(G))\rangle$; then $|F| = p|\Omega_{n-1}(Z(G))|$ and the subgroup F is abelian of exponent p^{n-1}. By Theorem 10.1, there is in G a normal abelian subgroup K of exponent $< p^n$ containing $\Omega_{n-1}(Z(G))$ as a subgroup of index p. Note that $K \nleq Z(G)$ since $\exp(K) = p^{n-1}$ and $|K| = |F| > |\Omega_{n-1}(Z(G))|$. In that case, there is in the set $G - C_G(K)$ an element w; then the subgroup $L = \langle w, K\rangle$ is nonabelian. By Theorem 285.1, applied to $K < L$, there is $b \in K$ such that the subgroup $S = \langle w, b\rangle$ is minimal nonabelian so $S \in \{M_p(n, n), M_p(n, n, 1)\}$, by hypothesis. This contradicts assertion $(*)$ since $o(b) \leq \exp(K) < p^n$. Thus, $\Omega_{n-1}(G) \leq Z(G)$. □

Note that if, in part (c), $n = 2$ for $p = 2$, then Theorem 10.1 is not applicable.

In particular, if G is as in Theorem A.133.1 (c), then $Z(G) \geq Z(S)$ for each minimal nonabelian subgroup $S \leq G$. It follows from Theorem 10.28 that the group $G/\Omega_{n-1}(G)$ is generated by abelian subgroups of type (p, p).

Problem 1. Study the nonabelian p-groups all of whose minimal nonabelian subgroups are metacyclic with noncyclic center.

Problem 2. Study the nonabelian p-groups all of whose minimal nonabelian subgroups have noncyclic centers. (In Theorem 319.1 a partial case of this problem is solved.) Is it true that the center of some minimal nonabelian subgroup of our G is contained in $Z(G)$?

Problem 3. Classify the nonabelian p-groups all of whose minimal nonabelian subgroups, except one, are metacyclic with noncyclic center.

Problem 4. Study the nonabelian p-groups G such that the quotient group $G/K_3(G)$ is minimal nonabelian.

Problem 5. Study the nonabelian groups of exponent p and order $> p^3$ such that $S \cap T > \{1\}$ for any two distinct minimal nonabelian $S, T < G$.

Problem 6. Study the nonabelian p-groups G such that $Z(S) \leq Z(G)$ for each minimal nonabelian $S \leq G$. (The groups of Theorem A.133.1 satisfy the above condition.)

Appendix 134 On irregular p-groups $G = \Omega_1(G)$ without subgroup of order p^{p+1} and exponent p

Let $G = \Omega_1(G)$ be an irregular p-group. Assume, in addition, that G is not of maximal class; then $|G| > p^{p+1}$ (Theorem 7.1 (b)). In that case, there is in G a normal subgroup R of order p^p and exponent p (Theorem 13.5) and suppose, in addition, that G has no subgroup of order p^{p+1} and exponent p. There is in R a G-invariant subgroup $L \cong \mathrm{E}_{p^2}$. Set $H = \mathrm{C}_G(L)$. We claim that $H < G$. Assume that this is false. Then in the set $G - R$ there is an element x of order p. In that case, the subgroup $F = \langle x, R\rangle$ of order p^{p+1} is not of maximal class. It follows that F is regular (Theorem 7.1 (b)) so that $\exp(F) = p$ since $\Omega_1(F) = F$ and $\mathrm{cl}(F) < p$ (Theorem 7.2 (b)), which is a contradiction. Thus, $H < G$ so $H \in \Gamma_1$. By Theorem 12.1 (b), the subgroup H is not absolutely regular (otherwise, $|\Omega_1(G)| = p^p < |G|$). Then, by Theorem 13.5, there is in H a G-invariant subgroup F of order p^p and exponent p since, in addition, H is not of maximal class. As the subgroup FL, being of class $< p$, is regular (Theorem 7.2 (b), it is of exponent p, and we conclude that $L \le F$ (otherwise, $|FL| \ge p^{p+1}$). A similar argument shows that the set $H - R$ has no elements of order p, which implies that $\Omega_1(H) = R$ is of order p^p and exponent p. In the case under consideration, if $y \in G - H$ is of order p, then the subgroup $\langle x, R\rangle$ is of maximal class and order p^{p+1}.

Thus, the following result holds:

Proposition A.134.1. *Let an irregular p-group $G = \Omega_1(G)$, which is not of maximal class, have no subgroup of order p^{p+1} and exponent p. If $L \cong \mathrm{E}_{p^2}$ is a normal subgroup of G (in the case when L does not exist, G is a 2-group of maximal class, by Lemma 1.4), then $\Omega_1(\mathrm{C}_G(L))$ is of order p^p and exponent p.*

Exercise 1. If, in the notation of Proposition A.134.1, there is in $\mathrm{C}_G(L)$ an irregular subgroup M of maximal class, then $|\Omega_1(M)| = p^{p-1}$.

Hint. If $M < \mathrm{C}_G(L)$ is irregular of maximal class, then $M \cap L = \mathrm{Z}(M)$ is of order p (otherwise, it follows that $ML = M \times L$ contains a subgroup of order p^{p+1} and exponent p). In that case, $ML = M \times Z$ for $|Z| = p$. Since the set $M \times Z$ has no subgroup of order p^{p+1} and exponent p, we conclude that $|\Omega_1(M)| = p^{p-1}$. □

Exercise 2. In the notation of Proposition A.134.1, the number of subgroups of order p in G that are not contained in $\mathrm{C}_G(L)$ is a multiple of p^p.

Hint. Use Theorem 13.2 (a) and the hint to the previous exercise. □

Exercise 3. In the notation of Proposition A.134.1, for any subgroup S of G of order p^p and exponent p not contained in $\mathrm{C}_G(L)$ the subgroup RS is of maximal class and order p^{p+1}.

https://doi.org/10.1515/9783110533149-095

Solution. We claim that such an S exists. Indeed, if $x \in G - R$ is of order p and $x \in T$, where T is maximal in $\langle x\rangle R$, then $\Omega_1(T) = T$ by the modular law and $|T| = p^p$ so that $\exp(T) = p$. Set $H = RS$. Let $R < U \leq H$, where $|U : R| = p$. Then $T = R(U \cap S)$, by the modular law. Therefore, $T = \langle x, R\rangle$ for some $x \in H - S$ of order p. Since $\exp(T) > p$ and $|T| = p^{p+1}$, it follows that T is of maximal class. By Exercise 10.10, the subgroup H is of maximal class. Since $R \lhd H$ is of order p^p and exponent p, we get $|H| = p^{p+1}$ (Theorem 9.6 (c)) so $H = T$ is of maximal class. □

Proposition A.134.2. *Suppose that $G = \Omega_1(G)$ is an irregular p-group of order $> p^{p+1}$. If each subgroup of G of order p^{p+2} has no section of order p^{p+1} and exponent p, then G is of maximal class.*

Proof. Assume that there is in G an irregular subgroup A of order p^{p+1}; then A is of maximal class. Let $A < B \leq G$, where $|B : A| = p$. It follows from

$$|B/\mho_1(B)| < p^{p+1}$$

that B is of maximal class (Theorem 12.12 (b)). In that case, G is of maximal class (Exercise 10.10).

Next we assume that all subgroups of G of order p^{p+1} are regular. If G has no normal subgroup of order p^p and exponent p, it is of maximal class (Theorem 12.1 (b)). Suppose that G is not of maximal class. Therefore, there is $R \lhd G$ of order p^p and exponent p. Let $x \in G - R$ be of order p. Set $M = \langle x, R\rangle$; then $|M| = p^{p+1}$ and $\Omega_1(M) = M$. By hypothesis, $\exp(M) > p$ so that M is irregular (Theorem 7.1 (b)), contrary to the result of the previous paragraph. Thus, G is of maximal class. □

In particular, if a p-group $G = \Omega_1(G)$ of exponent $> p$ has no sections of order p^{p+1} and exponent p, it is of maximal class.

Problem 1. Study the p-groups $G = \Omega_1(G)$ of order $> p^{p+1}$ containing a normal subgroup R of order p^p and exponent p such that for each $x \in G - R$ of order p the subgroup $\langle x, R\rangle$ is of maximal class. (See Proposition A.134.1 and Exercise 10.10.)

Problem 2. Let G be a group of order $> p^p$ and exponent $p > 2$. Is it true that the number of subgroups of maximal class and order p^p in G is a multiple of p? (This is true for $p = 3$, by Theorem 10.4.)

Problem 3. Let G be a group of order $> p^p$ and exponent $p > 2$. Is it true that the number of $D \lhd G$ such that G/D is of maximal class and order p^p is divisible by p? (This is not true for two-generator G and $p = 3$.)

Problem 4. Study the nonabelian p-groups all of whose minimal nonabelian subgroups are quasinormal.

Problem 5. Study the primary $\mathcal{A}_n$-groups G such that $S \cap T > \{1\}$ for any two distinct minimal nonabelian subgroups $S, T < G$. Consider in detail the following two cases: (i) $\exp(G) = p$, (ii) $G = \Omega_1(G)$.

Problem 6. Let S be a minimal nonabelian p-group. Describe the group of those automorphisms of S that fix all elements of order p from S.

Problem 7. Study the p-groups G such that $G/\mathrm{K}_3(G) \cong \mathrm{M}_p(m.n)$. Consider in detail the case $m = n = 2$.

Appendix 135 Nonabelian 2-groups of given order with maximal possible number of involutions

Let G be a 2-group and let $\vartheta_2(G) = |\{x \in G \mid x^2 = 1\}|$ be the number of solutions of the equation $x^2 = 1$ in G. Then $\mathrm{inv}(G)$, the number of involutions in G, is equal to $\vartheta_2(G) - 1 = c_1(G)$, the number of subgroups of order 2 in G. Below we find $\mathfrak{I}_n$, the maximal possible number of involutions in a nonabelian group of order 2^n. Let $\mathrm{T}(G) = \sum_{\chi \in \mathrm{Irr}(G)} \chi(1)$ be the sum of degrees of all irreducible characters of G.

If A, B are groups, then

$$\mathrm{inv}(A \times B) = (\mathrm{inv}(A) + 1) \times (\mathrm{inv}(B) + 1) - 1.$$

Therefore, one has

$$\mathrm{inv}(\mathrm{D}_8 \times \mathrm{E}_{2^{n-3}}) = 3 \cdot 2^{n-2} - 1 \implies \mathfrak{I}_n \geq 3 \cdot 2^{n-2} - 1 \tag{1}$$

and

$$\mathrm{T}(G) = 3 \cdot 2^{n-2} = \mathrm{inv}(G) + 1, \quad \text{where } G = \mathrm{D}_8 \times \mathrm{E}_{2^{n-3}}. \tag{2}$$

We show that, in fact,

$$\mathfrak{I}_n = 3 \cdot 2^{n-2} - 1 = 2^n - 2^{n-1} - 1 \quad \text{and} \quad \mathrm{T}(G) = 3 \cdot 2^{n-2}. \tag{3}$$

It follows from the first equality in (3) that the minimal number of element order > 2 in a nonabelian 2-group of order 2^n is equal to 2^{n-1}.

In general, the set of 2-groups with given number of involutions is infinite (however, the maximal number of involutions in a nonabelian 2-group of given order 2^n, as Lemma A.135.1 shows, is equal to $3 \cdot 2^{n-2} - 1$). In some cases, it is possible to classify arbitrary 2-groups G with a given number of involutions (as we know, this is made for $\mathrm{inv}(G) \in \{1, 3\}$ and $\mathrm{inv}(G) \equiv 1 \pmod 4$; in the last case G is either cyclic or of maximal class).

We begin with the following exercises.

Exercise 1. One has $\mathfrak{I}_4 = 3 \cdot 2^2 - 1 = 11$. If a nonabelian 2-group G of order 2^4 satisfies the last equality, then $G = \mathrm{D}_8 \times \mathrm{C}_2$.

Solution. By (1), $\mathrm{inv}(\mathrm{D}_8 \times \mathrm{C}_2) = 11$ implies $\mathfrak{I}_4 \geq 11$. In view of Lemma 65.1, G is not minimal nonabelian (indeed, if G is minimal nonabelian 2-group, then $\mathrm{inv}(G) \in \{3, 7\}$). By Theorem 1.2, G is not of maximal class (if G is of maximal class and order 2^4, then $\mathrm{inv}(G) \in \{1, 5, 9\}$, by Theorem 1.2). Let $S < G$ be minimal nonabelian. Then, by Proposition 10.17, $G = S * \mathrm{C}_G(S)$ since G is not of maximal class. If $\mathrm{C}_G(S) \cong \mathrm{E}_4$, then $G = S \times \mathrm{C}_2$, and it follows easily that $S \cong \mathrm{D}_8$; then, as we have noted, $\mathrm{inv}(G) = 11$. Now let $\mathrm{C}_G(S) \cong \mathrm{C}_4$. Then $G = \mathrm{D}_8 * \mathrm{C}_4$ (central product) and we claim that $\mathrm{inv}(G) = 7 < 11$,

https://doi.org/10.1515/9783110533149-096

a contradiction. Indeed, one has $Z(G) \cong C_4$ and $G/Z(G) \cong E_4$. Let $A_1/Z(G)$, $A_2/Z(G)$, $A_3/Z(G)$ be all maximal subgroups of $G/Z(G)$; then A_1, A_2, A_3 are abelian of type $(4, 2)$. In that case,

$$\mathrm{inv}(G) = \mathrm{inv}(A_1) + \mathrm{inv}(A_2) + \mathrm{inv}(A_3) - 2\,\mathrm{inv}(Z(G)) = 3 + 3 + 3 - 2 = 7,$$

aa asserted. Thus, $\mathfrak{I}_4 = 11$. □

The following exercise is fairly nontrivial.

Exercise 2. One has $\mathfrak{I}_5 = 3 \cdot 2^3 - 1 = 23$. Is it true that if a nonabelian 2-group G of order 2^5 satisfies the last equality, then $G = D_8 \times E_4$?

If $\chi \in \mathrm{Irr}(G)$, then $\eta_2(G)$ is the Frobenius–Schur indicator of χ. By the Frobenius–Schur formula (see [BZ, § IV.6, (16)]), we have

$$1 + \mathrm{inv}(G) = \sum_{\chi \in \mathrm{Irr}(G)} \eta_2(\chi)\chi(1). \tag{4}$$

Since $\eta_2(\chi) \in \{0, -1, 1\}$ (see [BZ, § IV.6]), it follows from (4) that

$$1 + \mathrm{inv}(G) \le \mathrm{T}(G) \tag{5}$$

with equality if and only if $\eta_2(\chi) = 1$ for all $\chi \in \mathrm{Irr}(G)$; then all irreducible representations of G are realized over the field of real numbers.

We begin with the following:

Lemma A.135.1. *If G is a nonabelian 2-group of order 2^n, then* $\mathrm{inv}(G) \le 3 \cdot 2^{n-2} - 1$. *Thus, in view of* (1), *we have* $\mathfrak{I}_n = 3 \cdot 2^{n-2} - 1$. *The maximal value of* $\mathrm{T}(G)$ *for a group G of order 2^n is* $3 \cdot 2^{n-2}$.

Proof. As we know, the theorem holds for $n \le 4$ (Exercise 1). Next we assume that $n > 4$. In view of (1) and (5), it suffices to prove that $\mathrm{T}(G) \le 3 \cdot 2^{n-2}$.

We proceed by induction on n. Note that if X is minimal nonabelian 2-group, then $\mathrm{inv}(X) \in \{1, 3, 5, 7\}$ (Lemma 65.1). Therefore, if G is minimal nonabelian, then $|G| = 8$ and $\mathfrak{I}_3 = 5$ (in that case, $G \cong D_8$ and $\mathrm{T}(G) = 6 = 3 \cdot 2$, by Lemma 65.1). Next we assume that G is not minimal nonabelian. Let $H < G$ be a nonabelian subgroup of index 2 in G. By induction, $\mathrm{T}(H) \le 3 \cdot 2^{(n-1)-2} = 3 \cdot 2^{n-3}$. If $\chi \in \mathrm{Irr}(G)$, then one has either $\chi_H = \mu \in \mathrm{Irr}(H)$ or $\chi_H = \mu_1 + \mu_2$, where $\mu_1, \mu_2 \in \mathrm{Irr}(H)$ of degree $\frac{1}{2}\chi(1)$ (see [BZ, Exercise VII.3.7]). In the first case, $\mu^G = \chi + \chi_1$, where distinct $\chi, \chi_1 \in \mathrm{Irr}(G)$ have the same degree $\mu(1)$. In the second case, $\mu_1^G = \chi = \mu_2^G$ has degree $2\mu(1)$. That argument shows that, in the case under consideration, $\mathrm{T}(G) \le 2\mathrm{T}(H)$. Therefore, if $\mathrm{T}(G)$ is maximal possible, then, in view of (5), one has

$$3 \cdot 2^{n-2} \le \mathrm{T}(G) \le 2\,\mathrm{T}(H) \le 2 \cdot (3 \cdot 2^{n-3}) = 3 \cdot 2^{n-2}, \tag{6}$$

which implies that $\mathrm{T}(G) = 3 \cdot 2^{n-2}$, i.e., the maximal possible value of $\mathrm{T}(G)$ for a nonabelian group of order 2^n is $3 \cdot 2^{n-2}$. By (6), one has $\mathrm{inv}(G) \le \mathrm{T}(G) - 1 = 3 \cdot 2^{n-2} - 1$. Since $\mathrm{inv}(D_8 \times E_{2^{n-3}}) = 3 \cdot 2^{n-2} - 1$ (see formula (1)), we get $\mathfrak{I}_n = 3 \cdot 2^{n-2} - 1$, completing the proof. □

It follows from the proof of the lemma that if $\mathrm{T}(G) = 3 \cdot 2^{n-2}$ for a group G of order 2^n, then $\mathrm{T}(H) = 3 \cdot 2^{n-3}$ for any nonabelian $H \in \Gamma_1$. This can be used for the description of the structure of G by induction. However, a description of the structure of G satisfying $\mathrm{T}(G) = 3 \cdot 2^{n-2}$ is very difficult. Note that any minimal nonabelian group G of order 2^n satisfies $\mathrm{T}(G) = 3 \cdot 2^{n-2}$. Indeed, $\mathrm{cd}(G) = \{1, 2\}$, by Ito's theorem on degrees [BZ, Theorem 7.2.3]. In that case,

$$\mathrm{T}(G) = |\mathrm{Lin}(G)| + 2|\mathrm{Irr}_1(G)| = 2^{n-1} + 2\frac{|G| - |G/G'|}{2^2} = 2^{n-1} + 2^{n-2} = 3 \cdot 2^{n-2}.$$

Exercise 3. Classify the nonabelian 2-groups G containing two elementary abelian subgroups of index 2. Is it true that $\mathrm{inv}(G) = 3 \cdot 2^{n-2} - 1$ for some $n > 2$?

Solution. Suppose that the group G of order 2^n satisfies the condition. Let $A, B \in \Gamma_1$ be distinct elementary abelian. Then $A \cap B = \mathrm{Z}(G)$. Let $\mathrm{Z}(G) < C \in \Gamma_1 - \{A, B\}$. Then C is also abelian. Since $\exp(G) > 2$, it follows that $\exp(C) = \exp(G) = 4$; moreover, we get $\Omega_1(C) = \mathrm{Z}(G)$ so that $\mathrm{inv}(C) = 2^{n-2} - 1$. Therefore, we have

$$\begin{aligned}\mathrm{inv}(G) &= \mathrm{inv}(A) + \mathrm{inv}(B) + \mathrm{inv}(C) - 2\,\mathrm{inv}(\mathrm{Z}(G)) \\ &= (2^{n-1} - 1) + (2^{n-1} - 1) + (2^{n-2} - 1) - 2(2^{n-2} - 1) \\ &= 3 \cdot 2^{n-2} - 1.\end{aligned}$$

Moreover, it is easy to show that $G = D \times E$, where $D \cong \mathrm{D}_8$ and $\exp(E) \le 2$. □

Exercise 4. Let G be a nonabelian group of order 2^n satisfying $\mathrm{inv}(G) = 3 \cdot 2^{n-2} - 1$. If all minimal nonabelian subgroups of G are isomorphic to D_8, then $G = \mathrm{D}_8 \times \mathrm{E}_{2^{n-3}}$.

Hint. This follows easily from Theorem 10.33. □

Exercise 5. Prove that the maximal value of $\mathrm{T}(G)$ for a nonabelian group G of order p^n is equal to $(2p - 1)p^{n-2}$.

Solution. The result of Exercise 5 is not new. Indeed, if S is a minimal nonabelian p-group, then $\mathrm{f}(S) = \frac{\mathrm{T}(S)}{|S|} = \frac{2p-1}{p^2}$ (check!). By [BZ, Lemma XI.6 (b)], $\mathrm{f}(G) \le \mathrm{f}(S) = \frac{2p-1}{p^2}$, which implies $\mathrm{T}(G) \le (2p - 1)p^{n-2}$. Since

$$\mathrm{T}(\mathrm{S}(p^3) \times \mathrm{E}_{p^{n-3}}) = \mathrm{T}(\mathrm{S}(p^3)) \cdot \mathrm{T}(\mathrm{E}_{p^{n-3}}) = (2p - 1)p^{n-2},$$

the result follows. □

If G is a nonabelian group of order 2^n satisfying $\mathrm{inv}(G) = 3 \cdot 2^{n-2} - 1$, then

$$\mathrm{T}(G) = 3 \cdot 2^{n-2},$$

which is maximal possible for the nonabelian groups of order 2^n (Exercise 5). However, the equality $\mathrm{T}(G) = 3 \cdot 2^{n-2}$ for a group G of order 2^n does not implies the equality $\mathrm{inv}(G) = 3 \cdot 2^{n-2}$ (indeed, a minimal nonabelian 2-group of arbitrary order has at most seven involutions).

Problem 1. Find the maximal (minimal) number of subgroups of order p^3 in a nonabelian group of order p^n and exponent p.

Problem 2. Describe the nonabelian 2-groups G of order 2^n such that $\mathrm{inv}(G)$ is maximal possible subjecting $\mathrm{inv}(G) < 3 \cdot 2^{n-2} - 1$.

Problem 3. Find the maximal number of elements of order p in an irregular p-group G of order p^n, $p > 2$.

Problem 4. Find the maximal number of cyclic subgroups of order p^k, $k > 1$, in a group of order p^n.

Problem 5. Study the p-groups all of whose maximal subgroups, except one, have exponent p.

Problem 6. Classify the groups G of order $2^n > 2^2$ satisfying $\mathrm{T}(G) = 3 \cdot 2^{n-2}$.

Problem 7. Is it possible to produce a character-free proof of Lemma A.135.1?

Problem 8. Find the maximal (minimal) number of abelian subgroups of type (p, p) in a nonabelian group of order p^n and exponent p.

Appendix 136 On metacyclic p-groups

Let G be a metacyclic p-group. Then there is in G a normal cyclic subgroup C such that G/C is cyclic. In many cases there is no in G a subgroup B such that $G = BC$ and $B \cap C = \{1\}$. It appears that such an B exists if $p > 2$ and $|C| = \exp(G)$. We prove that such an B also exists for $p = 2$ and $|C| = \exp(G)$ whenever G has no section isomorphic to Q_8.

Let $G = M_{p^n}$ with $n > 3$. Then $\Phi(G) = Z(G) \cong C_{p^{n-2}}$. Let $C < G$ be cyclic of index p. Since $c_{n-2}(G) = p$, there exists in G a normal cyclic subgroup $L \neq \Phi(G)$ of order p^{n-2}. Clearly, L is a maximal cyclic subgroup of G. Since $L \nleq \Phi(G)$, the quotient group G/L is cyclic. Assume that there is in G a subgroup M such that $G = LM$ with $L \cap M = \{1\}$. As $|M| = p^2$, the intersection $A \cap M$ is a central subgroup of order p. In that case, the noncyclic subgroup $\Omega_1(L) \times \Omega_1(M) \leq Z(G)$, which is a contradiction. Thus. L is not complemented in G.

Remark. A p-group G is modular if and only if all its subgroups are quasinormal. A p-group is modular if and only if it has no sections isomorphic to $S(p^3)$ for $p > 2$ and D_8 for $p = 2$. In particular, metacyclic p-groups for $p > 2$ are modular.

Our main result is the following:

Theorem A.136.1. *Let G be a metacyclic p-group of exponent p^e, $p > 2$. If $C < G$ is cyclic of order p^e, then $G = BC$ with $B \leq G$ is cyclic and $B \cap C = \{1\}$.*

Proof. All subgroups of our group G are quasinormal (see the remark). In what follows we shall use permutability of any two subgroups of G. In view of the basic theorem on abelian p-groups (see Introduction, Exercise 4), one may assume that the group G is nonabelian. By Lemma 1.4, the subgroup $R = \Omega_1(G) \cong E_{p^2}$. Set $\overline{G} = G/R$. If $\overline{G}$ is cyclic, then there is in G a cyclic subgroup of index p (indeed, in the case under consideration $R \nleq \Phi(G)$ so any maximal subgroup of G not containing R is cyclic), and then the result follows from the paragraph preceding the remark. Next we assume that the group G/R is noncyclic.

One has $e > 1$ since G is nonabelian. As G has no cyclic subgroup of index p, we get $|G| > p^3$ in view of the classification of groups of order p^3, $p > 2$. If $|G| = p^4$, then $G \cong M_p(2,2)$, and any cyclic subgroup of G of order p^2 is complemented in G. Therefore, in what follows, we assume that $|G| > p^4$. Using induction on $|G|$, one can suppose that for any proper section of G the theorem holds. Write $\exp(G) = p^e$. By assumption, $e > 2$ and $|G| > p^{e+1}$.

Let $C < G$ be cyclic of order p^e. We have to prove that C has a cyclic complement. Since $|Z \cap R| = p$ for any nonidentity cyclic $Z < G$, we get $\exp(\overline{G}) = p^{e-1}$ so that $\overline{C} = CR/R$ has order $p^{e-1} = \exp(\overline{G})$. Therefore, by induction, $\overline{G} = \overline{C}\overline{L}$, where $\overline{L}$ is cyclic and $\overline{C} \cap \overline{L} = \{\overline{1}\}$. We get $L = L_0R$, where L_0 is cyclic of index p in L. We get $L_0 \nleq C$

https://doi.org/10.1515/9783110533149-097

(otherwise $L = RL_0 \le CR$, which is a contradiction). It follows from $\overline{G} = \overline{CL}$ that

$$G = (CR)(L_0R) = (CL_0)R. \tag{1}$$

As the subgroup CL_0 (here we use the quasinormality of all subgroups of G) is noncyclic, we get $R < CL_0$, and (1) implies that $G = CL_0$. We claim that $C \cap L_0 = \{1\}$. To prove this, it suffices, in view of the product formula, to show that $|G| = |C||L_0|$. Indeed, one has $|L_0| = p|\overline{L}|$ so that

$$\frac{1}{p}|L_0| = |\overline{L}| = |\overline{G} : \overline{C}| = |G/R : CR/R| = |G : CR| = \frac{1}{p}|G : C| \implies |G : C| = |G : L_0|.$$

The proof is complete. □

Thus, if G is a metacyclic p-group, $p > 2$, then it is possible to choose a cyclic $C \le G$ such that $G = BC$, where $B \le G$ is cyclic and $B \cap C = \{1\}$. Theorem A.136.1 contains a more exact assertion.

Exercise 1. Suppose that a metacyclic 2-group G of exponent 2^e has no nonabelian section of order 8 (then G is modular). If C is a cyclic subgroup of G of order 2^e, then $G = BC$ with $B \le G$ is cyclic and $B \cap C = \{1\}$.

To solve the exercise, we have to repeat, word for word, the proof of Theorem A.136.1. Note that in a generalized quaternion group all nonidentity cyclic subgroups are not complemented.

Corollary A.136.2 (Theorem 207.2). *Suppose that G is a metacyclic p-group containing a normal cyclic subgroup C of order $p^e = \exp(G)$. Let, in addition, G have no nonabelian sections of order* 8. *Then $G = B \cdot C$, a semidirect product, with cyclic subgroup $B \le G$.*

The results of this appendix supplement some results obtained in Appendix 137.

Problem 1. Let G be a metacyclic p-group without nonabelian sections of order 8. Is it true that there is in G a normal cyclic subgroup C such that $G = BC$ with $B \cap C = \{1\}$ for some cyclic $B \le G$?

Problem 2. Study the nonabelian 2-groups all of whose minimal nonabelian subgroups have no nonabelian sections of order 8.

Appendix 137 Alternate proof of Lemma 207.1

Below we offer an alternate proof of the following result.

Proposition A.137.1. *If a metacyclic group G of exponent p^e contains a normal cyclic subgroup A of order p^e, then the quotient group G/A is cyclic.*

Proof. Assume that G is a counterexample of minimal order. Then $e > 1$ so that $|A| = p^e > p$, G is nonabelian (indeed, in abelian group of exponent p^e any cyclic subgroup of order p^e has a complement which is, in our case, is cyclic; see Introduction to Volume 1, Lemma 4 (b)). We claim that $|G| > p^4$. Indeed, if $|G| = p^4$, then $e = 2$ and G/C is cyclic for any normal $C \cong C_{p^2}$ since, in the case under consideration, $C \neq \Phi(G) = \Omega_1(G)$. Thus, $\exp(G) > p^2$ so that $|A| \geq p^3$. By Proposition 10.19, G has no nonabelian subgroup of order p^3.

(i) Assume that $G/A \cong E_{p^2}$. Then $A = \Phi(G) = \mho_1(G)$. Since A is cyclic, it is contained in a cyclic subgroup of order $p|A|$, contrary to maximality of $|A|$.

(ii) Assume that $|G : A| > p^2$.

(ii1) Suppose that the group G/A contains a subgroup $S/A \cong E_{p^2}$. By (i), S contains a cyclic subgroup L of index p so that $|L| > |A|$, a contradiction.

(ii2) Now let a group G/A have no subgroup isomorphic to E_{p^2}. Then $p = 2$ and G/A being, by assumption, noncyclic is a generalized quaternion group (Lemma 1.4 and Theorem 1.2). As G is not of maximal class (Theorem 1.2), there is in G a normal subgroup $\Omega_1(G) = R \cong E_4$ (Lemma 1.4). In that case, $\exp(G/R) = 2^{e-1}$ since each cyclic subgroup of G of order > 2 has with R the intersection of order 2. Set $\overline{G} = G/R$. As any cyclic subgroup $D < G$ of order > 2 satisfies

$$|\overline{D}| = |DR/R| = \frac{1}{2}|D|,$$

we get

$$\exp(\overline{A}) = \exp(\overline{G}) = 2^{e-1}.$$

Therefore, by induction, the quotient group

$$\overline{G}/\overline{A} \text{ is cyclic.} \tag{1}$$

One has

$$\overline{G}/\overline{A} \cong (G/R)/(AR/R) \cong G/AR \cong (G/A)/(AR/A). \tag{2}$$

Since G/A is a generalized quaternion group and $|AR/A| = 2$, the quotient group $(G/A)/(AR/A)$ cannot be isomorphic to the cyclic group $\overline{G}/\overline{A}$ (see (2)). This contradiction completes the proof. □

The following theorem contains some additional results on metacyclic p-groups.

https://doi.org/10.1515/9783110533149-098

Theorem A.137.2. *Suppose that a nontrivial normal subgroup A of a nonabelian metacyclic p-group G has a complement B. Then the following statements hold:*
(a) *The subgroup B is cyclic.*
(b) *The subgroup A is either cyclic or $p = 2$ and $|G : A| = 2$. In the second case, G is dihedral or semi-dihedral. If G is semi-dihedral, then A is cyclic or generalized quaternion. If G is dihedral, then A is any maximal subgroup of G.*
(c) *If a nontrivial normal subgroup A_1 of our group G has a complement, then $|A_1| = |A|$.*

Proof. (a)–(b) Obviously, $A \nleq \Phi(G)$ which implies that $d(G/A) < d(G) = 2$, and we conclude that the quotient group G/A is cyclic. As $G/A \cong B$, it follows that the subgroup B is cyclic. Assume that there is in A a G-invariant subgroup $R \cong E_{p^2}$. Then the subgroup $R\Omega_1(B)$ of order p^3 is generated by elements of order p, so, being a subgroup of a metacyclic group, it is nonabelian, and we conclude that $R\Omega_1(B) \cong D_8$. By Proposition 10.19, G is a 2-group of maximal class. It follows that $G \cong D_8$, and for this group assertions (a) and (b) hold. Now let A have no G-invariant subgroup isomorphic to E_{p^2} and assume that A is noncyclic. Then A is a 2-group of maximal class (Lemma 1.4). In that case, A contains a nonabelian subgroup of order 8, and Proposition 10.17 implies that G is a 2-group of maximal class. By Theorem 1.2, G is either dihedral or semi-dihedral, and for this G assertion (b) holds as can be checked easily.

(c) If A_1 is noncyclic, then $p = 2$ and $|G : A_1| = 2 = |G : A|$, by (b) so $|A_1| = |A|$. Now assume that A_1 is cyclic and has a complement B_1. Note that $G' \leq A \cap A_1$. It follows from $B \cap A = \{1\}$ that $B \cap G' = \{1\}$. Similarly, we have $B_1 \cap G' = \{1\}$. It follows that $B \cap A_1 = \{1\} = B_1 \cap A$ hence $|B_1| \leq |G : A| = |B|$ and $|B| \leq |G : A_1| = |B_1|$, and we conclude that $|B| = |B_1|$, which implies that $|A_1| = |A|$. □

Appendix 138 Subgroup characterization of a p-group of maximal class with an abelian subgroup of index p

We begin with the following definition.

Definition 1. A p-groups G is said to be an $\mathcal{M}_p$ group if it is nonabelian and any two its normal subgroups of different orders are incident.

It is trivial that a nonabelian group of order p^3 is an $\mathcal{M}_p$-group. All p-groups of maximal class are $\mathcal{M}_p$-groups. An extraspecial p-group of order $> p^3$ is not an $\mathcal{M}_p$-group.

Lemma A.138.1. *Suppose that G is an $\mathcal{M}_p$-group. Then the following statements hold:*
(i) *Nonabelian epimorphic images of G are $\mathcal{M}_p$-groups.*
(ii) *If $T \lhd G$ is of index p^2, then $T = \Phi(G)$. In particular, $\mathrm{d}(G) = 2$ and T is a unique G-invariant subgroup of index p^2 in G. It follows that $G' = \Phi(G)$.*
(iii) *If G is minimal nonabelian, then $|G| = p^3$.*
(iv) *If $H \lhd G$ is nonabelian, then $C_G(H) < H$.*
(v) *If $|G'| = p$, then $|G| = p^3$.*

Proof. (i) This is obvious.

(ii) All members of the set Γ_1 contain T so that $T = \Phi(G)$. Hence $\exp(G/G') = p$ so that $G' = \Phi(G)$.

(iii) By (i) and Lemma 65.1, we have

$$|G| = |G : G'||G'| = p^2 \cdot p = p^3.$$

(iv) One has $C_G(H) \lhd G$. Assume that $C_G(H) \nleq H$. Let $U \leq C_G(G)$ be G-invariant of minimal order not contained in H. Then $HU = H * U$ is a central product of order $p|H|$. As $H \cap U \leq Z(H)$ and $|H : Z(H)| \geq p^2$ since H is nonabelian and $|U : (H \cap U)| = p$, we get $|U| < |H|$. It follows that $H * U$ is not an $\mathcal{M}_p$-group, a contradiction.

(v) This follows from (ii). □

It follows from Lemma A.138.1 (ii) that any $\mathcal{M}_2$-group is of maximal class. If $p > 2$, then any $\mathcal{M}_p$-group containing an abelian subgroup of index p is of maximal class since any nonabelian subgroup of such a p-group is of maximal class.

Definition 2. A p-group G is said to be an $\mathcal{L}_p$-group if it is nonabelian and any its nonabelian subgroup H is an $\mathcal{M}_p$-group.

Of course, any $\mathcal{L}_p$-group is an $\mathcal{M}_p$-group but the converse is not true.

It is trivial that a nonabelian group of order p^3 is an $\mathcal{L}_p$-group. As we have noted, all 2-groups of maximal class are $\mathcal{L}_2$-groups and all extraspecial p-groups of order $> p^3$ are not $\mathcal{L}_p$-groups. Obviously, the property $\mathcal{L}_p$ is inherited by nonabelian subgroups and epimorphic images.

https://doi.org/10.1515/9783110533149-099

Our main result is the following:

Theorem A.138.2. *A p-group G is an $\mathcal{L}_p$-group if and only if it is a group of maximal class with abelian subgroup of index p.*

Proof. Let $S \leq G$ be minimal nonabelian. Then one has $|S| = p^3$ (Lemma A.138.1 (iii)). By Lemma A.138.1 (iv), $C_G(S) < S$ so that G is of maximal class (Proposition 10.17). It remains to prove that the fundamental subgroup G_1 of G is abelian. Assume that this is false. Then $|G_1| > p^3$. Let $T < G_1$ be minimal nonabelian. Then, by Lemma A.138.1 (iv), $C_{G_1}(T) < T$, so that G_1 is of maximal class since $|T| = p^3$, by the above, a contradiction. Thus, T does not exist so that G_1 is abelian.

Let G be a p-group of maximal class containing an abelian subgroup A of index p. We know that any proper nonabelian subgroup H of G is of maximal class and $H \cap A$ is abelian; then, by induction, H is an $\mathcal{M}_p$-group, and therefore G is an $\mathcal{L}_p$-group. □

We offer another proof of Theorem A.138.2 using induction on $|G|$. Let $R < G'$ be of index p. By Lemma A.138.1 (v), we have $|G/R| = p^3$ since $|G : G'| = p^2$. One may assume that G is not minimal nonabelian (otherwise, $|G| = p^3$). By induction, all nonabelian maximal subgroups of G, being $\mathcal{M}_p$-groups, are of maximal class. But not all members of the set Γ_1 are of maximal class (indeed, if $K \lhd G$ is of order p^2, then any maximal subgroup T of G such that $K < T \leq C_G(K)$ is not of maximal class). Thus, there is in the set Γ_1 an abelian member A. Then it follows that from Lemma 1.1 that

$$|Z(G)| = \frac{1}{p}|G : G'| = p.$$

By induction, the group $G/Z(G)$ is either abelian or of maximal class. If $G/Z(G)$ is abelian, then $|G| = p^3$ (Lemma A.138.1 (v)). If $G/Z(G)$ is of maximal class, so is F since $|Z(G)| = p$.

Problem 1. Is it true that any $\mathcal{M}_p$-group G is of maximal class?

To solve this problem, it suffices to prove that $Z(G)$ is cyclic. Indeed, assume that $Z(G)$ is cyclic but G is not of maximal class. If $|Z(G)| = p$, then, by induction on $|G|$, the quotient group $G/Z(G)$ is either abelian or of maximal class. In the first case, $|G| = p^3$, by Lemma A.138.1 (v), so G is of maximal class. In the second case, G is of maximal class since $|Z(G)| = p$. As we have noted, the problem is solved positively for $p = 2$. Now let $Z(G) \cong C_{p^k}$, $k > 1$. Then $G/\Omega_1(Z(G))$ is of maximal class, by induction, hence $k = 2$. By Lemma 1.4, there is in G a normal subgroup $R \cong E_{p^2}$. One has $R \cap Z(G) = \Omega_1(Z(G))$ and $RZ(G)/\Omega_1(Z(G)) \cong E_{p^2}$ coincides with the center of $G/\Omega_1(Z(G))$, a contradiction.

Problem 2. Classify the nonabelian p-groups all of whose factors of the lower central series besides the first one are cyclic.

Research problems and themes VI

3373. Classify the non-Dedekindian p-groups in which the normal closure of any nonnormal maximal cyclic subgroup is (i) of maximal class, (ii) metacyclic, (iii) abelian, (iv) minimal nonabelian.

3374. Study the nonabelian p-groups G such that $Z(H) \leq Z(G)$ for any nonabelian (minimal nonabelian) $H \leq G$.

3375. Classify the nonabelian p-groups G in which any two normal subgroups of different orders are incident. (Example: any p-group of maximal class.)

3376. Study the nonabelian p-groups G such that $|S : (S \cap G')| = p^2$ for any minimal nonabelian $S \leq G$.

3377. Study the 2-groups G such that, whenever $x \in G$ of order > 2 is real, then $\langle x \rangle \lhd G$.

3378. Does there exist a p-group covered by proper subgroups of maximal class?

3379. Study the nonmetacyclic p-groups all of whose maximal metacyclic subgroups are complemented.

3380. Study the two-generator p-groups G with elementary abelian Frattini subgroup.

3381. Does there exist an irregular p-group, $p > 2$, containing no normal maximal regular subgroup?

3382. Does there exist a nonabelian group of exponent p all of whose normal subgroups (maximal subgroups) are characteristic?

3383. Study the p-groups G such that (i) $G/\mho_2(G)$ is minimal nonabelian, (ii) $G/K_3(G)$ is minimal nonabelian, (iii) $G/\mho_2(G)$ is minimal nonmetacyclic of order 2^5.

3384. If a p-group G contains a subgroup of order p^{p+1} and exponent p, then it contains a normal subgroup of order p^p and exponent p (Theorems 9.6 and 13.5). Is it true that a p-group containing a subgroup isomorphic to $E_{p^{p+1}}$ possesses a normal subgroup isomorphic to E_{p^p}?

3385. Study the p-groups G such that $\Omega_1(G)$ ($\Omega_2(G)$) is special. (See [GMS1].)

3386. (i) Study the irregular p-groups all of whose maximal subgroups of exponent p are normal. (ii) Does there exist a p-group $G = \Omega_1(G)$ in which all maximal subgroups of exponent p have pairwise distinct orders?

3387. Study the p-groups G satisfying $\mathrm{cl}(H^G) = \mathrm{cl}(H)$ ($\mathrm{dl}(H^G) = \mathrm{dl}(H)$) for any $H < G$.

3388. Study the nonabelian p-groups G all of whose proper isolated subgroups have order p. (Example for $p = 2$: $G = M_2(2, 2)$. If $p > 2$, then $M_p(2, 2)$ has no proper isolated subgroups.)

https://doi.org/10.1515/9783110533149-100

3389. Study the p-groups $G = \Omega_1(G)$ containing no subgroup of order p^{p+1} and exponent p.

3390. Classify the non-Dedekindian 2-groups all of whose non-Dedekindian subgroups are normal. (For a partial case, see [FA].)

3391. Classify the p-groups all of whose subgroups are either abelian, or minimal nonabelian, or normal.

3392. Does there exist a p-group G such that, provided nonabelian $U, V \in \Gamma_1$ are distinct, then U is an $\mathcal{A}_i$-subgroup, V is an $\mathcal{A}_j$-subgroup for some $i \neq j$?

3393. Study the p-groups covered by nonnormal (nonnormal abelian, nonnormal regular) subgroups.

3394. Study the 2-groups generated by two subgroups isomorphic to (i) D_8, (ii) Q_8, (iii) D_8 and C_2, (iv) C_4, (v) D_8 and Q_8.

3395. Study the irregular p-groups G, $p > 2$, with regular Hughes subgroup $H_p(G)$. (By [HogK], $|G : H_p(G)| = p$ if $H_p(G)$ is abelian.)

3396. Study the non-Dedekindian p-groups G in which the normalizer of any nonnormal cyclic subgroup coincides with its centralizer.

3397. Estimate the number of p-groups that are lattice isomorphic to a given (i) p-group of maximal class, (ii) minimal nonabelian p-group, (iii) prime power $\mathcal{A}_2$-group.

3398. Classify the p-groups G all of whose subgroups non-incident with G' are nonnormal.

3399. Study the p-groups G with $\exp(\Omega_1(G)) > p$ but $\exp(\Omega_1(H)) = p$ for any $H \in \Gamma_1$.

3400. Given $n > 1$, study the p-groups $G = \Omega_n^{\#}(G)$, all of whose cyclic subgroups of order p^n are (i) normal, (ii) nonnormal.

3401. Describe the nonmetacyclic p-groups all of whose maximal cyclic subgroups have metacyclic normalizers.

3402. Study the nonmetacyclic p-groups all of whose two distinct maximal metacyclic subgroups have cyclic intersection.

3403. Let M be a minimal nonabelian p-group, $p > 3$. Describe the structure of M if it is isomorphic to a subgroup of some p-group of maximal class.

3404. Study the p-groups G such that H is normal in H^G for all $H \le G$.

3405. Suppose that all minimal nonabelian subgroups of a p-group G are quasinormal. Is it true that there is in G a normal minimal nonabelian subgroup?

3406. Study the p-groups G such that $|H^G : H| \le p$ for all $H \le G$.

3407. Does there exist a 2-group G of exponent > 2 such that $\exp(\Phi(\Phi(G)) = \exp(G)$?

3408. Study the p-groups G such that a Sylow p-subgroup of $\mathrm{Aut}(G)$ possesses an abelian subgroup of index p.

3409. Study the p-groups G of exponent $> p > 2$ such that, whenever $H \le G$, then $\mho_1(H) = \Phi(H)$.

3410. Study the irregular p-groups all of whose normal regular subgroups have exponent p.

3411. Let $\delta(G)$ be the minimal degree of a representation of a group G by permutations. Classify the p-groups G such that (i) $\delta(G) = \frac{1}{p}|G|$, (ii) $\delta(G) \le \frac{p^2+p-1}{p^3}|G|$.

3412 (Isaacs, private communication). Does there exist a p-group G with the property that $|A|^2 < |G : G'|$ for any abelian $A < G$?

3413. Given a p-group A of class $n > 1$, does there exist a p-group G of class $n + 1$ such that $Z_n(G) \cong A$?

3414. Study the p-groups G such that $\chi(1)^3 \mid |G|$ for all $\chi \in \mathrm{Irr}(G)$.

3415. Study the p-groups G such that, whenever $A, B < G$ are non-incident, then $|G : N_G(A \cap B)| \le p$.

3416. Study the irregular p-groups G such that $R^G \in \Gamma_1$ for any maximal regular $R < G$.

3417. Study the p-groups G such that, whenever $|H|^2 < |G|$ for $H < G$, then H is either cyclic or G-invariant.

3418. Study the nonabelian p-groups G such that, whenever $S, T < G$ are minimal nonabelian, there exists $\alpha \in \mathrm{Aut}(G)$ with $S^\alpha = T$.

3419. Study the nonmetacyclic p-groups all of whose maximal metacyclic subgroups are normal.

3420. Study the p-groups G such that, whenever $\{1\} < C < S < G$, where C is nonnormal cyclic in a minimal nonabelian S, then $|G : C^G| = p$.

3421. Classify the p-groups G containing a subgroup R of order p^2 such that there is only one maximal chain connecting R with G.

3422. Study the p-groups G such that there is only one maximal chain connecting any minimal nonabelian (maximal cyclic) subgroup with G.

3423. Study the p-groups all of whose subgroups of order p^4 are isomorphic. (This was solved by the second author; see § 303.)

3424. Suppose that a nonabelian p-group G has an abelian subgroup of index p. Is it true that $|G| \le |\mathrm{Aut}(G)|_p$?

3425. Study the p-groups of exponent $> p$ all of whose maximal normal subgroups of exponent p are isolated.

3426. Does there exist a p-group G of exponent $> p$ such that $\mathrm{H}_p(G) = \mho_1(G)$?

3427. Study the p-groups all of whose minimal nonabelian subgroups of the least order are isolated.

3428. Describe the automorphism group of the 2-group $G = \mathrm{Q}_{2^m} * \mathrm{C}_{2^n}$ of order 2^{m+n-1}.

3429. Is it true that the derived length of a product of pairwise permutable cyclic p-subgroups is bounded?

3430. Study the irregular p-groups containing a maximal absolutely regular subgroup of order p^p.

3431. Study the nonabelian p-groups containing exactly one metacyclic (nonmetacyclic) $\mathcal{A}_1$-subgroup.

3432. Study the irregular 3-groups in which all minimal nonabelian subgroups are maximal regular. (Example: any 3-group of maximal class.)

3433. Study the p-groups G such that $\exp(\mathrm{M}(G)) \geq \exp(G)$, where $\mathrm{M}(G)$ is the Schur multiplier of G. (See [BZ, Chapter 6, Theorem 38].)

3434. Study the p-groups G such that the derived subgroup G' is isolated.

3435. Study the p-groups such that any their nonnormal nonabelian subgroup contains an abelian subgroup of index p.

3436. Let H be a proper normal subgroup of order $\geq p^p$ of a p-group G. Study the structure of G such that each normal subgroup of G containing H as a subgroup of index p is of maximal class.

3437. Study the non-Dedekindian p-groups all of whose maximal nonnormal subgroups have index p^2.

3438. Study the p-groups G with $|H : H_G| \leq p$ for any $H < G$. (For $p = 2$, see [Wil7].)

3439. Describe the p-groups H such that $\mathrm{cd}(H) = \mathrm{cd}(G)$ and $\mathrm{k}(H) = \mathrm{k}(G)$, where G is (i) an $\mathcal{A}_2$-group, (ii) metacyclic, (iii) of maximal class.

3440. Study the p-groups G such that, whenever $H < G$ is nonabelian, then $\mathrm{C}_G(A) = A$ for some maximal abelian A in H.

3441. Study the irregular p-groups in which all maximal subgroups of exponent p are maximal regular.

3442. Study the non-Dedekindian p-groups G such that $C^G \in \Gamma_1$ for any nonnormal maximal cyclic $C < G$. (This is solved in § 282.)

3443. Classify the nonabelian p-groups G such that $\psi(1)^2 = |H : Z(H)|$ for all nonabelian $H \leq G$ and all $\psi \in \mathrm{Irr}_1(H)$.

3444. Estimate the number of p-groups of maximal class and order p^p and exponent (i) p, (ii) p^2.

3445. Study the nonabelian p-groups G such that $|G : S^G| = p$ for any $\mathcal{A}_1$-subgroup $S < G$.

3446. Study the non-Dedekindian p-groups G such that, whenever $H < G$ is nonnormal, then $|G : H^G| \leq p^2$.

3447. Study the p-groups of exponent $> p$ all of whose maximal elementary abelian subgroups coincide with their centralizers.

3448. Study the nonabelian p-groups G such that each their nonnormal subgroup is complemented in their normalizer.

3449. Study the p-groups G such that $G/Z(G)$ is special and $Z(G) \leq \Phi(G)$.

3450. Study the two-generator p-groups G such that $G/Z(G)$ is of maximal class.

3451. Study the p-groups G such that $G/Z(G)$ is absolutely regular and $Z(G) \leq \Phi(G)$. Estimate the number $|\Omega_1(G)|$. (According to Remark 7.2, due to Mann, the group G is regular.)

3452. Study the two-generator p-groups G containing a normal subgroup H of maximal class such that G/H is cyclic of order $> p$.

3453. Classify the metacyclic p-groups all of whose maximal subgroups are pairwise non-isomorphic. Is it true that then $p = 2$?

3454. Study the nonabelian p-groups G in which any nonabelian two-generator subgroup is either minimal nonabelian or metacyclic.

3455. Let $\mu_p(G)$ be the number of subgroups of order p^p of maximal class in a p-group G of maximal class, $p > 2$. Find all possible values of $\mu_p(G)$ (mod p).

3456. Study the p-groups G such that $C_G(L)$ is minimal nonabelian for any cyclic $L < G$ of order p^2.

3457. Classify the non-modular p-groups all of whose subgroups of order $> p$ are quasinormal. (See Theorem 1.25.)

3458. Classify the p-groups G admitting an automorphism ϕ of order p such that all subgroups of G (all members of the set Γ_1) are ϕ-invariant.

3459. Study the nonabelian p-groups G such that, whenever $S \leq G$ is minimal nonabelian, then all subgroups of S are quasinormal in G.

3460. Study the p-groups G such that for any their subgroup $E \cong \mathrm{E}_{p^2}$ the subgroup $\mathrm{C}_G(E)$ is metacyclic minimal nonabelian.

3461. Given an integer $t > 0$, a p-group G is said to be $\mathcal{D}_t$-group if, whenever $H < G$ and $|H|^t < G$, then $H \lhd G$. The $\mathcal{D}_1$-groups are Dedekindian. The $\mathcal{D}_2$-groups also classified. Is it true that if G is a $\mathcal{D}_t$-group, $t > 2$, then bounded its (i) class, (ii) derived length.

3462. Study the p-groups containing a minimal nonabelian subgroup $A \lhd G$ such that (i) G/A is cyclic, (ii) $G/A \cong \mathrm{E}_{p^2}$.

3463. Study the p-groups containing an extraspecial subgroup of index p.

3464. Study the two-generator p-groups containing a subgroup of maximal class $M \lhd G$ and index p^2. (See Theorem 12.12 (a).)

3465. Study the p-groups containing two distinct extraspecial (special) subgroups of index p.

3466. Study the p-groups of order $p^m > p^3$ in which all nonnormal subgroups have the same order.

3467. Study the p-groups in which every proper subgroup possesses a metacyclic subgroup of index p.

3468. Study the p-groups in which any maximal subgroup is a non-trivial direct product of two factors.

3469. Classify the p-groups G such that whenever $H \in \Gamma_1$ and A is quasinormal in H then A is quasinormal in G.

3470. Study the nonabelian p-groups G such that, whenever $A < G$ is maximal abelian, then $A/\mathrm{Z}(G)$ is maximal abelian in $G/\mathrm{Z}(G)$.

3471. Classify the two-generator p-groups G such that $\mathrm{Z}(G) \leq \Phi(G)$ and the quotient group $G/\mathrm{Z}(G)$ is minimal nonabelian (metacyclic).

3472. Study the p-groups $G = AB$, where (i) A and B are minimal nonabelian, (ii) A is metacyclic and B is cyclic, (iii) A and B are metacyclic, (iv) A is abelian and B is metacyclic, (v) A is minimal nonabelian and B is metacyclic, (vi) A and B are of class ≤ 2.

3473. It is easy to prove that if G is a nonabelian group of order 2^n, then the number of involutions in G does not exceed $3 \cdot 2^{n-2} - 1$. Classify the nonabelian groups of order 2^n containing exactly $3 \cdot 2^{n-2} - 1$ involutions.

3474. Study the p-groups all of whose maximal subgroups, except one, have exponent p.

3475. Study the 2-groups G admitting an involutory automorphism ϕ such that the group $\mathrm{C}_G(\phi)$ has only one involution.

3476. Describe the structure of Aut(H), where H is a homocyclic p-group.

3477. Study the p-groups G containing a maximal subgroup H such that, whenever an abelian (maximal abelian) $A < G$ is not contained in H, then $A \lhd G$.

3478. Let G be a p-group. Is it true that $|G|$ $(\exp(G))$ is restricted provided p^n does not divide $\exp(\mathrm{Aut}(G))$?

3479. Given a p-group H, estimate the number of p-groups G having the same number of elements of each order as H.

3480. Classify the p-groups G such that $|H^G : H| \le p^2$ for any $H < G$.

3481. Study the Schur multipliers and representation groups of p-groups of maximal class, $p > 2$.

3482. Study the p-groups all of whose outer automorphisms have order p.

3483. Suppose that A is the abelian p-group of given type. Find $|\exp(\mathrm{Aut}(A))|_p$ and $\mathrm{d}(P)$, where $P \in \mathrm{Syl}_p(\mathrm{Aut}(A)))$.

3484. Estimate the number of special p-groups G such that $|\mathrm{Z}(G)| = p^z$ and $\mathrm{d}(G) = d$.

3485. Does there exist a p-group G such that $A \cap B$ is special for any two distinct $A, B \in \Gamma_1$? In particular, does there exist a two-generator p-group G with special $\Phi(G)$?

3486. Study the non-Dedekindian p-groups G such that for any two nonnormal cyclic subgroups A, B of equal order (i) there is $\phi \in \mathrm{Aut}(G)$ such that $A^\phi = B$, (ii) A and B are conjugate (this problem was solved by the second author; see Problem 275).

3487. Let a p-group $G = ABC$, where A, B, C are pairwise permutable cyclic (abelian) subgroups. Estimate $\mathrm{dl}(G)$.

3488. Study the non-Dedekindian p-groups G such that $|G : \mathrm{N}_G(L)| \le p$ for any cyclic (abelian) $L < G$. (See Theorem 122.1.)

3489. Given $k > 1$, find the minimal n such that any group of order p^n and class $n - k$ is irregular.

3490. Let G be a two-generator nonabelian p-group. Given $n > 2$, find all possible values (mod p) of the number of $N \lhd G$ such that the quotient group G/N is minimal nonabelian of order p^n.

3491. Find all possible values (mod p) of the number of subgroups of maximal class and order p^k, $k \in \{3, \dots, p\}$, in a p-group of order $> p^p$.

3492. Study the p-groups all of whose subgroups of index p^2 have cyclic derived subgroups. (See § 139.)

3493. Classify the special p-groups G having special representation groups.

3494. Study the non-Dedekindian p-groups G such that $N_G(A)$ contains an abelian subgroup of index p for any nonnormal abelian $A < G$.

3495. Let $\mathrm{km}(G)$ ($\mathrm{kr}(G)$) be the number of conjugate classes of maximal metacyclic (maximal regular) subgroups in a nonmetacyclic (irregular) p-group G; then $\mathrm{km}(G) \ge p+1$ ($\mathrm{kr}(G) \ge p+1$) since the normal closures of such subgroups cover G. Study the nonmetacyclic (irregular) p-groups with $\mathrm{km}(G) = p+1$ ($\mathrm{kr}(G) = p+1$). Does there exist a nonmetacyclic (irregular) p-group G satisfying $\mathrm{km}(G) = p+2$ ($\mathrm{kr}(G) = p+2$)?

3496. Study the irregular p-groups G such that $|\Omega_1(G)| = |G/\mho_1(G)| = p^p$.

3497. Study the p-groups G such that $\alpha_1(G/L) \ge \alpha_1(G)$ for some $L \lhd G$ of order p.

3498. Let $\mathrm{T}(G)$ be the sum of degrees of irreducible characters of a group G. If G is a nonabelian group of order p^n, then $\mathrm{T}(G) \le (2p-1)p^{n-2}$, and the value $(2p-1)p^{n-2}$ is attained provided G is minimal nonabelian [BZ, Chapter 11]. Classify the nonabelian groups G of order p^n satisfying $\mathrm{T}(G) = (2p-1)p^{n-2}$.

3499. Study the irregular p-groups G such that, whenever $A < G$ is maximal regular, then any subgroup of G containing A as a subgroup of index p is two-generator.

3500. Study the irregular p-groups all of whose maximal regular subgroups are two-generator.

3501. Study the nonmetacyclic p-groups generated by any two distinct maximal metacyclic subgroups.

3502. Study the 2-groups G such that if $A, B \lhd G$ and $|A| < |B|$, then $A < B$. (For a partial case, see Appendix 138.)

3503. Study the irregular p-groups G such that, whenever $R < G$ is maximal regular, then all subgroups containing R as a subgroup of index p, are minimal irregular.

3504. Study the irregular p-groups G such that, whenever $A, B < G$ are distinct maximal regular, then $A \cap B$ is (i) abelian, (ii) G-invariant, (iii) absolutely regular, (iv) of exponent p.

3505. Study the 2-groups G such that $c_1(G) > 7$ but $c_1(H) \le 7$ for each $H \in \Gamma_1$. State a similar problem for $p > 2$.

3506 (Blackburn [Bla13]). Classify the 2-groups containing an involution x such that $C_G(x)/\langle x\rangle$ contains only one subgroup of order 2.

3507. Find $\alpha_1(\mathrm{B}(4,2))$, where $\mathrm{B}(4,2)$ is the two-generator group of exponent 4 and order 2^{12} (Burnside).

3508. Study the p-groups A, B such that A, B are characteristic in $A * B$. (Example: $A = Q_8$ and $B \cong C_4$ are characteristic in $A * B$ of order 2^4.)

3509 (see Miller's Problem 1906). Study the p-groups G such that a Sylow p-subgroup of Aut(G) is abelian. (See [KY].)

3510. Suppose that G is a nonabelian p-group all of whose minimal nonabelian subgroups of exponent $> p$ are isolated (assuming that G has such subgroups).

3511. Let G be a nonabelian p-group of exponent $> p$. Let K be generated by all minimal nonabelian subgroups of G of order p^3 and let L be generated by all minimal nonabelian subgroups of order $> p^3$. Then $G = KL$, by Theorem 10.28. Find conditions guaranteing that one of subgroups K, L coincides with G?

3512. Study the irregular p-groups all of whose maximal absolutely regular subgroups are complemented (normally complemented).

3513. Study the 2-groups all of whose maximal metacyclic subgroups have index at most 4.

3514. Study the p-groups all of whose proper noncyclic normal subgroups of equal order are isomorphic.

3515. Study the p-groups with two distinct isolated subgroups of index p.

3516. Study the p-groups which are isolated (critical) in a Sylow p-subgroup of their holomorph.

3517. Study the irregular p-groups, $p > 2$, containing only one normal subgroup of each of indices $p^2, p^3, \dots, p^p$.

3518. Suppose that a p-group G contains a metacyclic subgroup M of index p. Describe the structure of G provided all minimal nonabelian subgroups of G not contained in M have order p^3.

3519. Study the irregular p-groups all of whose maximal absolutely regular subgroups are isolated. (This is solved in Theorem A.118.1.)

3520. Study the p-groups all of whose $\mathcal{A}_2$-subgroups are isolated.

3521. Study the irregular p-groups $G = \Omega_1(G)$, $p > 2$, all of whose minimal nonabelian subgroups of exponent p are isolated.

3522. Classify the p-groups all of whose nonnormal subgroups are two-generator.

3523. Classify the nonabelian p-groups of exponent $p^e > p$ all of whose maximal abelian subgroups of exponent p^e are isolated.

3524. Study the irregular p-groups $G = \Omega_1(G)$ all of whose maximal subgroups of exponent p have order p^p. (See § 300.)

3525. Study the special 2-groups G such that Aut(G) is a 2-group.

3526. Given an abelian p-group A, is it true that there is a two-generator p-group G such that $Z(G) \cong A$? Estimate the minimum of $|G|$.

3527. Does there exist a p-group G such that, whenever distinct $A, B \in \Gamma_1$, then $A \cap B$ is of maximal class?

3528. Study the nonabelian p-groups all of whose minimal nonabelian subgroups (i) are complemented, (ii) have cyclic complements, (iii) have normal cyclic complements, (iv) have abelian complements.

3529 (see Miller's Problem 1906). Study the 2-groups with Hamiltonian automorphism group.

3530. Classify the p-groups G such that, for any maximal abelian $A < G$ there is an abelian $B < G$ such that $G = AB$.

3531. Let $\nu_k(X)$ be the number of normal subgroups of order p^k in a p-group X. Study the p-groups G such that $\nu_k(G) = \nu_k(M)$ for all k, where M is a given metacyclic p-group.

3532. Study the p-groups G such that $\nu_k(G) = \nu_k(A \times B)$ for all k, where A, B are p-groups of maximal class (see Problem 3531).

3533 (Janko). Study the p-groups in which any abelian subgroup of exponent $> p$ is of rank 2.

3534. Study the p-groups in which all nonnormal nonabelian subgroups are metacyclic.

3535. Study the irregular p-groups in which any minimal irregular subgroup is isolated.

3536. Classify the irregular p-groups all of whose maximal subgroups of exponent p have index $\leq p^2$.

3537. Study the irregular p-groups all of whose nonnormal subgroups have exponent p.

3538. Study the nonabelian p-groups G such that $A \cap B$ is isolated in A and B provided distinct $A, B \in \Gamma_1$.

3539. Study the p-groups G containing a normal subgroup N of order p such that $k(G) = k(G/N) + p - 1$.

3540. Classify the nonabelian p-groups G all of whose minimal nonabelian subgroups are conjugate in the holomorph of G. (Examples: D_{2^n} and Q_{2^n} for $n \geq 4$.)

3541. Describe the holomorphs of the p-groups with a cyclic subgroup of index p.

3542. Does there exist a p-group G such that G is isomorphic to a Sylow p-subgroup of $\mathrm{Aut}(M(G))$, where $M(G)$ is the Schur multiplier of G?

3543. Study the p-groups not covered by abelian subgroups of exponent $> p$.

3544. Study the 2-groups G with cyclic G' all of whose subgroups of order 2, except one, are not maximal cyclic. (Example: $M_2(2, 2)$.)

3545. Suppose that a p-group G has a nonnormal subgroup of order $> p$. Describe the structure of G provided the normal closures of all such subgroups have index p in G.

3546. Study the nonabelian p-groups G such that $M(G) \cong M(H)$ for all $H \in \Gamma_1$.

3547. (i) (See § 287.) Study the nonabelian p-groups in which the intersection of any two distinct subgroups of equal order is abelian (normal). (ii) Study the p-groups G such that $N_G(H)/H$ is cyclic for any nonabelian $H < G$.

3548. Does there exist a nonabelian p-group all of whose maximal abelian subgroups are characteristic?

3549. Study the special 2-groups G such that (i) $Z(G) < \Omega_1(G) < G$, (ii) $\Omega_1(G) = Z(G)$, (iii) $\Omega_1(G) = G$.

3550 (Old problem). Let $p^{r(X)}$ ($p^{e(X)}$) be the maximal order of elementary abelian subgroup (subgroup of exponent p) of a p-group X. Let a p-group $G = AB$, where $A, B < G$. Estimate $r(G)$ ($e(G)$) in terms of $r(A)$ and $r(B)$ ($e(A)$ and $e(B)$). In particular, find $r(AB)$ ($e(AB)$), where A and B are (i) abelian, (ii) regular, (iii) absolutely regular, (iv) metacyclic, (v) of maximal class.

3551. Classify the nonmetacyclic p-groups all of whose minimal nonmetacyclic subgroups are isolated.

3552. Study the p-groups all of whose maximal metacyclic (absolutely regular) subgroups have index $\leq p^2$.

3553. Classify the nonabelian p-groups G of order p^n, containing an abelian subgroup of index p and satisfying $\alpha_1(G) = p^{n-3}$.

3554. Study the irregular p-groups, $p > 2$, in which any two distinct maximal regular subgroups have absolutely regular intersection.

3555. Study the nonabelian p-groups with abelian centralizers of all nonabelian subgroups.

3556. Does there exist a special p-group, $p > 2$, whose automorphism group is a p-group?

3557. Describe the groups G of given order and exponent $p^e > p$ such that

$$c_1(G) + c_2(G) + \cdots + c_e(G)$$

is maximal possible for a given e.

3558. A 2-group G is said to be *generalized Dedekindian* if $G = Q_8 \times A$, where $A' = \{1\}$. Classify the 2-groups all of whose maximal subgroups are either abelian or generalized Dedekindian.

3559. Study the p-groups with a proper isolated minimal nonabelian subgroup (i) isomorphic to $M_p(n, n)$, (ii) M_{p^n}?

3560 (Janko). A p-group G is said to be *quasiabelian* if each its maximal abelian subgroup has index $\leq p$. The minimal nonabelian p-groups are quasiabelian. Subgroups of quasiabelian p-groups are quasiabelian. (i) Classify the minimal non-quasiabelian p-groups. (ii) Classify the p-groups all of whose maximal subgroups, except one, are quasiabelian.

3561. Classify the nonnilpotent groups all of whose minimal nonabelian (minimal nonnilpotent) subgroups are isolated.

3562. Study the p-groups G of exponent $> p$ in which all cyclic subgroups of orders $> p$ have normal closures of index p in G. (Compare with §§ 63, 282.)

3563. Classify the p-groups all of whose maximal abelian subgroups have index $\leq p^2$.

3564. Study the two-generator 2-groups G such that $\mathrm{Aut}(G)$ is a 2-group.

3565. Classify the p-groups all of whose maximal subgroups are either special or have cyclic derived subgroups.

3566. Study the p-groups G such that, whenever $|H|^2 \geq |G|$ for a subgroup $H < G$, then $H \triangleleft G$.

3567. Study the p-groups G such that, whenever $H \in \Gamma_1$ and $x \in G - H$, then $C_H(x)$ is cyclic. (This is solved in § 283.)

3568. Study the irregular p-groups G such that $|\Omega_1(G)| = p^p$ and $d(\Omega_1(G)) = 2$.

3569. Classify the p-groups G all whose cyclic subgroups are half-normal (i.e., their normalizers have index $\leq p$ in G).

3570. Study the non-Dedekindian p-groups G such that, whenever $H < G$ with $H_G = \{1\}$, then $|H| \leq p$.

3571. Classify the non-Dedekindian p-groups G such that if $A/Z(G) \leq G/Z(G)$ is Dedekindian, so is A. (This was solved by the second author).

3572. Study the nonabelian p-groups all of whose maximal abelian subgroups (maximal abelian subgroups of exponent $> p$, maximal regular subgroups, maximal metacyclic subgroups) are isolated in their normalizers (normal closures).

3573. Study the non-Dedekindian p-groups in which the normalizer of every nonnormal cyclic subgroup is metacyclic.

3574. Study the p-groups G such that (i) $C_G(x)$ is isolated for every $x \in G$, (ii) $N_G(A)$ is isolated for every $A \leq G$.

3575. Study the non-Dedekindian p-groups G such that $N_G(L)/C_G(L)$ is isomorphic to a Sylow p-subgroup of $\mathrm{Aut}(L)$ for any nonnormal $L < G$.

3576. Study the minimal non-pyramidal p-groups (see § 8). Is it true that such groups are minimal irregular? (ii) Study the p-groups all of whose maximal subgroups, except one, are pyramidal.

3577. Study the p-groups G with exactly two conjugate classes of nonnormal maximal cyclic subgroups. (This problem was solved by the second author; see § 277.)

3578. Study the nonabelian p-groups G such that, whenever $A < G$ is maximal abelian (minimal nonabelian, maximal metacyclic) and $A < B \leq G$ with $|B : A| = p$, then A is isolated in B.

3579. Let $\mathcal{M}$ be the set of maximal abelian and minimal nonabelian subgroups of a p-group G. Study the structure of G if, whenever distinct $A, B \in \mathcal{M}$, then $A \cap B \trianglelefteq G$.[1]

3580. Study the non-modular p-groups all of whose non-quasinormal (cyclic, abelian, minimal nonabelian) subgroups are conjugate (four problems).

3581. Study the nonabelian p-groups all of whose two distinct maximal abelian subgroups have intersection of exponent $\leq p^2$.

3582. Does there exist a pair of p-groups $N \trianglelefteq G$ such that $\alpha_2(G/N) > \alpha_2(G)$?

3583. Study the p-groups G such that, whenever $|H|^2 < |G|$, then $|G : N_G(H)| \leq p$.

3584. Study the irregular p-groups all of whose maximal regular subgroups have index $\leq p^2$.

3585. Study the p-groups all of whose two maximal abelian subgroups are permutable.

3586 (Old problem). Estimate the minimal class number of a group of order p^n.

3587. Find $c_1(G)$, where G is the standard wreath product of (i) E_{p^m} with the active factor E_{p^n}, (ii) C_{p^m} with the active factor E_{p^n}.

3588. Study the two-generator p-groups G such that $G/Z(G) \cong \Sigma_{p^3}$.

3589. Study the subgroup and normal structure of a two-generator p-group G such that $Z(G) = G'$.

---

1 Remark of the second author: If $S < G$ is minimal nonabelian, then all its maximal subgroups are G-invariant. It follows that $S \trianglelefteq G$, and we conclude that G is metahamiltonian, by Theorem 10.28; such groups G are classified in [FA].

3590. Classify the p-groups of exponent p containing a self-centralizing abelian subgroup of order p^3.

3591. Study the p-groups, $p > 2$, without elementary abelian subgroups of order p^4.

3592. Describe the subgroup structure of metacyclic p-groups with cyclic center.

3593. Describe the Schur multipliers and representation groups of Sylow subgroups of all simple groups.

3594. Study the p-groups of class 2 with trivial Schur multiplier.

3595. Describe the Schur multipliers and the representation groups of minimal nonabelian p-groups.

3596. Describe the representation groups of an abelian p-group G. Consider in detail the case when G is homocyclic.

3597. Study the p-groups in which the normal closures of all elementary abelian subgroups are abelian.

3598. Describe the representation groups of the extraspecial p-groups. When are these representation groups special?

3599. Estimate the maximal order of the Schur multiplier of a special group of order p^m with the center of order p^z.

3600. Study the p-groups G such that $G/\mho_2(G)$ is of (i) maximal class, (ii) maximal class with abelian subgroup of index p.

3601. Study the nonabelian p-groups in which any abelian subgroup is contained in a subgroup of maximal class. (This is a partial case of Problem 3378.)

3602. Study the nonabelian 2-groups G with $H = \Omega_2^{\#}(H)$ for any nonabelian $H \leq G$.

3603. Study the nonabelian primary $\mathcal{A}_n$-groups G, $n > 1$, such that each their minimal nonabelian (maximal abelian, maximal metacyclic) subgroup is contained in a proper isolated subgroup of G.

3604. Estimate the number of p-groups of maximal class and order p^n, $p > 2$, (i) containing an abelian subgroup of index p, (ii) with abelian $\Phi(G)$.

3605. Study the p-groups all of whose cyclic subgroups are TI-subgroups.

3606. Present a nonabelian group G of order p^n such that $|\mathrm{M}(G)|$ is maximal possible. Is it true that G is elementary abelian?

3607. Given $n > p^2$, does there exist a p-group G such that $\alpha_1(G) = n$? (See § 76.)

3608. Study the nonabelian p-groups G all of whose minimal nonabelian subgroups are nonmetacyclic.

3609. Study the p-groups all of whose sections of order p^4 are abelian.

3610. Study the nonabelian p-groups G all of whose proper nonabelian subgroups are two-generator.

3611. Study the p-groups all of whose subgroups that are not complemented are contained in $\Phi(G)$.

3612. Classify the 2-groups with exactly two conjugate classes of abelian subgroups of type $(2, 2)$.

3613. Classify the p-groups with at most three conjugate classes of nonnormal cyclic subgroups.

3614. Study the p-groups G of class > 2 such that, whenever $A < G$ is of class 2, then $\mathrm{cl}(A^G) = 2$.

3615. Study the p-groups whose Frattini subgroup is abelian of type (p^n, p).

3616. Describe the Schur multipliers and representation groups of the (i) primary $\mathcal{A}_2$-groups, (ii) automorphism groups of 2-groups of maximal class.

3617. Study the non-Dedekindian p-groups G such that, whenever $A, B < G$ are non-incident nonnormal, then $A^G \cap B^G$ is abelian.

3618. Describe the automorphism groups of the representation groups of homocyclic p-groups.

3619. Describe the automorphism group of the standard wreath product of two cyclic p-groups.

3620 (Mann). Classify the p-groups G satisfying $\mathrm{Aut}(G) \cong G$. Moreover, study the p-groups G that are isomorphic to a Sylow p-subgroup of $\mathrm{Aut}(G)$.

3621. Study the p-groups G of exponent $> p > 2$ satisfying $\mho_1(G) = \mho_1(\Phi(G))$.

3622. Let G be a p-group. Set

$$\mho^1(G) = \mho_1(G), \quad \mho^2(G) = \mho_1(\mho^1(G)), \quad \mho^3(G) = \mho^1(\mho^2(G)), \quad \text{etc.}$$

Given n, does there exist a p-group G of exponent p^2 such that $\mho^n(G) > \{1\}$?

3623. Let G be a nonabelian p-group of order $p^n > p^3$. Study the structure of G provided for any nonabelian $H \le G$ of order $p^k > p^3$ $(k \le n)$, one has $\alpha_1(H) = p^{k-4}$.

3624. Given $e > 1$, does there exist a p-group G of exponent p^e such that the subgroups $\Omega_1^{\#}(G), \Omega_2^{\#}(G), \dots, \Omega_e^{\#}(G)$ are pairwise non-incident?

3625. Study the p-groups G all of whose subgroups of index p^2 are two-generator.

3626. Study the p-groups G with $\Phi(G) \cong \mathrm{E}_{p^2}$.

3627. Classify the nonabelian groups G of exponent p and order $> p^3$ such that for any minimal nonabelian $M < G$ one has $C_G(M) \cong E_{p^2}$.

3628. Study the p-groups G such that, for some $n > 1$, the subgroup $Z_n(G)$ ($\mho_n(G)$) is (i) an $\mathcal{A}_1$-subgroup, (ii) an $\mathcal{A}_2$-subgroup.

3629. Study the irregular p-groups G such that any regular $R < G$ is either abelian (absolutely regular) or $\exp(R) = p$.

3630. Classify the non-Dedekindian p-groups G such that $A \le Z(N_G(A))$ for any non-normal abelian $A < G$.

3631. Describe $\mathrm{Aut}(G)$ and find $k(G)$, where G is a 2-group of order $> 2^4$ from Theorem 282.1. Study the 2-groups that are lattice isomorphic to that G.

3632. Characterize the p-groups with exactly two non-trivial characteristic subgroups.

3633. Study the p-groups G all of whose proper characteristic subgroups are abelian.

3634. Let A be a critical p-group. Does there is a p-group G containing a proper critical subgroup $\cong A$.

3635. Given a p-group A, present a condition guaranteing that A is characteristic in a Sylow p-subgroup of its holomorph.

3636. Present a condition guaranteing that $\mho_1(G)$ ($\Phi(G)$, G', $Z(G)$) is the unique G-invariant subgroup of its order in G.

3637. Let $G = A \times B$, where A and B are minimal nonabelian p-groups. Find the number of minimal nonabelian subgroups of order $|A|$ in G.

3638. A p-group G is of type $\mathcal{S}$ if $G = S \times E$, where S is a nonabelian group of order p^3 and E is elementary abelian p-group. Classify the p-groups all of whose maximal subgroups, except one, are either abelian or of type $\mathcal{S}$. (The second author has solved this problem in §§ 306–307.)

3639 (Mann, personal communication). Study the p-groups G such that the Schur multiplier $M(G)$ is non-characteristic in a representation group of G.

3640. Study the p-groups G all of whose maximal pyramidal subgroups have index p in G. (See § 8.)

3641. Study the p-groups G all of whose (i) critical subgroups are isolated, (ii) isolated subgroups are critical.

3642. Study the p-groups G of exponent $p^e > p$ such that $\Omega_e^{\#}(G)$ is minimal nonabelian.

3643. Study the nonabelian p-groups G all of whose members of the upper central series are isolated.

3644. Classify the special 2-groups G such that $c_2(G) \equiv 2 \pmod 4$.

3645. Study the nonabelian p-groups G all of whose maximal abelian subgroups are normal. Describe the minimal nonabelian subgroups of these groups. Estimate $cl(G)$ provided $G = A_1 \dots A_n$ is an irredundant product of normal maximal abelian subgroups of G. (If $p > 2$, then G is regular.)

3646. Study the p-groups G all of whose epimorphic images of order $\frac{1}{p}|G|$ are special.

3647. Classify the p-groups with exactly $p + 1$ conjugate classes of maximal abelian subgroups.

3648. Classify the p-groups without non-trivial isolated subgroups.

3649. Study the p-groups G with exactly one non-trivial critical subgroup.

3650. Classify the regular p-groups all of whose representation groups are of maximal class.

3651. Study the non-Dedekindian p-groups G such that $N_G(H) = H^G$ for all nonnormal subgroups $H < G$.

3652. Study the p-groups containing a critical subgroup of index p.

3653. Study the irregular 2-groups G such that, whenever A, B are nonnormal non-incident, then $A \cap B$ is G-invariant. (See § 287.)

3654 (Old problem). Describe the automorphism groups of (i) metacyclic p-groups, (ii) special p-groups.

3655 (Old problem). Study the noncyclic p-groups all of whose maximal subgroups are characteristic. Is it possible that such group has exponent p? Is it possible that $p > 2$?

3656. Suppose that a p-group G contains two distinct isolated maximal abelian subgroups A, B of exponent $> p$ such that $AB = G$. Study the structure of G.

3657. Find the number of groups G of order p^n with $|G'| = p$.

3658. Study the p-groups G such that $G/Z(H) \cong H$ for some $H \in \Gamma_1$.

3659. Study the p-groups G such that $G/Z(G)$ is powerful. (See § 26.)

3660. Study the structure of p-groups G from Theorem 231.1 if in addition $|G'| = p$.

3661. Study the p-groups in which the normalizers of all cyclic subgroups are isolated. (This was solved by the second author.)

3662. Study the p-groups in which the normalizers of all minimal nonabelian subgroups are isolated.

3663. Study the automorphism groups of the critical p-groups G such that $\Phi(G) = Z(G)$.

3664. Study the p-groups in which any critical subgroup is special.

3665. Study the nonabelian p-groups whose automorphism groups are metabelian.

3666 (Old problem). Study the p-groups G such that $|\mathrm{Aut}(G)/\mathrm{Inn}(G)|_p = p$.

3667. Classify the p-groups all of whose minimal nonabelian subgroups are critical in their normalizers.

3668. Given $n > 2$, study the primary $\mathcal{A}_n$-groups covered by $\mathcal{A}_2$-subgroups.

3669. Classify the nonmetacyclic 2-groups all of whose minimal nonmetacyclic subgroups have order (i) 2^4, (ii) 2^5.

3670. Does there exist a p-group G such that $|\mathrm{Aut}(H)| \geq |\mathrm{Aut}(G)|$ for all $H \in \Gamma_1$?

3671. Study the p-groups all of whose nonnormal subgroups are absolutely regular. (For $p = 2$, see § 16.)

3672. Classify the nonabelian p-groups $G = \Omega_1(G)$ such that $|G : C^G| = p$ for any nonnormal $C < G$ of order p.

3673. Study the p-groups with $\leq 2p + 1$ maximal abelian subgroups.

3674. Study the nonabelian p-groups G such that $|\bigcap_{S \in \mathcal{A}_1(G)} S| > p$ (here $\mathcal{A}_1(G)$ is the set of all $\mathcal{A}_1$-subgroups of G).

3675. Study the nonabelian groups G of exponent p and order p^n with maximal possible $\alpha_1(G)$.

3676. Study the p-groups all of whose nonnormal nonabelian subgroups are minimal nonabelian.

3677. Study the p-groups G such that for all $H \in \Gamma_1$ one has (i) $\mathrm{cd}(H) \cap \mathrm{cd}(G) = \{1\}$, (ii) $\mathrm{cd}(H) = \mathrm{cd}(G)$.

3678. Study the nonabelian p-groups such that, whenever $S < G$ is minimal nonabelian and $A < S$ is maximal, there is $H \in \Gamma_1$ with $S \cap H = A$.

3679. Study the p-groups G such that $\mathrm{cd}(G) = \mathrm{cd}(G/R)$ for all subgroups $R \leq G' \cap \mathrm{Z}(G)$ of order p.

3680. Study the p-groups G of order $> p^{n+1}$ such that $|\mathrm{Z}_n(G)| = p^n$, Consider in detail the cases $n = p, p + 1$.

3681. Estimate the number of two-generator irregular groups of order p^n. (Old problem) Consider in detail the case $n = p + 1$.

3682. Study the p-groups G such that for any abelian $A < G$ there is $N \triangleleft G$ such that $G/N \cong A$.

3683. Given $k > 1$, study the p-groups such that (i) $\mathrm{Z}_k(G)$ is special, (ii) whenever $\mathrm{Z}_k(G) < G$, then $\mathrm{cl}(\mathrm{Z}_k(G)) = k$.

3684. Study the p-groups G such that, whenever $H < G$, then H' is cyclic if and only if $(H^G)'$ is.

3685. Let $\mathcal{A}_1(X)$ denote the set of all minimal nonabelian subgroups of a nonabelian p-group X. Let G, G_0 be two nonabelian p-groups and let

$$\mathcal{A}_1(G) = \{A_1, \ldots, A_n\}, \quad \mathcal{A}_1(G_0) = \{B_1, \ldots, B_n\}.$$

Compare the structures of the groups G and G_0 provided $A_i \cong B_i$ and $A_i \cap A_j \cong B_i \cap B_j$ for all $i, j = 1, \ldots, n$.

3686. Study the p-groups G such that all overgroups of G of order $p|G|$ have the same class as G. (Examples: $G \cong \mathrm{SD}_{2^n}$, $G = \Sigma_{p^2}$, $p > 2$.)

3687. Classify the nonabelian 2-groups G all of whose noncyclic abelian subgroups are normal.

3688. Let $a(G)$ be the number of maximal abelian subgroups in a nonabelian p-group G. Find all possible values of $a(G) \pmod p$.

3689. (i) Is it true that the number of pairwise non-isomorphic minimal nonabelian subgroups in a metacyclic p-group, $p > 2$ is bounded? (ii) Study the p-groups all of whose minimal nonabelian subgroups are pairwise non-isomorphic.

3690. Study the subgroup $V(G) = \bigcap_{B \in \mathcal{A}(G)} \mathrm{N}_G(B)$, where G is a nonabelian p-group.

3691. Find the class number of the Sylow p-subgroup of the holomorph of the cyclic group of order p^n.

3692. Is it true that the derived length of the p-groups generated by two minimal nonabelian subgroups, is unbounded? (The class of such p-groups is not bounded as p-groups of maximal class with abelian subgroup of index p show: these groups are generated by two minimal nonabelian subgroups.)

3693. Let G be a representation group of the elementary abelian group of order p^n. Describe the group $\mathrm{Aut}(G)$.

3694. Let A be a maximal normal abelian subgroup of a p-group G. Study the structure of G provided for any $a \in A - \mathrm{Z}(G)$ and $x \in G - A$ the subgroup $\langle a, x\rangle$ is minimal nonabelian. (This was solved by the second author; see Theorem 170.1.)

3695. Study the p-groups G such that $\mathrm{cd}(H) \in \{1, p\}$ for all $H < G$.

3696. Study the p-groups G all of whose proper two-generator subgroups have cyclic (i) derived subgroups, (ii) Frattini subgroups.

3697. Study the p-groups G such that (i) $|G/\mathrm{Z}(G)| = p^3$. (ii) $|H/\mathrm{Z}(H)| \le p^3$ for all $H < G$.

3698. Find all possible values of $|G/\mho_1(G)|$, where G is a minimal irregular p-group, $p > 2$.

3699. Describe $\mathrm{Aut}(A \times B)$, where A is nonabelian of order p^3 and B is elementary abelian.

3700. Estimate the number $|\mathrm{Aut}(G)|$, where G is a group of maximal class and order p^{p+1}.

3701. Describe the structure of a p-group G satisfying $C_G(E)) = E$, where $E \cong E_{p^3}$, $p > 2$. (For $p = 2$, see § 51.)

3702. Study the nonabelian p-groups G such that $C_G(x) < H$ for all $H \in \Gamma_1$ and $x \in H - Z(G)$. (This is solved by the second author; see § 263.)

3703. Describe the group $\mathrm{Aut}(\mathrm{Aut}(G))$, where G is a 2-group of maximal class. (See Theorem 34.8.)

3704. Study the nonabelian p-groups G in which the intersection of any two distinct maximal abelian subgroups (i) is normal, (ii) coincides with $Z(G)$).

3705. Study the irregular p-groups G with absolutely regular derived subgroup G'. (See Appendix 8.)

3706. Classify the p-groups G such that $\Phi(H) \le H_G$ for all nonabelian subgroups $H \le G$.

3707. Study the primary $\mathcal{A}_n$-groups G, $n > 2$, such that, whenever $H \in \Gamma_1$ is nonabelian, then all $\mathcal{A}_1$-subgroups of H are G-conjugate.

3708. Study the nonabelian p-groups all of whose minimal nonabelian subgroups, except one, are (i) metacyclic, (ii) nonmetacyclic.

3709. Study the nonabelian p-groups G such that $\mathrm{Aut}(G)$ acts on the set $\mathcal{A}_1(G)$ of all minimal nonabelian subgroups of G transitively.

3710. Estimate the order of the p-groups G of exponent p such that $p^3 \nmid \exp(\mathrm{Aut}(G))$.

3711. Find the groups G of order p^p and exponent p such that $|\mathrm{Aut}(G)|_p$ is minimal possible.

3712. Does there exist a p-group of maximal class, $p > 2$, all of whose $p + 1$ maximal subgroups are pairwise non-isomorphic?

3713. Let G be a p-group of order $> p^{p+1}$. Find all possible numbers of subgroups $D \lhd G$ such that G/D is of maximal class and order p^{p+1}.

3714. Let G be an irregular minimal non-absolutely regular p-group (in that case G is of maximal class and order p^{p+1}). (i) Describe $\mathrm{Aut}(G)$. (ii) Estimate the number of minimal non-absolutely regular p-groups.

3715. Study the p-groups in which intersection of any two proper nonincident subgroups is abelian. (This was solved by the second author; see Appendix 123.)

3716. Study the non-Dedekindian p-groups all of whose nonnormal subgroups are complemented.

3717. Study the nonmetacyclic p-groups all of whose minimal nonmetacyclic subgroups are pairwise non-isomorphic.

3718. Study the nonabelian p-groups G such that A is isolated in A^G for any maximal abelian (minimal nonabelian, maximal metacyclic) subgroup $A < G$.

3719. Study the nonabelian p-groups G such that, whenever $H \leq G$ has an abelian subgroup of index p, then the same holds for H^G.

3720. Study the p-groups all of whose nonabelian maximal subgroups are special. (See [Cos] and Problem 3747.)

3721. Study the p-groups G such that, provided C is critical in G and $C \leq H \in \Gamma_1$, then C is critical in H.

3722. Study the nonabelian p-groups G such that, whenever $L < S \leq G$, where S is minimal nonabelian and L is nonnormal cyclic subgroup in S, then $|G : \mathrm{N}_G(L)| = p$.

3723. Study the p-groups G such that, whenever $H \in \Gamma_1$ and $x \in G - H$, then one has $\mathrm{C}_H(x) = \mathrm{Z}(H)$.

3724. Describe the abelian p-groups G such that a Sylow p-subgroup of $\mathrm{Aut}(G)$ is regular.

3725. Classify the non-Dedekindian p-groups of exponent $> p$ in which any nonnormal cyclic subgroup of order $> p$ is self-centralizing.

3726. Is it true that any p-group can be embedded in the Frattini subgroup of a two-generator p-group? (It is known that any p-group can be embedded in the Frattini subgroup of an appropriate p-group.)

3727. Study the nonabelian p-groups containing exactly $p + 2$ conjugate classes of maximal abelian subgroups.

3728. Study the p-groups G satisfying $|\mathrm{Aut}(G)|_p < |G|$.

3729. Describe the subgroup structure of the wreath product $W = \mathrm{C}_{p^m} \wr \mathrm{C}_{p^n}$. Find $\mathrm{Aut}(W)$, $\mathrm{cl}(W)$, $\Omega_1(W)$, $W/\mho_1(W)$, all types of minimal nonabelian subgroups of W.

3730. Study the primary p-groups G such that, whenever $A < B \leq G$, where A is maximal abelian in G and $|B : A| = p$, then $\Omega_1(B) = B$.

3731. Describe the Sylow 3-subgroups of the automorphism groups of 3-groups of maximal class.

3732. Study the nonabelian p-groups all of whose noncyclic maximal abelian subgroups are elementary abelian. (This is solved by the second author; see § 273.)

3733. Study the nonabelian p-groups all of whose characteristic abelian subgroups are cyclic.

3734. Does there exist a p-group admitting a non-trivial partition all of whose components are nonabelian?

3735. Study the p-groups G all of whose normal subgroups of equal order are isomorphic.

3736. Study the nonabelian p-groups (special p-groups) all of whose minimal nonabelian subgroups are isomorphic.

3737. Describe $\mathrm{Aut}(\Sigma_{p^n})$, where $\Sigma_{p^n} \in \mathrm{Syl}_p(S_{p^n})$. Consider in detail the case $p = 2$.

3738. Classify the non-Dedekindian p-groups G such that, whenever $A < G$ is nonnormal, there is only one maximal chain connecting A with G.

3739. Study the nonabelian p-groups all of whose nonabelian subgroups have trivial Schur multiplier.

3740. Study the nonabelian p-groups G such that, whenever $A/Z(G) \leq G/Z(G)$ is abelian, then either A is abelian or minimal nonabelian. (See Problem 3571.)

3741. Let A be a proper minimal nonmetacyclic (minimal nonabelian) subgroup of a p-group G. Suppose that, whenever $B < A$ is maximal, any maximal metacyclic (maximal abelian) subgroup of G containing B has index $\leq p^2$ in G. Study the structure of G.

3742. Study the p-groups of class > 2 without normal subgroup of class 2. (Example: any p-group of maximal class and order $\geq p^5$ containing an abelian subgroup of index p.)

3743. Describe the p-groups all of whose subgroups of index p^2 are complemented.

3744. Suppose that R be a subgroup of order p^p and exponent p of a p-group $G = \Omega_1(G)$, $|G| > p^{p+1}$. Suppose that the subgroup $\langle x, R\rangle$ is of maximal class for every $x \in G - R$ of order p. Study the structure of G.

3745. Study the nonabelian p-groups G containing an abelian subgroup A of index p such that, whenever $B/Z(G) < G/Z(G)$ is of order p^2 and $B \not\leq A$, then B is minimal nonabelian.

3746. Study the special p-groups all of whose nonabelian subgroups are either minimal nonabelian or special.

3747. Study the p-groups all of whose subgroups of index p^2 or p^3 are special. (See [Cos].)

3748. Study the nonabelian p-groups in which the intersection of any two distinct maximal abelian subgroups contains a cyclic subgroup of index p.

3749. Study the p-groups without metacyclic subgroups of order p^4 and exponent p^2. Consider the case $p = 2$ in detail.

3750. Study the p-groups G such that, whenever $M \triangleleft G$, then there is $N \triangleleft G$ such that $G/N \cong M$.

3751. Study the p-groups G such that, whenever H is an epimorphic image of G, there is $N \triangleleft G$ such that $N \cong H$.

3752. Classify the p-groups G such that $\mathrm{Aut}(G)$ is metacyclic.

3753. Study the p-groups G such that $|\mathrm{N}_G(S) : S| = p$ for any minimal nonabelian (maximal metacyclic) subgroup $S < G$.

3754. A p-group G is said to be *p-central* if $\Omega_1(G) \leq \mathrm{Z}(G)$ for $p > 2$ and $\Omega_2(G) \leq \mathrm{Z}(G)$ for $p = 2$ (see § 152). Describe the minimal non-p-central p-groups. (This was solved by the second author; see § 309; see also Appendix 129.)

3755. Study the nonabelian p-groups all of whose subgroups of class 2 are minimal nonabelian.

3756. Study the p-groups G such that, whenever $A < G$ is minimal nonabelian, then all subgroups of G containing A as a subgroup of index p, are two-generator.

3757. Study the p-groups G such that, whenever $H \in \Gamma_1$ and $R < H$ is not G-invariant, then $\mathrm{N}_G(R) \leq H$.

3758. Estimate the number of subgroups of order p^2 (of index p^2) in a group of maximal class of order p^p and exponent p.

3759. Let ϕ be a nonidentity p'-automorphism of a nonabelian p-group G and suppose that, whenever $S \leq G$ is minimal nonabelian, then ϕ acts trivially on some maximal subgroup of S. Study the structure of G.

3760. Study the p-groups G such that (i) $\mathrm{Aut}(G)$ is minimal nonabelian, (ii) minimal nonnilpotent.

3761. Study the nonabelian p-groups all of whose subgroups of class 2 are critical.

3762. Study the non-Dedekindian p-groups all of whose subgroups are normal in their normal closures?

3763. Study the p-groups $G = \Omega_1(G)$ of exponent $> p$ with the property that $\Omega_1(H) < H$ for all $H \in \Gamma_1$.

3764. Study the non-Dedekindian p-groups with at most p conjugate classes of nonnormal subgroups.

3765. Study the special p-groups G such that G/A is extraspecial for any maximal subgroup A in $\mathrm{Z}(G)$. In particular, classify the special p-groups G such that any their nonabelian epimorphic image is special.[2] (Ito, personal communication) Is it true that in the last case the order of $Z(G)$ is bounded?

3766. Study the irregular p-groups G of maximal class such that $G/Z(G)$ has an abelian subgroup of index p. (Any 3-group of maximal class satisfies the above condition.)

3767. Study the p-groups G such that, whenever $A^G = B^G$ for nonnormal $A, B < G$, then $|A| = |B|$. Consider the above condition only for A, B cyclic.

3768 (Mann). Classify the p-groups G such that $|\mathrm{Aut}(G)| = |G|$. Is it true that then $p = 2$? Moreover, study the p-groups G such that $|\mathrm{Aut}(G)|_p = |G|$.

3769. Classify the p-groups such that, whenever $A, B < G$ are non-incident and $|A| = p^2$, $|B| = p^3$, then $A \cap B > \{1\}$. Consider the case $\exp(G) = p$ in detail.

3770. Let G be an abelian p-group of given type and let

$$U = \langle \phi \in \mathrm{Aut}(G) \mid \phi_{\Omega_1(G)} = \mathrm{id} \rangle, \quad V = \langle \phi \in \mathrm{Aut}(G) \mid \phi \text{ acts trivially on } G/\mho_1(G) \rangle.$$

Study the structure of groups U, V.

3771. Let A be a maximal normal abelian subgroup of a nonabelian p-group G. Study the structure of G if any minimal nonabelian subgroup of G is of the form $\langle a, x \rangle$ for appropriate $a \in A$ and $x \in G - A$. (See Problem 3694 and Lemma 57.1.)

3772. Study the nonabelian p-groups, $p > 2$, with derived subgroup of index p^2. (For $p = 2$. see Proposition 1.6.)

3773. Classify the 2-groups all of whose maximal subgroups, except one, are Dedekindian. (This problem is solved by the second author; see § 257.)

3774. Classify the p-groups containing an elementary abelian subgroup of index p.

3775. Classify the 2-groups that are lattice isomorphic to the group $Q_8 \times A$, where A is an abelian 2-group.

3776. Classify the 2-groups G in which the intersection of any two distinct maximal subgroups is Dedekindian.

3777. Describe the group $\mathrm{Aut}(Q_8 \times A)$, where A is an abelian 2-group. Consider in detail the case when A is cyclic.

3778. Study the p-groups G such that, whenever $H \in \Gamma_1$ contains a non-G-invariant cyclic subgroup, then all such subgroups of H are G-conjugate.

3779. Study the irregular p-groups all of whose minimal irregular subgroups have order p^{p+1}. (For $p = 2$, see § 90.)

---

2 The nonabelian normal Sylow subgroup of minimal nonnilpotent group satisfies the above condition; in that case, $d(G)$ is even.

3780. Study the irregular p-groups in which the intersection of any two distinct maximal regular subgroups is (i) absolutely regular, (ii) of exponent p, (iii) abelian.

3781. Does there exist a p-group G such that $\mathrm{Aut}(G)$ is isomorphic to (i) $\mathrm{Aut}(\Phi(G))$, (ii) $\mathrm{Aut}(G')$?

3782. Given an abelian p-group A, find $\mathrm{d}(P)$, where (i) $P \in \mathrm{Syl}_p(\mathrm{Aut}(A))$, (ii) P is a Sylow p-subgroup of the holomorph of A.

3783. Classify the p-groups covered by proper isolated subgroups.

3784. Study the nonabelian p-groups G containing a maximal subgroup M such that the cyclic subgroups of G not contained in M constitute p conjugate classes.

3785. Study the p-groups all of whose non-trivial characteristic subgroups contain (i) $\mathrm{Z}(G)$, (ii) $\Phi(G)$. (See [Gor1, Exercise 5.3].)

3786. Describe the absolutely regular p-groups G, $p > 2$, such that a Sylow p-subgroup of $\mathrm{Aut}(G)$ is irregular.

3787. Classify the p-groups G of maximal class and order p^{p+1} such that $\mathrm{Aut}(G)$ is a regular p-group.

3788. Study a Sylow p-subgroup P of $\mathrm{Aut}(G)$, where G is an irregular p-group of maximal class. Estimate $\mathrm{d}(P)$ and $\mathrm{cl}(P)$.

3789. Study the p-groups of maximal class (i) that contain only one characteristic subgroup of index p, (ii) all of whose maximal subgroups are characteristic.

3790. Study the p-groups G all of whose maximal abelian and minimal nonabelian subgroups have the same order. Is it true that there is in G a normal minimal nonabelian subgroup?

3791. Describe the representation groups of (i) nonmetacyclic minimal nonabelian p-groups, (ii) primary $\mathcal{A}_2$-groups.

3792. Study the p-groups G such that, whenever $A/\mathrm{M}(G) \le \Gamma/\mathrm{M}(G)$ is abelian for any representation group Γ of G, then A is also abelian.

3793. Given $n > 1$, how many representation groups has the abelian group of type $(p, p^2, \dots, p^n)$?

3794. Given $d > 1$ and $e \ge 1$, how many representation groups has the homocyclic group of exponent p^e and rank d?

3795. Study the p-groups G such that the Schur multiplier $\mathrm{M}(G)$ is isolated in a representation group of G.

3796. Describe the group $\mathrm{Aut}(G \times G)$ in the terms of a p-group G.

3797. Describe the representation groups of Q, where Q is a nonabelian Sylow p-subgroup of a minimal nonnilpotent group.

3798. Describe the Schur multiplier and the representation groups of a Sylow 2-subgroup of the simple Suzuki group $\mathrm{Sz}(2^{2m+1})$.

3799. Study the nonabelian p-groups all of whose proper nonabelian subgroups are characteristic. (Example: SD_{16}.)

3800. Classify the nonabelian p-groups G such that $C_G(H) = Z(G)$ for any nonabelian $H \leq G$.

3801. Study the p-groups G of maximal class one of whose representation groups is of maximal class. Is it true that all representation groups of G are of maximal class?

3802. Study the regular p-groups whose representation groups are irregular.

3803. Study the p-groups covered by $p + 1$ abelian and minimal nonabelian subgroups.

3804. Classify the special p-groups G all of whose maximal abelian subgroups have order $p|Z(G)|$. (See Theorem 20.4 due to Mann.)

3805. Let $a(X)$ be the number of maximal abelian subgroups of a nonabelian p-group X. Describe all residues of $a(X) \pmod p$.

3806. Classify the non-Dedekindian p-groups G such that, whenever $S \leq G$ is minimal nonabelian, then if (i) $p = 2$, then either $|S| = 2^3$ or $S \cong M_{2^n}$, (ii) $p > 2$, then $S \in \{S(p^3), M_{p^n}\}$.

3807. Classify the p-groups all of whose maximal subgroups, except one, are special. (See [Cos].)

3808. Study the p-groups G such that, whenever $A < S < G$, where S is minimal nonabelian, there is in G a subgroup $H < G$ nonincident with S such that $S \cap H = A$.

3809. Study the special p-groups G of exponent $> p$ all of whose subgroups of order p, except one, are maximal cyclic.

3810. Study the p-groups in which the maximal cyclic subgroups have intersection $> \{1\}$. (See [Bla7].)

3811. Given a p-group H of class $\leq k$, does there exist a p-group G of class $> k$ such that $Z_k(G) \cong H$?

3812. Let S be a proper special subgroup of a p-group G and suppose that any subgroup of G containing S as a subgroup of index p is special. Study the structure of G.

3813. Classify the nonabelian 2-groups G such that if $S < G$ is minimal nonabelian, then $S \in \{D_8, Q_8, M_2(2.2)\}$. (See §§ 90, 92.)

3814. Is it true that, provided a p-group, $p > 2$, contains a subgroup $\cong E_{p^p}$, then it contains a normal subgroup $\cong E_{p^p}$? (This is true for $p \in \{3, 5\}$, by Theorems 10.4, 10.5 and [Kon1].) Consider this problem for groups of exponent p.

3815. Suppose that a nonabelian metacyclic subgroup H of order p^4 and exponent p^2 is a maximal metacyclic subgroup of a two-generator nonmetacyclic p-group G. Describe the structure of G.

3816. Study the non-Dedekindian p-groups G such that $|G : N_G(A)^G| = p$ for each nonnormal $A < G$.

3817. Study the p-groups all of whose subgroups containing an abelian subgroup of index p have class ≤ 2.

3818. Classify the nonabelian metacyclic subgroups all of whose minimal nonabelian subgroups (i) are isomorphic, (ii) have equal order.

3819. Let $G \cong \Sigma_{p^n} \in \mathrm{Syl}_p(S_{p^n})$, let $\mathcal{S}$ be a collection of pairwise non-incident subgroups $S_1, S_2, \ldots$ of G such that $S_i \cap S_j > \{1\}$ for each pair i, j. How much can $|\mathcal{S}|$ be?

3820. Study the p-groups G such that $A_G > \{1\}$ for any $A < G$ of order $> p$.

3821. Let G be a p-group such that $H = H_p(G) < G$. Study the structure of H provided $C_H(x)$ is of exponent p for all $x \in G - H$.

3822. Study the p-groups all of whose characteristic subgroups are waists.[3]

3823. Study the irregular p-groups G such that $R \cap Q \leq Z(G)$ for any two distinct maximal subgroups $R, S < G$ of exponent p.

3824. (i) Given a metacyclic p-group M, estimate the number of non-isomorphic metacyclic p-groups containing a maximal subgroup isomorphic to M. (ii) Study the metacyclic p-groups M such that if M is maximal in a p-group G, then G is nonmetacyclic. (Example: $M \cong SD_{2^n}$.)

3825. Study the p-groups G such that $\mathrm{Aut}(G)$ acts trivially on $G/\Phi(G)$. (Example: $G \cong SD_{2^n}$.) Is it true that then $p = 2$?

3826. Does there exist a p-group G with $d(G) > 2$ all of whose maximal subgroups are pairwise non-isomorphic?

3827. Study the irregular p-groups (nonabelian groups of exponent p) all of whose maximal subgroups are isomorphic. Estimate the orders of such p-groups.

---

3 A *waist* N of a p-group G is its normal subgroup N such that any normal subgroup of G of order $|N|$ coincides with N. Example: G' is a waist of an extraspecial p-group G. A nonabelian p-group is of maximal class if and only if all its normal subgroups of index $> p$ are waists, by Lemma 9.1 (b).

3828. Classify the nonabelian p-groups that are not generated by noncyclic abelian subgroups.

3829. Study the p-groups all of whose subgroups of index p^2 are isomorphic.

3830. Study the p-groups all of whose subgroups of order p^4 are two-generator.

3831. Study the nonabelian p-groups G all of whose proper subgroups are either of class ≤ 2 or metacyclic.

3832. Study the p-groups having only one representation group.

3833. Study the p-groups G such that $C_G(x)$ is of order p^3 for some $x \in G$.

3834 (Inspired by [GM]). Classify the p-groups containing a maximal elementary abelian subgroup of order p^2.

3835. Study the p-groups containing a self-centralizing cyclic subgroup of order (i) p^3, (ii) p^4.

3836 (A partial case of a known unsolved problem). Is it true that a group of order $> p$ and exponent p possesses an outer automorphism of order p?

3837. Classify the p-groups all of whose subgroups of index $\geq p^4$ are normal.

3838. Classify the p-groups G such that, whenever $p^2|H|^2 < |G|$ for $H < G$, then $H \trianglelefteq G$.

3839. Study the nonabelian p-groups G such that the subgroup $Z_2(G)$ is critical.

3840. Study the irregular p-groups G such that $R \cap Q \leq Z(G)$ for any two distinct maximal subgroups $R, S < G$ of exponent p.

3841. Given $n > 2$, estimate the maximal possible derived length of a group of order p^n and exponent p.

3842. Study the nonabelian p-groups G such that, whenever H is a nontrivial normal subgroup of G and $\phi \in \mathrm{Irr}^{\#}(H)$, all irreducible constituents of ϕ^G have the same degree.

3843. Study the nonabelian p-groups all of whose maximal abelian subgroups, except one, are either cyclic or elementary abelian.

3844 (Isaacs–Passman). Study the p-groups that have no irreducible character of degree p.

3845. Classify the non-p-central p-groups G all of whose maximal p-central subgroups have index p in G.

3846. Given $n > 3$, study the p-groups G of class n such that $Z(Z_i(G)) = Z(G)$ for $i \in \{2, \dots, n-1\}$.

3847. Study the p-groups all of whose maximal abelian subgroups are homocyclic.

3848. Study the p-groups G such that $\mathrm{M}(G) = \mathrm{Z}(\Gamma)$, where Γ is a representation group of G.

3849. Study the p-groups G such that all proper critical subgroups of any $H \in \Gamma_1$ are critical in G.

3850. Study the p-groups G such that, whenever $H \leq G$ is nonabelian, then $\mathrm{Z}(H) = \mathrm{Z}(G)$.

3851. Study the nonabelian p-groups all of whose $\mathcal{A}_1$-subgroups ($\mathcal{A}_2$-subgroups) have elementary abelian (cyclic) centralizers (four problems).

3852. Study the p-groups G such that any their outer p-automorphism has at most p^2 fixed points. (See Problem 2008.)

3853. Classify the nonabelian p-groups, $p > 2$, having only one subgroup isomorphic to (i) $\mathrm{S}(p^3)$, (ii) M_{p^3}.

3854. A subgroup H of a p-group G is said to be *weakly isolated* if it is isolated in $\mathrm{N}_G(H)$. Study the nonabelian p-groups all of whose (i) maximal abelian subgroups are weakly isolated, (ii) minimal nonabelian subgroups are weakly isolated. (These two problems were solved by the second author.)

3855. Study the nonmetacyclic p-groups all of whose maximal metacyclic subgroups are weakly isolated (see Problem 3854).

3856. Classify the nonabelian p-groups G such that, if $S < G$ is minimal nonabelian, then $S \in \{\mathrm{M}_{p^3}, \mathrm{D}_8, \mathrm{S}(p^3), \mathrm{M}_p(n, 1, 1),\ n > 1\}$. (This allows to classify the p-groups of exponent $> p$ containing exactly p maximal abelian subgroups of exponent $> p$.)

3857. Classify the p-groups all of whose nonnormal subgroups are abelian.

3858 (G. Freiman). Given k, study the p-groups G such that $|S^2| < k^2$ for any k-element subset S of G. Here $S^2 = \{xy \mid x, y \in S\}$.

3859. Classify the nonabelian p-groups G such that, whenever $x \in G - \mathrm{Z}(G)$, then $|G : \langle x\rangle^G| = p$. (This is solved in § 282.)

3860. Study the 2-groups all of whose subgroups $H < G$ with $\mathrm{d}(H) = 3$ are minimal nonmetacyclic.

3861. Study the non-Dedekindian p-groups G such that, whenever $A, B < G$ are nonnormal cyclic, then $A < \mathrm{N}_G(B)$ and $B < \mathrm{N}_G(A)$.

3862 (N. Blackburn [Bla13]). Classify the p-groups containing an element x of order p such that the centralizer $\mathrm{C}_G(x)$ is abelian of type (p^r, p) for $r > 1$. (For $r = 1$, see Proposition 1.8. See also Problem 3506.)

3863. Study the p-groups G such that $\mathrm{N}_G(F) = F * H$ for any nonnormal nonabelian $F \leq G$ and some $H < G$.

3864. Study the irregular but not minimal irregular p-groups containing exactly $p+1$ maximal regular subgroups. (Example: $\Sigma_{p^2} \times C_p$, where $\Sigma_{p^2} \in \mathrm{Syl}_p(S_{p^2})$.)

3865. Study the irregular p-groups G covered by proper minimal irregular subgroups.

3866. Study the p-groups of derived length 3 all of whose proper subgroups are metabelian.

3867. Study the p-groups G of maximal class and order $> p^{p+1}$, $p > 3$, such that $d(G_1) = p - 1$, where G_1 is the fundamental subgroup of G.

3868. Study the nonabelian p-groups G of exponent $> p$ such that, whenever $C < G$ is maximal cyclic, then either $|C| = p$ or C is a maximal abelian subgroup of G.

3869. Study the nonabelian groups of exponent p all of whose two subgroups of order p^3 are permutable.

3870. Study the p-groups containing exactly one proper irregular subgroup.

3871. Classify the 2-groups containing a normal subgroup of maximal class and index 4.

3872. Does there exist a p-group G all of whose maximal subgroups are critical?

3873. Classify the p-groups all of whose abelian subgroups of order $> p^2$ are normal.

3874. Study the irregular p-groups, $p > 2$, all of whose nonabelian regular subgroups are two-generator.

3875. Study the nonabelian p-groups all of whose minimal nonabelian subgroups are isomorphic to $M_p(m, m)$, where m is fixed.

3876. Given $m > 5$, does there exist a p-group of order p^m containing minimal nonabelian subgroups of orders $p^3, p^4, \ldots, p^{m-1}$? If the answer is "yes", is it true that m is bounded?

3877. Let G be an irregular p-group all of whose maximal subgroups of exponent p are normal. Is it true that the number of such subgroups is bounded? The same problem for maximal elementary abelian subgroups.

3878. Study the p-groups all of whose maximal subgroups are non-trivial (i) direct products, (ii) central products.

3879. Classify the nonabelian p-groups G such that, whenever $A, B < G$ are distinct minimal nonabelian and $A \cap B > \{1\}$, then $A \not\cong B$. (This is solved by the second author; see § 266.)

3880. A p-group G is said to be an ECF-*group* if $G' = \Phi(G)$ and all indices of its central series below G' are equal to p. Find the minimal n such that all d-generator ECF-groups with $|G'| = p^n$ are irregular. (If $d = 2$, then $n = p - 1$, by Theorem 9.5.)

3881. Study the nonabelian p-groups G such that, whenever $A < S < G$, where S is minimal nonabelian and $|S : A| = p$, then $C_G(A)$ is abelian.

3882. Let G be a nonabelian p-group. Let M be generated by all metacyclic minimal nonabelian subgroups of G and let N be generated by all nonmetacyclic minimal nonabelian subgroups of G. Study the structure of G provided $M < G$ and $N < G$.

3883. Study the p-groups G of class > 2 such that the normal closure of any minimal nonabelian subgroup in G is of class 2.

3884. Study the nonabelian p-groups G such that, whenever $H \leq G$ is nonabelian, then $Z(G) \leq H$.

3885. Let G be an irregular ECF-group (see Problem 3880) with $\exp(\Phi(G)) = p$. Is it true that $|\Phi(G)|$ is bounded? (If $d(G) = 2$, then $|\Phi(G)| = p^{p-1}$, by Theorem 9.5.)

3886. Study the p-groups G containing a subgroup N of order p such that there are $\leq p + 1$ maximal chains of subgroups of G containing N.

3887. Study the p-groups G containing exactly one minimal nonabelian subgroup of exponent $> p$. (This was solved by the second author; see Theorem 257.1.)

3888. Study the p-groups G containing a nonabelian subgroup S of order p^3 such that $N_G(S)$ is extraspecial.

3889. Let A, B be isomorphic minimal nonabelian p-groups such that $A \cap B = Z(A)$ and let $G = A * B$ be the central product. Describe all minimal nonabelian subgroups of G.

3890. Given a metacyclic minimal nonabelian p-group S with noncyclic center, does there exist a p-group G such that $\Phi(G) = S$? (See Theorem 294.1.)

3891. Is it true that $|\Omega_1(G)|$ is bounded provided all subgroups of order $p > 2$ of a p-group G are characteristic?

3892. Find the minimal order of a group of exponent p containing as subgroups all isomorphism types of p-groups of exponent p of orders $\leq p^4$.

3893. Classify the special p-groups G without proper characteristic subgroup of order $> |Z(G)|$.

3894. Study the nonabelian p-groups all of whose minimal nonabelian subgroups, except one, have order p^3.

3895. Find the minimal n such that any group of order p^n and exponent p with cyclic center contains two distinct subgroups of order p^3 with trivial intersection.

3896. Study the p-groups G admitting an automorphism α of order p such that all (all normal) subgroups of G are α-invariant.

3897. Study the p-groups all of whose normal abelian subgroups have exponent p. (Compare with § 216.)

3898. Given a p-group G of exponent $p^e > p$, study the structure of $G/\Omega_e^{\#}(G)$.

3899. Suppose that a p-group G contains an abelian subgroup of rank 2 and order $p^n > p^2$. Find a sufficient condition for G to contain a normal abelian subgroup of rank 2 and order p^n.

3900. Classify the 2-groups G such that $\mathrm{Aut}(G)$ is a 2-group of maximal class.

3901. Classify the 2-groups G such that $\mathrm{Aut}(G) \cong \mathrm{Aut}(M)$, where M is a 2-group of maximal class.

3902. Study the p-groups covered by extraspecial (special) subgroups.

3903. Classify the special p-groups all of whose minimal nonabelian subgroups have order p^5.

3904. Does there exist a p-group covered by irregular subgroups of maximal class?

3905. Classify the p-groups all of whose maximal abelian subgroups (minimal nonabelian subgroups), except one, are isolated.

3906. Study the two-generator irregular p-groups G, $p > 2$, such that $\Omega_1(G) \leq \mathrm{Z}(G)$.

3907. Study the nonabelian p-groups in which any two distinct minimal nonabelian subgroups with non-trivial intersection are non-isomorphic.

3908. Study the nonmetacyclic p-groups G such that each maximal metacyclic subgroup $M < G$ is isolated in $\mathrm{N}_G(M)$. (This was solved by the second author.)

3909. Does there exist a nonabelian (irregular) p-group all of whose maximal abelian (regular) subgroups have pairwise distinct orders?

3910. Classify the nonabelian p-groups that are not covered by nonabelian members of the set Γ_1.

3911. Does there exist an irregular p-group such that $\langle S, T\rangle$ is regular for any two minimal nonabelian $S, T < G$?

3912. Find the ranks and types of maximal abelian subgroups of the group $\mathrm{UT}(n, p)$.

3913. Does there exist a p-group of exponent $> p$ whose Frattini subgroup (derived subgroup) is nonabelian critical?

3914. How many pairwise non-conjugate self-centralizing cyclic subgroups in a p-group are possible?

3915. Study the p-groups of order p^n containing $\leq p + 1$ normal subgroups of any order.

3916 (Old problem). Given $n > 2$, estimate the minimal order of p-groups having the derived length n. (See Appendix 6.)

3917. Study the Schur multiplier and the representation groups of a nonabelian critical p-group G with cyclic $Z(G)$.

3918. Describe the two-generator p-groups containing a subgroup $\cong M_p(n, n)$ of index p.

3919. Estimate the number of groups of order 2^n without subgroup isomorphic to E_8.

3920. Study the non-Dedekindian p-groups in which the normalizer of any nonnormal maximal cyclic subgroup is minimal nonabelian.

3921. Study the groups of exponent p all of whose nonabelian subgroups of order p^3 are nonnormal.

3922. Study the p-groups G such that, whenever $A < B < G$, then $A^G < B^G$.

3923. State a sufficient condition for a p-group G to be characteristic in a Sylow p-subgroup of its holomorph.

3924. (i) Given n, does there exist a $\mathcal{D}_n$-group of order 2^{n+4}? (ii) Classify $\mathcal{D}_n$-groups, $n \in \{2, 3\}$. (See § 245.)

3925. Using Lemma 57.1, analyze the structure of the p-groups G of exponent $> p^2$, all of whose minimal nonabelian subgroups have order p^3. (See § 284.)

3926. Classify the p-groups G with a maximal normal abelian subgroup $A < G$ satisfying $\alpha_1(G) = c(G/A) + 1$. (Here $c(X)$ is the number of pairwise distinct nonidentity cyclic subgroup of a group X. See § 250.)

3927. Suppose that a 2-group G has no normal subgroup $\cong E_{2^k}$. Find the upper bound of the ranks of abelian subgroups of G. (By Theorem 50.1, if $k < 3$, then 4 is the desired upper bound.)

3928. Study the nonabelian p-groups all of whose nonnormal nonabelian subgroups are minimal nonabelian.

3929. Study the p-groups G that satisfy $\exp(G) = \exp(G/\Omega_i^{\#}(G))\exp(\Omega_i^{\#}(G))$ for all $p^i \leq \exp(G)$.

3930. Classify the p-groups containing only one metacyclic minimal nonabelian subgroup. (See Mann's commentary to Problem 115.)

3931. Classify the 2-groups G such that $A \cap B \trianglelefteq G$ for any non-incident $A, B < G$. (This is solved in § 287 for $p > 2$. See also § 241.)

3932. Let D be the intersection of all minimal nonabelian subgroups of a nonabelian p-group G. Study the structure of G provided $|D| > p$.

3933. Study the nonabelian p-groups in which normal closures of all maximal abelian (minimal nonabelian) subgroups are isolated.

3934. Study the p-groups G containing $H \in \Gamma_1$ such that, whenever $A < G$ with $A \nleq H$, then $A^G \in \Gamma_1$. (Example: G is of maximal class and $H = G_1$.)

3935. Classify the special p-groups of exponent p^2 all of whose cyclic subgroups of order p^2 are normal.

3936. Study the p-groups all of whose minimal nonabelian subgroups are normalizers of cyclic subgroups.

3937. Study the p-groups G that are $\mathcal{A}_n$-groups, $n > 1$, such that, whenever $A < S < G$, where S is minimal nonabelian and A is not G-invariant, then $|G : A^G| = p$.

3938. Study the p-groups G all of whose maximal subgroups are normal closures of non-G-invariant subgroups. (Here we do not assert that the normal closure of any non-G-invariant subgroup has index p in G; see § 282.)

3939. Study the class of p-groups G such that $|C^G : C| \leq p$ for any cyclic subgroup $C < G$.

3940. Estimate the number of groups of maximal class and order p^n (i) containing an abelian subgroup of index p, (ii) whose Frattini subgroup is abelian.

3941. Describe the subgroups of exponent p and minimal nonabelian subgroups in the irregular p-groups G with the cyclic $\mho_1(G)$.

3942. Study the irregular p-groups without regular subgroups of order p^{p+2}.

3943. Study the p-groups in which any member of the upper and lower central series of G is a unique G-invariant subgroup of its order.

3944. Classify the extraspecial p-groups with only one non-trivial characteristic subgroup.

3945. Study the p-groups all of whose proper epimorphic images of equal order are isomorphic.

3946. Study the p-groups all of whose minimal nonabelian subgroups are metacyclic.

3947. Study the non-Dedekindian p-groups all of whose proper nonnormal nonabelian subgroups have equal order.

3948. Find $w(n)$, the minimal order of a nonabelian p-group containing all abelian subgroups of order $\leq p^n$. Estimate $\frac{w(n+1)}{w(n)}$. (See Exercise 293.238.)

3949. Find $m(n)$, the minimal order of a p-group containing all metacyclic subgroups of order $\leq p^n$. Estimate $\frac{m(n+1)}{m(n)}$.

3950. Given $n > 2$, estimate $\nu(n)$, the minimal order of a p-group containing all minimal nonabelian subgroups of order $\leq p^n$.

3951. Study the p-groups G such that for any maximal cyclic $C < G$ there is only one subgroup of G of order $p|C|$, containing C.

3952. Classify the p-groups G such that $N_G(C) \cong M_p(n, n)$, $n > 1$, for any maximal cyclic $C < G$ of order $> p$?

3953. Study the p-groups, $p > 2$, without absolutely regular subgroups of order p^{p+1}.

3954. Describe the special 2-groups that are generated by involutions.

3955. Study the p-groups all of whose maximal cyclic subgroups are nonnormal. (Example: Σ_{p^2} for $p > 2$.)

3956. Study the p-groups G such that the Schur multipliers of G and all its maximal subgroups are trivial.

3957. Let E be a special p-group such that $\Omega_1(E) = Z(E) \cong E_{p^2}$. Describe the automorphism group of E.

3958. Study the p-groups of maximal class, $p > 2$, whose representation groups are also of maximal class. State a sufficient conditions for a p-group of maximal class to have (i) the trivial Schur multiplier, (ii) the Schur multiplier of order p.

3959. Classify the p-groups all of whose maximal subgroups have cyclic (metacyclic) Frattini (derived) subgroup.

3960. Study the p-groups G such that H/H_G is a TI-subgroup of a group G/H_G for all nonnormal $H < G$.

3961. Study the p-groups G such that $\alpha_1(G) - \alpha_1(H) \leq p$ for all nonabelian $H \in \Gamma_1$. (See § 76.)

3962. Study the irregular p-groups G such that $c_k(G) = c_k(R)$ for some regular p-group R and all k.

3963. Study the p-groups $G = \Omega_1(G)$ such that $\Omega_1(H) < H$ for exactly one $H \in \Gamma_1$.

3964. Study the 2-groups G such that $\Omega_2(G)$ is isomorphic to a Sylow 2-subgroup of the simple Suzuki group $Sz(2^{2m+1})$.

3965. Study the irregular p-groups G such that any their maximal regular subgroup R satisfies the following condition: whenever $R < S \leq G$, where $|S : R| = p$, then R is isolated in S.

3966. Study the p-groups G of order p^m such that, for any $i \leq m$, there is a characteristic subgroup in G of order p^i. Is it true that there is in G a maximal chain all of whose members are characteristic?

3967. Study the nonabelian p-groups G containing a maximal subgroup H such that all minimal nonabelian subgroups of G not contained in H, are conjugate. (Example: G is a 2-group of maximal class, $H \in \Gamma_1$ is nonabelian.)

3968. Given a p-group G, let $\mathcal{M}$ be the set of those $H < G$ that $\mathrm{Irr}(1_H^G) = \mathrm{Irr}(G)$. Study the structure of the maximal members of the set $\mathcal{M}$.

3969. Suppose that an involution t acts on a p-group G, $p > 2$. Is it true that if G possesses a subgroup $E \cong \mathrm{E}_{p^4}$, then there is in G a t-invariant subgroup $\cong E$?

3970. Given an irregular p-group H, describe all regular p-groups G such that, for all k, one has $\mathrm{c}_k(G) = \mathrm{c}_k(H)$ and $\mathrm{s}_k(G) = \mathrm{s}_k(H)$.

3971. Study the non-Dedekindian p-groups such that, whenever $C, C_1 < G$ are maximal nonnormal cyclic subgroups, there is $\alpha \in \mathrm{Aut}(G)$ with $C_1 = C^\alpha$.

3972. Estimate the minimal possible number of maximal chains $\mathrm{ch}(p, n)$ in a nonabelian group of exponent p and order p^n.

3973. Estimate the number of maximal chains in a special p-group G such that $|\mathrm{Z}(G)| = p^z$ and $\mathrm{d}(G) = d$.

3974. Study the nonabelian p-groups G admitting a p'-automorphism that has no fixed points on the set $G - \mathrm{Z}(G)$.

3975. Given an extraspecial group E of order p^{2n+1}, describe the p-groups of order p^{2n+1} having the same number of maximal chains as E.

3976. Estimate the maximal possible number of maximal chains in metacyclic p-group of order p^n and describe the groups for which that number is attained.

3977. Let H be a nonabelian normal subgroup of a p-group G and let $c(G/H)$ be the number of non-identity cyclic subgroups of the quotient group G/H. Is it true that $\alpha_1(G) \geq c(G/H) + \alpha_1(H)$? Describe the pairs $H \triangleleft G$ satisfying

$$\alpha_1(G) = c(G/H) + \alpha_1(H).$$

Is it true that, in the case under consideration, G/H is of exponent p (may be, even elementary abelian)? (See § 245.)

3978. Study the p-groups all of whose nonabelian minimal nonmetacyclic subgroups are conjugate.

3979. By § 76, if $\alpha_1(G) \leq 1 + p + p^2$, then all subgroups of index p^3 in such a p-group G are abelian. Study the structure of p-groups G such that $\alpha_1(G) = (1 + p + p^2) + 1$.

3980. Study the p-groups G such that $\exp(A \cap B) = p$ for any distinct $A, B \in \Gamma_1$.

3981. Describe the p-groups G of exponent p and order p^n with maximal possible number $\alpha_1(G)$. (For the extraspecial G, see Example 76.1.)

3982. Study the p-groups containing two non-conjugate self-centralizing cyclic subgroups.

3983. Study the p-groups G with the property that any abelian subgroup of G is contained in an $\mathcal{A}_2$-subgroup.

3984. Let $H \nleq \Phi(G)$ be a subgroup of a p-group G. Describe the structure of H if the intersection of H with any maximal subgroup of G not containing H, is (i) abelian, (ii) absolutely regular, (iii) of maximal class, (iv) of exponent p.

3985. Classify the nonabelian p-groups all of whose nonabelian subgroups (nonabelian maximal subgroups) have cyclic centers. (This was solved in § 238.)

3986. A normal series $H_0 < H_1 < \cdots < H_{n-1} < H_n = G$ is said to be an *upper abelian series* if, for $i \in \{1, \dots, n\}$, all $H_i \trianglelefteq G$ and H_i/H_{i-1} is a maximal abelian normal subgroup of G/H_{i-1}. Study such groups G satisfying $\mathrm{dl}(G) = n$.

3987. Study the irregular p-groups all of whose maximal absolutely regular subgroups are normal.

3988. Study the nonabelian groups G of exponent p such that, whenever $S_1, S_2 < G$ are minimal nonabelian, then $|\langle S_1, S_2\rangle| \le p^6$.

3989. Study the nonabelian p-groups G in which every maximal abelian subgroup A is contained in an irregular subgroup of order $p|A|$.

3990. Study the p-groups G with $\le p + 1$ conjugate classes of minimal nonabelian subgroups. Consider in detail the case $\exp(G) = p$.

3991. Study the p-groups, $p > 2$, containing two distinct maximal abelian subgroups isomorphic to E_{p^3}.

3992. Study the p-groups containing two distinct normal subgroups of order p^p and exponent p but not containing a subgroup of order p^{p+1} and exponent p.

3993. Study the p-groups such that, whenever $A \trianglelefteq G$ is of order $\ge p^p$, then every G-invariant subgroup containing A as a subgroup of index p is of maximal class.

3994. Study the non-Dedekindian p-groups G such that $\mathrm{cl}(H) = \mathrm{cl}(\mathrm{N}_G(H))$ for any nonnormal $H < G$.

3995. Study the class of p-groups G such that (i) $G/\mho_2(G)$ is extraspecial, (ii) $\Omega_2(G)$ is critical.

3996. Classify the non-Dedekindian p-groups of exponent $> p$ all of whose nonnormal cyclic subgroups have the same order.

3997. Study the p-groups G of exponent $p^e > p$ such that $\mathrm{cl}(G/\mho_i(G)) = i$ for all $i \le e$.

3998. Describe $\mathrm{Aut}(G)$, where $G = \mathrm{E}_{p^m} \wr \mathrm{E}_{p^n}$. Find $\mathrm{cl}(G)$ and $\mathrm{k}(G)$.

3999. Let G be a two-generator nonabelian group of exponent p with noncyclic center. Is it true that $Z(G)$ contains a subgroup of order p that is characteristic (not characteristic) in G?

4000. Study the p-groups G all of whose normal subgroups are characteristic. Consider in detail the case when G is metacyclic.

4001. Describe the p-groups G such that, whenever $A < G$ is abelian, then G possesses a normal abelian subgroup of order $|A|$.

4002. Classify the non-Dedekindian 2-groups all of whose maximal Dedekindian subgroups have index 2.

4003. Study the p-groups of class > 2 all of whose maximal subgroups of class 2 have index p.

4004. Study the 2-groups all of whose maximal nonnormal abelian subgroups are elementary abelian.

4005. Let a p-group P, $p > 2$, contain a normal subgroup $E \cong E_{p^k}$, $k \in \{4, 5\}$, and let an involution t act on P. Is it true that P contains a t-invariant (t-invariant normal) subgroup $Q \cong E_{p^k}$?

4006. Study the p-groups G, $p > 2$, all of whose minimal nonabelian subgroups, not contained in $\Phi(G)$, have order p^3.

4007. Let S be a minimal nonabelian subgroup of a p-group G. Suppose that for any maximal subgroup A of S the subgroup $C_G(A)$ is (i) minimal nonabelian, (ii) either abelian or minimal nonabelian. Study the structure of G.

4008. A subgroup A of a p-group X is said to be *half-normal* provided $|X : N_X(A)| \le p$. Study the p-groups G such that, whenever $H < G$ and $|H|^2 \le |G|$, then H is half-normal in G. (See § 122.)

4009. Study the p-groups all of whose cyclic subgroups are half-normal.

4010. Study the p-groups all of whose non-half-normal subgroups are conjugate.

4011. Study the two-generator p-groups G such that $\mathrm{Aut}(G)$ is a p-group. (For a number of interesting results connected with automorphisms of p-groups, see the paper [Hor1] by M. V. Horosevskii.)

4012. Study the groups of exponent p all of whose nonabelian subgroups (minimal nonabelian subgroups) are half-normal.

4013. Given n, study the p-groups G of maximal class and order p^n such that $\mathrm{cl}(G_1) = n - 3$.

4014. Study the p-groups G such that $\Omega_1(G)$ ($\Omega_2^{\#}(G)$) is irregular of order p^{p+2}.

4015. Study the p-groups G of order $> p^{p+1}$ such that the subgroup $\Omega_1(G)$ is of maximal class of order p^p.[4]

4016. Study the nonmetacyclic 2-groups all of whose minimal nonmetacyclic subgroups have order 2^5.

4017. Study the p-groups G such that $\sum_{S \in \mathcal{A}_1(G)} |S| \leq |G|$.

4018. Study the p-groups G such that, whenever $x \leq S \leq G$, where S is minimal nonabelian, then x is half-central in G, i.e., $|G : \mathrm{C}_G(x)| \leq p$. (This was solved by the second author; see Theorem 310.1.)

4019. Classify the p-groups G of maximal class such that $|G_1'| = p$.

4020. Classify the p-groups containing exactly p (exactly $p + 1$) pairwise distinct maximal abelian subgroups of exponent $> p$.

4021. Study the p-groups G such that $|G : \mathrm{N}_G(C)| \leq p^2$ for any abelian (cyclic) $C < G$. (See § 122.)

4022. Study the nonabelian p-groups G all of whose subgroups of class 2 have the derived subgroup of order p.

4023. Study the nonabelian p-groups all of whose minimal nonabelian subgroups except one, have order p^4.

4024. Study the p-groups G such that all elements of the set $G - \Phi(G)$ $(G - \mho_1(G)$, $G - G'$, $G - \mathrm{Z}_2(G))$ are half-central.

4025. Given an $\mathcal{A}_2$-group S, find $\mathrm{Aut}(S)$.

4026. Study the p-groups G such that all subgroups non-incident with $\Phi(G)$ are half-normal.

4027. Study the p-groups G such that $\Omega_1(G)$ is an $\mathcal{A}_2$-subgroup.

4028 (Old problem). Classify the groups of maximal class and exponent p.

4029. Study the representation groups of special p-groups.

4030. Given $p > 2$, does there exist an irregular p-group all of whose subgroups of order p^p are abelian?

---

4 One has $p > 2$. By Theorem 9.6 (c), G is not of maximal class. Since $|\mathrm{Z}(\Omega_1(G))| = p$, it follows that $\Omega_1(G) \nleq \Phi(G)$, by Lemma 1.4. Therefore, there is $H \in \Gamma_1$ such that $\Omega_1(G) \nleq H$, and we conclude that $|\Omega_1(H)| = p^{p-1}$. By Theorem 12.1, H is either absolutely regular or of maximal class. Let $R \lhd G$ be abelian of type (p, p). Then $\Omega_1(G) \nleq \mathrm{C}_G(R) \in \Gamma_1$ so that $\mathrm{C}_G(R)$ is absolutely regular. In that case, $|G/\mho_1(G)| = p^p$. It follows that all members of the set Γ_1 not containing $\Omega_1(G)$ are absolutely regular (Theorem 12.12).

4031. Study the p-groups containing a regular subgroup of index p^2 but not containing a normal regular subgroup of index p^2.

4032. Study the p-groups of exponent p containing a subgroup $E \cong \mathrm{E}_{p^p}$ but not containing a normal subgroup $\cong E$.

4033. Study the 2-groups G containing a subgroup $L \cong \mathrm{C}_4$ such that $\mathrm{C}_G(L)$ is (i) abelian of type $(4, 4)$, (ii) abelian of type $(2^n, 2)$ (see § 77).

4034. Study the 2-groups G containing a subgroup A of (i) type $(4, 2)$ such that $\mathrm{C}_G(A)$ is abelian of type $(2^n, 2)$, (ii) type $(4, 4)$ such that $\mathrm{C}_G(A)$ is abelian of type $(2^n, 2^2)$. State a similar problem for $p > 2$.

4035. Study the primary $\mathcal{A}_n$-groups G, $n > 2$, covered by normal $\mathcal{A}_1$- and $\mathcal{A}_2$-subgroups. (See Problem 860.)

4036. Study the non-Dedekindian p-groups G such that, whenever $A < G$ is nonnormal, then $\alpha_1(G) - \alpha_1(A^G) = p - 1$. (See Lemma 76.5.)

4037. Study the non-Dedekindian p-groups G such that, whenever $A < G$ is nonnormal, then A is normalized by all nonnormal subgroups of G. (See § 274, containing the solution of this problem by the second author.)

4038. Describe $\mathrm{Aut}(\Sigma_{p^2})$.

4039. Describe the automorphism group of the p-group from Theorem 13.7 (c).

4040. Study the p-groups G of exponent p such that for all abelian $A < G$ one has $|A^G : A| \leq p$.

4041. Two subgroups C_1, C_2 of a p-group G are said to be C-*equivalent* if $C_1^G = C_2^G$. Classify the non-Dedekindian p-groups all of whose nonnormal maximal cyclic subgroups are C-equivalent. Study the p-groups with exactly two classes of nonnormal C-equivalent subgroups (cyclic subgroups).

4042. Classify the nonabelian p-groups G containing an abelian subgroup A of index p such that, for any cyclic $L \leq A$, one has either $\mathrm{N}_G(L) = A$ or $L \leq \mathrm{Z}(G)$.

4043. Classify the p-groups G such that $\mathrm{N}_G(A)$ is normal in G for all maximal subgroups A in minimal nonabelian subgroups of G.

4044. Classify the nonabelian p-groups G of exponent p and order $> p^3$ such that $\mathrm{N}_G(A) \cong A \times \mathrm{C}_p$ for any nonabelian $A < G$ of order p^3.

4045. Given $n > 1$, study the p-groups G such that $\mathrm{Z}_n(G) \cong \mathrm{E}_{p^n}$. Consider in detail the groups G of exponent p.

4046. Describe the automorphism groups of the p-groups of maximal class with abelian subgroup of index p.

4047. Study the non-Dedekindian p-groups G such that $\mathrm{cl}(\mathrm{N}_G(H)) = \mathrm{cl}(H)$ for any nonnormal $H < G$.

4048. Study the p-groups all of whose maximal two-generator subgroups have index p.

4049. Study the special p-groups which are products of two abelian subgroups.

4050. Classify the non-Dedekindian p-groups G such that for each nonnormal cyclic $C < G$ the normal closure C^G is of (i) maximal class, (ii) metacyclic, (iii) minimal nonabelian.

4051. Study the nonabelian p-groups all of whose two minimal nonabelian subgroups are permutable.

4052. Study the nonabelian p-groups G all of whose maximal abelian subgroups (maximal regular subgroups, minimal nonabelian subgroups) are components of an irredundant covering of G.

4053. Study the special p-groups G such that $\mathrm{Aut}(G)$ is a p-group.

4054. Let H be a proper noncyclic subgroup of a p-group G. Describe the structure of H if $H \cap S$ is cyclic for any minimal nonabelian $S < G$ non-incident with H.

4055. Classify the p-groups G satisfying $\mathrm{f}(G) = \frac{1}{p}$. (It is easy to show that then $p = 2$.)

4056. Is it true that if $\mathrm{k}(H)$ is the same for all maximal subgroups of a p-group G, then all maximal subgroups of G are abelian?

4057. Let G_0 be a p-group of maximal class and order p^n. Describe the p-groups G of order p^n such that $\mathrm{k}(G) = \mathrm{k}(G_0)$. Consider in detail the cases when (i) G_0 has an abelian subgroup of index p, (ii) $p \le 3$.

4058. Classify the 2-groups G such that all elements of all their minimal nonabelian subgroups are real in G.

4059. Study the 2-groups G containing a maximal abelian subgroup all of whose elements are real in G.

4060. Study the p-groups G such that $\exp(A \cap B) \le p$ for any two distinct (i) minimal nonabelian subgroups $A, B < G$, (ii) $A, B \in \Gamma_1$.

4061. Let a p-group G be irregular with $|G/\mho_1(G)| = p^p$. Is it true that orders of subgroups of exponent p in G are bounded?

4062. Suppose that a 2-group G has no normal subgroup $\cong \mathrm{E}_{2^4}$. Find the exact upper bound of ranks of abelian subgroups of G.

4063. Let $A \in \Gamma_1$ be nonabelian. Study the structure of the p-group G provided all of its minimal nonabelian subgroups non-incident with A, are (i) metacyclic, (ii) of order p^3.

4064. Study the p-groups G such that the set $G - \Phi(G)$ is covered by minimal nonabelian subgroups.

4065. Study the p-groups containing an isolated homocyclic maximal abelian subgroup.

4066. Study the nonabelian p-groups G such that, whenever $A < G$ is maximal abelian and $x \in A - Z(G)$, then $C_G(x) = A$.

4067. Study the p-groups G such that $|H/\mho_1(H)| > |G/\mho_1(G)$ for any $H \in \Gamma_1$. (Example: $G = M_2(2, 2, 1)$.)

4068. Is it true that the maximal number of independent generators of abelian subgroups of a two-generator p-group is bounded?

4069. Classify the p-groups G such that $|M(G)| \geq \prod_{H \in \Gamma_1} |M(H)|$, where $M(X)$ is the Schur multiplier of X.

4070. Study the nonabelian p-groups G such that, whenever $A < G$ is minimal nonabelian and $A < N \leq G$ with $|N : A| = p$, then $cl(N) = 2$.

4071. Study the irregular p-groups G such that, whenever $R < G$ is nonnormal, then $|N_G(R) : R| = p$.

4072. Study the primary $\mathcal{A}_n$-groups, $n > 1$, in which any $\mathcal{A}_i$-subgroup is contained in an $\mathcal{A}_{i+1}$-subgroup of G for all $i < n$.

4073. Study the p-groups in which the intersection of any two distinct maximal subgroups is isolated.

4074. Study the nonabelian p-groups, $p > 2$, in which the normal closure of each minimal nonabelian subgroup is of maximal class.

4075. Study the two-generator p-groups of exponent $> p$ such that $G/\mho_2(G) \cong B(2, p)$, where $B(2, p)$ is the two-generator Burnside group of exponent p.

4076. Study the class of p-groups G containing a proper normal subgroup $H \nleq Z(G)$ such that, whenever (i) $x \in H - Z(G)$, then $C_G(x) \leq H$, (ii) $A < H$ is nonnormal, then $N_G(A) \leq H$.

4077. Study the irregular p-groups G of order $> p^{p+2}$ such that $|\Omega_2(G)| = p^{p+2}$. (For $p = 2$, this is solved in § 52.)

4078. Let G be an L_n-group (see §§ 17, 19). Estimate $|G/\mho_1(G)|$. Consider the case $n = p + 1$ in detail.

4079. Does there exist a p-group of maximal class (of exponent p), $p > 2$, containing an abelian subgroup of index p^4 but not containing a normal abelian subgroup of the same index?

4080. Let a p-group G be such that $|G/\mho_1(G)| = p^p$. Is it true that $|H/\mho_1(H)|$ is bounded for all $H < G$?

4081. Describe the two-generator extensions of a cyclic 2-group by a 2-group of maximal class. Find the automorphism groups of those extensions.

4082. Classify the p-groups, $p > 2$, covered by subgroups $\cong M_{p^n}$, $n = 3, 4, \ldots$.

4083. Study the nonabelian groups of exponent p all of whose subgroups $\cong S(p^3)$ are nonnormal.

4084. Study the nonabelian p-groups G such that, whenever $N \triangleleft G$ is nonabelian, then N has exactly one G-invariant subgroup of index p.

4085. Study the p-groups G of exponent $p^e > p^2$ such that $p^{e-1} \nmid \exp(\mathrm{Aut}(G))$.

4086. Study the nonabelian p-groups $G = \Omega_1(G)$, $p > 2$, that are not generated by subgroups $\cong S(p^3)$.

4087. Study the 2-groups that are not generated by real elements.

4088. Study the non-Dedekindian p-groups G such that, whenever a minimal nonabelian $A < G$ is nonnormal, then A is contained in exactly one member of the set Γ_1. (Compare with Theorem 28.1.)

4089. Suppose that a p-group G, $p > 2$, has a maximal nonnormal elementary subgroup of order p^2. Is it true that G has no subgroup of order p^{p+1} and exponent p? (Compare with [GM].)

4090. Suppose that a p-group G, $p > 2$, has a maximal nonnormal elementary abelian subgroup of order p^3. Is it true that the order of maximal elementary abelian (maximal subgroup of exponent p) is bounded? (Compare with [GM].)

4091. Classify the nonabelian p-groups G such that $H^G = N_G(H)$ for all nonnormal subgroups $H < G$.

4092. Given $i > 1$, study the p-groups all of whose subgroups of orders $p^2, \ldots, p^i$ are normal. (See § 298 and Appendix 132.)

4093. Classify the metacyclic p-groups (p-groups of maximal class, of exponent p) all of whose maximal subgroups are isomorphic (pairwise non-isomorphic).

4094. Let G be an irregular p-group of maximal class. Classify the p-groups H such that $s_n(H) = s_n(G)$ for all positive integers n. The similar problem in the case when G is (i) minimal nonabelian, (ii) metacyclic, (iii) a group of exponent p, (iv) minimal nonmetacyclic.

4095. Describe the p-groups that are lattice isomorphic to minimal nonabelian p-groups (p-groups of maximal class).

4096. Describe the 2-groups G of exponent > 4 such that $G/\mho_2(G) \cong \mathrm{B}(2,4)$, the two-generator Burnside group of exponent 4 (of order 2^{12}). (See Problem 4075.)

4097. Study the p-groups G, $p > 2$, such that all elements of the set $G - (M \cup N)$ have order p for some fixed distinct $M, N \in \Gamma_1$. (For $p = 2$, see Theorem 43.12.)

4098. Classify the nonabelian p-groups all of whose nonabelian (nonabelian regular) subgroups are two-generator.

4099. Study the nonabelian p-groups all of whose minimal nonabelian subgroups are nonmetacyclic.

4100. Classify the p-groups all of whose subgroups, except one, are p-central.

Bibliography

[Alle] R. B. J. T. Allenby, Normal subgroups contained in Frattini subgroups are Frattini subgroups, *Proc. Amer. Math. Soc.* **78** (1980), 315–318.

[Alp1] J. L. Alperin, On a special class of regular p-groups, *Trans. Amer. Math. Soc.* **106** (1963), 77–99.

[Alp3] J. L. Alperin, Centralizers of abelian normal subgroups of p-groups, *J. Algebra* **1** (1964), 110–113.

[Alp2] J. L. Alperin, Large abelian subgroups of p-groups, *Trans. Amer. Math. Soc.* **117** (1965), 10–20.

[AK1] B. Amberg and L. Kazarin, On the rank of a finite product of two p-groups, in: *Groups – Korea '94*, Walter de Gruyter, Berlin (1995), 1–8.

[AHZ] L. J. An, R. F. Hu and Q. H. Zhang, Finite p-groups with a minimal nonabelian subgroup of index p (IV), *J. Algebra Appl.* **14** (2015), no. 3, Article ID 1550020.

[ALQZ] L. J. An, L. L. Li, H. P. Qu and Q. H. Zhang, Finite p-groups with a minimal nonabelian subgroup of index p (II), *Sci. China Math.* **56** (2013), 1–17.

[AZ] L. J. An and Q. H. Zhang, Finite metahamiltonian p-groups, *J. Algebra* **442** (2015), 23–35.

[AS] M. Aschbacher and S. D. Smith, *The Classification of Quasithin Groups 1*, Math. Surveys Monogr. 111, American Mathematical Society, Providence, 2004.

[BW] A. H. Baartman and J. J. Woeppel, The automorphism group of a p-group of maximal class with abelian maximal subgroup, *Fund. Math.* **93** (1976), no. 1, 41–46.

[Bae2] R. Baer, Partitionen endlicher Gruppen, *Math. Z.* **75** (1961), 333–372.

[BY] G. Ban and S. Yu, Minimal abelian groups that are not automorphisms groups, *Arch. Math.* **70** (1998), 427–434.

[Bei2] B. Beisigel, Die Automorphismengruppen homozyklisher p-Gruppen, *Arch. Math.* **29** (1977), no. 4, 363–366.

[Ber43] Y. Berkovich, *Groups of Prime Power Order. Vol. 1*, Walter de Gruyter, Berlin, 2008.

[Ber54] Y. Berkovich, Finite groups all of whose small subgroups are normal, *J. Algebra Appl.* **13** (2014), no. 6, Article ID 1450006.

[Ber53] Y. Berkovich, Some remarks on subgroups of maximal class of a finite p-group, *J. Algebra Appl.* **14** (2015), Article ID 1550043.

[Ber59] Y. Berkovich, Finite p-groups all of whose $\mathcal{A}_k$-subgroups for a fixed k are metacyclic, *J. Algebra Appl.* **16** (2017), no. 3, Article ID 1750055.

[Ber58] Y. Berkovich, On the number of minimal nonabelian subgroups of a nonabelian finite p-group, *J. Algebra Appl.* **16** (2017), no. 3, Article ID 1750054.

[Ber25] Y. Berkovich, Finite Groups. Selected Topics. Parts I, II, in preparation.

[BJ1] Y. Berkovich and Z. Janko, *Groups of Prime Power Order. Vol. 2*, Walter de Gruyter, Berlin, 2008.

[BJ3] Y. Berkovich and Z. Janko, *Groups of Prime Power Order. Vol. 3*, Walter de Gruyter, Berlin, 2011.

[BJ4] Y. Berkovich and Z. Janko, *Groups of Prime Power Order. Vol. 4*, Walter de Gruyter, Berlin, 2016.

[BJ5] Y. Berkovich and Z. Janko, *Groups of Prime Power Order. Vol. 5*, Walter de Gruyter, Berlin, 2016.

[BZhang] Y. Berkovich and Q. H. Zhang, $\mathcal{A}_1$- and $\mathcal{A}_2$-subgroups of finite p-groups, *J. Algebra Appl.* **13** (2013), no. 2, Article ID 1350095.

[BZ] Y. Berkovich and E. Zhmud, *Characters of Finite Groups. Parts 1 and 2*, Transl. Math. Monogr. 172 and 181, American Mathematical Society, Providence, 1998, 1999.

https://doi.org/10.1515/9783110533149-101

[Bla3] N. Blackburn, On a special class of p-groups, *Acta Math.* **100** (1958), 45–92.

[Bla5] N. Blackburn, Generalizations of certain elementary theorems on p-groups, *Proc. Lond. Math. Soc.* **11** (1961), 1–22.

[Bla7] N. Blackburn, Finite groups in which the nonnormal subgroups have nontrivial intersection, *J. Algebra* **3** (1966), 30–37.

[Bla12] N. Blackburn, On centralizers in p-groups, *J. Lond. Math. Soc. (2)* **9** (1975), 478–482.

[Bla13] N. Blackburn, Groups of prime-power order having an abelian centralizer of type $(r, 1)$, *Monatsh. Math.* **99** (1985), 1–18.

[Bla6] N. Blackburn, Automorphisms of finite p-groups, *J. Algebra* **3** (1988), 28–29.

[BlaH] N. Blackburn and L. Hethelyi, Some further properties of soft subgroups, *Arch. Math.* **65** (1997), no. 5, 365–371.

[Blac2] S. R. Blackburn, Groups of prime power order with derived subgroup of prime order, *J. Algebra* **219** (1999), 625–657.

[Bra] R. Brandl, Groups with few non-normal subgroups, *Comm. Algebra* **23** (1995), no. 6, 2091–2098.

[Carl] J. Carlson, Maximal elementary abelian subgroups of rank 2, *J. Group Theory* **10** (2007), 5–13.

[Che] Y. Cheng, On finite p-groups with cyclic commutator subgroups, *Arch. Math.* **39** (1982), 295–298.

[CM] D. Chillag and A. Mann, Nearly odd order and nearly real finite groups, *Comm. Algebra* **26** (1998), no. 7, 2041–2064.

[CT] G. Corsi Tani, Automorphisms fixing every normal subgroup of a p-group, *Boll. Unione Mat. Ital. Ser. A* **64** (1985), 245–252.

[Cos] J. Cossey, Special maximal subgroups of p-groups, *Bull. Aust. Math. Soc.* **89** (2014), no. 3, 415–419.

[Cre1] E. Crestani, On 3-generated 2-groups whose maximal subgroups are 2-generated, *Comm. Algebra* **39** (2001), 881–887.

[Cre2] E. Crestani, *Monotone 2-groups*, Ph.D. Thesis, University of Padova, 2009.

[CrMe] E. Crestani and F. Menegazzo, Monotone 2-groups, *J. Group Theory* **15** (2012), 359–383.

[Ear] B. E. Earnley, *On finite groups whose groups of automorphisms are abelian*, Ph.D. Thesis, Wayne State University, 1975.

[ERNS] S. Evans-Riley, M. F. Newman and C. Schneider, On the soluble length of groups with prime-power order, *Bull. Aust. Math. Soc.* **59** (1999), 343–346.

[FA] X. G. Fang and L. J. An, The classification of finite metahamiltonian p-groups, preprint.

[FT] W. Feit and J. G. Thompson, Solvability of groups of odd order, *Pacific J. Math.* **13** (1963), 775–1029.

[GMS1] N. Gavioli, A. Mann and C. Scoppola, Two applications of the Hughes subgroup of finite group, in: *Ischia Group Theory 2006*, World Scientific, Singapore (2007), 138–146.

[GMS2] N. Gavioli, V. Monti and C. Scoppola, Pro-p-groups with waists, *J. Algebra* **351** (2012), 130–137.

[Gil] J. D. Gillam, A note on finite metabelian p-groups, *Proc. Amer. Math. Soc.* **25** (1970), 189–190.

[Gla1] G. Glauberman, Large abelian subgroups of finite p-groups, *J. Algebra* **196** (1997), 301–338.

[Gla2] G. Glauberman, Large abelian subgroups of groups of prime exponent, *J. Algebra* **237** (2001), 735–768.

[Gla5] G. Glauberman, Large subgroups of small class in finite p-groups, *J. Algebra* **272** (2004), 128–153.

[Gla7] G. Glauberman, Existence of normal subgroups in finite p-groups, *J. Algebra* **319** (2008), 800–805.

[Gla8] G. Glauberman, A p-group with no normal large abelian subgroup, in: *Character Theory of Finite Groups*, Contemp. Math. 524, American Mathematical Society, Providence (2010), 61–65.

[GM] G. Glauberman and N. Mazza, p-groups with maximal elementary abelian subgroup of rank 2, *J. Algebra* **323** (2010), 1729–1737.

[Gol] Y. A. Golfand, On groups all of whose subgroups are nilpotent, *Dokl. Akad. Nauk SSSR* **125** (1948), 1313–1315.

[Gor1] D. Gorenstein, *Finite Groups*, Harper and Row, New York, 1968.

[Gor3] D. Gorenstein, *Reviews on Finite Groups*, American Mathematical Society, Providence, 1974.

[GLS] D. Gorenstein, R. Lyons and R. Solomon, *The Classification of the Finite Simple Groups. Part 1. Chapter G: General Group Theory*, American Mathematical Society, Providence, 1995.

[GLS5] D. Gorenstein, R. Lyons and R. Solomon, *The Classification of the Finite Simple Groups. Number 5*, American Mathematical Society, Providence, 2002.

[GLS6] D. Gorenstein, R. Lyons and R. Solomon, *The Classification of the Finite Simple Groups. Number 6*, American Mathematical Society, Providence, 2005.

[Grov2] J. R. J. Groves, On direct products of regular p-groups, *Proc. Amer. Math. Soc.* **37** (1973), 377–379.

[Grov1] J. R. J. Groves, On minimal irregular p-groups, *J. Aust. Math. Soc.* **16** (1973), 78–89.

[Grov3] J. R. J. Groves, Some criteria for the regularity of a direct product of regular p-groups, *J. Aust. Math. Soc. Ser. A* **24** (1977), 35–49.

[Grü4] O. Grün, Einige Sätze über Automorphismen abelschen p-Gruppen, *Abh. Math. Semin. Univ. Hambg.* **24** (1960), 54–58.

[Hal] M. Hall, *The Theory of Groups*, Macmillan, New York, 1959.

[HS] M. Hall and J. K. Senior, *On Groups of Order 2^n, ($n \le 6$)*, Macmillan, New York, 1964.

[Hal1] P. Hall, A contribution to the theory of groups of prime power order, *Proc. Lond. Math. Soc.* **36** (1933), 29–95.

[Hal2] P. Hall, On a theorem of Frobenius, *Proc. Lond. Math. Soc.* **40** (1936), 468–501.

[Hal8] P. Hall, The Eulerian functions of a group, *Quart. J. Math.* **7** (1936), 134–151.

[Hal5] P. Hall, On groups of automorphisms, *J. Math.* **182** (1940), 194–204.

[Hal3] P. Hall, The classification of prime power groups, *J. Reine Angew. Math.* **182** (1940), 130–141.

[Hal4] P. Hall, *Nilpotent Groups*, Queen Mary College, London, 1957.

[HH] P. Hall and G. Higman, On the p-length od p-solvable groups and the reduction theorem for Burnside's problem, *Proc. Lond. Math Soc.* **3** (1956), 1–42.

[Har] K. Harada, On some 2-groups of normal 2-rank 2, *J. Algebra* **20** (1972), no. 1, 90–93.

[Haw] T. O. Hawkes, On the automorphism group of a 2-group, *Proc. Lond. Math. Soc.* **26** (1973), 207–225.

[H] P. V. Hegarty, Minimal abelian automorphism groups of finite groups, *Rend. Semin. Mat. Univ. Padova* **94** (1995), 121–135.

[Hei3] H. Heineken, Nilpotente Gruppen, deren sämtliche Normalteiler charakteristisch sind, *Arch. Math.* **33** (1979/80), 497–503.

[HL1] H. Heineken and H. Liebeck, On p-groups with odd order automorphism groups, *Arch. Math.* **24** (1973), 465–471.

[Hel1] G. T. Helleloid, *Automorphism groups of finite p-groups: Structure and applications*, Ph.D. Thesis, Stanford University, 2007.

[Het5] L. Hethelyi, Soft subgroups of p-groups, *Ann. Univ. Sci. Budapest Eótvős Sect. Math.* **27** (1984), 81–85.

[Het2] L. Hethelyi, On subgroups of p-groups having soft subgroups, *J. Lond. Math. Soc. (2)* **41** (1990), 425–427.

[Het4] L. Hethelyi, Some remarks on 2-groups having soft subgroups, *Studia Sci. Math. Hungar.* **27** (1992), 295–299.

[Het3] L. Hethelyi, Some additional properties of subgroups of p-groups having soft subgroups, *Studia Sci. Math. Hungar.* **29** (1994), 317–320.

[HK] L. Hethelyi and B. Külshammer, Characters, conjugacy classes and centrally large subgroups of p-groups of small rank, *J. Algebra* **340** (2011), 199–210.

[HKS1] L. Hethelyi, B. Külshammer and B. Sambale, Conjugacy classes and characters of finite p-groups, *Comm. Algebra* **39** (2011), 657–685.

[Hig] G. Higman, Finite groups in which every element has prime power order, *J. Lond. Math. Soc.* **32** (1957), 337–342.

[HP] W. M. Hill and D. B. Parker, The nilpotence class of the Frattini subgroup, *Israel J. Math.* **15** (1973), 211–215.

[HW] W. M. Hill and C. R. B. Wright, Normal subgroups contained in the Frattini subgroup, *Proc. Amer. Math. Soc.* **35** (1972), 413–415.

[HogK] G. T. Hogan and W. P. Kappe, On the H_p-problem for finite p-groups, *Proc. Amer. Math. Soc.* **20** (1969), 450–454.

[Hop2] C. Hopkins, Non-abelian groups whose groups of automorphisms are abelian, *Ann. of Math.* **29** (1927/28), 508–520.

[Hor1] M. V. Horosevskii, The automorphism groups of finite p-groups (in Russian), *Algebra Logika* **10** (1971), 81–86.

[Hor2] M. V. Horosevskii, The automorphism group of wreath products of finite groups, *Sibirsk. Mat. Zh.* **14** (1973), 651–659.

[Hor3] M. V. Horosevskii, Automorphisms of finite groups (in Russian), *Mat. Sb. (N.S.)* **93** (1974), 576–587.

[HTh] D. R. Hughes and J. G. Thompson, The H_p-problem and the structure of H_p-groups, *Pacific J. Math.* **9** (1959), 1097–1101.

[Hum] K. Hummel, The order of the automorphism group of a central product, *Proc. Amer. Math. Soc.* **47** (1975), 37–40.

[Hup1] B. Huppert, *Endliche Gruppen. Band. 1*, Springer, Berlin, 1967.

[HupB] B. Huppert and N. Blackburn, *Finite Groups II*, Springer, Berlin, 1982.

[Isa1] I. M. Isaacs, *Character Theory of Finite Groups*, Academic Press, New York, 1976.

[Isa6] I. M. Isaacs, Commutators and commutator subgroup, *Amer. Math. Monthly* **84** (1977), 720–722.

[Isa5] I. M. Isaacs, *Algebra: A Graduate Course*, Brooks/Cole, Pacific Grove, 1994.

[Isa19] I. M. Isaacs, *Finite Group Theory*, American Mathematical Society, New York, 2008.

[IsP1] I. M. Isaacs and D. S. Passman, A characterization of groups in terms of degrees of degrees of their characters I, II, *Pacific J. Math.* **15** (1965), 877–903; ibid. **24** (1968), 468-510.

[Ish] K. Ishikawa, On finite p-groups which have only two conjugacy lengths, *Israel J. Math.* **129** (2002), 119–123.

[Ito1] N. Ito, On the degrees of irreducible representations of a finite group, *Nagoya Math. J.* **3** (1951), 5–6.

[Ito10] N. Ito, Über eine Frattini-Gruppe duale Bildung, *Nagoya Math. J.* **9** (1955), 123–127.

[IM] N. Ito and A. Mann, Counting classes and characters of groups of prime exponent, *Israel J. Math.* **156** (2006), 205–220.

[Iwa] K. Iwasawa, Über die endlichen Gruppen und die Verbände ihrer Untergruppen, *J. Univ. Tokyo* **4** (1941), 171–199.

[JRY] V. K. Jain, P. K. Rai and M. K. Yadav, On finite groups with abelian automorphism groups, *Internat. J. Algebra Comput.* **23** (2013), 1063–1077.

[Jan8] Z. Janko, Finite 2-groups with exactly four cyclic subgroups of order 2^n, *J. Reine Angew. Math.* **566** (2004), 135–181.

[Jan27] Z. Janko, Finite 2-groups all of whose maximal cyclic subgroups of composite order are self-centralizing, *J. Group Theory* **10** (2007), 1–4.

[Jan47] Z. Janko, Finite nonabelian p-groups all of whose nonabelian maximal subgroups are either metacyclic or minimal nonabelian, *Arch. Math.* **96** (2011), 105–107.

[Jan42] Z. Janko, Finite nonabelian p-groups having exactly one maximal subgroup with a noncyclic center, *Arch. Math* **96** (2011), 101–103.

[Jan48] Z. Janko, Finite p-groups all of whose maximal subgroups, except one, have its derived subgroup of order $\le p$, *Glas. Mat.* **47** (2012), no. 2, 325–332.

[Jan38] Z. Janko, Finite p-groups with many minimal nonabelian subgroups, *J. Algebra* **367** (2012), 363–370.

[Jan54] Z. Janko, Finite p-groups of exponent p^e all of whose cyclic subgroups of order p^e are normal, *J. Algebra* **416** (2014), 274–286.

[Jan51] Z. Janko, Finite p-groups with $A \cap B$ being maximal in A or B for any two nonincident subgroups A and B, *Glas. Mat. Ser. III* **49** (2014), no. 1, 113–118.

[Jan58] Z. Janko, Classification of finite p-groups with cyclic intersection of any two distinct conjugate subgroups, *Glas. Mat Ser. III* **50(70)** (2015), no. 1, 101–161.

[Jan49] Z. Janko, Finite p-groups all of whose maximal subgroups, except one, have cyclic derived subgroup, *J. Algebra Appl.* **14** (2015), no. 1, Article ID 1450080.

[Jan57] Z. Janko, Finite p-groups with some isolated subgroups, *J. Algebra* **465** (2016), 41–61.

[Jan56] Z. Janko, Finite 2-groups with some prescribed minimal nonabelian subgroups, *J. Algebra Appl.* **16** (2017), no. 8, Article ID 1750053.

[Jan36] Z. Janko, Finite nonabelian 2-groups all of whose minimal nonabelian subgroups are isomorphic to $M_{2^{n+1}} = \langle a, t \mid a^{2^n} = t^2 = 1, a^t = a^{1+2^{n-1}} \rangle$, $n \ge 3$ fixed, manuscript.

[Jan39] Z. Janko, Finite p-groups all of whose nonabelian maximal subgroups have the largest possible centers, manuscript.

[Jan50] Z. Janko, Finite p-groups with exactly two sizes of conjugate classes, manuscript.

[Jan55] Z. Janko, Classification of finite p-groups with cyclic intersection of any two distinct conjugate subgroups, *Glas. Mat. Ser. III*, to appear.

[Jan37] Z. Janko, Finite nonabelian 2-groups all of whose minimal nonabelian subgroups are metacyclic and have exponent 4, *J. Algebra*, to appear.

[John] D. L. Johnson, A property of finite p-groups with trivial multiplicator, *Amer. J. Math.* **98** (1976), 105–108.

[Kar1] G. Karpilovsky, *The Schur Multiplier*, Clarendon Press, Oxford, 1986.

[Kaz1] L. S. Kazarin, Groups with certain conditions for normalizers of subgroups (in Russian), *Uch. Zap.Perm Univ.* **218** (1969), 268–279.

[Kaz2] L. S. Kazarin, On some classes of finite groups, *Soviet Math. Dokl.* **12** (1971), no. 2, 549–553.

[Kaz3] L. S. Kazarin, Groups with restrictions on normalizers of subgroups (in Russian), *Izv. Vuzov (Mathematics)* **2** (1973), 41–50.

[Kaz4] L. S. Kazarin, On product of two nilpotent groups (in Russian), *Vopr. Teor. Grupp Gomologicheskoj Algebry* **1981** (1981), 62–66; ibid. **1982** (1982), 47-49.

[KazSi] L. S. Kazarin and V. Sidel'nikov, On the automorphism group of the Suzuki p-algebra, *Math. Notes* **80** (2006), 500–508.

[Keg2] O. H. Kegel, Die Nilpotenz der H_p-Gruppen, *Math. Z.* **75** (1960), 373–376.
[Khu1] E. I. Khukhro, *Nilpotent Groups and Their Automorphisms*, De Gruyter, Berlin, 1993.
[Khu2] E. I. Khukhro, *p-Automorphisms of Finite p-Groups*, Cambridge University, Cambridge, 1998.
[Kim2] I. Kiming, Structure and derived length of finite p-groups possessing an automorphism of p-power order having exactly p fixed points, *Math. Scand.* **62** (1988), 153–172.
[Kin1] B. W. King, Normal subgroups of groups of prime-power order, in: *Proceedings of the Second International Conference on the Theory of Groups*, Lecture Notes in Math. 372, Springer, Berlin (1973), 401–408.
[KY] R. D. Kitture and M. K. Yadav, Finite groups with abelian automorphism groups: A survey, to appear.
[Klu] F. L. Kluempen, The power structure of 2-generator 2-groups of class two, *Algebra Colloq.* **9** (2002), no. 3, 287–302.
[Kno] H. G. Knoche, Über den Frobeniusschen Klassenbegriff in nilpotenten Gruppen. I, II, *Math. Z.* **55** (1951), 71–83; ibid **59** (1953), 8–16.
[Kon1] M. Konvisser, Embedding of abelian subgroups in p-groups, *Trans. Amer. Math. Soc.* **153** (1971), 469–481.
[Kur] J. Kurdics, Automorphisms of a minimal nonabelian p-group with odd p. I, *Glas. Math. Ser. III* **46(66)** (2011), no. 2, 367–383.
[KS] H. Kurzweil und B. Stellmacher, *Theorie der endlichen Gruppen. Eine Einführung*, Springer, Berlin, 1998.
[Laf1] T. J. Laffey, The minimum number of generators of a finite p-group, *Bull. Lond. Math. Soc.* **5** (1973), 288–290.
[Laf3] T. J. Laffey, A lemma on finite p-groups and some consequences, *Proc. Cambridge Philos. Soc.* **75** (1974), 133–137.
[Laf4] T. J. Laffey, Centralizers of elementary abelian subgroups in finite p-groups, *J. Algebra* **51** (1978), 88–96.
[Laf2] T. J. Laffey, Bounding the order of a finite p-group, *Proc. Roy. Irish Acad. Sect. A* **80** (1980), no. 2, 131–134.
[Lan] G. L. Lange, Two-generator Frattini subgroups of finite p-groups, *Israel J. Math* **29** (1978), no. 4, 357–360.
[Leon2] Y. K. Leong, Odd order nilpotent groups of class two with cyclic centre, *J. Aust. Math. Soc.* **17** (1974), 142–153.
[Leon1] Y. K. Leong, Finite 2-groups of class two with cyclic centre, *J. Aust. Math. Soc. Ser. A* **27** (1979), 125–140.
[Lew] M. L. Lewis, On a p-group Camina pairs, *J. Group Theory* **15** (2012), 469–483.
[LQC] L. L. Li, H. P. Qu and G. Y. Chen, Central extension of inner abelian p-groups (I), *Acta Math. Sinica (Chin. Ser.)* **53** (2010), no. 4, 675–684.
[LP] P. J. Li, A classification of finite p-groups whose proper subgroups are of class ≤ 2 (I), (II), *J. Algebra Appl.* **12** (2013), no. 3, Article ID 1250170; ibid. **12** (2013), no. 3, Article ID 1250171.
[LW] H. G. Liu and Y. L. Wang, The automorphism group of a generalized extraspecial p-group, *Sci. China Math.* **53** (2010), 315–334.
[LM] P. Longobardi and M. Maj, On p-groups of breadth two, *Algebra Colloq.* **6** (1999), 121–124.
[LubM] A. Lubotzky and A. Mann, Powerful p-groups. 1, *J. Algebra* **105** (1987), 484–505.
[Macd10] I. D. Macdonald, On cyclic commutator subgroups, *J. Lond. Math. Soc.* **38** (1963), 419–422.

[Macd1] I. D. Macdonald, Generalizations of a classical theorem on nilpotent groups, *Illinois J. Math.* **8** (1964), 556–570.

[Macd14] I. D. Macdonald, A computer application to finite p-groups, *J. Aust. Math. Soc. Ser. A* **17** (1974), 102–112.

[Macd5] I. D. Macdonald, The breadth of finite p-groups, *Proc. Roy. Soc. Edinburgh Sect. A* **78** (1977/78), 1–39.

[Macd3] I. D. Macdonald, Commutators and their products, *Amer. Math. Monthly* **93** (1986), 440–444.

[Mal2] I. Malinowska, p-automorphisms of finite p-groups: Problems and questions, in: *Advances in Group Theory*, Aracne, Rome (2002), 111–127.

[Mal6] I. Malinowska, On the structure of the automorphism group of a minimal nonabelian p-group (metacyclic case), *Glas. Mat. Ser. III* **47** (2012), no. 67, 153–164.

[Mal7] I. Malinowska, Finite groups all of whose small subgroups are pronormal, *Acta Math. Hungar.* **147** (2015), no. 2, 324–337.

[Man18] A. Mann, Finite groups with maximal normalizers, *Illinois J. Math.* **12** (1968), 67–75.

[Man2] A. Mann, Regular p-groups, *Israel J. Math* **10** (1971), 471–477.

[Man3] A. Mann, Regular p-groups. II, *Israel J. Math.* **14** (1973), 294–303.

[Man6] A. Mann, Regular p-groups and groups of maximal class, *J. Algebra* **42** (1976), 136–141.

[Man5] A. Mann, The power structure of p-groups I, II, *J. Algebra* **42** (1976), 121–135; ibid **318** (2007), 953–956.

[Man7] A. Mann, Conjugacy classes in finite groups, *Israel J. Math.* **31** (1978), 78–84.

[Man4] A. Mann, Regular p-groups. III, *J. Algebra* **70** (1981), 89–101.

[Man15] A. Mann, Generators of p-groups, in: *Groups*, London Math. Soc. Lecture Note Ser. 121, Cambridge University Press, Cambridge (1986), 273–281.

[Man24] A. Mann, On the power structure of some p-groups, *Rend. Circ. Mat. Palermo (2)* **23** (1990), 227–235.

[Man16] A. Mann, Some applications of powerful p-groups, in: *Groups St. Andrews 1989. Volume 2*, London Math. Soc. Lecture Note Ser. 160, Cambridge University Press, Cambridge (1991), 370–385.

[Man20] A. Mann, Some questions about p-groups, *J. Aust. Math. Soc. Ser. A* **67** (1999), 356–379.

[Man21] A. Mann, The derived length of p-groups, *J. Algebra* **224** (2000), 263–267.

[Man11] A. Mann, The number of generators of finite p-groups, *J. Group Theory* **8** (2005), 317–338.

[Man29] A. Mann, On characters-classes duality and orders of centralizers, *Contemp. Math.* **402** (2006), 215–217.

[Man34] A. Mann, On skew p-groups, *Publ. Math. Debrecen* **69** (2006), 353–360.

[Man31] A. Mann, On the exponent of the product of two groups, *Rend. Semin. Mat. Univ. Padova* **115** (2006), 205–207.

[Man35] A. Mann, The exponents of central factor and commutator subgroups, *J. Group Theory* **10** (2007), 435–436.

[Man36] A. Mann, Corrigendum "The number of generators of finite p-groups", *J. Group Theory* **14** (2011), 329–331.

[Man22] A. Mann, Finite p-Groups, in preparation.

[MM] A. Mann and C. Martinez, The exponent of finite groups, *Arch. Math.* **67** (1996), 8–10.

[MP-F] A. Mann and F. Poznick-Fradkin, Subgroups of powerful groups, *Israel J. Math.* **138** (2003), 19–28.

[Maz2] V. D. Mazurov, 2-groups with an automorphism of odd order fixing all involutions (in Russian), *Algebra Logika* **8** (1969), no. 6, 874–885.

[McK2] S. McKay, On the structure of a special class of p-groups, *Quart. J. Math. (2)* **38** (1987), 489–502.

[McK1] S. McKay, *Finite p-Groups*, Queen Mary Math. Notes 18, University of London, London, 2000.

[MS] U. Meierfrankenfeld and B. Stellmacher, The generic groups of p-type, preprint (1997).

[Men] F. Menegazzo, Automorphisms of p-groups with cyclic commutator subgroup, *Rend. Semin. Mat. Padova* **90** (1993), 81–101.

[Mie1] R. J. Miech, Metabelian p-groups of maximal class, *Trans. Amer. Math. Soc.* **152** (1970), 331–373.

[Mie2] R. J. Miech, Some p-groups of maximal class, *Trans. Amer. Math. Soc.* **189** (1974), 1–47.

[Mie3] R. J. Miech, On p-groups with a cyclic commutator subgroup, *J. Aust. Math. Soc.* **20** (1975), 178–198.

[Mie4] R. J. Miech, The metabelian p-groups of maximal class, *Trans. Amer. Math. Soc.* **236** (1978), 93–119.

[Mie5] R. J. Miech, The metabelian p-groups of maximal class. II, *Trans. Amer. Math. Soc.* **272** (1982), 465–484.

[Mil9] G. A. Miller, The groups in which every subgroup is either abelian or Hamiltonian, *Trans. Amer. Math. Soc.* **8** (1907), 25–29.

[Mil8] G. A. Miller, On the holomorph of the cyclic group of order p^m, *Trans. Amer. Math. Soc.* **9** (1908), 232–236.

[MilM] G. Miller and H. Moreno, Non-abelian groups in which every subgroup is abelian, *Trans. Amer. Math. Soc.* **4** (1903), 398–404.

[Mill] W. H. Mills, The automorphisms of the holomorph of a finite abelian group, *Trans. Amer. Math. Soc.* **85** (1956), 1–34.

[MLC] E. Morgado Morales and M. Lazo Cortis, On the Sylow p-groups of the automorphism group of a finite homocyclic p-group, *Rev. Cienc. Mat.* **6** (1985), 35–44.

[Mori4] M. Morigi, On p-groups with abelian automorphism group, *Rend. Semin. Mat. Univ. Padova* **92** (1994), 47–58.

[Mor5] M. Morigi, On minimal number of generators of finite non-abelian p-groups having an abelian automorphism group, *Comm. Algebra* **23** (1995), 2045–2065.

[Mori2] M. Morigi, Power automorphisms of finite p-groups, *Comm. Algebra* **70** (1999), 4853–4877.

[Mul] O. Müller, On p-automorphisms of finite p-groups, *Arch. Math.* **32** (1979), 533–538.

[NRSB] L. A. Nazarova, A. V. Roiter, V. V. Sergeichuk and V. M. Bondarenko, Application of modules over a dyad to the classification of finite p-groups that have an abelian subgroup of index p and to the classification of pairs of mutually annihilating operators (in Russian), *Zap. Nauchn. Sem. LOMI* **28** (1972), 69–92.

[Nek2] K. G. Nekrasov, On some 2-groups with a small noncyclic Frattini subgroup, in: *Algebraic and Logical Constructions* (in Russian), Tver (1994), 53–65.

[Nin] Y. Ninomiya, Finite p-groups with cyclic subgroups of index p^2, *Math. J. Okayama Univ.* **36** (1994), 1–21.

[Pas] D. S. Passman, Nonnormal subgroups of p-groups, *J. Algebra* **15** (1970), no. 3, 352–370.

[Pas1] D. S. Passman, Character theory and group rings, in: *Character Theory of Finite Groups*, Contemp. Math. 524, American Mathematical Society, Providence 2010, 139–148.

[Paz] G. Pazdersky, Prime power groups which are cyclic extensions of elementary Abelian groups, *Math. Nachr.* **97** (1980), 57–68.

[Qu1] H. P. Qu, Finite p-groups all of whose maximal normal abelian subgroups are soft, *Sci. China Math.* **53** (2010), no. 11, 3037–3040.

[QH2] H. P. Qu and R. F. Hu, A central extension of minimal non-abelian p-groups (III), *Acta. Math. Sin.* **53** (2010), no. 6, 1051–1064.

[QXA] H. P. Qu, M. Y. Xu and L. J. An, Finite p-groups with a minimal nonabelian subgroup of index p (III), *Sci. China Ser. A* **58** (2015), no. 4, 763–780.

[QYXA] H. P. Qu, S. S. Yang, M. Y. Xu and L. J. An, Finite p-groups with a minimal nonabelian subgroup of index p (I), *J. Algebra* **358** (2012), 178–188.

[QZ2] H. P. Qu and Q. H. Zhang, A central extension of minimal non-abelian p-groups (II), *Acta. Math. Sin.* **53** (2010), no. 5, 933–944.

[QZGA] H. P. Qu, L. P. Zhao, J. Gao and L. J. An, Finite p-groups with a minimal nonabelian subgroup of index p (V), *J. Algebra Appl.* **13** (2014), no. 7, Article ID 1450032.

[QZ] H. P. Qu and L. F. Zheng, A central extension of minimal non-abelian p-groups (IV) (in Chinese), *Acta. Math. Sin.* **54** (2011), no. 5, 739–752.

[Rea1] E. W. Read, On the Schur multiplier of a wreath product, *Illinois J. Math.* **20** (1976), 456–466.

[Rea2] E. W. Read, On the centre of a representation group, *J. Lond. Math. Soc. (2)* **16** (1977), 43–50.

[Red1] L. Redei, Das schiefe Produkt in der Gruppentheorie, *Comment. Math. Helvet.* **20** (1947), 225–267.

[Red2] L. Redei, Die endlichen einstufig nichtnilpotenten Gruppen, *Publ. Math. Debrecen* **4** (1956), 303–324.

[Rock] D. M. Rocke, p-groups with abelian centralizers, *Proc. Lond. Math. Soc. (3)* **30** (1975), 55–75.

[Roi] M. Roitman, On Zsigmondy primes, *Proc. Amer. Math. Soc.* **125** (1997), 1913–1919.

[Ron] C. Ronse, On centralizers of involutions in 2-groups, *Math. Proc. Cambridge Philos. Soc.* **86** (1979), 199–204.

[Rus2] D. J. Rusin, The 2-groups of rank 2, *J. Algebra* **149** (1992), 1–31.

[Sam2] B. Sambale, Exponent and p-rank of finite p-groups and applications, *Arch. Math.* **103** (2014), 11–20.

[ST] J. Sangroniz and J. Tent, 2-groups with few rational conjugate classes, *J. Algebra* **338** (2011), 114–121.

[Schm3] P. Schmid, On the automorphism group of extraspecial 2-groups, *J. Algebra* **234** (2000), 492–506.

[Sch3] O. Schmidt, Groups having only one class of nonnormal subgroups (in Russian), *Mat. Sb.* **33** (1926), 161–172.

[Sch4] O. Schmidt, Groups with two classes of nonnormal subgroups, in: *Proceedings of a Seminar on the Theory of Groups* (in Russian), GONTI, Moscow (1938), 7–26.

[Scot] W. R. Scott, *Group Theory*, Prentice Hall, Upper Saddle River, 1964.

[Sh] S. Shahriari, On normal subgroups of capable groups, *Arch. Math.* **48** (1987), 193–198.

[Sha1] A. Shalev, The structure of finite p-groups: Effective proof of coclass conjectures, *Invent. Math.* **115** (1994), 315–345.

[Sha2] A. Shalev, Finite p-groups, in: *Finite and Locally Finite Groups*, Kluwer Academic, Dodrecht (1995), 401–450.

[SU] V. M. Sitnikov and A. D. Ustjuzaninov, Finite groups with three classes of non-invariant subgroups (in Russian), *Ural Gos. Univ. Mat. Zap.* **6** (1967), no. 1, 94–102.

[SQG] Q. W. Song, H. P. Qu and X. Y. Guo, Finite p-groups of class 3 all of whose proper subgroups have class at most 2, *Algebra Colloq.* **17** (2010), 191–201.

[SS] E. G. Straus and G. Szekeres, On a problem of D. R. Hughes, *Proc. Amer. Math. Soc.* **9** (1958), 157–158.

[Suz2] M. Suzuki, *Structure of a Group and the Structure of its Lattice of Subgroups*, Ergeb. Math. Grenzgeb. (3) 10, Springer, Berlin, 1956.

[Suz1] M. Suzuki, *Group Theory I, II*, Springer, Berlin, 1982, 1986.

[Sze] G. Szekeres, On finite metabelian p-groups with two generators, *Acta Sci. Math. (Szeged)* **21** (1960), 270–291.
[Tes] L. Teschke, Über die Normalteiler der p-Sylowgruppe der symmetrische Gruppe vom Grade p^m, *Math. Nachr.* **87** (1979), 197–212.
[Tho2] J. G. Thompson, Finite groups with fixed-point-free automorphisms of prime order, *Proc. Natl. Acad. Sci. USA* **45** (1959), 578–581.
[Tho4] J. G. Thompson, Fixed points of p-groups acting on p-groups, *Math. Z.* **86** (1964), 12–13.
[Tho5] J. G. Thompson, Centralizers of elements in p-groups, *Math. Z.* **96** (1967), 292–293.
[Tho3] J. G. Thompson, Nonsolvable finite groups all of whose local subgroups are solvable. 1, *Bull. Amer. Math. Soc.* **74** (1968), 383–437.
[Tua1] H. F. Tuan, A theorem about p-groups with abelian subgroup of index p, *Acad. Sinica Sci. Record* **3** (1950), 17–23.
[V-LF1] A. Vera-Lopez and G. A. Fernandez-Alcober, Centralizers of small order in p-groups of maximal class, *Comm. Algebra* **20** (1992), no. 4, 1051–1059.
[V-LF2] A. Vera-Lopez and G. A. Fernandez-Alcober, The conjugacy vector of a p-group of maximal class, *Israel J. Math.* **86** (1994), 233–255.
[V-LLO] A. Vera-Lopez and L. Ortiz de Elguea, The conjugacy-vectors of all relative holomorphs of an elementary abelian group of order 16, *Port. Math.* **47** (1990), no. 3, 243–257.
[V-LS] A. Vera-Lopez and J. Sangroniz, Classification of all relative holomorphs of an abelian group of type $C_4 \times C_4$ (in Spanish), *Rev. Acad. Cienc. Zaragoza (2)* **43** (1988), 7–12.
[Ver2] L. Verardi, A class of finite groups of exponent p in which every normal subgroup is characteristic, *Boll. Unione Mat. Ital. Ser. B 4* **6** (1988), 307–317.
[Ver1] L. Verardi, On groups whose non-central elements have the same finite number of conjugates, *Boll. Unione Mat. Ital. Ser. A 2* **7** (1988), 391–400.
[Waa5] R. W. van der Waall, A note on modular p-groups, *Indag. Math.* **35** (1973), 47–48.
[Waa1] R. W. van der Waall, On finite p-groups whose commutator subgroups are cyclic, *Indag. Math.* **35** (1973), 342–345.
[Waa3] R. W. van der Waall, On modular p-groups, *J. Algebra* **25** (1973), 125–127.
[Waa2] R. W. van der Waall, Certain normal subgroups of the Frattini subgroup of a finite group, *Indag. Math.* **36** (1974), 382–386.
[Waa4] R. W. van der Waall, On modular p-groups II, *Indag. Math.* **37** (1975), 79–82.
[WaaN] R. W. van der Waall and C. H. W. M. de Nijs, On the embedding of a finite group as a Frattini subgroup, *Bull. Belg. Math. Soc. Simon Stevin* **2** (1995), 519–527.
[Wal] G. E. Wall, On Hughes' H_p-problem, in: *Proceedings of the International Conference on the Theory of Groups* (Canberra 1965), Gordon and Breach, New York (1967), 357–362.
[Wam] J. W. Wamsley, The deficiency of metacyclic groups, *Proc. Amer. Math. Soc.* **24** (1970), 724–730.
[WZ] L. Wand and Q. Zhang, Finite p-groups whose non-abelian subgroups have the same center, *J. Group Theory* **17** (2014), 689–703.
[Web3] U. H. M. Webb, The occurrence of groups as automorphisms of nilpotent p-groups, *Arch. Math.* **37** (1981), no. 6, 481–498.
[Web4] U. H. M. Webb, On the rank of a p-group of class 2, *Canad. Math. Bull.* **26** (1983), 101–105.
[Web2] U. H. M. Webb, The number of stem covers of an elementary abelian p-group, *Math. Z.* **182** (1983), no. 3, 327–337.
[Web5] U. H. M. Webb, The Schur multiplier of a nilpotent group, *Trans. Amer. Math. Soc.* **291** (1985), 755–763.
[Web6] U. H. M. Webb, The using graphs to investigate the automorphism groups of nilpotent groups, in: *Graphs and Applications*, Wiley, New York (1985), 223–247.

[Weir2] A. Weir, The Sylow subgroups of the symmetric groups, *Proc. Amer. Math. Soc.* **6** (1955), 534–541.

[Wie3] J. Wiegold, The multiplicator of direct product, *Quart. J. Math. Oxford (2)* **22** (1971), 103–105.

[Wie2] J. Wiegold, The Schur multiplier: An elementary approach, in: *Groups - St. Andrews 1981*, London Math. Soc. Lect. Notes 71, Cambridge University Press, Cambridge (1982), 137–154.

[Wil3] B. Wilkens, 2-groups of breadth 3, *J. Algebra* **318** (2007), 202–224.

[Wil5] B. Wilkens, Finite p-groups not characteristic in any p-group in which they are properly contained, *Israel J. Math.* **166** (2008), 97–112.

[Wil6] B. Wilkens, p-groups without noncharacteristic normal subgroups, *Israel J. Math.* **172** (2009), 357–369.

[Wil7] B. Wilkens, More on core-2 2-groups, *J. Group Theory* **20** (2017), no. 2, 193–225.

[Wol] B. Wolf, A note on p'-automorphisms of p-groups P of maximal class centralizing the center of P, *J. Algebra* **190** (1997), no. 1, 163–171.

[Wri] D. Wright, Degree of minimal permutation representation of covering groups of abelian groups, *Amer. J. Math.* **96** (1974), 578–592.

[Xu3] M. Y. Xu, A theorem on metabelian p-groups and some consequences, *Chin. Ann. Math. Ser. B* **5** (1984), 1–6.

[Xu1] M. Y. Xu, The power structure of finite p-groups, *Bull. Aust. Math. Soc.* **36** (1987), no. 1, 1–10.

[Xu4] M. Y. Xu, Regular p-groups and their generalizations, manuscript.

[XAZ] M. Y. Xu, L. J. An and Q. H. Zhang, Finite p-groups all of whose nonabelian proper subgroups are generated by two elements, *J. Algebra* **319** (2008), 3603–3620.

[XZA] M. Y. Xu, Q. H. Zhang and L. J. An, Finite p-groups all of whose proper non-abelian subgroups are generated by two elements, *J. Algebra* **319** (2008), no. 9, 3603–3620.

[Yin3] J. H. Ying, On finite groups whose automorphism groups are nilpotent, *Arch. Math.* **29** (1977), 41–44.

[Yin2] J. H. Ying, Finite groups with restricted quotient structures, *Bull. Inst. Math. Acad. Sinica* **8** (1980), 1–14.

[Zap1] G. Zappa, Finite groups in which all nonnormal subgroups have the same order (in Italian), *Atti Accad. Naz. Lincei Rend. Lincei Mat. Appl. (9)* **14** (2002), no. 1, 4–16.

[Zap2] G. Zappa, Finite groups in which all nonnormal subgroups have the same order II (in Italian), *Atti Accad. Naz. Lincei Rend. Lincei Mat. Appl. (9)* **14** (2003), no. 1, 13–21.

[[Zha] L. H. Zhang, Intersection of maximal subgroups which are not minimal nonabelian of finite p-groups, *Comm. Algebra* **45** (2017), no. 8, 3321–3330.

[Zha3] Q. H. Zhang, A characterization of smallest Suzuki 2-group, *Acta Math. Sin. (Engl. Ser.)* **24** (2008), no. 12, 2011–2014.

[Zha2] Q. H. Zhang, A characterization of metacyclic p-groups by counting subgroups, in: *Proceedings of the International Conference on Algebra 2010: Advances in Algebraic Structures*, World Scientific, Hackensack (2011), 713–720.

[ZGQX] Q. H. Zhang, X. Q. Guo, H. P. Qu and M. Y. Xu, Finite groups which have many normal subgroups, *J. Korean Math. Soc.* **46** (2009), no. 6, 1165–1278.

[ZL2] Q. H. Zhang and P. J. Li, Finite p-groups with a cyclic subgroup of index p^3, *J. Math. Res. Appl.* **32** (2012), no. 5, 505–529.

[ZLS] Q. H. Zhang, X. X. Li and M. J. Su, Finite p-groups whose nonnormal subgroups have orders $\le p^3$, manuscript.

[ZS] Q. H. Zhang and M. J. Su, Finite 2-groups whose nonnormal subgroups have orders $\le 2^3$, *Front. Math. China* **7** (2012), no. 5, 971–1003.

[ZW] Q. H. Zhang and J. J. Wei, The intersection of subgroups of finite groups, *Arch. Math.* **96** (2011), 9–17.

[ZZLS] Q. H. Zhang, L. B. Zhao, M. M. Li and Y. Q. Shen, Finite p-groups all of whose subgroups of index p^3 are abelian, manuscript.

[ZG1] X. H. Zhang and X. Y. Guo, Finite p-groups whose non-normal cyclic subgroups have small index in their normalizers, *J. Group Theory* **15** (2012), 641–659.

[ZG2] L. B. Zhao and X. Y. Guo, Finite p-groups with exactly one $\mathcal{A}_1$-subgroup of given structure of order p^3, *Acta Math. Sin. (Engl. Ser.)* **29** (2013), no. 11, 2099–2110.

[Zhm5] E. M. Zhmud, The isomorphisms of the lattice of normal subgroups of a finite nilpotent group (in Russian), *Vestnik Kharkov Univ.* **26** (1967), 3–17.

Author index

https://doi.org/10.1515/9783110533149-102

Subject index

All chapters with numbers < 257 and all appendices with numbers < 110 are located in the previous volumes.

https://doi.org/10.1515/9783110533149-103

O

P

De Gruyter Expositions in Mathematics

Volume 63
Yakov Berkovich, Lev S. Kazarin, Emmanuel M. Zhmud'
Characters of Finite Groups. Volume 1, 2nd edition, 2017
ISBN 978-3-11-022406-1, e-ISBN 978-3-11-022407-8

Volume 62
Yakov Berkovich, Zvonimir Janko
Groups of Prime Power Order, 2016
ISBN 978-3-11-029534-4, e-ISBN 978-3-11-029535-1

Volume 61
Yakov Berkovich, Zvonimir Janko
Groups of Prime Power Order, 2016
ISBN 978-3-11-028145-3, e-ISBN 978-3-11-028147-7

Volume 60
Benjamin Fine, Anthony Gaglione, Alexei Myasnikov, Gerhard Rosenberger, Dennis Spellman
The Elementary Theory of Groups, 2014
ISBN 978-3-11-034199-7, e-ISBN 978-3-11-034203-1

Volume 59
Friedrich Haslinger,
The d-bar Neumann Problem and Schrödinger Operators, 2014
ISBN 978-3-11-031530-1, e-ISBN 978-3-11-031535-6

Volume 56
Yakov Berkovich, Zvonimir Janko
Groups of Prime Power Order, 2011
ISBN 978-3-11-020717-0, e-ISBN 978-3-11-025448-8

Volume 42
Helmut Strade
Simple Lie Algebras over Fields of Positive Characteristic, 2nd Edition, 2017
ISBN 978-3-11-051676-0, e-ISBN 978-3-11-051760-6

Volume 38
Helmut Strade
Simple Lie Algebras over Fields of Positive Characteristic, 2nd edition, 2017
ISBN 978-3-11-051516-9, e-ISBN 978-3-11-051544-2

www.degruyter.com